Table of Aton

DN-12

	Symbol	Atomic Number	Atomic Weight
Mercury	Hg	80	200.59
Molybdenum	Mo	42	95.94
Neodymium	Nd	60	144.24
Neon	Ne	10	20.179
Neptunium	Np	93	237.0482
Nickel	Ni	28	58.71
Niobium	Nb	41	92.9064
Nitrogen	N	7	14.0067
Nobelium	No	102	(254)
Osmium	Os	76	190.2
Oxygen	O	8	15.9994
Palladium	Pd	46	106.4
Phosphorus	P	15	30.9738
Platinum	Pt	78	195.09
Plutonium	Pu	94	(242)
Polonium	Po	84	(210)
Potassium	K	19	39.102
Praseodymium	Pr	59	140.9077
Promethium	Pm	61	(147)
Protactinium	Pa	91	231.0359
Radium	Ra	88	226.0254
Radon	Rn	86	(222)
Rhenium	Re	75	186.2
Rhodium	Rh	45	102.9055
Rubidium	Rb	37	85.4678
Ruthenium	Ru	44	101.07

	Symbol	Atomic Number	Atomic Weight
Samarium	Sm	62	150.4
Scandium	Sc	21	44.9559
Selenium	Se	34	78.96
Silicon	Si	14	28.086
Silver	Ag	47	107.868
Sodium	Na	11	22.9898
Strontium	Sr	38	87.62
Sulfur	S	16	32.06
Tantalum	Ta	73	180.9479
Technetium	Tc	43	98.9062
Tellurium	Te	52	127.60
Terbium	Tb	65	158.9254
Thallium	Tl	81	204.37
Thorium	Th	90	232.0381
Thulium	Tm	69	168.9342
Tin	Sn	50	118.69
Titanium	Ti	22	47.90
Tungsten	W	74	183.85
Uranium	U	92	238.029
Vanadium	V	23	50.9414
Xenon	Xe	54	131.30
Ytterbium	Yb	70	173.04
Yttrium	Y	39	88.9059
Zinc	Zn	30	65.37
Zirconium	Zr	40	91.22

The Elements of Chemistry

Second Edition

The Elements of Chemistry

Second Edition

LAWRENCE P. EBLIN

OHIO UNIVERSITY

HARCOURT, BRACE & WORLD, INC.

New York / Chicago / San Francisco / Atlanta

ISBN: 0-15-522071-3

Library of Congress Catalog Card Number: 70-105695

Printed in the United States of America

Preface

The second edition of this text, like the first, is intended primarily for students in a one-semester or two-quarter course who do not necessarily intend to become chemists. I hope, however, that now and then a student will decide upon chemistry as a profession because of the interest which this book has helped to develop. I have attempted to write in a style that the uninitiated can understand, and yet to deal competently with the fundamentals. The student is entitled to an opportunity to acquire a correct concept of the nature of chemistry, and he should not be handicapped by an inadequate introduction in case he decides to explore the subject further.

It remains my conviction that there can be no real appreciation of the nature of chemistry without the ability to read and write chemical equations. This kind of basic chemical literacy must be one of the fundamental objectives of any introductory course. Since the interpretation of chemical equations requires an understanding of their stoichiometric implications, I have not attempted to keep this text completely free of arithmetic. Sample problems and solutions are employed in the presentation of each new type of calculation throughout the text. Such arithmetical topics as significant figures, exponential numbers, and units of measurement are discussed in the Appendixes for the benefit of those who need to refer to them at any time. There is also a table of logarithms, together with an explanation of how to use it.

The sequence of chapters has been carefully planned and reviewed, as has the sequence of topics within each chapter. I am convinced that no chemical topic is inherently too difficult to be understood by a beginning student if prerequisite concepts have previously been developed. The idea that atoms exist should be developed before the structure of atoms is considered; methods of expressing concentrations must be presented before the occasion arises to use them in connection with equilibrium constants.

In this text atomic structure is introduced in Chapter 3, and the Aufbau Principle is fully developed in Chapter 4. Electronegativity is considered when covalent bonding is first discussed; this makes it possible to present the concept of oxidation state understandably at an early point in the course. Oxidation and reduction are defined in terms of changes in oxidation state when they are first mentioned. The mole concept is introduced in Chapter 2 and is used consistently in calculations based on chemical equations, titrations, and Faraday's Law. A knowledge of mole fractions is assumed to be necessary for a real understanding of Dalton's Law of Partial Pressures.

Some topics have been developed historically because this seemed to be the most effective approach. Other topics are discussed without historical development; for example, the carbon-12 standard for atomic weights is presented without reference to previous atomic weight scales.

I have again chosen to avoid the term *valence* almost entirely, referring instead to the oxidation number or to the number of covalent bonds, as the case may be. I have also chosen to avoid equivalents and normality; instead, the presentation is consistently in terms of moles and molarity.

I have named inorganic compounds by using Roman numerals in parentheses to indicate oxidation states. I have chosen to use torr instead of mm Hg, Celsius instead of centigrade, and Kelvin instead of absolute. I have used a hexagon to represent cyclohexane, and a hexagon with an inscribed circle to represent benzene.

New topics in the second edition include colloids, osmosis, Brønsted-Lowry and Lewis concepts of acidity, kinetics, thermodynamics, and biochemistry. The treatment of thermodynamics is an informal one that enables the student to understand why natural events, including chemical reactions, proceed in one direction rather than the other. One of the most important reasons for studying chemistry is the pathway it opens for an effective attack on the problems of biology. It is for this reason that the second edition includes a chapter dealing with carbohydrates, lipids, proteins, and nucleic acids.

Each chapter concludes with a list of important terms introduced in the chapter, giving again their definitions; a list of exercises; and a list of references, each with a brief description that will help the student to use it most effectively. Answers for selected exercises are in the Appendix; answers for all the exercises are provided in a separate booklet available to instructors using the text.

The second edition of my laboratory manual, *The Elements of Chemistry in the Laboratory* (published concurrently), follows the organization of this edition of the text.

Many have helped, directly or indirectly, in the preparation of this book. I am especially indebted to my son, James, who read each chapter critically and provided answers for the numerical exercises, and to my wife, Geraldine, who typed the manuscript with meticulous care and played the role of lector for proofreading.

LAWRENCE P. EBLIN

Contents

The Elements of Chemistry

Second Edition

1

Chemical Concepts and the Atomic Theory

1-1 The Concept of Chemical Change

The progress that man has made toward becoming master of the earth has involved a growing capacity for utilizing the materials of his surroundings, such as wood, water, air, stone, and metals. *Chemistry* is the science concerned with the composition of all materials, wherever they may occur, and the transformations that they undergo when they are subjected to changing conditions.

Matter is whatever occupies space and has mass. *Materials* are the different varieties of matter, and the *properties* of a material are the characteristics that distinguish it from other materials. Two specimens of matter are presumed to be of identical composition if each of them has all the properties of the other. A change of conditions may cause some of the properties of a material to undergo a temporary change without causing the material to lose its identity; for example, water freezes when it is sufficiently cooled, and the ice melts to give liquid water that is identical with the original liquid in every respect. Such occurrences, in which there is no change in composition, are examples of *physical change.*

When sugar is heated, an odor develops and a black residue is obtained that does not have the sweet taste of sugar and will not dissolve in water as sugar does. This transformation involves a more profound change than the freezing of water or the melting of ice. After the change has occurred, the sugar is no longer present; in its place are other substances with properties markedly different from those of sugar. This type of transformation, in which there is a change in composition, is called *chemical change*, and a particular instance of it is called a *chemical reaction.*

One of man's earliest important achievements was chemical in nature—the discovery of fire and how to control it for cooking his food and for keeping himself warm. Other chemical arts that gradually developed include the tanning of leather, the making of glass, the extraction of dyes and medicines from plants, and the winning of such metals as iron and copper from their ores. These skills developed independently of one another, by the slow method of trial and error, over many centuries. Discoveries in one of the chemical arts meant little to those engaged in the other arts. There was no set of general principles to provide the intellectual command that characterizes the chemical profession today.

1-2 Varieties of Materials

Mixtures are materials of variable composition; their properties vary with the relative amounts of their ingredients. It is possible to separate the components of a mixture without depending on chemical reactions. For example, when a mixture of sand and sugar is treated with water, only the sugar dissolves in the water. The result is an intimate mixture of the sugar and water. Since it is not obvious to the eye that more than one substance is present, this mixture is called a *homogeneous mixture* or a *solution.* Mixtures composed of different kinds of particles large enough to be individually distinguished and observed are known as *heterogeneous mixtures.* A mixture of sand and sugar, for example, is a heterogeneous mixture.

The term *pure substance* is used to designate any material of which all specimens have the same properties and the same composition. Pure substances can be grouped in two classes, chemical compounds and elementary substances, according to whether they can be formed from simpler substances or broken down into simpler substances in a chemical reaction.

The starting materials of a chemical reaction are called *reactants*, and the substances formed are referred to as *products* of the reaction. A chemical reaction in which two or more reactants combine to form a single product is known as a *chemical combination*. For example, when mercury, a liquid metal with a silvery appearance, is heated in air, it combines chemically with oxygen, an important component of air, to form a red, powdery product that chemists have named "mercuric oxide."

When mercuric oxide is heated strongly, it ceases to exist as the red, powdery substance and in its place are the separate substances mercury and oxygen. This type of reaction, in which a single substance decomposes to form two or more different products, is known as *chemical decomposition*.

Any substance that can be formed by the chemical combination of other substances or can be decomposed to form two or more products is called a *chemical compound*. A compound, unlike a mixture, has an invariable composition and a fixed set of properties. A substance that cannot be decomposed and cannot be formed by the chemical combination of other substances is called an *elementary substance*. Thus, mercury and oxygen are elementary substances because neither can be decomposed to form other substances and neither can be formed by the combination of other substances, but mercuric oxide is a chemical compound because, on the one hand, it can be formed by the chemical combination of mercury and oxygen and, on the other hand, it can decompose to form mercury and oxygen.

The materials of our physical surroundings appear to be numberless. Indeed, it would be difficult to comprehend the nature of these materials without the concepts of mixture, chemical compound, elementary substance, combination, and decomposition. The usefulness of these distinctions is apparent from Figure 1-1, which charts the varieties of materials.

1-3 Alchemy and the Chemical Elements

Robert Boyle first clearly stated the basic concept that underlies all of modern chemistry in a book entitled *The Sceptical Chymist*, published in London in 1661. Boyle proposed the powerful idea that substances of all kinds are composed of a limited number of distinct varieties of matter, the *chemical elements*. A chemical compound is a decomposable substance because it is composed of at least two chemical elements. An elementary substance cannot be decomposed because it consists of only one chemical element.

FIGURE 1-1

Varieties of materials

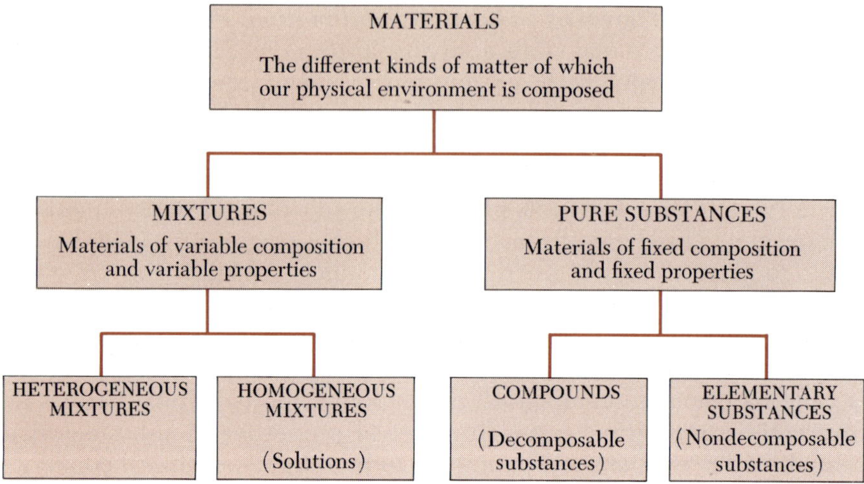

Prior to acceptance of the concept of chemical elements, it was generally assumed that the properties of different materials were only incidental qualities and that in changing a material into another material one merely divested it of certain properties and bestowed other properties upon it. We now recognize that each substance has a definite set of properties and that we change the identifying characteristics of a substance only by transforming it into one or more other substances through chemical change. Each product of a chemical reaction has its own unique properties, as does each of the reactants; but the products contain the same chemical elements as the reactants.

Alchemy was the medieval art concerned with efforts to perform such profound transformations as the conversion of base metals into gold and silver. Such efforts failed because the alchemists were attempting the transmutation of one chemical element into another. This is a difficult feat and was not accomplished until the present century. Nevertheless, the alchemists made many important contributions to scientific progress. Among their achievements were the discoveries of three of our most important acids (nitric, sulfuric, and hydrochloric) and five elements not previously known (zinc, bismuth, antimony, arsenic, and phosphorus).

Alchemy evolved into chemistry as magic and secrecy gave way to the combination of intelligence with honesty and open communication. Robert Boyle himself was an alchemist whose achievements did much to hasten the change.

1-4 Physical and Chemical Properties

The *physical properties* of a substance are those characteristics that may be observed or measured without transforming it into other materials. Common physical properties include color, odor, taste, conductivity for heat and electricity, density, ductility, tensile strength, melting point, boiling point, and solubility.

Many substances are liquid at ordinary temperatures, many others are solid, and some are gaseous. Most liquid substances become gaseous if sufficiently heated and solid if sufficiently cooled. Most substances that are ordinarily solid become liquid if sufficiently heated and in turn become gaseous at still higher temperatures. Substances that are gaseous at ordinary temperatures become liquid if sufficiently cooled and in turn become solid at still lower temperatures.

We speak of the liquid, solid, and gaseous conditions as the three *states of matter.* A *solid* object has a size and a shape that are characteristic of that particular object. A *liquid* specimen conforms to the shape of its container but has a volume that is characteristic of that particular specimen. A *gaseous* specimen not only assumes the shape of its container but occupies all the space that the container makes available to it; it exerts a pressure that is characteristic of that particular specimen in the available volume at the prevailing temperature.

Some of the important physical properties of substances refer to their tendency to undergo a *change of state*, that is, to change from one of the three states of matter to another. Water is characterized by its freezing point: 0° on the Celsius (centigrade) scale and 32° on the Fahrenheit scale. When water is cooled to this temperature it forms ice (solid water). When ice is heated at 0°C, it becomes liquid water. The change of state from liquid to solid, as when water becomes ice, is referred to as *freezing, solidification*, or *crystallization*; the change from solid to liquid, as when ice becomes liquid water, is called *melting* or *fusion.* The conversion of gas to liquid, as when ordinary air changes to liquid air at extremely low temperatures, is called *liquefaction.* When the vapor of a liquid returns to the liquid state, as when water vapor becomes dew during the night, the change is called *condensation.* The change from liquid to gas is called *volatilization*, *vaporization*, or *evaporation.* A liquid that readily changes into gas (vapor) is said to be *volatile.* Some solids vaporize without melting; this change of state is called *sublimation.* Dry Ice (carbon dioxide) sublimes.

In addition to its physical properties, every substance possesses other characteristics that are appropriately called *chemical properties.* These include chemical stability toward heat, light, and shock and behavior with other substances under varying conditions. Briefly stated, the chemical properties of a substance are those characteristics associated with its susceptibility to chemical change. For example, we say that nitroglycerin is highly explosive, gasoline is combustible, iron has the property of being converted into "iron rust" upon exposure to moist air, and grape juice has the property of producing alcohol by fermentation. All these statements are descriptions of chemical properties.

1-5 Mass, Weight, Volume, and Density

The *mass* of any selected portion of matter—for example, a solid object, or the liquid in a bottle, or the gas inside a flask—is a measure of the amount of matter present. Ordinarily the mass of a portion of matter is determined by the familiar process called weighing. When we determine the *weight* of an object, as with a spring scale, we measure the gravitational attractive force acting between the object and the earth. The magnitude of this force depends upon the mass of the object, the mass of the earth, and the distance between the center of the object and the center of the earth. Thus, unlike the mass, which is constant, the weight of an object varies from one point to another on the earth's surface.

The instrument with which we determine the mass of a specimen of matter in the laboratory is a chemical balance (Figure 1-2). Regardless of the external appearances of modern balances, many of which are of the "one-pan" variety, the principle of a chemical balance is that when we weigh the specimen we balance its attraction to the earth against the attraction the earth has for a set of "weights." For example, a 1-gram "weight" is an object that is known to have a mass of 1 gram. The amount of matter present in such an object is 1 gram, regardless of where the object is. The attraction between it and the earth changes with altitude, and therefore its weight—but not its mass—changes with altitude. However, the earth's attraction for a "weight" with a mass of 1 gram varies with altitude in the same manner as its attraction for any other 1-gram specimen of matter. Therefore, the weight of an object as determined by a chemical balance is a measure of its mass. Chemists often refer to the mass of a specimen as its weight because it is traditional and conventional to do so. Even a chemist who is careful to say that he has determined the mass of a specimen may also say that he has weighed it!

The *density* of a specimen of matter is the mass per unit volume. The specimen is weighed and its volume is measured; its density is then calculated by dividing the mass by the volume:

$$D = \frac{M}{V}$$

FIGURE 1-2

The balance

When the pointer is at the midpoint of the scale, the mass of the object is equal to the mass of the weights.

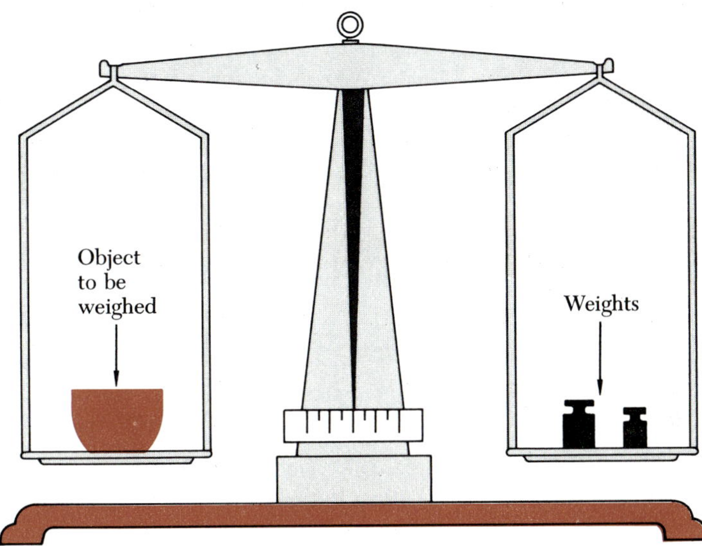

where D is the density, M is the mass, and V is the volume. In scientific laboratory work, the mass is usually expressed in grams and the volume in milliliters (cubic centimeters) so that the density is expressed in grams per milliliter (g/ml) or grams per cubic centimeter (g/cm^3).

The volume of a given mass of a substance changes with temperature, so the density must always refer to a specified temperature. For gases it is also necessary to specify the pressure, since the volume of a given mass of a gas is highly dependent on the pressure, as we shall see in Chapter 6.

1-6 The Law of Conservation of Mass

When faith in magic was widespread, many believed that matter could be created and destroyed. It appeared that some natural phenomena, such as the growth of a tree, involved creation of matter and that matter disappeared when a candle was burned. Almost 2500 years ago, however, the Greek philosopher Democritus stated that things cannot be formed from nothing and cannot be reduced to nothing. Nevertheless, it was not until the eighteenth century, when fairly sophisticated means of weighing reacting substances were available, that scientists seriously investigated this subject.

The principle that there is no detectable gain or loss of mass during any physical or chemical change is now accepted by all scientists. This principle is known as the *Law of Conservation of Mass.*

Although the conservation of mass has been assumed to be true for the last two centuries, it has been subjected to rigorous experimental tests within the present century. In connection with the development of atomic energy, mass has been converted into energy. Energy is hardly "nothing," however, and with modification for this conversion, the Law of Conservation of Mass is thought to be completely valid.

1-7 Air and Combustion

The development of chemistry into a quantitative science (one that answers the question, How much?) was not possible until the chemical balance was sufficiently developed for experimental work to include mass determinations of considerable precision. Prior to this development, most chemical discoveries were limited to qualitative observations (determinations of the materials present but not their precise relative quantities).

The alchemists were well acquainted with the change that many metals undergo when they are heated in air. They gave to this transformation the name "calcination," derived from *calx*, the alchemical term for the crumbly residue that remains. The alchemists concluded correctly that calcination must be related to the burning that takes place when combustible materials such as wood are heated in air. They knew that when wood burns it disappears and ashes remain and that when a metal undergoes complete calcination it likewise disappears and the calx remains.

The *phlogiston theory of combustion* was proposed by Johann Becker in 1667 and developed into an elaborate system of chemical thought by his student Georg Stahl. The theory attributed combustion to the escape of phlogiston, a "fire substance" or "element of combustibility" assumed to be present in all combustible substances. According to this theory, when wood burns, phlogiston escapes, leaving ashes; therefore, it was concluded, wood is composed of ashes and phlogiston. The smelting of ores by heating with charcoal was explained as a transfer of phlogiston from the charcoal to the ore, and respiration was explained in terms of phlogistic concepts to the satisfaction of all but a few. In spite of the faulty premise, much chemical progress was made during the period that the phlogiston theory was in vogue.

One weakness of the phlogiston theory was that it could not explain why air must be present for burning to occur. Worse yet, it could not explain the fact that the calx of a metal, assumed to be formed by the escape of phlogiston from the metal, weighed more than the original metal.

In 1774 Antoine Laurent Lavoisier, in France, was performing experiments on what happens when a metal is heated in air. He placed a weighed sample of tin in a glass retort, which he sealed and weighed and then heated. The weight of the sealed retort did not change, but the tin itself (covered with calx) gained weight. Since the weight of the empty retort had not changed, Lavoisier concluded that the weight gained by the tin came from the air that was in the retort when it was sealed. He therefore tried, by heating the calx, to remove the part of air that had combined with the tin, but he could not induce the calx to decompose. However, Joseph Priestley, in England, strongly heated the calx of mercury with a large burning glass (Figure 1-3) and succeeded in liberating a gas. He collected the gas over water and found that it supported combustion better than did ordinary air. He found that mice could live in it much longer than in an equal volume of ordinary air. He even ventured to breathe the gas himself, noting that "my breast felt peculiarly light and easy for some time afterwards." This was the gas

FIGURE 1-3

Schematic diagram of apparatus used by Priestley in his discovery of oxygen

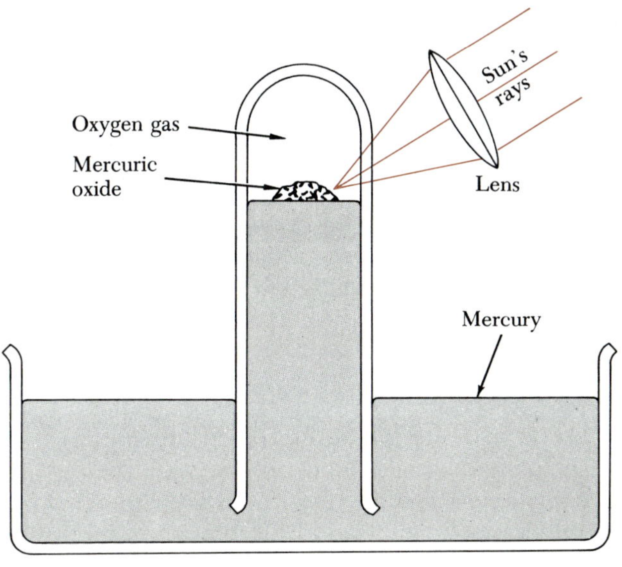

now known as oxygen, and Priestley is credited with its discovery in 1774.

Lavoisier repeated Priestley's experiments, making accurate quantitative measurements. He heated mercury in a retort and produced the calx, noting what part of the volume of the air in the retort was consumed and the amount of calx produced. Then he heated the calx more strongly and found that the volume of gas generated was equal to the volume that had previously disappeared. He also noted that after part of the air was removed by combining it with mercury, the gas remaining was unfit for respiration and unable to support combustion. Yet the gas produced by strongly heating the calx was better than air in both respects.

Lavoisier gave the name *azote* (from the Greek, meaning "without life") to the portion of ordinary air that is incapable of supporting combustion. The element is still known as azote in the French language, but in English it is known as *nitrogen* ("niter former") because of its presence in niter (which is now called potassium nitrate). Although Lavoisier was probably the first to show that nitrogen is an element, it had been observed earlier by Daniel Rutherford, a student at the University of Edinburgh, as the residual gas remaining after burning carbonaceous material in air. Rutherford is credited with the discovery of nitrogen in 1772 because he reported his results in a medical dissertation in that year. The gas was found independently at about the same time by Priestley and others.

Lavoisier gave the name *oxygen* ("acid former") to the active component of air, the gas that supports combustion and respiration and combines with metals when they undergo calcination. His assumption that this element is a constituent of all acids was incorrect.

The quantitative experiments of Lavoisier, and his highly competent interpretation of the results, were largely responsible for the overthrow of the phlogiston theory and its replacement by the *oxygen theory of combustion.* His findings left little doubt that combustion and calcination involve combination with oxygen. The new ideas were ably set forth in Lavoisier's textbook, *Traité élémentaire de chimie*, published in 1789. Its appearance marked the culmination of the Chemical Revolution and the dawn of the new chemical era in which we are now living.

1-8 Number and Abundance of the Elements

The gas now known as hydrogen has been known for several centuries as a flammable gas that could be produced by bringing certain substances together and allowing them to react. Robert Boyle knew that this gas is produced when iron is added to acids. Henry Cavendish, in

England, collected quantities of the gas and studied it systematically; he is credited with the discovery of hydrogen in 1766.

In 1781 Cavendish proved that water is the *only* substance formed when hydrogen burns in air or in pure oxygen. This indicated that hydrogen is one of the chemical elements and that water is a compound composed of hydrogen and oxygen. Lavoisier therefore, in establishing a new system of chemical nomenclature, gave the name *hydrogen* ("water former") to the element. (Cavendish had called it "inflammable air.")

The discovery that air is a mixture mainly of nitrogen and oxygen and the proof that water is a compound of hydrogen and oxygen were highlights of the Chemical Revolution that set the stage for the discovery of additional chemical elements. Lavoisier's first list of elements, which was included in his 1789 textbook, contained 33 entries. It included seven metals known even in prehistoric times: gold, silver, copper, tin, lead, mercury, and iron. The list was suddenly lengthened, in 1807 and 1808, when Sir Humphry Davy used electric current to obtain specimens of five metals never before seen by man: sodium, potassium, calcium, strontium, and barium. Since then the list of elements has been extended until the number now stands at 104.

On the earth the most abundant element is oxygen. Almost half the mass of the earth's crust, including the oceans and the atmosphere, is due to oxygen (Figure 1-4). A little more than half of the remainder is due to silicon. The next most abundant elements are six metals: alumi-

FIGURE 1-4

Relative abundance of elements in the earth's crust

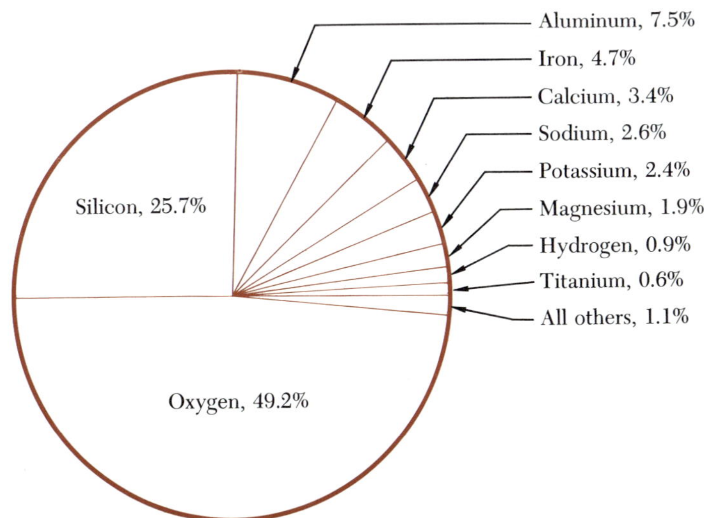

num, iron, calcium, sodium, potassium, and magnesium. These eight elements constitute approximately 98 percent of our environment. The remainder consists of hydrogen, titanium, and small quantities of other elements.

About three-fourths of the elements are classified as *metals*, characterized by so-called metallic properties. These properties include high conductivity for heat and electricity, ductility, malleability, and the peculiar reflective power commonly called "metallic luster." Most of the remaining elements do not have these properties and are called *nonmetals*; well-known examples are oxygen, sulfur, chlorine, nitrogen, and hydrogen. A few of the elements are not easily classified as metals or nonmetals and are known as *metalloids*; among them are silicon, germanium, and arsenic.

1-9 Atoms and Molecules

The atomic theory was fully developed only a little more than a century and a half ago. True, certain Greek philosophers, notably Leucippus and Democritus, concluded 2500 years ago that all matter is composed of invisibly small, separate, and indestructible particles that they called *atoms* (Greek: *atomos*, indivisible). We are not sure what led these philosophers to this view of the physical world. Not all philosophers accepted the atomistic view of the structure of matter, and it was largely forgotten until the seventeenth century.

As we have seen, Robert Boyle and others developed the idea that the many different varieties of matter are composed of only a few chemical elements. This concept led in turn to the idea that there are only as many different kinds of atoms as there are chemical elements. In this form the atomic theory was utilized by John Dalton, an English scientist, to explain the laws of chemical combination that had been formulated in the century following Boyle. His explanations were presented in his book, *New System of Chemical Philosophy*, published in 1808.

Dalton's assumptions were (1) that all the atoms of a particular element are alike and have the same mass and (2) that chemical combination involves the chemical bonding (by unspecified means) of definite numbers of atoms of each of the combining elements to form *molecules* of the resulting compound. This theory offers a simple explanation for what initially appears to be a set of highly complicated facts. For example, the Law of Conservation of Mass applies to chemical reactions as well as to merely physical transformations because the atoms involved in either type of process retain their identities. In a chemical reaction, the molecules of the products contain the same atoms that were originally present in the molecules of the reactants. Each individual atom

FIGURE 1-5

Molecules of two compounds

A molecule of a compound consists of more than one kind of atom.

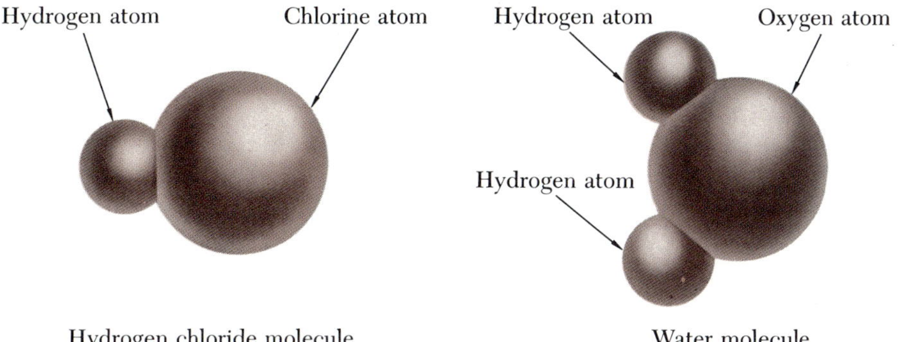

FIGURE 1-6

Molecules of two elements

A molecule of an element consists of one or more atoms of the same kind.

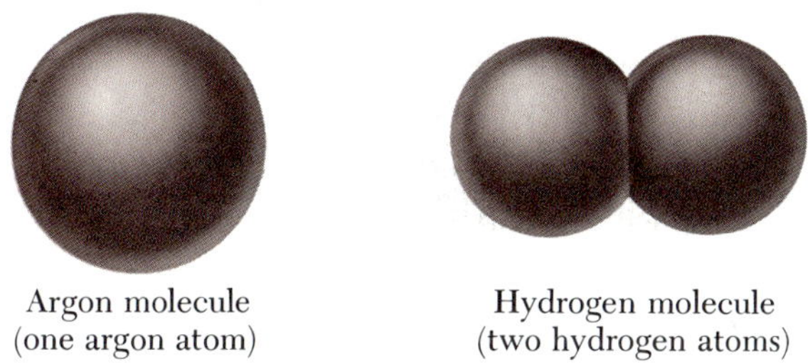

retains its original mass, so the sum of the masses of the molecules of the products is the sum of the masses of the molecules of the reactants.

A molecule can be characterized as a minute, uncharged particle of a substance that moves about as a whole so that its component parts do not become detached during the motions. A molecule of a compound consists of more than one kind of atom (Figure 1-5), whereas a molecule of an element consists of one or more atoms of the same kind (Figure 1-6). From the standpoint of chemistry, a molecule may be thought of as *the smallest particle of a substance as it normally exists.*

1-10 The Law of Constant Composition

The Law of Constant Composition is a direct consequence of the fact that each molecule of a substance consists of a definite number of atoms of each of its constituent elements, each atom having a fixed and characteristic mass. It is a fact of nature, then, that two elements uniting do not combine in fortuitous quantities but in definite proportions. The *Law of Constant Composition* states that the proportion by weight of the elements present in any pure compound is always the same. This law was not discovered by any one man but came about by gradual development as chemistry became an increasingly exact science. In fact, the law has been tacitly assumed by most chemists since about 1630. However, its validity could not be firmly established until the chemical balance was developed into an instrument of precision.

It has been found by experiment that 1.000 gram of oxygen combines with 1.520 grams of magnesium or with 4.086 grams of zinc. When quantities other than these are involved, the ratio is nevertheless the same. Thus, 1.000 kilogram of oxygen combines with 1.520 kilograms of magnesium or with 4.086 kilograms of zinc; likewise, 1.000 pound of oxygen combines with 1.520 pounds of magnesium or with 4.086 pounds of zinc. Even if magnesium oxide is obtained in ways other than by direct union of magnesium with oxygen, it nevertheless always contains 1.520 parts (by weight) of magnesium for 1.000 part of oxygen. Likewise, regardless of the method used to prepare zinc oxide, it always contains 4.086 parts (by weight) of zinc for 1.000 part of oxygen.

1-11 The Law of Simple Multiple Proportions

In many instances the same elements are found to combine in different, but still definite, proportions to form two or more different compounds. It is well known that carbon and oxygen combine to give two different compounds, called carbon dioxide and carbon monoxide. Carbon monoxide is 42.88 percent carbon and 57.12 percent oxygen by weight; carbon dioxide is 27.27 percent carbon and 72.73 percent oxygen. A direct comparison of these percentages reveals nothing significant. However, if we divide the percentage of oxygen by the percentage of carbon (that is, grams of oxygen per 100 grams of oxide by grams of carbon per 100 grams of oxide) for each of these compounds, we obtain the number of grams of oxygen combined with 1 gram of carbon (or the number of pounds of oxygen combined with 1 pound of carbon, etc.).

FOR CARBON MONOXIDE: $$\frac{57.12 \text{ g oxygen}}{42.88 \text{ g carbon}} = 1.33 \text{ g oxygen per g carbon}$$

FOR CARBON DIOXIDE: $$\frac{72.73 \text{ g oxygen}}{27.27 \text{ g carbon}} = 2.66 \text{ g oxygen per g carbon}$$

Clearly, in carbon dioxide twice as much oxygen is combined with a given amount of carbon as in carbon monoxide.

If we choose, we may divide the percentage of carbon by the percentage of oxygen for each of these compounds and obtain the number of grams of carbon combined with 1 gram of oxygen.

FOR CARBON MONOXIDE: $$\frac{42.88 \text{ g carbon}}{57.12 \text{ g oxygen}} = 0.750 \text{ g carbon per g oxygen}$$

FOR CARBON DIOXIDE: $$\frac{27.27 \text{ g carbon}}{72.73 \text{ g oxygen}} = 0.375 \text{ g carbon per g oxygen}$$

Thus, in carbon dioxide only half as much carbon is combined with a given amount of oxygen as in carbon monoxide.

This example is one of many that might be used to illustrate the *Law of Simple Multiple Proportions.* This law states that if there exists a series of two or more compounds containing the same elements, there is a ratio of small whole numbers between the different weights of one element that are combined with any chosen weight of another element. In the case of carbon dioxide and carbon monoxide, the ratio is expressed as 2:1 or 1:2, depending on how the calculation is made. The law is excellently illustrated by the five oxides of nitrogen, in which the quantity of oxygen combined with 1 gram of nitrogen is 0.571, 1.142, 1.713, 2.284, and 2.855 grams, respectively. If each of these numbers is divided by the smallest of the set, the results are 1, 2, 3, 4, and 5.

The Law of Simple Multiple Proportions can be easily accounted for by the atomic theory. Thus we picture the two oxides of carbon as having the chemical formulas CO (carbon monoxide) and CO_2 (carbon dioxide), where C is the symbol representing one carbon atom, O represents one oxygen atom, and the subscript 2 attached to the symbol for oxygen in the second formula represents the fact that there are two atoms of oxygen combined with one carbon atom. There is twice as much oxygen per unit mass of carbon in the second oxide as there is in the first, simply because there are two atoms of oxygen per atom of carbon in the second oxide but only one atom of oxygen per atom of carbon in the first.

1-12 The Law of Definite Proportions

Another law dealing with weight relations in chemical reactions is the *Law of Definite Proportions*, which states that in any chemical reaction there is a definite ratio between the weight of each reactant or product and the weight of any other reactant or product (Figure 1-7). As an illus-

FIGURE 1-7

The Law of Definite Proportions

In any chemical reaction, the ratio of the amounts of any two substances, whether reactants or products, remains the same even though the amounts themselves are different.

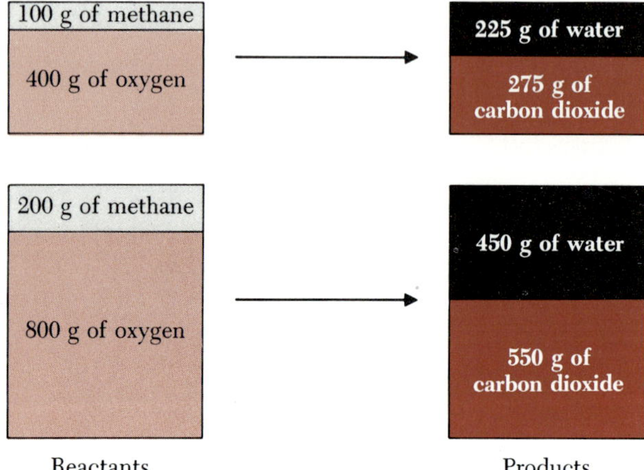

tration of this law, when methane (the chief component of natural gas) is burned, the products are carbon dioxide and water:

$$\underset{100\ \text{g}}{\text{Methane}} + \underset{400\ \text{g}}{\text{oxygen}} \longrightarrow \underset{275\ \text{g}}{\text{carbon dioxide}} + \underset{225\ \text{g}}{\text{water}}$$

In this reaction, the weight of oxygen involved is 4 times the weight of methane being burned, the weight of carbon dioxide formed is 2.75 times the weight of methane burned, and the weight of water produced is 2.25 times the weight of methane consumed. As indicated by the numbers written beneath each reactant and product, the burning of 100 grams of methane requires 400 grams of oxygen and produces 275 grams of carbon dioxide and 225 grams of water.

The Law of Definite Proportions is a necessary corollary to the Law of Constant Composition and is readily understood in terms of atoms and molecules:

1. A molecule of any substance consists of definite numbers and kinds of atoms.
2. A definite number of molecules of each reactant is needed to provide the number of atoms of each kind that are to constitute the molecules of the reaction products.
3. Since each molecule of each reaction product consists of definite numbers and kinds of atoms, a definite number of molecules of each product will be obtained.

1-13 Chemical Symbols and Molecular Formulas

The elements of which a compound is constituted and the number of atoms of each element in one of its molecules are indicated by its *molecular formula.* The molecular formula consists of the *symbols* of the constituent elements with subscripts to indicate the number of atoms of each element present in one molecule. For example, the molecular formula for table sugar is $C_{12}H_{22}O_{11}$. This formula means that one sugar molecule contains 12 carbon atoms, 22 hydrogen atoms, and 11 oxygen atoms.

Obviously, chemical formulas mean nothing unless the symbols of the elements can be recognized. Although more than a hundred elements are now known, many of them are scarce and do not occur in substances with which we are familiar. The names and symbols for some of the more common metals and nonmetals are listed in Tables 1-1 and 1-2. Note that the symbols are often the initial letter or first two letters

TABLE 1-1

Common elements having symbols derived from their English names

Metals		*Nonmetals and metalloids*	
Aluminum	Al	Argon	Ar
Barium	Ba	Arsenic	As
Beryllium	Be	Boron	B
Bismuth	Bi	Bromine	Br
Cadmium	Cd	Carbon	C
Calcium	Ca	Chlorine	Cl
Cesium	Cs	Fluorine	F
Chromium	Cr	Helium	He
Cobalt	Co	Hydrogen	H
Lithium	Li	Iodine	I
Magnesium	Mg	Krypton	Kr
Manganese	Mn	Neon	Ne
Molybdenum	Mo	Nitrogen	N
Nickel	Ni	Oxygen	O
Platinum	Pt	Phosphorus	P
Titanium	Ti	Selenium	Se
Uranium	U	Silicon	Si
Vanadium	V	Sulfur	S
Zinc	Zn	Xenon	Xe

TABLE 1-2

Elements having symbols derived from their Latin names

English name	*Latin name*	*Symbol*
Antimony	Stibium	Sb
Copper	Cuprum	Cu
Gold	Aurum	Au
Iron	Ferrum	Fe
Lead	Plumbum	Pb
Mercury	Hydrargyrum	Hg
Potassium	Kalium	K
Silver	Argentum	Ag
Sodium	Natrium	Na
Tin	Stannum	Sn

of the English names but that some symbols are derived from the Latin names. The symbol for one of the elements, tungsten (W), is derived from its German name (*Wolfram*). The first letter of a chemical symbol is always capitalized, and the second letter is never capitalized.

Formulas for compounds do not always contain subscripts. The subscript is always understood to be 1 if no subscript is written. The compound formed by combining hydrogen with chlorine consists of molecules containing one atom of each element; the formula is written HCl, meaning H_1Cl_1. Water contains two hydrogen atoms for every oxygen atom, and for this reason its formula is written H_2O (meaning H_2O_1).

There is no strict rule concerning the sequence in which symbols should appear in a formula. However, custom dictates that the formula for water should be written H_2O rather than OH_2, the symbol for oxygen should appear last in the formula for any oxide, the symbol for chlorine should be last in the formula for any chloride, and the symbol for the metal should appear first in the formula for a compound consisting of a metal and a nonmetal.

The formulas of some compounds are indicated rather clearly by their names. For example, *carbon dioxide* refers to the compound whose formula is CO_2, and *carbon monoxide* refers to the compound with the formula CO.

The formula for an elementary substance indicates the number of atoms present in one molecule of the substance. Thus, the formula for elementary hydrogen, H_2, indicates that a molecule of hydrogen consists of two hydrogen atoms; the formula for argon gas, Ar, indicates that an argon molecule consists of a single argon atom.

1-14 The Scope of Chemistry

Chemistry, in general, deals with the composition, structure, and properties of individual substances and with the chemical reactions in which they take part. Since the days of the alchemists we have progressed so far that not only has the disciplined science of chemistry come into being, but it has become so complex that we divide it into various fields.

Analytical chemistry is concerned with the separation, identification, and determination of the composition of individual substances and with the principles involved in these operations. *Qualitative analysis* deals primarily with the identification of the ingredients of mixtures and the constituents of chemical compounds and with the separation processes that are often required for successful identifications. *Quantitative analysis* is concerned with procedures for ascertaining the quantity of each ingredient or constituent present.

Physical chemistry is concerned with the laws and theories that underlie all of chemistry and its applications. In its search for understanding of the structure of matter and its transformations through chemical change, physical chemistry relies upon the fundamental principles of physics.

Organic chemistry deals with the chemistry of carbon compounds, and *inorganic chemistry* is concerned with all the other elements. *Biochemistry*, or *biological chemistry*, strives for an understanding of the chemical structure of living materials and the chemistry of the life processes.

The chemistry with which this introductory text is concerned may be described as *general chemistry*: the basic concepts, principles, and laws that are universally recognized to be the foundations of chemistry, together with factual material that is either inherently important or relevant as illustrative material. The main objective in this first chapter has been to show the development of the basic chemical laws and their explanation by the atomic theory. The atom is not the simple thing that Dalton probably believed it to be, but our main purpose here has been to show why chemists came to believe in the *existence* of atoms long before anything was known about their structure or about how they are bonded together in molecules.

NEW TERMS

Alchemy: the medieval art that aspired to transmute base metals into gold, discover a universal cure for disease, and prolong life indefinitely. Chemistry evolved from alchemy.

Atom: one of the unit particles of elements, of which molecules are composed.

Chemical change: change in composition.

Chemical combination: a chemical reaction in which two or more different substances combine to form one substance.

Chemical decomposition: the type of chemical reaction in which one substance produces two or more other substances.

Chemistry: the science concerned with the composition of substances and the transformations they undergo.

Compound: a distinct substance with a fixed set of properties, composed of two or more constituent elements chemically combined in definite proportions.

Density: the ratio of the mass of a specimen of matter to its volume.

Element (chemical element): one of a limited number of distinct varieties of matter of which all substances are composed. A chemical compound is composed of at least two chemical elements, whereas an elementary substance is composed of a single chemical element.

Elementary substance: a substance that cannot be decomposed and cannot be formed by the chemical combination of other substances.

Freezing (solidification, crystallization): the change from the liquid state to the solid (crystalline) state.

Gaseous state: the state of matter in which a substance takes the shape of its container and fills all parts of the container to the same degree.

Liquefaction: the conversion of a gas to a liquid.

Liquid state: the state of matter in which a substance takes the shape of its container but has a volume of its own. The molecules are free to move among themselves but do not have the tendency to separate from one another that gas molecules have.

Mass (of a body): the amount of matter in a body.

Materials: the different kinds of matter of which the universe is composed. Some materials are mixtures, for example, soil and paint; some are pure substances, for example, water and sulfur.

Matter: something that occupies space and possesses mass.

Melting (fusion): the change from the solid to the liquid state.

Metalloids: chemical elements that have both metallic and nonmetallic properties.

Metals: chemical elements that, when not chemically combined with other elements, have a peculiar luster and are good conductors of heat and electricity.

Mixture: a material composed of two or more ingredients that are not chemically combined. A homogeneous mixture is one that appears to consist of only one substance; a heterogeneous mixture is one in which the different particles are large enough to be individually observed.

Molecular formula: the combination of symbols and subscripts representing the actual number of atoms of each element in one molecule of a substance; for example, the molecular formulas for elementary oxygen and for water are O_2 and H_2O, respectively.

Molecule: one of the smallest particles of a substance, consisting of one or more atoms of each constituent element.

Nonmetals: chemical elements that lack the luster and conductivity that are characteristic of metals.

Oxide: a compound containing oxygen and one other element.

Physical change: change in form or state with no change in composition.

Products: the substances formed in a chemical reaction.

Properties: the characteristics of a material that distinguish it from other materials. These characteristics are usually classified as physical and chemical properties.

Reactants: the starting materials in a chemical reaction.

Solid state: the state of matter in which a substance has no perceptible tendency to flow but has sufficient rigidity to maintain a definite shape independent of the shape of any container in which it is placed.

Solutions: homogeneous mixtures.

Sublimation: evaporation of a solid.

Substance: a material that has a fixed composition and a definite set of properties because it has only one ingredient. As used in chemistry, a substance is pure. It may be a compound or an elementary substance, but its molecules are all alike.

Symbol: a letter or a pair of letters used to represent an element or one atom of an element.

Transmutation: the conversion of one element into another. In ordinary physical and chemical processes, there is no transmutation of elements. Radioactivity involves transmutation.

Vaporization (evaporation): the change from the liquid to the gaseous (vapor) state.

Weight: as commonly used in chemical practice, weight is a measure of the mass of a specimen of matter, as determined with a chemical balance. Technically, the weight of a body is the force with which it is attracted toward the earth.

The Laws of Chemical Combination

Law of Conservation of Mass: As the law applies in chemical reactions, it states that the total mass of the products of a reaction is the same as the total mass of the reactants.

Law of Constant Composition: The proportion by weight of the elements present in any pure compound is always the same.

Law of Definite Proportions: In any chemical reaction there is a definite ratio between the weight of each reactant or product and the weight of any other reactant or product.

Law of Simple Multiple Proportions: If there exists a series of two or more compounds containing the same elements, there is a ratio of small whole numbers between the different weights of one element that are combined with a chosen weight of another element.

EXERCISES

1-1 Distinguish between the meanings of the following paired terms: materials and substances; alchemy and chemistry; gas and air; oxygen and nitrogen; combination and decomposition; elements and compounds; solutions and heterogeneous mixtures; physical and chemical properties; physical and chemical changes; metals and nonmetals; atoms and molecules; mass and weight; density and weight; freezing and melting; phlogiston theory and oxygen theory of combustion; organic chemistry and biochemistry; symbols and formulas; Co and CO.

1-2 What facts of nature are embodied in the following generalizations?
(a) Law of Constant Composition
(b) Law of Simple Multiple Proportions
(c) Law of Definite Proportions
(d) Law of Conservation of Mass

1-3 Name several materials that have been discovered or synthesized within the present century and several known since prehistoric times.

1-4 Name some materials whose uses depend on the following properties: electrical conductivity, low density, high electrical resistance, ease of stretching, difficulty of stretching, combustibility, noncombustibility, high melting point, solubility in water, insolubility in water, hardness, tensile strength.

1-5 Describe the differences among the three states of matter.

1-6 What is implied when we say that air is a mixture but water is a compound?

1-7 Classify the following properties as chemical or physical: color, flammability, compressibility, porosity, solubility.

1-8 Classify the following changes as chemical or physical: melting of ice, burning of coal, growth of a tree, digestion of food, production of light by an incandescent lamp, production of light by a burning candle.

1-9 Classify the following materials as elements, compounds, solutions, or heterogeneous mixtures: oxygen, water, air, mercury, milk.

1-10 Describe the difference between molecules of compounds and molecules of elementary substances.

1-11 A block of wood 1.00 meter long, 10.0 centimeters wide, and 5.0 centimeters thick weighs 36.0 kilograms. Calculate its density.

1-12 An empty graduate cylinder weighs 43.72 grams. With 25.0 milliliters of ethyl alcohol added, it weighs 63.47 grams. Calculate the density of ethyl alcohol.

1-13 Carbon dioxide is 27.27% carbon and 72.73% oxygen by weight. What weight of CO_2 will be formed by burning 100 grams of carbon? How much CO_2 will be formed from 100 pounds of carbon?

1-14 How many atoms are in one molecule of table sugar, $C_{12}H_{22}O_{11}$? In one molecule of carbon dioxide? In 100 molecules of carbon monoxide?

1-15 What is the formula of the compound whose name is nitrogen dioxide? Dinitrogen oxide? Dinitrogen pentoxide?

REFERENCES

Asimov, I., *A Short History of Chemistry*, Doubleday, Garden City, N.Y., 1965 (paperbound). Chemical understanding based on historical developments.

Dinga, G. P., "The Elements and the Derivation of Their Names and Symbols," *Chemistry*, **41**(2), 20 (1968). Includes date of discovery of each element, with name and nationality of discoverer.

Holton, G., and Roller, D. H. D., *Foundations of Modern Physical Science*, Addison-Wesley, Reading, Mass., 1958. The fundamental concepts of physics and chemistry presented in historical perspective.

Ihde, A. J., *The Development of Modern Chemistry*, Harper & Row, New York, 1964. The historical foundations of chemistry and the growth of the specialized areas.

Kesselman, B., "The Skeptical Chemist," *Chemistry*, **39**(1), 8 (1966). Robert Boyle's search for truth.

Kieffer, W. F., "Chemistry, Curiosity, and Culture," *Journal of Chemical Education*, **45**(9), 551 (1968). The importance to nonscientists of the part science plays in modern culture.

Leicester, H. M., *The Historical Background of Chemistry*, Wiley, New York,

1957. The development of chemical concepts from ancient times through the present.

Read, J., *Through Alchemy to Chemistry*, Harper & Row, New York, 1963 (paperbound). The rise and spread of alchemy and the evolution of chemistry.

Walton, H. F., "The Humanistic Values of Science," *Chemistry*, **38**(12), 8 (1965). Scientific exploration as an artistic enterprise.

Weeks, M. E., and Leicester, H. M., *Discovery of the Elements*, 7th ed., Chemical Education, Easton, Pa., 1968. Accounts of the discoveries of the chemical elements, with biographical sketches of the discoverers.

Winderlich, R. (transl. R. E. Oesper), "History of the Chemical Sign Language," *Journal of Chemical Education*, **30**(2), 58 (1953). Old and new symbolism for representing elements and compounds.

2

Stoichiometry and the Mole Concept

2-1 The Atomic Weights of the Elements

One of the highly significant events of the early years of the present century, because it contradicted one of the basic assumptions of Dalton's atomic theory, was the discovery that many chemical elements occur as mixtures of atoms having different masses. Atoms of the same element that have different masses are called *isotopes* of that element. One effect of the discovery of their existence was that it complicated the concept of *atomic weight*, which is the number that tells us how the mass of an atom of one element compares with the mass of an atom of another element.

Like any other system of measurement, the atomic weight scale is completely arbitrary and it has been changed from time to time. Since 1961 the standard for atomic weights has been the most abundant isotope of carbon, which was defined to have a mass of exactly 12 arbitrary units called *atomic mass units* (amu). This isotope of carbon is therefore referred to as "carbon-12" and is symbolized as C-12 or as ^{12}C. Another isotope of carbon has been found, by experimental measurement, to have a mass of 13.00344 amu. This isotope of carbon is referred to as "carbon-13" because its mass is so nearly 13. All isotopes have masses (isotopic weights) that are very nearly whole numbers of atomic mass units. The integer that approximates the isotopic weight is called the *mass number* of an isotope.

Individual atoms are so small that the tiniest piece of matter visible to our eyes contains more atoms than one of us could count in a lifetime. Chemical operations are usually carried out with large numbers of atoms – numbers so great as to make it very likely that any collection of atoms being worked with will contain the different isotopes of each element in the proportion corresponding to their natural distribution. For this reason it is the *average* mass of the atoms of a particular element that figures in the calculations of the chemist. The masses of the individual isotopes and their relative abundances determine what this average will be. As an example, 80.077 percent of the atoms in natural boron are B-11 atoms (mass 11.009 amu) and 19.923 percent are B-10 atoms (mass 10.013 amu). Therefore, the average mass of boron is

$$(0.80077 \times 11.009 \text{ amu}) + (0.19923 \times 10.013 \text{ amu}) = 10.811 \text{ amu}$$

This weighted average of the masses of the isotopes is usually referred to as the *atomic weight* of the element, although some chemists refer to this quantity as the "relative atomic mass" of the element. The isotopic weights and atomic weights of four elements are given in Table 2-1.

TABLE 2-1

Isotopic weights and atomic weights of four elements

Element	*Mass number*	*Isotopic weight* (amu)	*Abundance* (%)	*Atomic weight* (amu)
Hydrogen	1	1.007825	99.985	1.0080
	2	2.01410	0.015	
Boron	10	10.013	80.077	10.81
	11	11.009	19.923	
Carbon	12	12.00000	98.892	12.011
	13	13.00344	1.108	
Oxygen	16	15.99491	99.759	15.9994
	17	16.99914	0.037	
	18	17.99916	0.204	

The atomic weights of all the elements are listed inside the back cover of this book.

2-2 Molecular Weights

The mass of a molecule is simply the sum of the masses of the atoms it contains. The *molecular weight* of a substance is a number that tells us how the mass of an average molecule of the substance compares with the mass of a carbon-12 atom taken as the standard. In other words, molecular weights represent average masses of molecules in terms of the same atomic mass units (amu) that are used for atomic weights.

The molecular formula for a substance states how many atoms of each kind are present in one molecule of the substance. Therefore, the molecular weight of a substance can be computed by adding the atomic weights for all the atoms represented in the molecular formula.

EXAMPLE 1

Compute the molecular weight of table sugar, $C_{12}H_{22}O_{11}$.

$$\begin{aligned} 12 \text{ carbon atoms} \times 12.01 \text{ amu/atom} &= 144.1 \text{ amu} \\ 22 \text{ hydrogen atoms} \times 1.01 \text{ amu/atom} &= 22.2 \text{ amu} \\ 11 \text{ oxygen atoms} \times 16.00 \text{ amu/atom} &= 176.0 \text{ amu} \end{aligned}$$

The molecular weight is the total, 342.3 amu.

EXAMPLE 2

Compute the molecular weight of elementary oxygen, O_2.

The molecular formula means that elementary oxygen is diatomic; i.e., each molecule is composed of two oxygen atoms.

$$2 \text{ oxygen atoms} \times 16.00 \text{ amu/atom} = 32.00 \text{ amu}$$

Therefore, the molecular weight of O_2 is 32.00 amu.

EXAMPLE 3

What is the molecular weight of helium, He?

The fact that the molecular formula for helium is the same as the chemical symbol for the element means that helium is a monatomic gas; i.e., each molecule consists of a single atom. The atomic weight of helium is 4.00260 amu. Therefore, the molecular weight of helium is also 4.00260 amu.

In later chapters we shall discover some of the experimental procedures for determining molecular weights. These procedures make it possible to know the molecular weight of a substance even though its molecular formula is not known.

2-3 Cannizzaro's Principle and Atomic Weights

A molecule always contains a whole number of atoms of each constituent element. In a group of compounds containing a particular element, it is probable that there will be some compounds containing only one atom of that particular element in each molecule, others containing two atoms of the element in each molecule, and so on. The larger the group of compounds containing the element, the more likely it becomes that there will be at least one compound in the group that contains only one atom of the element in each molecule.

After a compound has been subjected to chemical analysis and has had its molecular weight determined, it is possible to calculate the portion of the molecular weight that is due to each of its constituent elements. For example, chemical analysis of sulfur dioxide shows that this compound is 50 percent sulfur and 50 percent oxygen by weight. Experimental determination of its molecular weight gives 64 amu as the approximate result. Therefore, the portion of the molecular weight due to each of its two constituent elements is approximately 50 percent of 64 amu, or 32 amu. From this it follows that neither the atomic weight of sulfur nor the atomic weight of oxygen could possibly be greater than 32 amu, because there must be at least one atom of each element in one molecule of the compound.

The value that can reasonably be assumed for the atomic weight of an element is the smallest contribution that element makes toward the molecular weight of any of its compounds (Figure 2-1). This principle was first clearly stated by the Italian chemist Stanislao Cannizzaro in 1858 and is known as *Cannizzaro's Principle.* For example, the least weight due to oxygen in the molecular weight of any oxygen compound is about 16 amu, and the smallest contribution of sulfur to the molecular weight of any sulfur compound is about 32 amu. Therefore, the atomic weight of oxygen is approximately 16 amu, and the atomic weight of sulfur is approximately 32 amu.

From the data presented in Table 2-2 it can be inferred, as an illustration of Cannizzaro's Principle, that the atomic weight of chlorine is about 35.5 amu. By dividing the weight of chlorine in one molecule of each compound by 35.5 amu per atom, we can find the number of

FIGURE 2-1

Cannizzaro's Principle

The weight of one atom of an element is the smallest weight due to that element in one molecule of any of its compounds. In these hydrogen compounds, the weight of one atom of hydrogen is the weight due to the one hydrogen atom in hydrogen chloride. (In each case the small balls represent hydrogen atoms.)

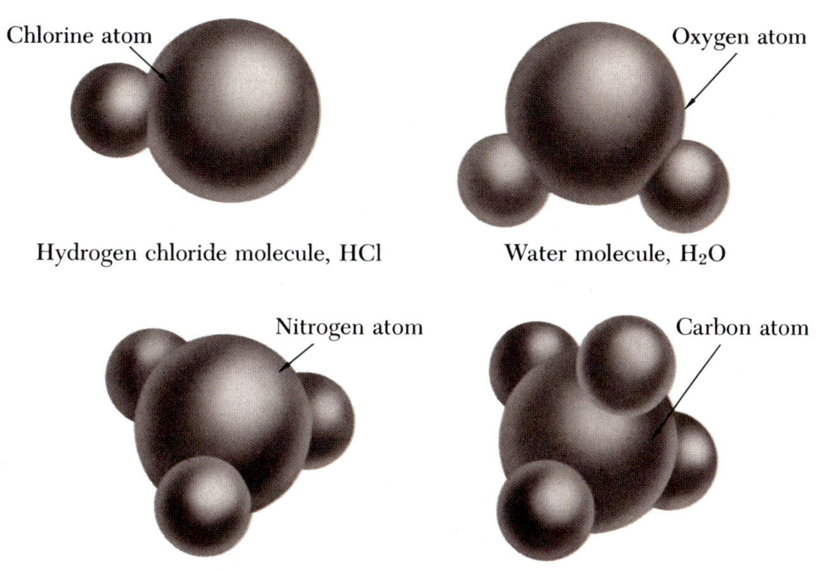

chlorine atoms in one molecule of each compound (1, 4, 2, 1, and 3, respectively).

TABLE 2-2

Approximate atomic weight of chlorine

Compound containing chlorine	*Approximate molecular weight* (amu), *experimental*	*Percent of chlorine (by chemical analysis)*	*Weight of chlorine in one molecule of compound* (amu)
1	36.5	97.24	35.5
2	154	92.19	142
3	135	52.51	70.9
4	64.5	54.96	35.4
5	119	89.10	106

2-4 Moles of Atoms

Since atoms are so small that we do not work with them individually in chemical operations, it is important to know when we are dealing with equal numbers of them. Comparisons between equal numbers of atoms are usually much more useful than comparisons between equal masses of substances since, as we have seen, chemical reactions involve definite numbers of atoms.

Knowledge of the atomic weights of the elements with which we are working enables us to know when we are dealing with equal numbers of atoms. For example, let us consider the fact that the atomic weights of carbon and oxygen are 12.011 amu and 15.9994 amu, respectively. If we were to take 12.011 grams (g) of carbon and 15.9994 g of oxygen, we would have masses that are in the same proportion as the average masses of the individual atoms; therefore, *the number of carbon atoms taken would be the same as the number of oxygen atoms.* When the mass expressed in grams is numerically equal to the atomic weight of an element, that amount is called a *mole of atoms.* For example, 12.011 g of carbon is 1 mole of carbon atoms and 15.9994 g of oxygen is 1 mole of oxygen atoms and both quantities contain the same number of atoms. (Some chemists use the term *gram-atom* to refer to a mole of atoms. The expression *gram-atomic weight* also is used to mean the same thing.)

The gram is firmly established by convention as the unit of mass in the chemical profession. When the gram seems to be an inappropriately large or excessively small unit for the purpose at hand, the milligram (mg) or the kilogram (kg) may be used (1 mg = 0.001 g; 1 kg = 1000 g. Units of the metric system are discussed in Appendix 7.) Likewise, the terms millimole and kilomole are sometimes encountered. For example, 12.011 mg of carbon is 1 millimole of carbon atoms and 15.9994 kg of oxygen is 1 kilomole of oxygen atoms. In the chemical industries where English units are still in use, we encounter such terms as "pound-mole" and "ounce-mole." One pound-mole of carbon atoms is 12.011 lb of carbon, and 1 ounce-mole of oxygen atoms is 15.9994 oz of oxygen.

2-5 Moles of Molecules

When the mass of a substance, expressed in grams, is numerically the same as its molecular weight, that amount of the substance is a *mole of molecules.* Most commonly, such a quantity of a substance is called, simply, a *mole* of the substance. For example, 1 mole of table sugar ($C_{12}H_{22}O_{11}$) is 342.3 g, 1 mole of elementary oxygen (O_2) is 32.00 g, and

1 mole of helium (He) is 4.0026 g. (Some chemists prefer to use the term *gram-molecular weight* for 342.3 g of table sugar, 32.00 g of O_2, etc.)

One mole of a molecular substance contains as many moles of atoms of each constitutent element as there are atoms of that element in one molecule of the substance. From the knowledge that the molecular formula for table sugar is $C_{12}H_{22}O_{11}$ we know that one molecule of this sugar contains 12 carbon atoms, 22 hydrogen atoms, and 11 oxygen atoms; therefore, 1 mole of the sugar contains 12 moles of carbon atoms, 22 moles of hydrogen atoms, and 11 moles of oxygen atoms. Similarly, 1 mole of elementary oxygen (O_2) molecules contains 2 moles of oxygen atoms. Since helium is monatomic, 1 mole of helium molecules is the same as 1 mole of helium atoms.

As with atoms, most molecules are so small that we are involved with large numbers of them in ordinary chemical operations. We know, however, that we are dealing with equal numbers of molecules whenever we are handling masses of substances that are in the same proportion as their molecular weights. Let us consider, for example, carbon dioxide and methane. The molecular formula CO_2 tells us that 1 mole of carbon dioxide is $12.01\ g + (2 \times 16.00)\ g = 44.01\ g$. The molecular formula CH_4 shows that 1 mole of methane is $12.01\ g + (4 \times 1.008)\ g = 16.04\ g$. If we were to take 44.01 g of carbon dioxide and 16.04 g of methane, we would have masses of the two compounds that are in the same proportion as their molecular weights (which are 44.01 amu and 16.04 amu, respectively). Therefore, the number of CO_2 molecules would be the same as the number of CH_4 molecules.

The number of molecules in 1 mole of any molecular substance is the same as the number of molecules in 1 mole of any other molecular substance. This is, in turn, the same as the number of atoms in 1 mole of atoms of any element.

2-6 Empirical Formulas

By definition, the molecular formula for a substance is a chemical formula that tells how many atoms of each kind are present in one molecule of the substance. Clearly, then, a molecular formula cannot be written for a substance unless it is a molecular substance. The compounds known as salts do not consist of molecules; instead, they are assemblages of charged particles called *ions*. For such compounds, the only chemical formulas that can be written are *empirical formulas*, sometimes called *simplest formulas*. Molecular substances can also have empirical formulas; in fact, they are a necessary first step in determining molecular formulas.

The empirical formula for a compound expresses, in terms of the smallest possible integers, the *ratio* of the numbers of atoms of each constituent element present in any chosen mass of the compound and therefore the ratio of the numbers of moles of these atoms.

EXAMPLE 1

Given that 1.000 g of bismuth combines with 0.2727 g of fluorine, write the empirical formula for bismuth fluoride.

We first determine how many moles of Bi and F atoms there are in the given weights. In the atomic weight table we find the atomic weights of bismuth and fluorine to be 209.0 amu and 19.00 amu, respectively (each rounded off here to four digits because the experimental data are four-digit numbers).

$$\frac{1.000 \text{ g}}{209.0 \text{ g/mole}} = 0.004785 \text{ mole of Bi atoms}$$

$$\frac{0.2727 \text{ g}}{19.00 \text{ g/mole}} = 0.01435 \text{ mole of F atoms}$$

The ratio of the number of moles of F atoms to the number of moles of Bi atoms is

$$\frac{0.01435}{0.004785} = 3.000 \text{ moles of F atoms for each mole of Bi atoms}$$

Therefore, the empirical formula is BiF_3.

EXAMPLE 2

The analysis of a pure sample of aluminum oxide shows that it is 52.9% aluminum and 47.1% oxygen. What is the empirical formula of aluminum oxide?

In 100.0 g of aluminum oxide there are 52.9 g of aluminum and 47.1 g of oxygen. The atomic weight of aluminum is 27.0 amu and the atomic weight of oxygen is 16.0 amu (each rounded off here to three digits because the analytical data are three-digit numbers).

$$\frac{52.9 \text{ g}}{27.0 \text{ g/mole}} = 1.96 \text{ moles of Al atoms}$$

$$\frac{47.1 \text{ g}}{16.0 \text{ g/mole}} = 2.94 \text{ moles of O atoms}$$

$$\frac{2.94}{1.96} = 1.50 \text{ moles of O atoms for each mole of Al atoms}$$

The ratio 1.50:1.00 is the same as 3.00:2.00. The empirical formula for aluminum oxide is Al_2O_3.

EXAMPLE 3

Analysis of calcium nitrate shows that it is 24.4% calcium, 17.1% nitrogen, and 58.5% oxygen. Calculate its empirical formula.

$$\frac{24.4\ \text{g}}{40.1\ \text{g/mole}} = 0.61 \text{ mole of Ca atoms}$$

$$\frac{17.1\ \text{g}}{14.0\ \text{g/mole}} = 1.22 \text{ moles of N atoms}$$

$$\frac{58.5\ \text{g}}{16.0\ \text{g/mole}} = 3.66 \text{ moles of O atoms}$$

$$\frac{1.22}{0.61} = 2.00 \text{ moles of N atoms for each mole of Ca atoms}$$

$$\frac{3.66}{0.61} = 6.00 \text{ moles of O atoms for each mole of Ca atoms}$$

Therefore, the empirical formula for calcium nitrate is CaN_2O_6 or $Ca(NO_3)_2$.

As the above examples show, every chemical compound must be subjected to elemental analysis before its empirical formula can be determined. Once this has been done and the empirical formula has been established, we can use the formula to calculate the percentage composition.

EXAMPLE

The empirical formula for calcium chloride is $CaCl_2$. Calculate its percentage composition.

From the table of atomic weights, 1 mole of Ca atoms weighs 40.08 g and 2 moles of Cl atoms weigh 2 × 35.453 g, or 70.906 g. Since 40.08 g + 70.906 g = 110.99 g,

$$\%\ \text{calcium} = \frac{40.08\ \text{g}}{110.99\ \text{g}} \times 100\% = 36.11\%$$

$$\%\ \text{chlorine} = \frac{70.906\ \text{g}}{110.99\ \text{g}} \times 100\% = 63.89\%$$

2-7 Determination of Molecular Formulas

In order to arrive at the molecular formula for a substance it is necessary, first of all, to perform a chemical analysis upon it so that its empirical formula can be calculated. In addition, it is necessary to measure the molecular weight of the substance in order to know whether the molecular formula is the same as the empirical formula or an integral multiple of it.

EXAMPLE 1

Chemical analysis shows that acetylene and benzene have the same percentage composition, 92.26% carbon and 7.74% hydrogen. Other laboratory experiments indicate that there are 26 g in 1 mole of acetylene and 78 g in 1 mole of benzene. What are the molecular formulas of these two compounds?

In 100.00 g of either compound there are 92.26 g of carbon and 7.74 g of hydrogen.

$$\frac{92.26 \text{ g}}{12.01 \text{ g/mole}} = 7.68 \text{ moles of C atoms}$$

$$\frac{7.74 \text{ g}}{1.008 \text{ g/mole}} = 7.68 \text{ moles of H atoms}$$

Each compound has equal numbers of carbon and hydrogen atoms, and they have the same empirical formula, CH (or HC). If the molecular formula of either of these compounds were CH, the molecular weight of that compound would be 13 amu (the sum of the atomic weights of carbon and hydrogen). If the molecular weights of the two compounds were unknown, we would still be able to say that their molecular weights can be represented as $13n$ amu, where n is an integer equal to the number of C and H atoms in one molecule of compound. Since the molecular weights are known from experimental measurements to be 26 amu for acetylene and 78 amu for benzene, we can derive their molecular formulas as follows:

For acetylene, $13n = 26$; $26 = 13 \times 2$; therefore $n = 2$, and the molecular formula for acetylene is C_2H_2.

For benzene, $13n = 78$; $78 = 13 \times 6$; therefore $n = 6$, and the molecular formula for benzene is C_6H_6.

See Figure 2-2.

EXAMPLE 2

In ethane, 1.0000 g of hydrogen is combined with 3.9721 g of carbon. By experiment the molecular weight of ethane is found to be approximately 30 amu. What is the molecular formula for ethane?

FIGURE 2-2

Two compounds with the same empirical formula

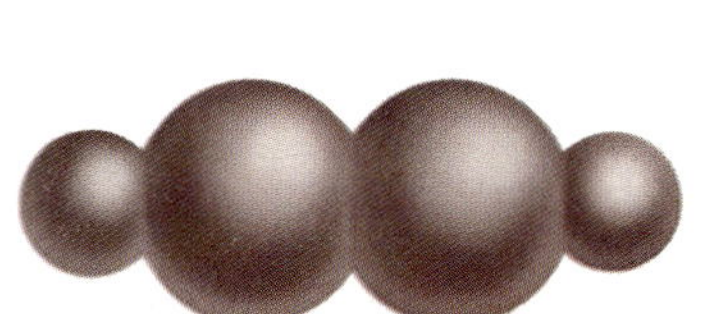

Acetylene
Molecular formula, C_2H_2
Empirical formula, CH

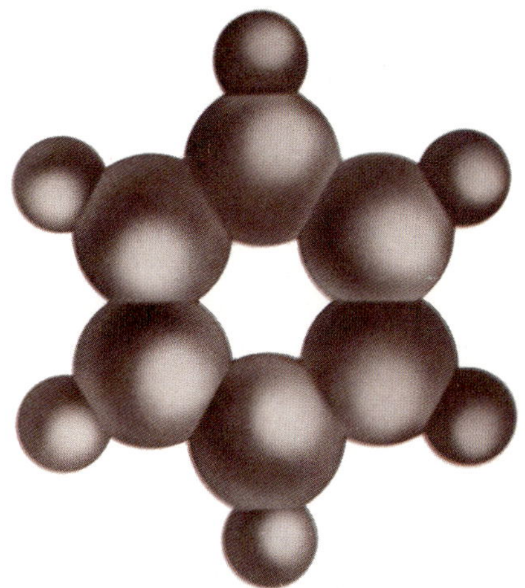

Benzene
Molecular formula, C_6H_6
Empirical formula, CH

$$\frac{3.9721\text{ g}}{12.011\text{ g/mole}} = 0.3307\text{ mole of C atoms}$$

$$\frac{1.0000\text{ g}}{1.0080\text{ g/mole}} = 0.9921\text{ mole of H atoms}$$

$$\frac{0.9921}{0.3307} = 3.000\text{ moles of H atoms for each mole of C atoms}$$

Therefore, the empirical formula is CH_3 (or H_3C) and the molecular weight is 15 amu or a multiple of 15 amu. Since the molecular weight is known by experiment to be 30 amu, we have $15n = 30$ from which $n = 2$. Therefore, the molecular formula is C_2H_6 (or H_6C_2).

EXAMPLE 3

Analysis of carbon dioxide shows that it is 27.3% carbon and 72.7% oxygen. Molecular weight measurements show that there are approximately 44 g in 1 mole. What is the molecular formula of this compound?

$$\frac{27.3 \text{ g}}{12.0 \text{ g/mole}} = 2.27 \text{ moles of C atoms}$$

$$\frac{72.7 \text{ g}}{16.0 \text{ g/mole}} = 4.54 \text{ moles of O atoms}$$

$$\frac{4.54}{2.27} = 2.00 \text{ moles of O atoms for each mole of C atoms}$$

Therefore, the empirical formula is CO_2 and the molecular weight can be represented as $44n$ amu, where n is an integer. Since the molecular weight is known, from experiments, to be approximately 44 amu, we have $44n = 44$, from which $n = 1$. The molecular formula is the same as the empirical formula, CO_2.

Even an approximate value for the molecular weight of a compound enables us to write the correct molecular formula if the empirical formula is already known. This is fortunate, because most experimental methods for determining molecular weights do not give precise results. Some of the experimental procedures available for determining the molecular weights of substances are described at appropriate points in this text.

2-8 General Application of the Mole Concept

The sum of the atomic weights of the atoms represented in the empirical formula for an ionic (nonmolecular) compound can be properly referred to as the *formula weight* of the substance, although it is *not* the molecular weight. In discussions of nonmolecular compounds the number of grams numerically equal to the formula weight is commonly referred to as 1 mole of the compound.

Let us consider, as an example, the ionic compound NaCl (table salt). The atomic weight of sodium is 23.0 amu, and for chlorine it is 35.5 amu. Therefore, the formula weight for NaCl is 58.5 amu, and 1 mole of NaCl is 58.5 g. Furthermore, 1 mole of NaCl contains 23.0 g of sodium ions and 35.5 g of chloride ions. More simply, 1 mole of NaCl consists of 1 *mole* of sodium ions and 1 *mole* of chloride ions. The formula NaCl represents a chemical unit as surely as if it were a molecule; in this instance the chemical unit is a pair of ions.

When using the term *mole* we should be careful to avoid ambiguities that could cause unnecessary confusion. For example, "1 mole of oxygen" is an undefined quantity. If we mean 32 g of elementary oxygen, O_2, we should say "1 mole of O_2 molecules," so as to avoid confusion with 1 mole of oxygen atoms (which weighs 16 g).

2-9 Avogadro's Number

We have seen (Section 2-4) that 1 mole of atoms of any element contains the same number of atoms as 1 mole of atoms of any other element. The number of atoms in 1 mole of atoms of any element is called *Avogadro's Number*, in honor of Amedeo Avogadro, the eminent Italian physicist whose pioneering work early in the nineteenth century eventually made it possible to obtain reliable values for the atomic weights of the elements. The value of Avogadro's Number has been experimentally determined to be 6.023×10^{23}. We can say, therefore, that 6.023×10^{23} atoms of an element constitute 1 mole of atoms of that element and that the mass of 6.023×10^{23} atoms of an element is W g, where W is the atomic weight of the element.

In most chemical operations it need not concern us that many of the chemical elements consist of two or more isotopes. The number of atoms in any working quantity of an element is almost always so great that we can confidently use the *average* value for the mass of its atoms. The table of atomic weights is used for carrying out chemical calculations involving masses of different elements because the values given in the table reflect the natural distribution of the various isotopes of each element.

One mole of carbon-12 atoms weighs exactly 12 g and 1 mole of carbon-13 atoms weighs 13.00344 g, but 1 mole of "natural carbon" atoms (i.e., the mixture of isotopes whose average mass is 12.011 amu) weighs 12.011 g. The number of carbon-12 atoms in exactly 12 g of carbon-12 is the same as the number of carbon-13 atoms in 13.00344 g of carbon-13. This number is, in turn, the same as the number of carbon atoms in 12.011 g of natural carbon.

We also have seen (Sections 2-5 and 2-8) that 1 mole of molecules of any substance contains the same number of molecules as 1 mole of molecules of any other substance and that this number is the same as the number of atoms in a mole of atoms of any element or the number of ions in 1 mole of ions of any kind.

We can now generalize and say that 1 mole is an Avogadro Number of particles, objects, or formula units of any species; for example, the unit may be an atom, a molecule, an ion, or an assemblage of ions (Figure 2-3).

2-10 Chemical Equations

When one or more starting substances are converted into one or more other substances, we say that a *chemical reaction* has occurred. Any chemical reaction can be described by a *chemical equation*, which is a statement employing chemical formulas instead of words.

FIGURE 2-3

Applications of the mole concept

One mole contains Avogadro's Number of formula units. The cubes are represented in the relative sizes needed to contain the quantities indicated.

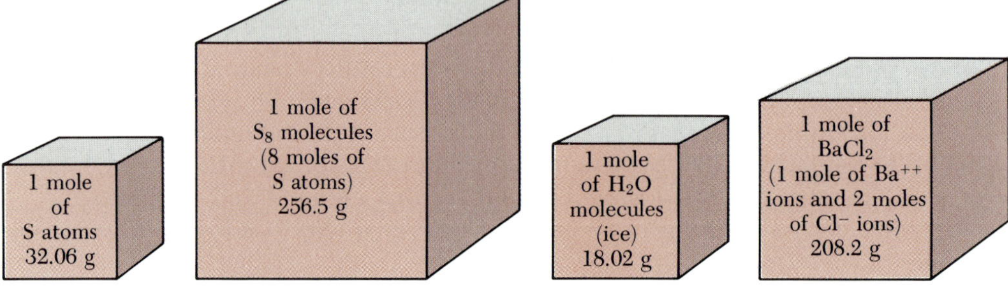

The burning of carbon is a reaction between carbon (C) and oxygen gas (O_2), and the product of the reaction is carbon dioxide gas (CO_2). This reaction can be represented by the following chemical equation:

$$C + O_2 \longrightarrow CO_2$$

The + sign means "and," and the arrow means "react to give." Since atoms are neither created nor destroyed in a chemical reaction, the same number of each kind of atom must be represented on both sides of a chemical equation. In the equation above, there are two oxygen atoms and one carbon atom on each side of the arrow, but the arrangement of the atoms after the reaction is different from their arrangement before the reaction.

Water (H_2O) is the product formed by allowing hydrogen gas (H_2) to combine with oxygen, as when it burns in air. Initially, the equation for this reaction is written simply with the formulas for the reactants and the product:

$$H_2 + O_2 \longrightarrow H_2O \qquad \text{(unbalanced equation)}$$

This equation is called a *skeleton equation.* In this example the skeleton equation is *unbalanced,* because there are two oxygen atoms on the left-hand side of the equation and only one on the right. To be correct, a chemical equation must be *balanced*; that is, it must be adjusted to show the same number of atoms of each element on both sides.

The skeleton equation that we have written for the burning of the H_2 gas may be balanced by recognizing that one O_2 molecule has twice as many oxygen atoms as are needed to convert one H_2 molecule into one molecule of H_2O. In other words, there are enough oxygen atoms in one O_2 molecule to convert two H_2 molecules into two H_2O molecules. Therefore, we balance the equation as follows:

$$2H_2 + O_2 \longrightarrow 2H_2O$$

In the balanced equation there are four hydrogen atoms and two oxygen atoms on each side of the arrow, but the arrangement of the atoms before and after the reaction is different (Figure 2-4).

In a balanced equation the numbers written in front of the formulas are called *coefficients*. The 2 before H_2 in the equation above means two H_2 molecules, and the 2 before H_2O means two H_2O molecules. The coefficient for O_2 is understood to be one.

Since 1 mole of any molecular substance contains the same number of molecules as 1 mole of any other molecular substance, the above equation also states that *2 moles* of H_2 react with *1 mole* of O_2 to form *2 moles* of H_2O. If the reaction represented by a chemical equation is a reaction involving ionic substances, the coefficients in the equation do not refer to numbers of molecules, but they still do refer to numbers of moles of the substances. For example, the equation

$$2NaF + CaCl_2 \longrightarrow CaF_2 + 2NaCl$$

states that 2 moles of NaF (an ionic compound) react with 1 mole of $CaCl_2$ (another ionic compound) to form 1 mole of CaF_2 and 2 moles of NaCl (both ionic compounds). As long as the chemical equation is balanced, the coefficients indicate the number of moles of reactants and products—whether the formulas are empirical or molecular.

Balancing equations comes easily with practice, but knowing the

FIGURE 2-4

The rearrangement of atoms in a chemical reaction

The reactants, the O_2 molecule and the two H_2 molecules, contain the same atoms as the products, the two H_2O molecules. The balanced chemical equation shows this relationship.

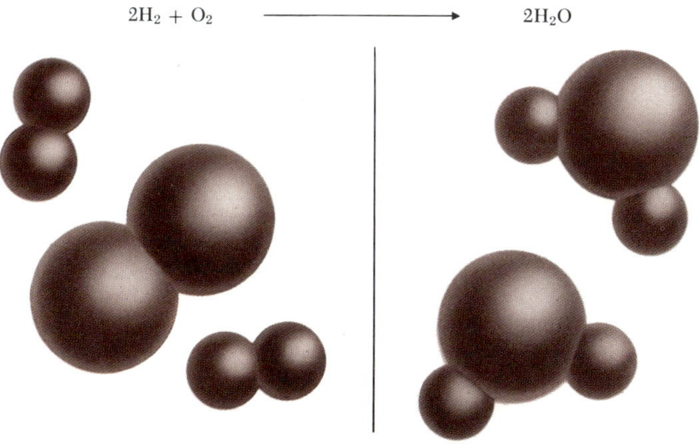

formulas of reactants and products is frequently more difficult. The steps in writing a chemical equation are as follows:

1. Write the formula(s) of the reactant(s) with + signs between them:

$$CO + O_2$$

2. Place an arrow after the reactants, and to the right of the arrow write the formula(s) of the product(s):

$$CO + O_2 \longrightarrow CO_2$$

3. Balance the equation by inserting the coefficients that are needed in front of the formulas. In other words, adjust the number of molecules so that the number of atoms of each element is the same on each side of the equation:

$$2CO + O_2 \longrightarrow 2CO_2$$

   In making this adjustment, never change the formula of any substance—*do not add or remove subscripts.*

Eventually you will be able to predict the formulas of the products of many chemical reactions. However, you should first learn to balance equations in which the formulas of all starting materials and products are given.

2-11 Weight Relations in Chemical Reactions

From the fact that a chemical equation has meaning in terms of moles, it follows that it also has meaning in terms of the weights (masses) of the reactants and products. The weight of 1 mole of each substance can be calculated from its formula, and the balanced equation states how many moles of each substance are involved in the reaction. One of the most valuable features of balanced chemical equations, then, is that they can be used to calculate the weights of reactants and products involved in chemical reactions. The use of equations in such calculations is called *stoichiometry* and is illustrated in the following examples.

EXAMPLE 1

How many moles of hydrogen gas (H_2) are needed to produce 10 moles of ammonia gas (NH_3)?

First we write the balanced equation:

$$N_2 + 3H_2 \longrightarrow 2NH_3$$

This tells us that 3 moles of H_2 are needed to produce 2 moles of NH_3, or 1.5 moles of H_2 for each mole of NH_3. Thus, 10 moles of NH_3 require 1.5 moles × 10 = 15 moles of H_2.

EXAMPLE 2

How many grams of nitrogen are needed to produce 10 moles of ammonia?

The equation (see Example 1) tells us that 1 mole of N_2 is needed to produce 2 moles of NH_3, or 0.5 mole of N_2 for each mole of NH_3. Thus, 10 moles of NH_3 require 0.5 mole × 10 = 5 moles of N_2. In the table of atomic weights we find that the atomic weight of nitrogen (rounded off to three digits) is 14.0 amu, so 1 mole of N_2 is 28.0 g. Thus, the required 5 moles of N_2 weigh 5 × 28.0 g = 140 g.

EXAMPLE 3

How many grams of ammonia will be produced from 100 g of hydrogen?

The atomic weight of hydrogen is 1.008 amu, so the weight of 1 mole of H_2 is 2 × 1.008 g = 2.016 g.

$$\frac{100 \text{ g}}{2.016 \text{ g/mole}} = 49.6 \text{ moles of } H_2$$

The equation (see Example 1) indicates that each mole of H_2 produces $\frac{2}{3}$ or 0.667 mole of NH_3, so 49.6 moles of H_2 will produce

$$0.667 \times 49.6 \text{ moles} = 33.1 \text{ moles of } NH_3$$

The weight of 1 mole of NH_3 is 14.0 g + 3.0 g = 17.0 g. Thus, the weight of the 33.1 moles of NH_3 is 33.1 × 17.0 g = 563 g.

EXAMPLE 4

How many moles of water will be produced if 10 moles of sulfuric acid, H_2SO_4, react with the necessary quantity of aluminum hydroxide, $Al(OH)_3$?

From the balanced equation

$$2Al(OH)_3 + 3H_2SO_4 \longrightarrow Al_2(SO_4)_3 + 6H_2O$$

it is apparent that 3 moles of H_2SO_4 produce 6 moles of H_2O, or 1 mole of H_2SO_4 produces 2 moles of H_2O. Thus, 10 moles of H_2SO_4 will produce 2 moles × 10 = 20 moles of H_2O.

EXAMPLE 5

In the reaction of aluminum hydroxide with sulfuric acid, how many grams of aluminum sulfate will be produced from 10 moles of sulfuric acid?

The equation (see Example 4) indicates that 3 moles of H_2SO_4 are required to produce 1 mole of $Al_2(SO_4)_3$. Thus, 10 moles of H_2SO_4 will produce $\frac{10}{3}$ moles of $Al_2(SO_4)_3$. The atomic weights of Al, S, and O are 26.98, 32.06, and 16.00 amu, respectively. Therefore, the weight of 1 mole of $Al_2(SO_4)_3$ is (2×26.98) g + $3[32.06 + (4 \times 16.00)]$ g = 53.96 g + (3×96.06) g = 53.96 g + 288.18 g = 342.1 g. The weight of $\frac{10}{3}$ moles of $Al_2(SO_4)_3$ is

$$\frac{10}{3} \times 342.1 \text{ g} = 1140 \text{ g}$$

In summary, stoichiometric calculations involve such problems as calculating the *mass* of one substance, A, required to react with a stated mass of another substance, B. Calculations of this sort follow a pattern consisting of three steps:

1. From the given number of *grams* of B, calculate the number of *moles* of B. (Divide the number of grams of B by the formula weight of B.)
2. From the balanced equation, calculate the number of moles of A required to react with the number of moles of B under consideration.
3. From the calculated number of *moles* of A, find the number of *grams* of A. (Multiply the number of moles of A by the formula weight of A.)

2-12 Incomplete Reactions

The basic concept in the stoichiometric approach to chemical problems is that the chemical equation, correctly interpreted, gives the correct relationship between the weights of the reacting substances.

We assume, in making a stoichiometric calculation, that *one* of the reactants is completely consumed. Indeed, if all reactants are present in the molar amounts required by the equation, we assume that all the reactants are completely transformed. However, most reactions appear to stop before the conversion of reactants to products is complete. This is the case if the reaction is *reversible* under the prevailing conditions; for example, if N_2 and H_2 gases are brought together they react in accordance with the equation

$$N_2 + 3H_2 \longrightarrow 2NH_3$$

to produce ammonia. However, at temperatures that permit N_2 and H_2 to react, NH_3 decomposes to give N_2 and H_2. Thus, when the reaction indicated by the above equation is occurring, a reverse reaction indicated by the following equation is occurring simultaneously:

$$2NH_3 \longrightarrow N_2 + 3H_2$$

Each reaction produces the substances that are the reactants in the other reaction; so if none of the substances is permitted to escape, neither reaction actually stops, and the equation can be written as follows:

$$N_2 + 3H_2 \rightleftharpoons 2NH_3$$

A state of *chemical equilibrium* is established in which the rate of the forward reaction equals the rate of the reverse reaction. In any such case of incomplete reaction, we cannot compute the weights of the products from the weights of the reactants unless we know how nearly the reaction has approached completion. This topic will be pursued further in Chapter 11.

NEW TERMS

Atomic mass unit: one-twelfth the mass of the carbon-12 atom.

Atomic weight: the weight of an average atom of an element on the scale $^{12}C = 12$ exactly. The atomic weight is the weighted average of the isotopic weights.

Avogadro's Number: the number of atoms in a mole of atoms, the number of molecules in a mole of molecules, the number of ions in a mole of ions, or the number of whatever chemical units one may wish to designate in a mole of such units. The number, N, is 6.023×10^{23}.

Cannizzaro's Principle: the weight of one atom of an element is the smallest weight due to that element in one molecule of any of its compounds.

Chemical equation: a statement representing a chemical reaction, employing chemical formulas instead of words.

Chemical equilibrium: the state that prevails when a chemical reaction is occurring at the same rate as its reverse.

Chemical formula: any collection of symbols and subscripts used to represent a pure substance. Empirical formulas and molecular formulas are examples of chemical formulas.

Empirical formula (simplest formula): the combination of symbols and subscripts used to represent the relative number of atoms of the constituent elements in a compound; for example, the empirical formulas for hydrogen peroxide and benzene are HO and CH, respectively, although their molecular formulas are H_2O_2 and C_6H_6.

Formula weight: the sum of the atomic weights of the atoms in a formula. If the formula is the molecular formula of a substance, the formula weight is also its molecular weight.

Ions: charged particles, especially atoms or molecules that have acquired electrical charges.

Isotopes: atoms of the same element that differ from one another in mass.

Isotopic weight: the weight of a particular isotope on the scale $^{12}C = 12$ exactly.

Mass number: the integer that most nearly approximates the mass of an isotope on the atomic weight scale.

Mole of atoms: the quantity of an element of which the mass in grams is numerically equal to the atomic weight of the element; for example, since the atomic weight of sodium is 22.9898 amu, a mole of sodium atoms is 22.9898 g.

Molecular weight: the weight of one molecule of a substance on the scale $^{12}C =$ 12 exactly. The molecular weight is the sum of the atomic weights of the constituent elements, taking into account the number of atoms of each element in one molecule of the substance.

Pound-mole of atoms: the quantity of an element the weight of which in pounds is numerically equal to the atomic weight of the element; for example, a pound-mole of sodium atoms is 22.9898 lb.

Reversible reaction: a reaction in which the products are able to interact to produce the starting materials.

Stoichiometry: computation based on the weight relations implicit in chemical equations.

EXERCISES

2-1 In 1 molecular weight of each of four compounds of element X, the weights of element X are found to be 32, 48, 64, and 80 amu. Is it correct to infer that the atomic weight of element X is 32 amu? Why or why not?

2-2 The least weight of carbon in a molecular weight of any known carbon compound is six-sevenths as much as the least weight of nitrogen in a molecular weight of any nitrogen compound. If it has been established that the atomic weight of carbon is approximately 12 amu, what is the atomic weight of nitrogen?

2-3 The standards formerly used for atomic weights include $H = 1$ exactly, $O = 100$ exactly, and $O = 16$ exactly. Calculate the atomic weight of sulfur on each of these scales.

2-4 Oxygen is a diatomic gas, O_2, and helium is a monatomic gas. How does the molecular weight of oxygen compare with that of helium?

2-5 What is the weight of 1.0 mole of phosphorus atoms? Of 4.2 moles of carbon atoms? Of 0.012 mole of uranium atoms? Of 3.5×10^{-16} mole of copper atoms?

2-6 How many moles of chlorine atoms are there in 100.0 g of chlorine atoms? In 100.0 g of Cl_2 molecules? In 1000.0 g of potassium perchlorate, $KClO_4$?

2-7 What is the weight of 1.0 mole of carbon dioxide molecules? Of 4.0 moles of water molecules?

2-8 How many moles of nitrogen molecules are there in 650.0 g of N_2?

2-9 In a mixture of 2.0 kg each of N_2 and CO_2, how many moles of molecules are there? How many moles of atoms?

2-10 How many moles of aluminum sulfate are there in 1.000 kg of $Al_2(SO_4)_3$?

2-11 Calculate the percentage composition of uranium hexafluoride, UF_6; limestone, $CaCO_3$; methane, CH_4; ethyl alcohol, C_2H_6O; formaldehyde, CH_2O; carbon disulfide, CS_2; potassium chlorate, $KClO_3$.

2-12 Calculate the empirical formulas of the compounds with the following percentage compositions:
(a) copper, 79.85%; sulfur, 20.15%
(b) phosphorus, 43.6%; oxygen, 56.4%
(c) carbon, 76.86%; hydrogen, 12.90%; oxygen, 10.24%
(d) potassium, 26.58%; chromium, 35.35%; oxygen, 38.07%
(e) potassium, 28.7%; hydrogen, 1.5%; phosphorus, 22.8%; oxygen, 47.0%
(f) sodium, 32.85%; aluminum, 12.85%; fluorine, 54.30%
(g) nicotine: carbon, 74.03%; hydrogen, 8.70%; nitrogen, 17.27%

2-13 Compute the molecular formulas of the following compounds:

Empirical formula	*Molecular weight* (amu)
HO	34
CH_2O	30
CH_2O	180
CH_2	28
CH_3	30
CN	52
P_2O_5	284

2-14 The molecular weight of an organic compound is known to be not less than 60 nor more than 64. The compound is 38.7% carbon, 9.7% hydrogen, and 51.6% oxygen. What is its molecular formula?

2-15 A compound has the following composition: 24.3% carbon, 4.1% hydrogen, 71.6% chlorine. The molecular weight of the compound is 99. What is its molecular formula?

2-16 Balance the following equations:
(a) $H_2 + O_2 \longrightarrow H_2O$
(b) $H_2 + Cl_2 \longrightarrow HCl$
(c) $Na + O_2 \longrightarrow Na_2O_2$
(d) $Al + O_2 \longrightarrow Al_2O_3$
(e) $P + O_2 \longrightarrow P_4O_{10}$
(f) $N_2 + H_2 \longrightarrow NH_3$
(g) $Al + Cl_2 \longrightarrow AlCl_3$

(h) $CO + O_2 \longrightarrow CO_2$
(i) $NO + O_2 \longrightarrow NO_2$
(j) $HgO \longrightarrow Hg + O_2$
(k) $KClO_3 \longrightarrow KCl + O_2$
(l) $Ag_2O \longrightarrow Ag + O_2$
(m) $NaNO_3 \longrightarrow NaNO_2 + O_2$
(n) $N_2O \longrightarrow N_2 + O_2$
(o) $H_2O_2 \longrightarrow H_2O + O_2$
(p) $Zn + HCl \longrightarrow ZnCl_2 + H_2$
(q) $Mg + H_2SO_4 \longrightarrow MgSO_4 + H_2$
(r) $Na + HOH \longrightarrow NaOH + H_2$
(s) $NaBr + Cl_2 \longrightarrow NaCl + Br_2$
(t) $KOH + H_2SO_4 \longrightarrow K_2SO_4 + HOH$
(u) $Ca(OH)_2 + HNO_3 \longrightarrow Ca(NO_3)_2 + HOH$
(v) $Al(OH)_3 + H_2SO_4 \longrightarrow Al_2(SO_4)_3 + HOH$
(w) $P_4O_{10} + H_2O \longrightarrow H_3PO_4$
(x) $N_2O_5 + H_2O \longrightarrow HNO_3$
(y) $Na_2O + H_2O \longrightarrow NaOH$
(z) $CaO + H_2O \longrightarrow Ca(OH)_2$

2-17 (a) How many moles of oxygen are required for the combustion of 5.0 moles of methane? The equation for the reaction is

$$CH_4 + 2O_2 \longrightarrow CO_2 + 2H_2O$$

(b) How many moles of oxygen are necessary to burn 500.0 g of methane?
(c) How many grams of oxygen are needed to burn 3.00 moles of methane?
(d) How many grams of oxygen are needed to burn 250.0 g of methane?
(e) What weight of water is produced if the weight of carbon dioxide obtained is 300.0 g?

2-18 Express the value of 1 amu in grams. How many atomic mass units make 1 g?

REFERENCES

Benson, S. W., *Chemical Calculations*, 2nd ed., Wiley, New York, 1963 (paperbound). An introduction to the use of mathematics in chemistry.

International Union of Pure and Applied Chemistry, "Atomic Weights," *Chemistry*, **40**(6), 19 (1967). Values adopted in 1965.

Kieffer, W. F., *The Mole Concept in Chemistry*, Reinhold, New York, 1962 (paperbound). The many uses of the mole concept in chemical arithmetic.

Labbauf, A., "The Carbon-12 Scale of Atomic Masses," *Journal of Chemical Education*, **39**(6), 282 (1962). Relationship of the former scale to the new scale, which became effective January 1, 1962.

Roberts, E., *A Programmed Sequence on Exponential Notation*, Freeman, San Francisco, 1962 (paperbound). Developed by Chemical Education Material Study (CHEM Study).

Sienko, M. J., *Chemistry Problems*, Benjamin, New York, 1967 (paperbound). Chapters 3 and 4 are concerned with the mole and calculations based on chemical equations.

Walker, R. A., and Johnston, H., *The Language of Chemistry*, Prentice-Hall, Englewood Cliffs, N.J., 1967 (paperbound). Deals with reading, writing, and understanding chemical equations.

Wichers, E., "Why the Carbon-12 Scale?" *Chemistry*, **37**(3), 12 (1964). An explanation by the chairman of the Commission on Atomic Weights of the International Union of Pure and Applied Chemistry.

Young, J. A., *Chemical Concepts*, Prentice-Hall, Englewood Cliffs, N.J., 1963 (paperbound), pp. 41–72. Programmed instruction. Chemical arithmetic based on the mole concept.

3

Atomic Numbers and the Periodic Law

3-1 Varying Properties of the Chemical Elements

The properties of the chemical elements vary within wide limits. Of the elements now known, all but 22 have properties that are characteristic of metals: conductivity, ductility, etc. (Section 1-8). Most of the metallic elements—for example, copper and silver—are solids at ordinary temperatures, but mercury is a liquid and several other metals will melt when they are placed in boiling water. The 22 nonmetals include 11 gases, 10 solids, and 1 liquid (bromine).

Both metals and nonmetals vary widely in chemical activity. Gold and platinum are metals that suffer no oxidation or corrosion under any ordinary conditions, but sodium metal reacts instantly with water, generating hydrogen gas. Argon is a chemically inert gas, but fluorine is a gas that combines with many other elements immediately upon contact.

The properties of corresponding compounds of the elements vary as widely as the properties of the elements themselves. For example, the binary compounds of oxygen, called oxides, include liquid water (hydrogen oxide), solid mercuric oxide, and the gaseous carbon oxides.

3-2 Classification of the Elements

The study of the chemical elements has been greatly aided by classifying them into a relatively small number of groups or families. As chemistry developed in the nineteenth century, it became evident that certain elements could be grouped together on the basis of apparent similarities. For example, anyone acquainted with the properties of the metals lithium, sodium, and potassium would easily note that they are much more like one another than one of them is like any of the other elements. They are considered to be members of the same family of elements, the alkali metals. Again, chlorine, bromine, and iodine are nonmetals that resemble one another to a significant degree and are regarded as members of the same family, the halogens.

The first attempt to establish a relationship between the properties of chemical elements and their atomic weights was made in 1829 by Johann Doebereiner, in Germany. He selected elements with similar properties and arranged them, in order of increasing atomic weights, in groups of three, which came to be known as *Doebereiner's triads.* For example, the atomic weight of sodium (23) is approximately equal to the average of the atomic weights of the elements it so closely resembles, lithium (6.9) and potassium (39). Doebereiner's discovery encouraged chemists to look for further systematic relationships among the chemical elements. In particular, other efforts were soon forthcoming to relate the atomic weights to the other properties of the elements, even though there was no particular reason for expecting such a correlation nor any logical explanation for it once it was found to exist.

3-3 Newlands' Octaves

After the middle of the nineteenth century, when reliable atomic weights first became available, attempts to correlate atomic weights with other properties began to show significant success. In 1865 John Newlands, in England, arranged the known elements according to increasing atomic weights and found that among those of lowest atomic weights each element after hydrogen strongly resembles the seventh element beyond it. (Helium, and the other elements of its family, had not yet been discovered.) His arrangement of these elements is shown at the top of the next page.

Li	Be	B	C	N	O	F
6.9	9.0	10.8	12.0	14.0	16.0	19.0
Na	Mg	Al	Si	P	S	Cl
23.0	24.3	27.0	28.1	31.0	32.1	35.5

This arrangement of these 14 elements became known as *Newlands' octaves* (the interval from an element in the first row to its related element in the second row forms an octave, or series of eight). Although the properties of each element in a row differ decidedly from those of the other elements in the same row, there is an unmistakable gradation in properties as the atomic weight increases within either row. The truly remarkable feature of the arrangement is revealed in this set of facts:

1. Sodium bears a strong resemblance to lithium in its physical and chemical properties. Both are highly reactive metals of low density and form compounds with similar formulas: LiOH, NaOH; LiCl, NaCl; etc.
2. Magnesium and beryllium have approximately the same density and form compounds with analogous formulas: $Be(OH)_2$, $Mg(OH)_2$; $BeCl_2$, $MgCl_2$; etc.
3. Aluminum and boron form compounds with similar formulas ($AlCl_3$, BCl_3, etc.), even though the similarities of their properties are not as marked, aluminum being a metal and boron a nonmetal.
4. Carbon and silicon are both nonmetals and form many compounds having comparable formulas: CCl_4, $SiCl_4$; CH_4, SiH_4; CO_2, SiO_2; etc. Many of the properties of these compounds with similar formulas are quite different, however.
5. The nonmetals nitrogen and phosphorus form some compounds that have similar formulas and properties: HNO_3, HPO_3; NH_3, PH_3; etc.
6. The chemistry of sulfur resembles the chemistry of oxygen in many ways. Similarities exist between H_2S and H_2O, between CS_2 and CO_2, etc.
7. Chlorine and fluorine are members of a family of nonmetals, the halogens, which resemble one another in many respects.

The limitation of Newlands' scheme is that this *law of octaves* does not apply much beyond the two rows shown here.

3-4 Mendeleev and the Periodic Law

Dmitri Mendeleev, in Russia, and Julius Lothar Meyer, in Germany, were the first chemists to arrange all the then-known elements in a concise *periodic table*. In 1869 Mendeleev published a table with the elements listed in horizontal rows and vertical columns, in order of increasing atomic weights.

The success of Mendeleev's table was due to the emphasis he placed on the repetition of both physical and chemical properties at regular intervals. He was so confident about the accuracy of the arrangement that he boldly insisted that (1) where the elements did not appear in the correct order according to their atomic weights, the accepted atomic weights were in error and (2) some of the places in his table should be left vacant, to be occupied later by elements yet to be discovered.

Using his table, Mendeleev predicted the properties of several elements that were not yet known. Later, when they were discovered, they fitted neatly into the periodic table, and their properties proved to be very much as predicted. Mendeleev's observation that properties of elements recur periodically throughout the sequence of elements is called the *Periodic Law.*

Although deserving much credit for pointing out independently the periodic variation of the properties of the elements, Meyer did not pursue the problem in the thoroughgoing manner of Mendeleev. It was Mendeleev who made extensive and fruitful predictions based on the periodic relationships.

Mendeleev's work stimulated the search for additional elements and renewed investigation of the elements that were "out of order" in his table. In most cases errors were found in the accepted atomic weights, as Mendeleev had predicted. However, there are still several instances in which the increasing atomic weight sequence is violated. For example, tellurium must be regarded as the fifty-second element and iodine as the fifty-third, even though the atomic weight of tellurium (127.60) is greater than that of iodine (126.90). The properties of iodine resemble those of chlorine and bromine so closely that iodine is properly situated beneath them in the table, and the properties of tellurium likewise require that it be placed beneath sulfur and selenium in the periodic table.

The reversal of tellurium and iodine in the periodic table can no longer be attributed to inaccurately determined atomic weights. We are forced to acknowledge a more fundamental characteristic of the atom than atomic weight. In Chapter 2 we mentioned the discovery of isotopes in the early years of the present century. The heavier isotopes of tellurium and the lighter isotopes of iodine are the most abundant, thus making the average atomic weight of the naturally occurring mixture of isotopes greater for tellurium than for iodine.

It is now known that nearly all elements have isotopes. The fact that isotopes exist and that the different isotopes of an element have chemical properties that are virtually identical means that the atomic weight of an element, being only an average of the isotopic weights, cannot be the important factor that Mendeleev thought it to be in determining the properties of an element. To understand the feature of atomic structure that determines the chemical behavior of the elements, we must first consider the nature of the subatomic particles of which atoms are composed.

3-5 Electrons and X-Rays

The first of the subatomic particles to be discovered was the *electron*. Knowledge of the electron began with the study of electrical discharges through a so-called cathode ray tube (Figure 3-1). This device consists of a glass tube with wires sealed into both ends, each wire attached to a metal disk. The air in the tube is drawn out with a pump through a glass side arm. Then, when the wires are connected to a source of high-voltage electricity, the metal disks act as electrodes, there is a discharge of electricity between them, and a greenish-yellow fluorescent glow appears on the glass wall of the tube. If an obstacle is placed between the cathode (negative terminal) and the glass wall opposite, a sharp shadow is cast on the wall. The rays responsible for the fluorescence thus are found to originate at the cathode; for this reason they are called *cathode rays*.

If an object is placed in the path of the cathode rays, it casts a shadow. A pinwheel placed in this position is caused to rotate. If a magnet is held near a cathode ray tube, the rays are deflected, indicating that they consist of electrically charged particles. Their deflection toward the positive side of an electric field indicates that they possess a negative charge. By measuring the amount of their deflection in electric and magnetic fields of known strengths, Sir J. J. Thomson in 1897 determined the ratio of charge to mass, e/m, for the cathode ray particles. He found that even when different metals are used as cathode material, the ratio remains

FIGURE 3-1

The cathode ray tube

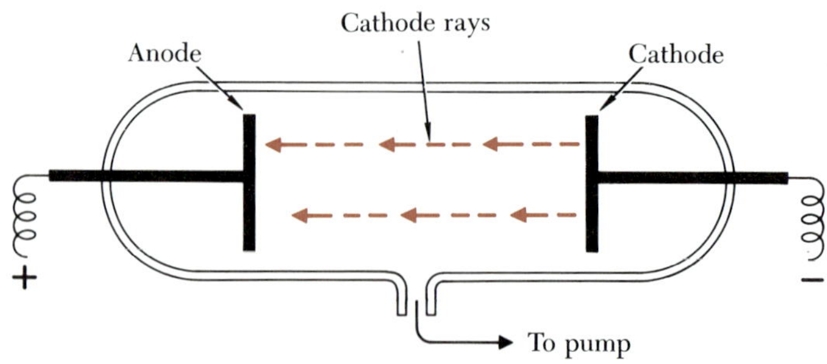

the same in every instance. Thus, the negatively charged particles issuing from the cathode came to be regarded as a species of subatomic particle occurring in all atoms. The name *electron* was given to this elementary unit of negative electrical charge. The value now accepted for the charge/mass ratio of the electron is 1.759×10^8 coulombs/g. (The coulomb is the amount of electrical charge conveyed by a current of 1 ampere in 1 second.)

Meanwhile, in late 1895 the German physicist Wilhelm Konrad Roentgen was studying the behavior of a cathode ray tube in a darkened room when he happened to notice a greenish glow from a square of cardboard lying on a nearby bench. This cardboard had been coated with crystals of a fluorescent compound in connection with a totally different experiment. On this particular occasion Roentgen had covered the cathode ray tube with black paper so that no light could enter or leave it. He concluded that the glow of the fluorescent crystals was caused by some kind of ray coming from the cathode ray tube, going through the black paper, and traveling to the bench where the crystals lay—even though, to the human eye, the room was utterly dark.

Roentgen found that the rays would expose photographic film and that they would pass through whatever objects he would hold between the tube and the film. His findings reached a climax when he placed a sheet of photographic film beneath his wife's hand and exposed it to the rays. When the film was developed it showed the bone structure of Frau Roentgen's hand.

Since Roentgen was unable to identify the rays coming from his tube, he called them *X-rays*. Within a few weeks of their discovery, they were put to use for observing broken bones in human bodies. Their use in the scientific study of matter has resulted in discoveries that are of the greatest importance in understanding the structures of materials, especially crystalline solids.

3-6 Charge of the Electron

During the first two decades of the present century, the American physicist Robert A. Millikan designed experiments to determine the charge, e, of an electron. He measured the quantity of electrical charge on small drops of oil. Oil droplets sprayed from an atomizer presumably obtain their initial charges by friction. When a droplet is situated between two charged plates (Figure 3-2), the gravitational pull on it can be counterbalanced by the required electrical "pull." The movement of the charged droplet is observed with a telescope. When the droplet is

FIGURE 3-2

Oil-drop method for determining the charge of an electron

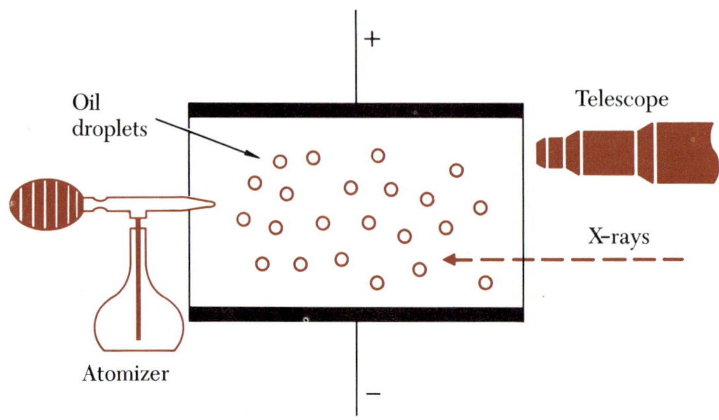

neither rising nor falling, i.e., when it is in equilibrium, the electrical "pull" is equal to the gravitational pull. The charge on the droplet can be changed to a new value by allowing a beam of X-rays to pass through the gas in which the drop is suspended. Millikan found that the charge is always an integral multiple of 1.602×10^{-19} coulomb. It is concluded, therefore, that the charge of one electron is 1.602×10^{-19} coulomb.

From the previously known value of the charge/mass ratio of the electron, it was now possible to calculate its mass:

$$\frac{1.602 \times 10^{-19} \text{ coulomb/electron}}{1.759 \times 10^{8} \text{ coulombs/g}} = 9.11 \times 10^{-28} \text{ g/electron}$$

The result shows that it would require 1837 electrons to have the same mass as one ordinary hydrogen atom. On the scale of atomic weights, the mass of an electron is 0.0005486 amu. This is equivalent to saying that the mass of 6.023×10^{23} electrons, that is, 1 mole of electrons, is 0.0005486 g.

3-7 Rutherford's Nuclear Theory of the Atom

Normally, all matter is electrically neutral. After the discovery that all atoms contain one or more electrons, which are negatively charged, scientists naturally searched for a positively charged particle. Toward the end of the nineteenth century a number of physicists studied positively charged rays in discharge tubes containing hydrogen. Their investigations led to the discovery of the *proton*, which has exactly the same quantity of electrical charge as the electron, but positive instead of negative. (The existence of two kinds of electricity has been known since 1733. In 1747, Benjamin Franklin proposed that the two varieties be referred to as *positive* and *negative*, and these terms are still in use.)

If you accept the idea that atoms contain (positive) protons and (negative) electrons, you will naturally wonder next about the arrangement of these particles inside the atom. The positive and negative particles might be uniformly mixed and occupy the space we call "the atom," or all the particles of one kind might be grouped together in a nucleus with the particles of the other kind entirely outside the nucleus. In 1911 Lord Rutherford performed an experiment that led him to conclude that the protons are clustered together in a small nucleus at the center of the atom (Figure 3-3).

The first strong evidence that atoms are really quite complex had been obtained in 1896, when Henri Becquerel, a French physicist, discovered radioactivity. The term *radioactivity* refers to the spontaneous disintegration of certain kinds of atoms, producing one or more of three kinds of radiations, designated as alpha (α), beta (β), and gamma (γ) rays.

Rutherford's experiment consisted in directing the positively charged alpha particles emitted by radium (a radioactive element) through a thin gold foil and onto a movable detector screen coated with zinc sulfide (Figure 3-4). Each time an alpha particle hit the zinc sulfide screen, a momentary flash, or scintillation, could be observed. With the movable screen it was possible to observe the scintillations caused by alpha particles that had been deflected from their original direction. In Ruther-

FIGURE 3-3

The nuclear concept of the atom

The nucleus has mass exceeding 99.9 percent of the total mass of the atom, and charge equal to the atomic number of the atom. The spherical region outside the nucleus contains rapidly moving electrons. The diameter of the atom is 10,000 or more times the diameter of the nucleus.

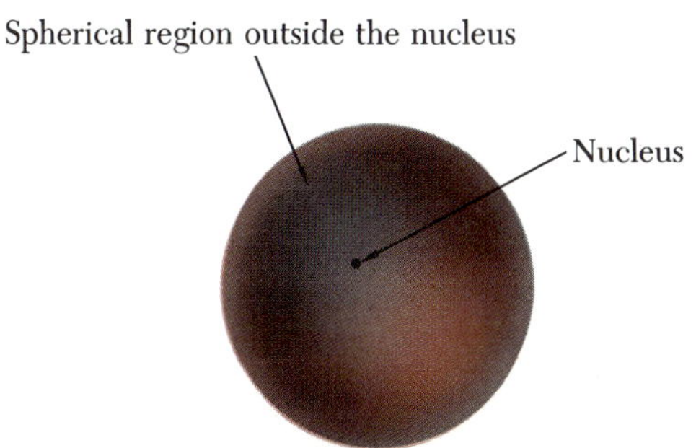

FIGURE 3-4

Rutherford's experiment with alpha particles and a thin gold foil

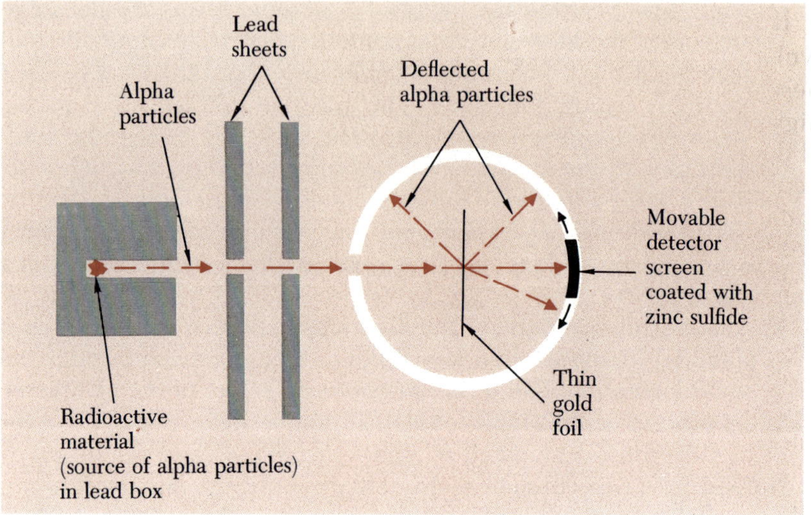

ford's experiment, nearly all the alpha particles passed through the gold foil as if it were not there.

A gold atom is about 3 angstrom units (Å), i.e., 3×10^{-8} cm, in diameter. The gold foil that Rutherford used was about 4×10^{-5} (4000×10^{-8}) cm thick, so the alpha particles had to travel through approximately a thousand gold atoms to reach the zinc sulfide screen—yet most of them reached the screen. From this, Rutherford concluded that a gold atom must be largely "empty space." In a very few instances an alpha particle would be deviated from its path; and occasionally (about once in 100,000 times) one of the particles would be deflected back at a large angle, indicating that it had experienced a very strong repulsive force. Rutherford concluded that there must be located within each gold atom a positively charged body capable of exerting an enormous repulsive force upon one of the positive alpha particles. This positively charged body at the center of the atom he termed the *nucleus*.

3-8 Nuclear Charge and Mass

From a mathematical analysis of the frequency with which the alpha particles were deflected, Rutherford, knowing that the alpha particle has twice the charge of a proton, calculated that the nucleus of the gold atom has the same charge as 79 protons. In the periodic table of the elements, gold is the seventy-ninth element, and it was referred to as the element with atomic number 79 even before Rutherford's research. The results of his investigation suggested that the *atomic number* of an element is something more than its serial number: it is the number of protons in the nucleus of an atom of that element.

In Rutherford's laboratory in England in 1913, the young physicist Henry Moseley compared the wavelengths of the characteristic X-rays of a number of elements and thus arrived at values for the nuclear charges of the atoms. He found that the wavelengths generally decrease as the atomic weights increase, but it was not possible to find any exact relationship between X-ray wavelengths and atomic weights. There was, however, an unfailing relationship between X-ray wavelengths and the ordinal positions of the elements in the periodic system (Figure 3-5).

Moseley's discovery provided proof that there is in the atom a fundamental quantity that increases by regular steps as we pass from one element in the periodic system to the next. This quantity is the *charge* of the nucleus, i.e., the number of protons in the nucleus.

The atom must contain another kind of particle in addition to protons and electrons, because the atomic weight of an element—and the mass

FIGURE 3-5

Relationship between X-ray wavelengths and atomic numbers

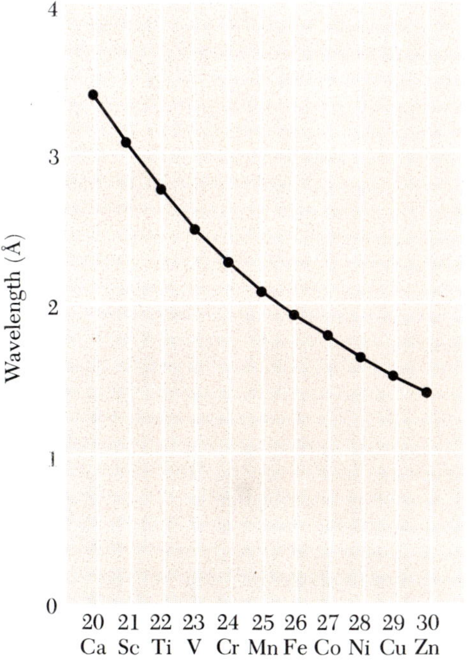

number of each isotope — exceeds the atomic number (except in the case of ^{1}H). According to Rutherford's concept of the nuclear atom, the space outside the nucleus is occupied solely by electrons. The third type of subatomic particle, therefore, must be in the nucleus with the protons — and it must not have much effect on the chemical behavior of the atom. In 1932 this particle was discovered by an English physicist, James Chadwick, and was named the *neutron*. The mass of a proton is 1.0073 amu, and the mass of a neutron is 1.0087 amu. You will recall that the mass of an electron is only 0.0005486 amu (Table 3-1).

The mass of an electron is so much less than the mass of a proton or the mass of a neutron that nearly all the mass of an atom is in the nucleus where the protons and neutrons are. The mass of a neutron, we have just seen, is only slightly in excess of 1 amu, and the same is true of the mass of a proton. Therefore, the mass of any atom or any atomic nucleus is very nearly an integer. We have seen (Section 2-1) that the integer which approximates the mass of an atom, expressed in atomic mass units, is called its mass number. Now it can be seen that *the mass number of an atom is the total number of protons and neutrons in its nucleus.* However, the actual mass of an atom is never exactly an integer (except that the mass of a ^{12}C atom is exactly 12 by definition).

The mass of each isotope of an element can be measured with an instrument called the *mass spectrograph.* In this apparatus the atoms of the element under investigation are given electrical charges, and those with different masses but the same charge travel in different paths as they pass through a magnetic field. When the charged particles, called *ions*, fall upon a photographic film (Figure 3-6), they strike the film at different places according to their masses. An electrical device for measuring the intensities of the ion beams may be substituted for the photographic film; the instrument is then known as a *mass spectrometer.*

In modern chemistry the behavior of an element is explained in terms of the number and arrangement of the electrons, which are outside

TABLE 3-1

Subatomic particles

Name	*Charge*[a]	*Mass*[b] (amu)
Proton	+1	1.0073
Neutron	0	1.0087
Electron	−1	0.0005486

[a] The unit of charge employed here is the amount of charge of the electron itself.
[b] The mass is expressed on the atomic weight scale, defined as $^{12}C = 12$ amu exactly.

FIGURE 3-6

The mass spectrograph

Gas molecules admitted to the evacuated space are given electrical charges. The ions emerge into a magnetic field which deflects the heaviest ions least.

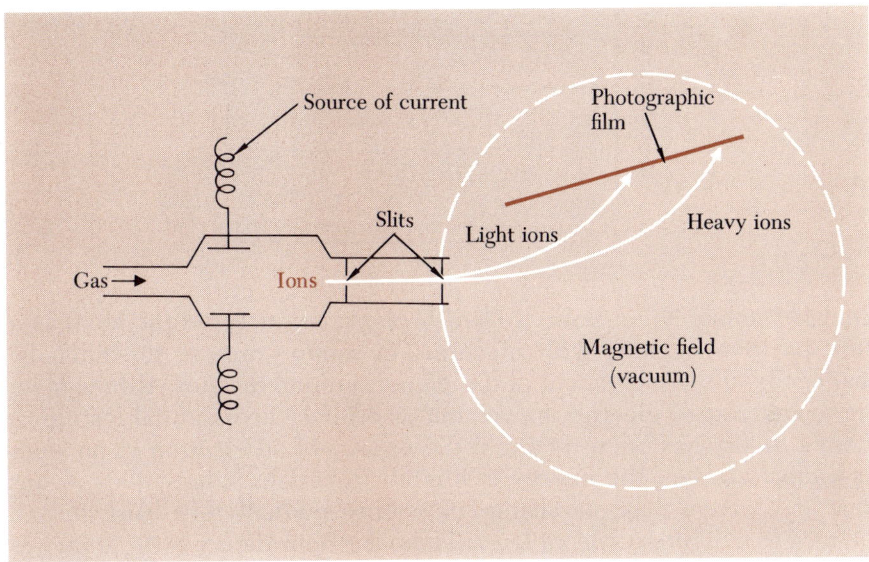

the nucleus of the atom. Since the number of electrons in a neutral atom is the same as the number of protons in its nucleus, which is in turn the atomic number of the element, we see that the chemical behavior of an element can still be said to depend on its atomic number.

An element can now be defined as *a substance all of whose atoms have the same atomic number.*

3-9 Arrangement of the Electrons in Atoms

Conclusions regarding the arrangement of extranuclear electrons have been based on the periodic table of the elements and on studies of atomic spectra.

If an electrical discharge is passed through a gas confined at low pressure in a sealed tube, the gas gives off light that is found to consist of certain wavelengths characteristic of its elements. If the gas consists of only one element, these wavelengths constitute an *atomic spectrum.* In 1913, to explain the existence of atomic spectra and the energy radi-

FIGURE 3-7

Energy levels of electrons in an atom

ated by atoms, Niels Bohr, a Danish physicist, pictured the atom as a tiny positive nucleus with electrons in motion around the nucleus, analogous to the movement of the planets around the sun. Although we no longer regard electron movement as orbital, the essential feature of Bohr's theory has been retained: the energy of an electron in an atom is *quantized*, i.e., the energy is limited to certain values. This means that an electron may not change its energy gradually but must shift all at once to a different *energy level*. Thus, the Bohr theory of the atom was attuned to the *quantum theory* of electromagnetic radiation, proposed in 1900 by the German physicist Max Planck. The different levels corresponding to large energy differences that are possible for electrons in atoms are designated by the numbers 1, 2, 3, 4, etc. (Figure 3-7). The number of the energy level, n, is called the *principal quantum number*.

The maximum number of electrons that can exist in an energy level is $2n^2$. Thus, for the first or lowest energy level, $n = 1$, so the maximum number of electrons for this level is 2. The second energy level, $n = 2$, can contain $2(2^2)$ or 8 electrons. The maximum population for the third level is 18; for the fourth level, 32; for the fifth level, 50; etc. These energy levels are sometimes referred to as *shells*. The first energy level is frequently called the *K* shell; the second, the *L* shell; the third, the *M* shell; the fourth, the *N* shell; the fifth, the *O* shell; etc.

Suppose you were building up an atom. Each electron enters the lowest energy level available, so the single electron in a hydrogen atom (atomic number 1) would be in the *K* shell. The two electrons in a helium atom (atomic number 2) would completely fill the *K* shell. With lithium (atomic number 3), two electrons would fill the *K* shell, and the third would have to go to the next lowest energy level, the *L* shell. Consider magnesium (atomic number 12): two electrons complete the *K* shell, the next eight electrons fill the second energy level (the *L* shell), and the last two electrons occupy the *M* shell. Table 3-2 lists the first 20 elements, with the number of electrons in each shell.

TABLE 3-2

Arrangement of the electrons in the atoms of the first twenty elements

Atomic number	*Name*	*Symbol*	*Shell*			
			K	*L*	*M*	*N*
1	Hydrogen	H	1			
2	Helium	He	2			
3	Lithium	Li	2	1		
4	Beryllium	Be	2	2		
5	Boron	B	2	3		
6	Carbon	C	2	4		
7	Nitrogen	N	2	5		
8	Oxygen	O	2	6		
9	Fluorine	F	2	7		
10	Neon	Ne	2	8		
11	Sodium	Na	2	8	1	
12	Magnesium	Mg	2	8	2	
13	Aluminum	Al	2	8	3	
14	Silicon	Si	2	8	4	
15	Phosphorus	P	2	8	5	
16	Sulfur	S	2	8	6	
17	Chlorine	Cl	2	8	7	
18	Argon	Ar	2	8	8	
19	Potassium	K	2	8	8	1
20	Calcium	Ca	2	8	8	2

If you arrange these first 20 elements according to Newlands' octaves (including helium, neon, and argon, which were not known in Newlands' time) and beneath each symbol note the number of electrons in the *outermost* shell, a periodic variation is evident:

H							He
1							2
Li	Be	B	C	N	O	F	Ne
1	2	3	4	5	6	7	8
Na	Mg	Al	Si	P	S	Cl	Ar
1	2	3	4	5	6	7	8
K	Ca						
1	2						

There is only one electron in the outermost shell of lithium, sodium, and potassium—hence their resemblance. The presence of seven electrons in the outermost shell of a chlorine atom causes it to behave much like

FIGURE 3-8

Periodic table of the elements

	IA	IIA	IIIB	IVB	VB	VIB	VIIB	VIII	
							TRANSITION ELEMENTS		
1	1 H Hydrogen								
2	3 Li Lithium	4 Be Beryllium							
3	11 Na Sodium	12 Mg Magnesium							
4	19 K Potassium	20 Ca Calcium	21 Sc Scandium	22 Ti Titanium	23 V Vanadium	24 Cr Chromium	25 Mn Manganese	26 Fe Iron	27 Co Cobalt
5	37 Rb Rubidium	38 Sr Strontium	39 Y Yttrium	40 Zr Zirconium	41 Nb Niobium	42 Mo Molybdenum	43 Tc Technetium	44 Ru Ruthenium	45 Rh Rhodium
6	55 Cs Cesium	56 Ba Barium	57 La* Lanthanum	72 Hf Hafnium	73 Ta Tantalum	74 W Tungsten	75 Re Rhenium	76 Os Osmium	77 Ir Iridium
7	87 Fr Francium	88 Ra Radium	89 Ac† Actinium						

*** Lanthanides**	58 Ce Cerium	59 Pr Praseodymium	60 Nd Neodymium	61 Pm Promethium	62 Sm Samarium
† Actinides	90 Th Thorium	91 Pa Protactinium	92 U Uranium	93 Np Neptunium	94 Pu Plutonium

a fluorine atom, which also has seven electrons in its outermost shell. The similarities among the other elements are accounted for in the same manner. This is the basis for the Periodic Law, which may be restated as follows: *Many of the physical and chemical properties of the elements vary periodically with their atomic numbers.* Note that the outermost

	IB	IIB	IIIA	IVA	VA	VIA	VIIA	0
								2 He Helium
			5 B Boron	6 C Carbon	7 N Nitrogen	8 O Oxygen	9 F Fluorine	10 Ne Neon
			13 Al Aluminum	14 Si Silicon	15 P Phosphorus	16 S Sulfur	17 Cl Chlorine	18 Ar Argon
28 Ni Nickel	29 Cu Copper	30 Zn Zinc	31 Ga Gallium	32 Ge Germanium	33 As Arsenic	34 Se Selenium	35 Br Bromine	36 Kr Krypton
46 Pd Palladium	47 Ag Silver	48 Cd Cadmium	49 In Indium	50 Sn Tin	51 Sb Antimony	52 Te Tellurium	53 I Iodine	54 Xe Xenon
78 Pt Platinum	79 Au Gold	80 Hg Mercury	81 Tl Thallium	82 Pb Lead	83 Bi Bismuth	84 Po Polonium	85 At Astatine	86 Rn Radon

NONMETALS

63 Eu Europium	64 Gd Gadolinium	65 Tb Terbium	66 Dy Dysprosium	67 Ho Holmium	68 Er Erbium	69 Tm Thulium	70 Yb Ytterbium	71 Lu Lutetium
95 Am Americium	96 Cm Curium	97 Bk Berkelium	98 Cf Californium	99 Es Einsteinium	100 Fm Fermium	101 Md Mendelevium	102 No Nobelium	103 Lr Lawrencium

shells of helium and neon, members of the noble gas family, are completely filled. Since the maximum population of the *M* shell is 18, you might expect argon—the next noble gas after neon—to have 18 electrons in its outermost energy level, but argon has only 8 electrons in the *M* shell. The explanation for this is considered in Chapter 4.

3-10 Periods and Groups in the Periodic Table

Several elements have been discovered since Mendeleev's periodic table was first constructed, but all of them have fitted into the table without difficulty. The imperfections that exist in any periodic table of the elements have motivated various people, including Mendeleev himself, to construct tables with new features. In 1895 Julius Thomsen, a Danish chemist, proposed a form of periodic table essentially like that shown in Figure 3-8.

In the table, the elements are arranged in order of increasing atomic number in 7 rows and 18 columns. The elements in any one row constitute a *series* or *period*; those in any one column constitute a *group*. Usually, the elements in a group consist of a *family* of elements of similar properties. For example, the Group IA elements (except hydrogen) are the alkali metals; the Group VIIA elements are the halogens; and those in Group 0 are the noble gas elements. This periodic recurrence of similar properties is attributed to the presence of identical numbers of electrons in the outermost shells of the atoms. Restricting ourselves momentarily to the first 20 elements, we recall that the halogens fluorine and chlorine both have seven outer electrons; that the alkali metal atoms lithium, sodium, and potassium have one electron in their outermost shell; etc.

3-11 The First Five Periods

The *first period* of the table contains only the elements hydrogen, $_1H$, and helium, $_2He$. (The numbers written as subscripts preceding the symbols are the atomic numbers.) The *second period* contains eight elements, beginning with lithium, $_3Li$, and ending with neon, $_{10}Ne$. The *third period* also contains eight elements, beginning with sodium, $_{11}Na$, and ending with argon, $_{18}Ar$.

If the sequence of periodic similarities in electronic configuration and chemical properties that occurs among the first 18 elements were to repeat itself once more, the next noble gas element after argon would be element 26. Actually, element 26 is iron, and the next noble gas element is krypton, element 36. Thus, the *fourth period* contains 18 elements, or 10 more than either the second or third period.

The first two elements in the fourth period are members of Groups IA and IIA, respectively. Potassium, $_{19}K$, is an alkali metal, as is sodium, $_{11}Na$; and calcium, $_{20}Ca$, belongs to the same family as magnesium, $_{12}Mg$. Similarly, the last six elements in the fourth period are members of Groups IIIA, IVA, VA, VIA, VIIA, and 0, respectively. The ten elements beginning with scandium, ${}_{21}Sc$, and ending with zinc, ${}_{30}Zn$, have no counterparts in the preceding period. They are known as *transition elements.*

The *fifth period* also contains 18 elements; each element has a counterpart immediately above it in the fourth period. The first two elements in the period are members of Groups IA and IIA, and the last six belong to Groups IIIA, IVA, VA, VIA, VIIA, and 0, respectively. The ten elements beginning with yttrium, ${}_{39}Y$, and ending with cadmium, ${}_{48}Cd$, are transition elements.

In the second and third periods there is a noticeable gradation in the properties of the elements. The first element of each period is a highly active metal; the last element is a noble gas; the next-to-last element is a highly active nonmetal; and the intervening five elements shift gradually in their properties from metallic to nonmetallic. The same is true of the A groups and noble gases in the other periods. Among the transition elements (the B groups), the differences between successive elements are much less marked than among the so-called *representative* elements (the A groups and the noble gases).

3-12 The Sixth and Seventh Periods

If the sequence of similarities that occurs among the elements of the fourth and fifth periods were to repeat itself once more, the next noble gas element after xenon, ${}_{54}Xe$, would be element 72. Actually, the next noble gas element is radon, element 86. Thus, the *sixth period* is an even longer period, consisting of 32 elements. As in the previous periods, the first two elements are members of Groups IA and IIA, and the last six are members of Groups IIIA, IVA, VA, VIA, VIIA, and 0, respectively. The 24 elements beginning with lanthanum, ${}_{57}La$, and ending with mercury, ${}_{80}Hg$, are transition elements. Of these 24 transition elements, only lanthanum and those beginning with hafnium, ${}_{72}Hf$, and ending with mercury have counterparts immediately above them in the fifth period. The 14 elements beginning with cerium, ${}_{58}Ce$, and ending with lutetium, ${}_{71}Lu$, have no counterparts in the preceding periods. These elements are very much alike and are known collectively as the *lanthanides*, since they immediately follow lanthanum, which they closely resemble. The lanthanides are *inner transition elements.* We have followed the custom of showing the lanthanides separately because their insertion between Groups IIIB and IVB makes the table inconveniently long.

Since the third period is the same length as the second (8 elements) and the fifth period is the same length as the fourth (18 elements), we would expect the seventh period to contain 32 elements, like the sixth. In other words, the next noble gas element after radon should be element 118. Of the elements now known, radon is the last noble gas element, but each of the elements beyond radon falls into place in the *seventh period* beneath its counterpart in the sixth period. The seventh period, like the preceding periods, begins with an alkali metal, francium, $_{87}Fr$; the second member of the period, radium, $_{88}Ra$, is the counterpart of barium. The 14 elements beginning with thorium, $_{90}Th$, and ending with lawrencium, $_{103}Lr$, are counterparts of the lanthanides and are listed beneath them. These elements are known as the *actinides*, since they immediately follow actinium, $_{89}Ac$. They constitute the second inner transition series.

3-13 Relationships Among the Elements

Elements in the different groups exhibit important features:

1. Members of a chemical family are found in the same column. For example,
    a. The noble gases, also known as the rare gases (and formerly as the inert gases), are all found in Group 0.
    b. Group VIIA, immediately preceding the noble gases, contains the halogens, the most active nonmetals.
    c. The elements of Group IA whose atomic numbers immediately follow those of the noble gas elements are the alkali metals, the most active metals.
2. The physical and chemical properties of the elements in a group show a more or less regular gradation with increasing atomic number. For example,
    a. The melting points of the alkali metals (on the Celsius scale): $_{3}Li$, 180°; $_{11}Na$, 98°; $_{19}K$, 64°; $_{37}Rb$, 39°; $_{55}Cs$, 29° (Figure 3-9).
    b. The radii of the atoms of the elements of Group VIA (in angstroms): $_{8}O$, 0.74; $_{16}S$, 1.04; $_{34}Se$, 1.14; $_{52}Te$, 1.37 (Figure 3-10).
    c. The solubilities of the sulfides of the metals of Group IIB (grams per 100 ml of water at 18°C): ZnS, 0.00069; CdS, 0.00013; HgS, 0.000001.
3. In general, elements in the same group form compounds with similar formulas.
    a. Consider the formulas for the highest oxides (those containing the most oxygen) of the Group VIA elements: SO_3, SeO_3, TeO_3.
    b. Consider the formulas for the highest oxides of the Group VB elements: V_2O_5, Nb_2O_5, Ta_2O_5.

FIGURE 3-9

Melting points of the alkali metals

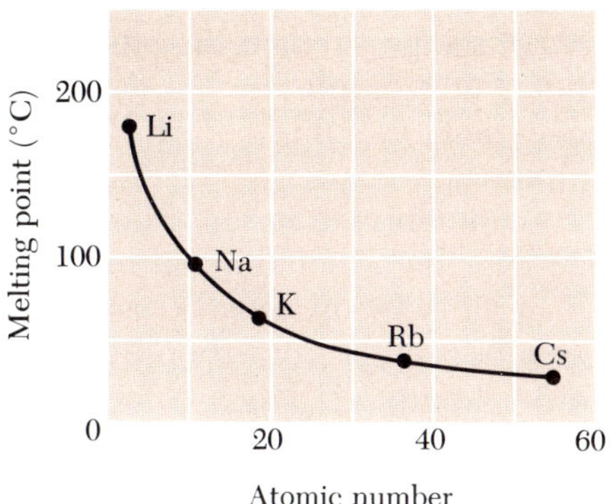

FIGURE 3-10

Radii of Group VIA atoms

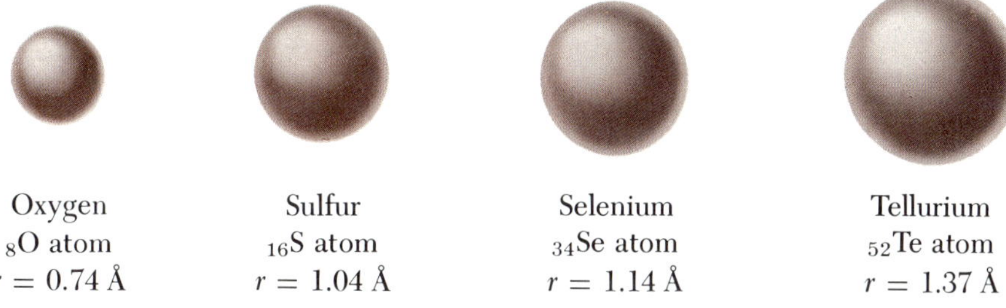

4. In the A groups, metallic properties are accentuated with increase in atomic number. (Metallic properties include thermal and electrical conductivity, malleability, ductility, luster, and the ability to form positive ions.)
5. In the A groups, nonmetallic properties diminish with increase in atomic number. For example, in Group VA the properties change from those of a gaseous element (nitrogen) to those of a lustrous solid (bismuth).
6. The most metallic elements are in the lower left region, and the most nonmetallic ones are in the upper right. All nonmetals are found in A groups. In Figure 3-8 a heavy line separates the metals from the nonmetals. However, most of the elements near this line are borderline elements, sometimes called *metalloids*, and are difficult to classify. For example, the commercially available forms of arsenic and selenium are sold under the labels "arsenic metal" and "selenium metal," but chemists generally agree that these

elements should be classified as nonmetals (unless we take refuge under the classification of metalloids).

7. The first member of a group is usually more different from the other members of the group than the other members are different from each other. For example,
   a. Fluorine does not resemble chlorine as much as chlorine resembles bromine.
   b. Boron is considerably different from the other elements of Group IIIA; it is the only nonmetal in the group.
8. The members of an A group usually have a few points of resemblance with members of the corresponding B group. For example,
   a. Manganese in Group VIIB and chlorine in Group VIIA form acids with similar formulas and properties: permanganic acid, $HMnO_4$, and perchloric acid, $HClO_4$.
   b. The properties of titanium compounds in many instances resemble those of the corresponding silicon compounds.

Grouping similar elements into the same columns of the table simplifies the study of the elements, but the gradation of properties within each group must be remembered. In some cases certain properties of one element are quite different from those of other elements in the same group. For example, chlorine, bromine, and iodine form oxyacids (acids containing oxygen) with the formulas $HClO_3$, $HBrO_3$, and HIO_3, respectively, but fluorine does not form an oxyacid.

3-14 Prediction of "New" Elements

When Mendeleev constructed his first table, only 62 elements were known, so he left a number of blank spaces for elements yet to be discovered. Mendeleev had the insight and courage to predict the properties of the missing elements, and his predictions provided substantial help in the search for these elements. For example, the properties he predicted in 1871 for element 32 (on the basis of its position immediately beneath silicon and to the left of arsenic) are much like the properties this element was found to have when it was discovered by the German chemist Clemens Winkler in 1886. Winkler named the element *germanium.*

None of the members of the noble gas family were known when Mendeleev constructed his table, but when argon was discovered in the earth's atmosphere in 1894, the existence of an entire group of elements previously unknown was implied. A search for the other members of the group was initiated, and by 1898 all the presently known noble gases had been discovered.

As recently as 1939 the spaces for elements 43, 61, 85, and 87 were blank, but since then all these elements have been discovered. Mendeleev did not predict the existence of the transuranium elements (elements beyond uranium); with the later inclusion of the noble gases, his table provided for only 92 elements. In connection with wartime research on atomic energy, elements 93 and 94 were obtained as products of nuclear reactions, and twelve transuranium elements have now been obtained by this procedure. Most of them have been obtained only in very small quantities and are too unstable to exist in nature. On the basis of periodic table relationships, the properties of each of these "synthetic" elements have been rather successfully predicted prior to their discovery. Element 103 is lawrencium ($_{103}Lr$), which was obtained in 1961 at the Lawrence Radiation Laboratory of the University of California. Little is known about its properties, but lawrencium apparently completes the actinide series. It is expected that any elements beyond 103 will occupy the main portion of the periodic table, with element 104 situated beneath hafnium ($_{72}Hf$), etc. Already, the preparation of element 104 has been claimed by both Russian and American scientists.

3-15 Value of the Periodic Table in Stimulating Research

Many investigations of the physical and chemical properties of the elements have been based on apparent discrepancies between the alleged properties and the properties expected on the basis of location in the periodic table. These investigations have sometimes resulted in the correction of serious errors.

Chemical research has frequently been based on familiarity with periodic table relationships. For example, if a certain element is useful as a catalyst for a particular type of reaction, the catalytic action of other elements selected according to their position in the table certainly merits investigation.

The periodic table has served as a useful guide in the determination of atomic weights. For example, when radium was discovered it was found to combine with chlorine in the ratio 113:35.5. If the formula for the compound is $RaCl$, the atomic weight of radium is 113; if the formula for the chloride is $RaCl_2$, the atomic weight of radium is 226. There is no way to determine the formula for the chloride as long as the atomic weight of radium is unknown; without the formula there is no way to know whether the atomic weight is 113 or a multiple of 113. However, in the periodic table Mendeleev left a blank space directly beneath barium. The other known properties of radium indicate that it belongs in this space. With radium located in this position, an atomic weight of

226 becomes reasonable—an atomic weight of 113 would require that radium be situated near cadmium and indium, where there is no available space.

NEW TERMS

Actinides: the 14 elements following actinium (second inner transition series) in the periodic table.

Atomic nucleus: the very small region at the center of an atom in which all the protons and neutrons are located. The diameter of an atom is 10,000 or more times the diameter of the nucleus, but the nucleus contains more than 99.9 percent of the mass of the atom.

Atomic number: the number of units of positive charge possessed by the nucleus of an atom. The atomic number is the same as the number of protons in the nucleus, and it determines the ordinal position of the element in the periodic table.

Atomic spectrum: a pattern of definite wavelengths caused by jumps of electrons between different energy levels.

Cathode rays: the invisible rays that move from the cathode (negative electrode) to the anode (positive electrode) in an evacuated tube when the electrodes are connected to a source of high voltage. Cathode rays consist of electrons.

Coulomb: the amount of electrical charge conveyed by an electric current of 1 ampere in 1 second.

Electron: a subatomic particle with a mass of 9.1×10^{-28} g (which is $\frac{1}{1837}$ of the mass of a hydrogen atom) and a negative charge of 1.602×10^{-19} coulomb. On the atomic weight scale, the mass of an electron is 0.0005486 amu. The charge of an electron is the unit usually employed to express the charges of subatomic particles and ions.

Element: a substance in which all the atoms have the same atomic number.

Energy level: one of a set of states of energy that may be assumed by an electron in an atom.

Group: the set of chemical elements appearing in any one column of the periodic table (Group IA, Group IIA, Group IIIB, etc.).

Inner transition elements: the elements in the sixth and seventh periods of the periodic table that have no counterparts in previous periods. The lanthanides (elements 58–71) constitute the first inner transition series; the actinides (elements 90–103), the second.

Lanthanides: the 14 elements following lanthanum (first inner transition series) in the periodic table.

Mass number: the integer that most nearly expresses the mass of an isotope on the atomic weight scale. The mass number of an atom is the total number of protons and neutrons in its nucleus.

Neutron: a subatomic particle, devoid of electrical charge, with a mass of 1.0087 amu.

Period: the set of chemical elements appearing in any one row of the periodic table (first period, second period, etc.).

Periodic Law: if the elements are considered in the order of their atomic numbers, similar properties recur periodically. In brief, most of the properties of the elements are periodic functions of the atomic numbers.

Periodic table: an arrangement of the chemical elements in rows and columns in the order of their atomic numbers, with similar elements in the same columns.

Principal quantum numbers: the numbers (1, 2, 3, etc.) used to designate the various principal energy levels an electron may occupy in an atom.

Proton: a subatomic particle with a positive charge of 1.602×10^{-19} coulomb and a mass of 1.0073 amu.

Radioactivity: the spontaneous disintegration of certain kinds of atoms, producing one or more of three kinds of radiations: alpha, beta, and gamma rays.

Representative elements: the nontransition elements, i.e., the A groups and the noble gases.

Shell: a term commonly used in referring to the electrons of a given principal energy level, for example, *K* shell, *L* shell, etc.

Transition elements: the B groups, i.e., the elements in the fourth and succeeding periods of the periodic table that have no counterparts in preceding periods.

X-rays: the highly penetrating electromagnetic radiation emanating from the spot where cathode rays strike a metal. X-rays are also known as Roentgen rays. Their wavelengths are expressed in angstroms (Å).

EXERCISES

3-1 Explain what is meant by the periodicity of the elements. By way of example, use a physical property and a chemical property.

3-2 Predict the missing properties of potassium and chlorine:

Element	*Atomic radius* (Å)	*Melting point* (°C)	*Boiling point* (°C)
Sodium	1.57	97.5	889
Potassium			
Rubidium	2.16	38.8	679
Fluorine	0.72	−218	−188
Chlorine			
Bromine	1.14	−7	59

3-3 Construct a blank periodic table and indicate the locations of the following sets of elements: alkali metals, halogens, noble gases, transition elements, representative elements, A and B subgroups, inner transition elements, lanthanides, actinides, metals, nonmetals.

3-4 Which two elements are most like selenium? Like bromine? Like rubidium? Like antimony? Like neon? Like strontium? Like tin?

3-5 Given the formulas $LiNO_3$, $Mg(NO_3)_2$, AlF_3, CS_2, H_2O, $SiCl_4$, predict the formulas for the following compounds: potassium nitrate, beryllium nitrate, boron fluoride, silicon sulfide, hydrogen sulfide, carbon fluoride.

3-6 How are isotopes alike? How are they different?

3-7 The atomic weight of argon (element 18) is 39.948, and the atomic weight of potassium (element 19) is 39.102. What can be said, on the basis of this information alone, regarding the relative abundance of the different isotopes of these two elements?

3-8 Find the instances in which atomic weights are not in the same order as atomic numbers.

3-9 Look up the atomic number of tin. How many protons are there in a tin atom? How many electrons? What additional information do you need before stating how many neutrons are in a tin atom?

3-10 What is the greatest number of electrons ever present in an M shell? In an N shell?

3-11 State, define, describe, or explain:
(a) the Periodic Law
(b) periods, groups, families
(c) metals, nonmetals, metalloids
(d) atomic number and mass number
(e) transition elements and representative elements

3-12 Discuss the basis of Mendeleev's predictions.

3-13 Do you think that an element presently unknown may be discovered and found to have an atomic number that places it between magnesium and aluminum in the periodic table? Justify your answer.

3-14 How can a piece of matter be electrically neutral if it consists of atoms that contain electrically charged particles?

3-15 The radius of a carbon nucleus is about 2.8×10^{-13} cm. (a) Compute the density of a carbon-12 nucleus and compare it with the density of diamond, 3.5 g/cm^3. (b) The radius of a carbon atom is 0.77 Å; compute the ratio of the volume of the atom to the volume of the nucleus.

REFERENCES

Andrade, E. N. da C., *Rutherford and the Nature of the Atom*, Doubleday, Garden City, N.Y., 1964 (paperbound). A short biography by a colleague.

Friend, J. N., *Man and the Chemical Elements*, 2nd ed., Scribner's, New York, 1961. Written from a historical and industrial viewpoint.

Mazurs, E. G., "Ups and Downs of the Periodic Table," *Chemistry,* **39**(7), 6 (1966). Many of the different forms of periodic tables are shown.

Nier, A. O. C., "The Mass Spectrometer," *Scientific American,* **188**(3), 68 (1953). An account of the modern version of the mass spectrograph and its many uses.

Reed, R. I., and Robertson, D. H., "Mass Spectrometry. Part 1: The Spectrometer," *Chemistry,* **42**(6), 7 (1969). How mass spectra are obtained.

Seaborg, G. T., and Valens, E. G., *Elements of the Universe,* Dutton, New York, 1962. Outgrowth of an educational television series on the chemical elements.

Van Spronsen, J. W., "The Prehistory of the Periodic System of the Elements," *Journal of Chemical Education,* **36**(11), 565 (1959). A consideration of the state of chemical theory at the time the Periodic Law was taking shape.

Van Spronsen, J. W., "The Priority Conflict between Mendeleev and Meyer," *Journal of Chemical Education,* **46**(3), 137 (1969). A critical review of the independent achievements of Mendeleev and Meyer.

4

Electronic Configurations of the Elements

4-1 Electron Orbitals in Atoms

The periodicity that results when the elements are arranged in order of increasing atomic number is due to the variations in the arrangement of the electrons in the atoms. The tendency of certain properties to "repeat" at particular intervals is correlated with repeating electronic configurations in the outer shells of the atoms. One situation that especially requires an explanation is the variation in the lengths of the periods in the table: a period of 2 elements ($_1H$ and $_2He$), followed by two periods of 8 elements ($_3Li$ through $_{10}Ne$ and $_{11}Na$ through $_{18}Ar$), then two periods of 18 elements ($_{19}K$ through $_{36}Kr$ and $_{37}Rb$ through $_{54}Xe$), a period of 32 elements ($_{55}Cs$ through $_{86}Rn$), and finally an incomplete period.

The quantum theory of the atom has undergone extensive modification in the last half-century. The original Bohr model of the atom showed each electron revolving in a definite orbit outside the nucleus, but we no longer think that an electron behaves like a planet in a fixed orbit.

The flat orbits of the Bohr model have been replaced by three-dimensional orbitals. An *orbital* is a region in the space surrounding the nucleus of an atom that can be occupied by no more than two electrons. The "shape" of an orbital is portrayed by drawing a figure in such a manner that there is, say, a 95 percent probability that, at a particular instant, the electron will be somewhere within the region represented by the figure.

When each electron is in the lowest energy level available to it (Section 3-9), the atom is in the *ground state.* If any electron in the atom is in a higher energy level, the atom is said to be in an *excited state.* In accounting for the arrangement of the elements in the periodic table, our first concern is with the electronic arrangement in the ground states of atoms. In the ground states of the known elements, electrons inhabit four types of orbitals referred to by the four letters s, p, d, and f. Each of these one-letter names corresponds to a particular value for the *orbital quantum number, l.* The symbol s corresponds to $l = 0$, p means $l = 1$, d represents $l = 2$, and f symbolizes $l = 3$. According to quantum theory, an electron whose principal quantum number is n can have values for its orbital quantum number equal to zero or any integer up to $n - 1$.

Without attempting to describe the motion of an electron that inhabits an s orbital, we nevertheless can say that the shape of the orbital itself is spherically symmetrical (Figure 4-1a). This means only that

FIGURE 4-1

Shapes of *s* and *p* orbitals

(a) *s* orbital. (b) *p* orbital.

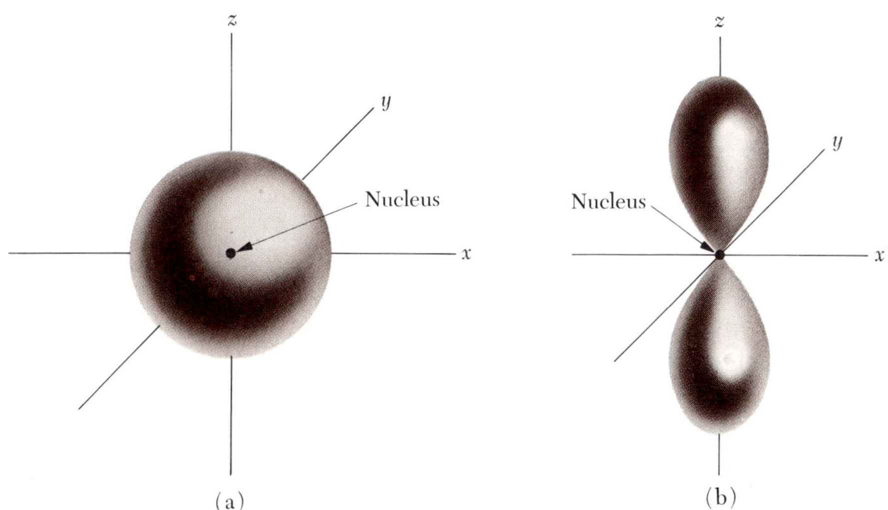

there is a spherical region surrounding the nucleus which has a 95 percent probability of "containing" the electron.

A *p* orbital has approximately the shape of a dumbbell, with two spheres situated on opposite sides of the nucleus (Figure 4-1b). An electron that inhabits a *p* orbital has equal probability of being in either half of it. The shapes of *d* and *f* orbitals are considerably more complicated.

Any orbital, whatever its type, can accommodate *no more than two* electrons. Thus, the electronic population of a particular orbital is always one of three numbers: 0, 1, or 2. According to theory, all electrons behave as though they were spinning and this spinning charge sets up a small magnetic field. An electron can be thought to spin about its axis in a clockwise or counterclockwise direction; these two directions of spin correspond to tiny magnets oriented in opposite directions. Two electrons can inhabit the same orbital only if they have opposite "spins."

4-2 Shells and Subshells of Electrons

When atoms of the known elements are in their ground states, all electrons are contained within seven shells designated by the principal quantum numbers (see Section 3-9). The average distance from the nucleus of electrons in the second shell is greater than for those in the first; for those in the third it is greater than for those in the second; and so on. The energy of electrons in successive shells increases (i.e., the electrons are held less tightly) as the distance between the shell and the nucleus increases. However, all the electrons in a given shell do not necessarily have exactly the same energy. Each shell beyond the *K* shell contains one or more *subshells*; electrons in different subshells do not have exactly the same energy (Figure 4-2).

Each shell contains one *s* orbital. The *K* shell contains only this one orbital, designated 1*s* (where 1 is the principal quantum number, *n*). In other words, the *K* shell is not divided into different subshells; the expressions "the 1*s* orbital," "the 1*s* subshell," "the first energy level," and "the *K* shell" all have the same meaning. The fact that there is only an *s* orbital in the shell for which the principal quantum number is 1 is implied by the statement (Section 4-1) that when $n = 1$ the only possible value for l, the orbital quantum number, is zero ($n - 1 = 1 - 1 = 0$).

The second (*L*) shell contains two subshells: one consisting of one *s* orbital, designated 2*s*, and a second subshell consisting of three *p* orbitals, each designated 2*p*. Each *p* orbital is perpendicular to the other two so that, considered together, they are distributed about the nucleus in a symmetrical manner.

The presence of both *s* and *p* orbitals in the second shell is in agree-

FIGURE 4-2

Energy of electrons in different subshells

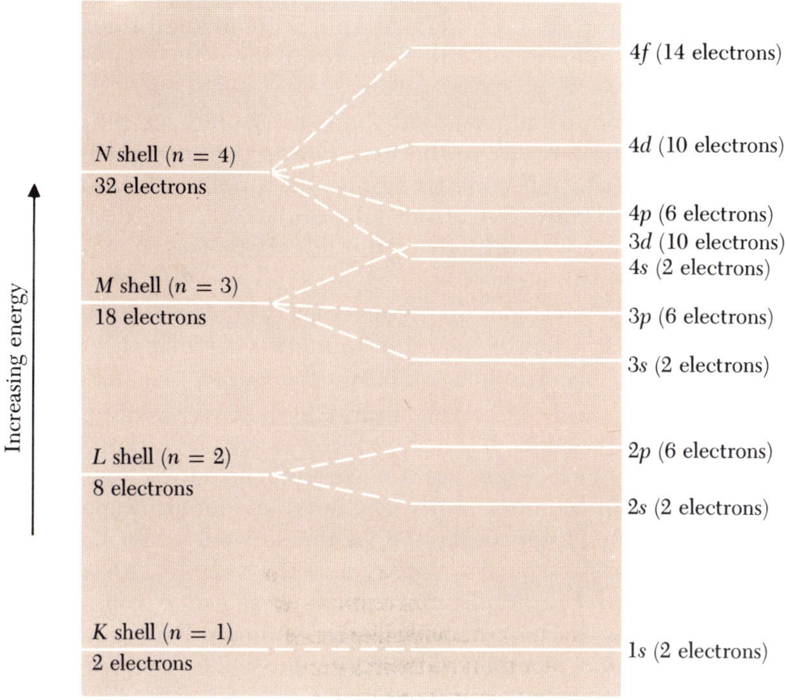

ment with the statement that when $n = 2$ the orbital quantum number may have values of 0 or 1 ($n - 1 = 2 - 1 = 1$). The *p* orbitals are the orbitals for which $l = 1$.

The number of orbitals of a particular kind in a given shell is governed by a third quantum number known as the *magnetic quantum number, m_l*. The possible values for m_l for a given value of l are the integers from $-l$ to $+l$, including 0. Thus the number of different possible values for m_l is $2l + 1$. When $l = 0$, the only possible value for m_l is 0, so there is only one *s* orbital in any shell. But when $l = 1$, the number of different possible values for m_l is 3 ($2l + 1 = 2 + 1 = 3$), so there are three *p* orbitals in any given shell.

The third (*M*) shell contains three subshells: one consisting of one *s* orbital, designated 3*s*; a second consisting of three *p* orbitals, each designated 3*p*; and a third consisting of five *d* orbitals, each designated 3*d*. The five *d* orbitals are spatially distributed so that the *d* subshell is symmetrical with respect to the nucleus.

The fourth (*N*) shell contains four subshells designated 4*s*, 4*p*, 4*d*, and 4*f*. An *f* subshell consists of seven *f* orbitals symmetrically distributed about the nucleus.

The energy of an electron in a given orbital in a particular shell is lower (i.e., the electrons are held more tightly) than that of an electron in an orbital of the same type in a shell with a higher number, i.e., $1s < 2s < 3s < 4s < 5s < 6s < 7s$. Within a given shell the *s* electrons have the lowest energy and the *f* electrons have the highest energy, i.e., $s < p < d < f$.

Each shell contains as many subshells as are indicated by the principal quantum number: one in the first, two in the second, three in the third, etc. Each subshell consists of an odd number of orbitals: one *s* orbital; three *p* orbitals; five *d* orbitals; and seven *f* orbitals. The net result is that there are n^2 orbitals in the *n*th shell: 1 in the first, 4 in the second, 9 in the third, etc.

The number of electrons in a particular orbital is restricted by the possible values for a fourth quantum number known as the *spin quantum number*, m_s. According to quantum theory, m_s can have either of the two values $+\frac{1}{2}$ and $-\frac{1}{2}$, corresponding to the two possible directions of electron spin (Section 4-1). Therefore, as we have noted, any one orbital, regardless of its type, has a maximum capacity of two electrons. Since the *n*th shell contains n^2 orbitals, the total electron capacity of the *n*th shell is $2n^2$, i.e., 2 for the first, 8 for the second, 18 for the third, 32 for the fourth, etc.

The fundamental principle that applies to the electronic configurations of atoms was discovered in 1925 by Wolfgang Pauli and is known as *Pauli's Exclusion Principle.* This principle states that *no two electrons in an atom ever have the same set of values for the four quantum numbers* n, l, m_l, and m_s.

4-3 Electronic Configurations of First Period Elements

The arrangement of the elements in the periodic table is the result of the principles of atomic structure considered in Section 4-2. The first period of the table contains only two elements, hydrogen and helium. In the first shell (the *K* shell) of any atom there is only one subshell, consisting of a single orbital, 1*s*, with a maximum capacity of two electrons. We may represent the electronic configuration of the hydrogen atom by using the abbreviated notation $1s^1$. The number 1 preceding the letter *s* indicates the principal quantum number ($n = 1$); *s* represents the *s* orbital; and the superscript 1 indicates that there is one electron in this orbital. The helium atom has two electrons, both in the same orbital; the notation for its electronic configuration is $1s^2$.

In Table 4-1, each electronic configuration is indicated by the abbreviated notation and also by a schematic representation in which a circle is used for each orbital. The presence of a lone electron in the 1*s* orbital of the hydrogen atom is indicated by an upturned arrow in the 1*s* circle. Since the entering helium electron fills the 1*s* orbital and has its spin in the opposite direction, it is symbolized by a downturned arrow in the same circle. The *K* shell, i.e., the 1*s* subshell, is complete at this point, and no more electrons can be added to it. In the next element, the third electron must go into the next higher energy level, which is the 2*s* subshell.

4-4 Electronic Configurations of Second Period Elements

Lithium is the first element of the second period. Two of its three electrons are in the first shell; the third is in the next higher level, the 2*s* orbital. In abbreviated notation its configuration is $1s^2 2s^1$. In beryllium, the next element, the fourth electron is also in the 2*s* orbital. With a $1s^2 2s^2$ configuration, the beryllium atom has a completed 2*s* orbital. The remaining elements of this series have electrons in the 2*p* orbitals. Remember that there are three of these, each with a capacity for two electrons.

In the boron atom, one electron is located in a 2*p* orbital, so the configuration may be represented as $1s^2 2s^2 2p^1$. In Table 4-1 you can see that the sixth electron of the carbon atom is not in the same 2*p* orbital as the fifth electron. Likewise, the seventh electron of the nitrogen atom is in the third 2*p* orbital. According to the *Principle of Maximum Multiplicity*, electrons enter each orbital of a given subshell singly so as to minimize electrical repulsion as long as this possibility exists. Only after each of the orbitals has one electron does a second electron enter any orbital of that subshell. This principle was first stated by Friedrich Hund, the German spectroscopist, and is often referred to as *Hund's Rule*.

In the remaining elements of this period, the 2*p* orbitals are progressively filled, so there is a $2p^4$ configuration for oxygen, $2p^5$ for fluorine, and $2p^6$ for neon. In the neon atom, both the *K* shell and the *L* shell are filled to their maximum capacity. Helium and neon are the only noble gases whose outer shells are filled to capacity.

4-5 An Alternative Notation for Electronic Configurations

In the helium atom the *K* shell contains as many electrons as it contains in an atom of any other element, however great the atomic number. To

TABLE 4-1

Electronic configurations of the first eighteen elements

	Element	*Abbreviated notation*	*Schematic representation* 1s	2s	2p	3s	3p
First period	$_{1}H$	$1s^1$	(↑)				
	$_{2}He$	$1s^2$	(↑↓)				
Second period	$_{3}Li$	$1s^22s^1$	(↑↓)	(↑)	○ ○ ○		
	$_{4}Be$	$1s^22s^2$	(↑↓)	(↑↓)	○ ○ ○		
	$_{5}B$	$1s^22s^22p^1$	(↑↓)	(↑↓)	(↑) ○ ○		
	$_{6}C$	$1s^22s^22p^2$	(↑↓)	(↑↓)	(↑) (↑) ○		
	$_{7}N$	$1s^22s^22p^3$	(↑↓)	(↑↓)	(↑) (↑) (↑)		
	$_{8}O$	$1s^22s^22p^4$	(↑↓)	(↑↓)	(↑↓) (↑) (↑)		
	$_{9}F$	$1s^22s^22p^5$	(↑↓)	(↑↓)	(↑↓) (↑↓) (↑)		
	$_{10}Ne$	$1s^22s^22p^6$	(↑↓)	(↑↓)	(↑↓) (↑↓) (↑↓)		
Third period	$_{11}Na$	$1s^22s^22p^63s^1$	(↑↓)	(↑↓)	(↑↓) (↑↓) (↑↓)	(↑)	○ ○ ○
	$_{12}Mg$	$1s^22s^22p^63s^2$	(↑↓)	(↑↓)	(↑↓) (↑↓) (↑↓)	(↑↓)	○ ○ ○
	$_{13}Al$	$1s^22s^22p^63s^23p^1$	(↑↓)	(↑↓)	(↑↓) (↑↓) (↑↓)	(↑↓)	(↑) ○ ○
	$_{14}Si$	$1s^22s^22p^63s^23p^2$	(↑↓)	(↑↓)	(↑↓) (↑↓) (↑↓)	(↑↓)	(↑) (↑) ○
	$_{15}P$	$1s^22s^22p^63s^23p^3$	(↑↓)	(↑↓)	(↑↓) (↑↓) (↑↓)	(↑↓)	(↑) (↑) (↑)
	$_{16}S$	$1s^22s^22p^63s^23p^4$	(↑↓)	(↑↓)	(↑↓) (↑↓) (↑↓)	(↑↓)	(↑↓) (↑) (↑)
	$_{17}Cl$	$1s^22s^22p^63s^23p^5$	(↑↓)	(↑↓)	(↑↓) (↑↓) (↑↓)	(↑↓)	(↑↓) (↑↓) (↑)
	$_{18}Ar$	$1s^22s^22p^63s^23p^6$	(↑↓)	(↑↓)	(↑↓) (↑↓) (↑↓)	(↑↓)	(↑↓) (↑↓) (↑↓)

emphasize the fact that elements beyond helium have the helium configuration in the *K* shell, we may modify our notation by writing (He)$2s^1$ for lithium, (He)$2s^2$ for beryllium, (He)$2s^22p^1$ for boron, etc. A similar procedure may be used for elements following the other noble gases, for example, (Ne)$3s^1$ for sodium.

4-6 Electronic Configurations of Third Period Elements

The first element of the third period, sodium, has one electron in the *M* shell, i.e., in the $3s$ orbital, and its electronic configuration may be represented as $1s^22s^22p^63s^1$ or (Ne)$3s^1$. The successive elements in this period have the $3s$ and $3p$ orbitals progressively filled in the same manner as the elements of the second period have the $2s$ and $2p$ orbitals progressively filled (see Table 4-1). This explains the positions of these elements in the periodic table: magnesium beneath beryllium, aluminum beneath boron, etc.

The third period ends with element 18, argon, which has the same outer-shell configuration, s^2p^6, as the neon atom. Argon, like neon, is a noble gas, even though the maximum capacity of the M shell is 18 rather than 8. In fact, all the noble gases except helium have eight electrons, i.e., the s^2p^6 configuration, in the outer shell.

4-7 The Aufbau Principle

In considering electronic configurations, we have assumed that each entering electron goes into the lowest energy level in which a vacancy exists. This is the *Aufbau Principle* (German: *Aufbau*, building up). With certain exceptions, each successive element has the configuration of the element preceding it plus one electron.

In the first three periods it is sufficient to use two general principles: (1) electrons in each successive shell (K, L, M, N, etc.) have increasing average energies; (2) electrons in orbitals of a given shell have increasing average energies in the order s, p, d, f. However, in the higher shells the orbitals overlap in energy. For instance, the $3d$ orbitals are of higher energy than the $4s$ (see Figure 4-2); the $4d$ orbitals are of higher energy than the $5s$; and the $4f$ orbitals are of higher energy than either the $6s$ or $5p$. In Figure 4-3 the energy associated with each type of orbital is represented diagrammatically.

A mnemonic chart for the order of filling of orbitals is provided in Figure 4-4. The orbitals are presented in the same order as in Figure 4-3. In using this idealized chart we assume that each successive element has the same configuration as the element preceding it, with one electron added. Exceptions will be pointed out.

4-8 Electronic Configurations of Fourth Period Elements

In accordance with the Aufbau Principle, we would expect the first element of the fourth period, potassium, to have one electron in the fourth (N) shell, i.e., in the $4s$ orbital, since the $3p$ orbitals are filled in argon,

FIGURE 4-3

Approximate relative energy of electrons in subshells

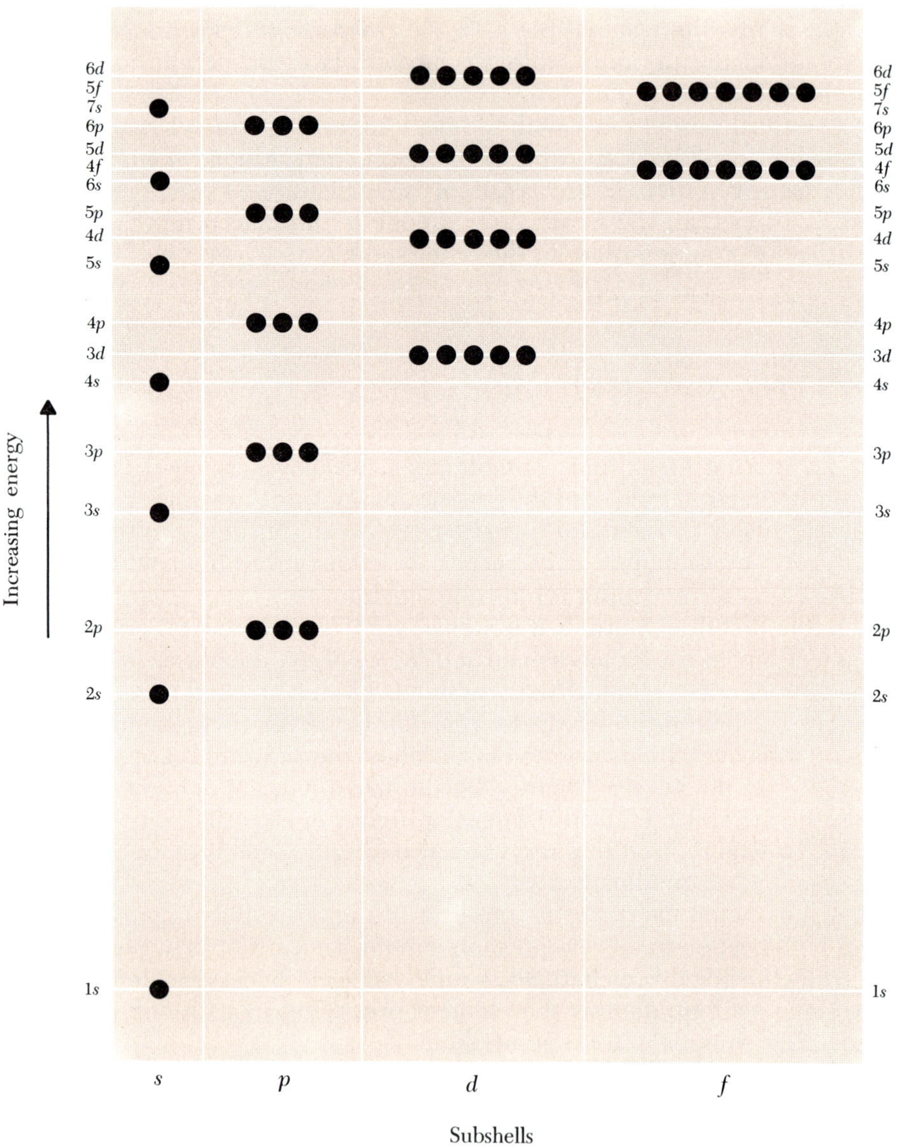

the element immediately preceding potassium (see Figure 4-4). The electronic configuration of potassium is indeed $1s^2 2s^2 2p^6 3s^2 3p^6 4s^1$ or, more briefly, $(Ar)4s^1$. The twentieth electron of the calcium atom also goes into the $4s$ orbital, filling it. The twenty-first element, scandium, shows why the fourth period is longer than the preceding periods—the $3d$ orbitals begin to fill up, as would be predicted from our mnemonic

FIGURE 4-4

Approximate order of entry of electrons into subshells

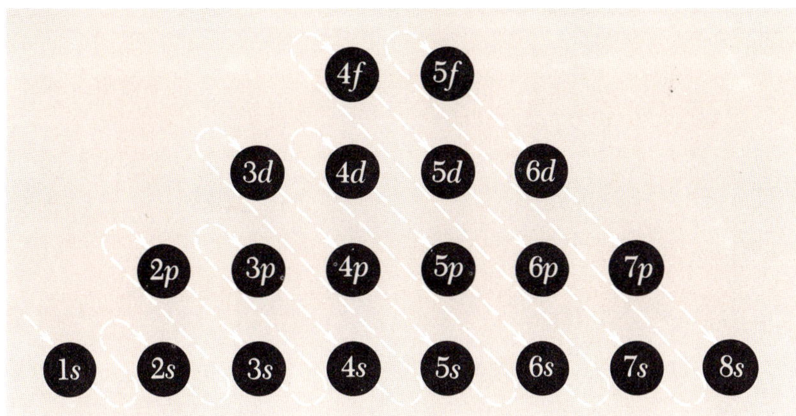

TABLE 4-2

Electronic configurations of fourth period elements

Element	*Abbreviated notation*	*Schematic representation of outer orbitals*				
		$3s$	$3p$	$3d$	$4s$	$4p$
$_{19}$K	(Ar)$4s^1$	⇅	⇅⇅⇅	○○○○○	↑	○○○
$_{20}$Ca	(Ar)$4s^2$	⇅	⇅⇅⇅	○○○○○	⇅	○○○
$_{21}$Sc	(Ar)$3d^14s^2$	⇅	⇅⇅⇅	↑○○○○	⇅	○○○
$_{22}$Ti	(Ar)$3d^24s^2$	⇅	⇅⇅⇅	↑↑○○○	⇅	○○○
$_{23}$V	(Ar)$3d^34s^2$	⇅	⇅⇅⇅	↑↑↑○○	⇅	○○○
$_{24}$Cr	(Ar)$3d^54s^1$	⇅	⇅⇅⇅	↑↑↑↑↑	↑	○○○
$_{25}$Mn	(Ar)$3d^54s^2$	⇅	⇅⇅⇅	↑↑↑↑↑	⇅	○○○
$_{26}$Fe	(Ar)$3d^64s^2$	⇅	⇅⇅⇅	⇅↑↑↑↑	⇅	○○○
$_{27}$Co	(Ar)$3d^74s^2$	⇅	⇅⇅⇅	⇅⇅↑↑↑	⇅	○○○
$_{28}$Ni	(Ar)$3d^84s^2$	⇅	⇅⇅⇅	⇅⇅⇅↑↑	⇅	○○○
$_{29}$Cu	(Ar)$3d^{10}4s^1$	⇅	⇅⇅⇅	⇅⇅⇅⇅⇅	↑	○○○
$_{30}$Zn	(Ar)$3d^{10}4s^2$	⇅	⇅⇅⇅	⇅⇅⇅⇅⇅	⇅	○○○
$_{31}$Ga	(Ar)$3d^{10}4s^24p^1$	⇅	⇅⇅⇅	⇅⇅⇅⇅⇅	⇅	↑○○
$_{32}$Ge	(Ar)$3d^{10}4s^24p^2$	⇅	⇅⇅⇅	⇅⇅⇅⇅⇅	⇅	↑↑○
$_{33}$As	(Ar)$3d^{10}4s^24p^3$	⇅	⇅⇅⇅	⇅⇅⇅⇅⇅	⇅	↑↑↑
$_{34}$Se	(Ar)$3d^{10}4s^24p^4$	⇅	⇅⇅⇅	⇅⇅⇅⇅⇅	⇅	⇅↑↑
$_{35}$Br	(Ar)$3d^{10}4s^24p^5$	⇅	⇅⇅⇅	⇅⇅⇅⇅⇅	⇅	⇅⇅↑
$_{36}$Kr	(Ar)$3d^{10}4s^24p^6$	⇅	⇅⇅⇅	⇅⇅⇅⇅⇅	⇅	⇅⇅⇅

FIGURE 4-5

Table showing periodicity in electron configurations of the elements

1 H $1s^1$								
3 Li $2s^1$	4 Be $2s^2$							
11 Na $3s^1$	12 Mg $3s^2$							
19 K $4s^1$	20 Ca $4s^2$	21 Sc $3d^14s^2$	22 Ti $3d^24s^2$	23 V $3d^34s^2$	24 Cr $3d^54s^1$	25 Mn $3d^54s^2$	26 Fe $3d^64s^2$	27 Co $3d^74s^2$
37 Rb $5s^1$	38 Sr $5s^2$	39 Y $4d^15s^2$	40 Zr $4d^25s^2$	41 Nb $4d^45s^1$	42 Mo $4d^55s^1$	43 Tc $4d^65s^1$	44 Ru $4d^75s^1$	45 Rh $4d^85s^1$
55 Cs $6s^1$	56 Ba $6s^2$	57 La* $5d^16s^2$	72 Hf $5d^26s^2$	73 Ta $5d^36s^2$	74 W $5d^46s^2$	75 Re $5d^56s^2$	76 Os $5d^66s^2$	77 Ir $5d^76s^2$
87 Fr $7s^1$	88 Ra $7s^2$	89 Ac† $6d^17s^2$						

*** Lanthanides**	58 Ce $4f^15d^16s^2$	59 Pr $4f^36s^2$	60 Nd $4f^46s^2$	61 Pm $4f^56s^2$	62 Sm $4f^66s^2$
† Actinides	90 Th $6d^27s^2$	91 Pa $5f^26d^17s^2$	92 U $5f^36d^17s^2$	93 Np $5f^46d^17s^2$	94 Pu $5f^67s^2$

chart. Note that the 3*d* orbitals, with a capacity of ten electrons, are completely filled before the 4*p* orbitals obtain any electrons (Table 4-2).

Since there are no *d* orbitals in the first and second shells, the ten elements from scandium through zinc have no counterparts in the preceding periods. They are referred to as *transition elements*. The 4*s* or-

									2 He $1s^2$
			5 B $2p^1$	6 C $2p^2$	7 N $2p^3$	8 O $2p^4$	9 F $2p^5$	10 Ne $2p^6$
			13 Al $3p^1$	14 Si $3p^2$	15 P $3p^3$	16 S $3p^4$	17 Cl $3p^5$	18 Ar $3p^6$
28 Ni $3d^84s^2$	29 Cu $3d^{10}4s^1$	30 Zn $3d^{10}4s^2$	31 Ga $4p^1$	32 Ge $4p^2$	33 As $4p^3$	34 Se $4p^4$	35 Br $4p^5$	36 Kr $4p^6$
46 Pd $4d^{10}$	47 Ag $4d^{10}5s^1$	48 Cd $4d^{10}5s^2$	49 In $5p^1$	50 Sn $5p^2$	51 Sb $5p^3$	52 Te $5p^4$	53 I $5p^5$	54 Xe $5p^6$
78 Pt $5d^96s^1$	79 Au $5d^{10}6s^1$	80 Hg $5d^{10}6s^2$	81 Tl $6p^1$	82 Pb $6p^2$	83 Bi $6p^3$	84 Po $6p^4$	85 At $6p^5$	86 Rn $6p^6$

63 Eu $4f^76s^2$	64 Gd $4f^75d^16s^2$	65 Tb $4f^96s^2$	66 Dy $4f^{10}6s^2$	67 Ho $4f^{11}6s^2$	68 Er $4f^{12}6s^2$	69 Tm $4f^{13}6s^2$	70 Yb $4f^{14}6s^2$	71 Lu $4f^{14}5d^16s^2$
95 Am $5f^77s^2$	96 Cm $5f^76d^17s^2$	97 Bk $5f^97s^2$	98 Cf $5f^{10}7s^2$	99 Es $5f^{11}7s^2$	100 Fm $5f^{12}7s^2$	101 Md $5f^{13}7s^2$	102 No $5f^{14}7s^2$	103 Lr $5f^{14}6d^17s^2$

bital is being filled in the two elements preceding them, and the $4p$ orbitals are being filled in the elements following them (beginning with $_{31}$Ga and ending with $_{36}$Kr). The filling of the fourth shell is interrupted to permit the $3d$ orbitals to fill. The transition elements of this period, $_{21}$Sc through $_{30}$Zn, are often referred to as the $3d$ elements.

TABLE 4-3

Electronic configurations of the elements in their ground states

Atomic number	*Element*	1(*K*)	2(*L*)	3(*M*)	4(*N*)	5(*O*)	6(*P*)	7(*Q*)
		s	*s p*	*s p d*	*s p d f*	*s p d f*	*s p d*	*s*
1	H	1						
2	He	2						
3	Li	2	1					
4	Be	2	2					
5	B	2	2 1					
6	C	2	2 2					
7	N	2	2 3					
8	O	2	2 4					
9	F	2	2 5					
10	Ne	2	2 6					
11	Na	2	2 6	1				
12	Mg	2	2 6	2				
13	Al	2	2 6	2 1				
14	Si	2	2 6	2 2				
15	P	2	2 6	2 3				
16	S	2	2 6	2 4				
17	Cl	2	2 6	2 5				
18	Ar	2	2 6	2 6				
19	K	2	2 6	2 6	1			
20	Ca	2	2 6	2 6	2			
21	Sc	2	2 6	2 6 1	2			
22	Ti	2	2 6	2 6 2	2			
23	V	2	2 6	2 6 3	2			
24	Cr	2	2 6	2 6 5	1			
25	Mn	2	2 6	2 6 5	2			
26	Fe	2	2 6	2 6 6	2			
27	Co	2	2 6	2 6 7	2			
28	Ni	2	2 6	2 6 8	2			
29	Cu	2	2 6	2 6 10	1			
30	Zn	2	2 6	2 6 10	2			
31	Ga	2	2 6	2 6 10	2 1			
32	Ge	2	2 6	2 6 10	2 2			
33	As	2	2 6	2 6 10	2 3			
34	Se	2	2 6	2 6 10	2 4			
35	Br	2	2 6	2 6 10	2 5			
36	Kr	2	2 6	2 6 10	2 6			
37	Rb	2	2 6	2 6 10	2 6	1		
38	Sr	2	2 6	2 6 10	2 6	2		
39	Y	2	2 6	2 6 10	2 6 1	2		
40	Zr	2	2 6	2 6 10	2 6 2	2		
41	Nb	2	2 6	2 6 10	2 6 4	1		
42	Mo	2	2 6	2 6 10	2 6 5	1		
43	Tc	2	2 6	2 6 10	2 6 6	1		
44	Ru	2	2 6	2 6 10	2 6 7	1		
45	Rh	2	2 6	2 6 10	2 6 8	1		
46	Pd	2	2 6	2 6 10	2 6 10			
47	Ag	2	2 6	2 6 10	2 6 10	1		
48	Cd	2	2 6	2 6 10	2 6 10	2		
49	In	2	2 6	2 6 10	2 6 10	2 1		
50	Sn	2	2 6	2 6 10	2 6 10	2 2		
51	Sb	2	2 6	2 6 10	2 6 10	2 3		
52	Te	2	2 6	2 6 10	2 6 10	2 4		

Atomic number	*Element*	1(*K*)	2(*L*)		3(*M*)			4(*N*)				5(*O*)				6(*P*)			7(*Q*)
		s	*s*	*p*	*s*	*p*	*d*	*s*	*p*	*d*	*f*	*s*	*p*	*d*	*f*	*s*	*p*	*d*	*s*
53	I	2	2	6	2	6	10	2	6	10		2	5						
54	Xe	2	2	6	2	6	10	2	6	10		2	6						
55	Cs	2	2	6	2	6	10	2	6	10		2	6			1			
56	Ba	2	2	6	2	6	10	2	6	10		2	6			2			
57	La	2	2	6	2	6	10	2	6	10		2	6	1		2			
58	Ce	2	2	6	2	6	10	2	6	10	1	2	6	1		2			
59	Pr	2	2	6	2	6	10	2	6	10	3	2	6			2			
60	Nd	2	2	6	2	6	10	2	6	10	4	2	6			2			
61	Pm	2	2	6	2	6	10	2	6	10	5	2	6			2			
62	Sm	2	2	6	2	6	10	2	6	10	6	2	6			2			
63	Eu	2	2	6	2	6	10	2	6	10	7	2	6			2			
64	Gd	2	2	6	2	6	10	2	6	10	7	2	6	1		2			
65	Tb	2	2	6	2	6	10	2	6	10	9	2	6			2			
66	Dy	2	2	6	2	6	10	2	6	10	10	2	6			2			
67	Ho	2	2	6	2	6	10	2	6	10	11	2	6			2			
68	Er	2	2	6	2	6	10	2	6	10	12	2	6			2			
69	Tm	2	2	6	2	6	10	2	6	10	13	2	6			2			
70	Yb	2	2	6	2	6	10	2	6	10	14	2	6			2			
71	Lu	2	2	6	2	6	10	2	6	10	14	2	6	1		2			
72	Hf	2	2	6	2	6	10	2	6	10	14	2	6	2		2			
73	Ta	2	2	6	2	6	10	2	6	10	14	2	6	3		2			
74	W	2	2	6	2	6	10	2	6	10	14	2	6	4		2			
75	Re	2	2	6	2	6	10	2	6	10	14	2	6	5		2			
76	Os	2	2	6	2	6	10	2	6	10	14	2	6	6		2			
77	Ir	2	2	6	2	6	10	2	6	10	14	2	6	7		2			
78	Pt	2	2	6	2	6	10	2	6	10	14	2	6	9		1			
79	Au	2	2	6	2	6	10	2	6	10	14	2	6	10		1			
80	Hg	2	2	6	2	6	10	2	6	10	14	2	6	10		2			
81	Tl	2	2	6	2	6	10	2	6	10	14	2	6	10		2	1		
82	Pb	2	2	6	2	6	10	2	6	10	14	2	6	10		2	2		
83	Bi	2	2	6	2	6	10	2	6	10	14	2	6	10		2	3		
84	Po	2	2	6	2	6	10	2	6	10	14	2	6	10		2	4		
85	At	2	2	6	2	6	10	2	6	10	14	2	6	10		2	5		
86	Rn	2	2	6	2	6	10	2	6	10	14	2	6	10		2	6		
87	Fr	2	2	6	2	6	10	2	6	10	14	2	6	10		2	6		1
88	Ra	2	2	6	2	6	10	2	6	10	14	2	6	10		2	6		2
89	Ac	2	2	6	2	6	10	2	6	10	14	2	6	10		2	6	1	2
90	Th	2	2	6	2	6	10	2	6	10	14	2	6	10		2	6	2	2
91	Pa	2	2	6	2	6	10	2	6	10	14	2	6	10	2	2	6	1	2
92	U	2	2	6	2	6	10	2	6	10	14	2	6	10	3	2	6	1	2
93	Np	2	2	6	2	6	10	2	6	10	14	2	6	10	4	2	6	1	2
94	Pu	2	2	6	2	6	10	2	6	10	14	2	6	10	6	2	6		2
95	Am	2	2	6	2	6	10	2	6	10	14	2	6	10	7	2	6		2
96	Cm	2	2	6	2	6	10	2	6	10	14	2	6	10	7	2	6	1	2
97	Bk	2	2	6	2	6	10	2	6	10	14	2	6	10	9	2	6		2
98	Cf	2	2	6	2	6	10	2	6	10	14	2	6	10	10	2	6		2
99	Es	2	2	6	2	6	10	2	6	10	14	2	6	10	11	2	6		2
100	Fm	2	2	6	2	6	10	2	6	10	14	2	6	10	12	2	6		2
101	Md	2	2	6	2	6	10	2	6	10	14	2	6	10	13	2	6		2
102	No	2	2	6	2	6	10	2	6	10	14	2	6	10	14	2	6		2
103	Lr	2	2	6	2	6	10	2	6	10	14	2	6	10	14	2	6	1	2

In Table 4-2 note the discrepancy between the configurations of chromium ($_{24}Cr$) and copper ($_{29}Cu$) and Figure 4-4. The $3d^5 4s^1$ arrangement gives to the chromium atom a *half-filled* 3*d* subshell, a more stable arrangement than the expected $3d^4 4s^2$. In the case of $_{29}Cu$ the $3d^{10} 4s^1$ arrangement gives a filled 3*d* subshell and thereby a filled *M* shell. The stability of filled and half-filled *p* and *d* subshells is a phenomenon that is frequently encountered.

In Figure 4-5 the electronic configurations of the elements are presented in abbreviated form. As you follow the order of entry of electrons with increasing atomic number, that is, across a period, you will notice that only those subshells are given in which electrons have just entered. Once a subshell is completed, it is omitted from the notation (except for the *s* subshells in the transition and inner transition series). In Table 4-3 the electronic configurations of the elements are given in full.

4-9 Electronic Configurations of Fifth Period Elements

Figure 4-4 indicates that after $_{36}Kr$, (Ar)$3d^{10} 4s^2 4p^6$, the next two electrons will enter the 5*s* orbital. Thus, $_{37}Rb$ has a (Kr)$5s^1$ arrangement and $_{38}Sr$ has a (Kr)$5s^2$ configuration. Consequently, these two elements fall beneath $_{19}K$ and $_{20}Ca$ in the periodic table. The entire fifth period is essentially a repetition of the fourth. The elements $_{39}Y$ through $_{48}Cd$ comprise the second transition series; they are the 4*d* elements. After $_{48}Cd$ the 5*p* orbitals fill, and the period ends with $_{54}Xe$, another noble gas element.

A comparison of the configurations listed in Table 4-3 with those expected on the basis of Figure 4-4 reveals that there is one more electron in the 4*d* subshell (and one less in the 5*s*) than expected in the series $_{41}Nb$ through $_{47}Ag$. The difference between the energies of 4*d* and 5*s* orbitals is not great (see Figure 4-3). We have here a delicate balance between the tendency of all competing orbitals to become filled or half-filled as soon as possible and the tendency of an entire inner shell to become completely filled as soon as competing tendencies permit.

4-10 Electronic Configurations in the Sixth Period

In the sixth period the entering electrons of the first two elements, $_{55}Cs$ and $_{56}Ba$, appear in the 6*s* orbital, thus following the pattern of the preceding series. In this period, however, a complication develops that has no parallel in any preceding period. According to Figure 4-4, the filling of the 6*s* orbital is followed by the occupancy of the 4*f*, 5*d*, and 6*p* orbitals, respectively. The energy levels 4*f* and 5*d* are so close (see Figure 4-3) that the chart is not completely reliable at this point, because in $_{57}$La, the fifty-seventh electron enters a vacant 5*d* orbital instead of a vacant 4*f* orbital. Thereafter, however, the 4*f* orbitals fill with almost uninterrupted regularity, followed by the filling of the 5*d* and then the 6*p* orbitals (see Table 4-3). Thus, elements 57–80 constitute the third transition series, and elements 58–71, the lanthanides, are the first inner transition series. The lanthanides are sometimes referred to as the 4*f* elements. The other elements of the third transition series (lanthanum and elements 72–80) are the 5*d* elements.

4-11 Electronic Configurations in the Seventh Period

The seventh period begins with the filling of the 7*s* orbital, as expected. In the remaining elements of this period, electrons enter the 6*d* and 5*f* orbitals in much the same manner as they entered the 5*d* and 4*f* orbitals. The difference between the energies of the 6*d* and 5*f* orbitals is so slight that there is some uncertainty as to the actual configurations of several of the actinides. In general, the actinides (the second inner transition series) are the 5*f* elements and thus are the analogs of the 4*f* elements, the lanthanides.

On the basis of Figure 4-5 (see also Table 4-3), we would expect elements 104–112 (assuming they are capable of existence) to complete a fourth transition series, the 6*d* elements. We may also speculate that elements 113–118 would be 7*p* elements and that the 8*s* subshell included for good measure in Figure 4-4 would not be occupied until we reached elements 119 and 120.

NEW TERMS

Aufbau Principle: the principle that each element has the electron configuration of the preceding element plus one electron that enters the lowest energy orbital in which a vacancy exists.

Excited state: any state of an atom in which at least one electron is in an energy level higher than the lowest possible for that electron.

Ground state: the lowest energy state of an atom, i.e., the state in which the electrons are in the lowest energy orbitals available to them.

Orbital: a region within an atom that can be occupied by electrons. Any orbital can accommodate only two electrons.

Pauli's Exclusion Principle: the principle according to which no two electrons in an atom ever have the same set of values for all four quantum numbers.

Principle of Maximum Multiplicity (Hund's Rule): the principle according to which no orbital of a given subshell contains two electrons unless every other orbital of that subshell contains at least one electron.

Subshell: a set of like orbitals within the same shell; for example, the 2*p* subshell consists of three *p* orbitals in the second (*L*) shell.

EXERCISES

4-1 State, define, describe, or explain
(a) orbitals, subshells, shells, principal quantum numbers
(b) ground state and excited state of an atom

4-2 What is the maximum number of electrons in the third shell? The fourth shell? The outermost shell?

4-3 What is the electronic configuration characteristic of the noble gases? The halogens? The alkali metals?

4-4 Construct a blank periodic table and indicate the location of the following: *s* elements, *p* elements, *d* elements, *f* elements.

4-5 What is the greatest number of orbitals ever present in a *d* subshell? In an *L* shell? In a shell with principal quantum number 3?

4-6 What is the greatest number of electrons ever present in a *p* subshell? In an *f* orbital? In a *K* shell? In a shell with principal quantum number 2?

4-7 Compare the electronic configurations of elements of Groups VIIA and VIIB, IA and IB, IVA and IVB, 0 and VIII.

4-8 On the basis of Hund's Rule, predict the number of paired electrons and the number of unpaired electrons in the *d* orbitals of the shell for which the principal quantum number is 3, given that there is a total of eight *d* electrons in this shell. Which element is this?

4-9 Show the electronic configurations of elements with the following atomic numbers: 7, 14, 21, 36, 55, 65, 79, 85, 100.

4-10 In terms of atomic structure, explain why there are 18 elements in the fourth period of the periodic table and only 8 in the third period.

4-11 Use the periodic table to find
(a) the atomic number of lead
(b) the element with electronic configuration $1s^2 2s^2 2p^6 3s^2 3p^6 3d^{10} 4s^2 4p^6 5s^2$
(c) the two elements most like selenium
(d) the two elements most like gadolinium
(e) the highest oxidation state for tungsten
(f) whether KCl should be ionic or covalent

4-12 Without the periodic table, ascertain
(a) which two elements should most resemble strontium
(b) the atomic numbers of the halogens
(c) how many alkali metals exist
(d) whether phosphorus is a metal or a nonmetal

4-13 Do you think hydrogen should be grouped with the elements of Group IA or VIIA? What are the arguments for and against your choice?

4-14 What are the electronic structures characteristic of transition elements? Inner transition elements?

4-15 If as many as 25 new elements were to be discovered, which one would be the alkali metal next after francium?

REFERENCES

Becker, C., "Geometry of the *f* Orbitals," *Journal of Chemical Education,* **41**(7), 358 (1964). Photographs of models.

Campbell, J. A., "Interpreting Electronic Structures," *Journal of Chemical Education,* **26**(9), 477 (1949). An interpretation of the periodicity of chemical properties in terms of atomic orbitals available for bonding; description of an *s*, *p*, *d*, *f* demonstration board.

Friedman, H. G., Choppin, G. R., and Feuerbacher, D. G., "The Shapes of the *f* Orbitals," *Journal of Chemical Education,* **41**(7), 354 (1964). Illustrations that are difficult to find in journals or textbooks.

Glasstone, S., *Sourcebook on Atomic Energy*, 3rd ed., Van Nostrand, Princeton, N.J., 1967. A very readable and authoritative book. Chapter 2 deals with fundamental particles, and Chapter 4 considers the structure of the atom.

Guillemin, Victor, *The Story of Quantum Mechanics*, Scribners, New York, 1968. An introductory account of the evolution of quantum physics and an interpretation of its philosophical implications.

Hoffman, B., *The Strange Story of the Quantum*, 2nd ed., Dover, New York, 1959 (paperbound). An account of the growth of our atomic knowledge for the general reader.

Moeller, T., *The Chemistry of the Lanthanides*, Reinhold, New York, 1963 (paperbound). Electronic configurations of lanthanide and actinide elements are presented on pages 14 and 100, respectively.

Pearson, R. G., *Chemical and Engineering News*, **37**(26), 73 (1959). Photographs of models of *d* orbitals.

5

Chemical Bonding and Oxidation States

5-1 Formation of Ions

As we have previously stated (Section 2-6), salts consist of ions instead of molecules. When we compare the formulas for different salts, we note that the ions have different amounts of charge. Sodium chloride (NaCl) consists of Na^+ ions and Cl^- ions; calcium chloride ($CaCl_2$), of Ca^{++} ions and Cl^- ions; sodium sulfide (Na_2S), of Na^+ ions and S^{--} ions; and calcium sulfide (CaS), of ions that are both doubly charged, Ca^{++} and S^{--}.

Why does a calcium ion have twice the charge of a sodium ion, and why does chlorine form an ion whose charge is only half as great as that of a sulfide ion? To answer such questions, we need to know how atoms become ions.

A negative ion is formed when one or more electrons from some outside source join an atom and become part of it. A negative charge is imparted because the species formed contains more electrons than there are protons in the nucleus (Figure 5-1).

$$\underset{\text{chlorine atom}}{Cl} + \underset{\text{one electron}}{e^-} \longrightarrow \underset{\text{chloride ion}}{Cl^-} \text{ (a mononegative ion)}$$

$$\underset{\text{sulfur atom}}{S} + \underset{\text{two electrons}}{2e^-} \longrightarrow \underset{\text{sulfide ion}}{S^{--}} \text{ (a binegative ion)}$$

A sulfide ion has twice the charge of a chloride ion because the sulfur atom acquires two extra electrons in becoming a sulfide ion, whereas the chlorine atom acquires only one extra electron. This difference exists because a chlorine atom has seven electrons in its outermost shell (one less than eight), whereas a sulfur atom has only six electrons in its outermost shell (two less than eight). Both the sulfide ion and the chloride ion have eight electrons in their outermost shells—the same number of electrons that neon and argon atoms have in their outermost shells. Neon and argon are two of the noble gas elements, which do not readily form either negative or positive ions. In general, an atom of a nonmetal, in forming a negative ion, will gain enough electrons so that the outermost shell of the ion produced will have a noble gas structure. Then all the *s* and *p* orbitals of the outermost shell are filled. This generalization is sometimes referred to as the "octet rule," since the *s* and *p* orbitals have a total capacity of eight electrons.

FIGURE 5-1

Negative ions

Atoms of nonmetals form negative ions.

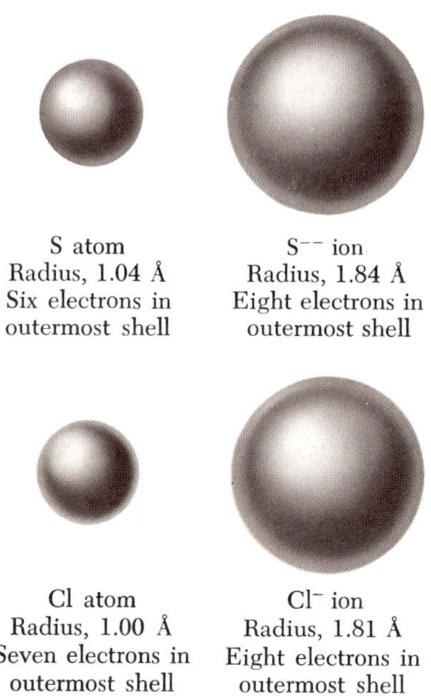

A positive ion is formed when an atom loses one or more electrons (Figure 5-2). The atom is neutral before the electrons are lost; consequently, the loss of electrons will convert an atom into a positive ion:

Na	$\longrightarrow$	Na^+ (a monopositive ion) +	e^-
sodium atom		sodium ion	one electron
Ca	$\longrightarrow$	Ca^{++} (a bipositive ion) +	$2e^-$
calcium atom		calcium ion	two electrons

By losing the two electrons in its outermost shell, calcium becomes a doubly charged positive ion. Sodium becomes an ion with one positive charge when it loses the single electron in its outermost shell. Neither atom loses electrons spontaneously, but either may serve as the "outside source" of electrons required in the formation of such negative ions as Cl^- and S^{--}.

5-2 Electronic Symbols

The number of electrons in the outermost shell of an atom is of such great importance in determining chemical behavior that there is need

FIGURE 5-2

Positive ions

Atoms of metals form positive ions.

Na atom
Radius, 1.57 Å
One electron in
outermost shell

Na^+ ion
Radius, 0.95 Å
A sodium atom that
has lost its outer-shell electron

Mg atom
Radius, 1.36 Å
Two electrons in
outermost shell

Mg^{++} ion
Radius, 0.65 Å
A magnesium atom that
has lost its outer-shell electrons

for a convenient method for indicating what the number is. An *electronic symbol* for an element consists of the chemical symbol surrounded by a certain number of dots. Each dot represents one electron of the *outermost* shell. These outermost electrons are often called *valence electrons* because the word "valence" was long used—and to some extent is still used—to refer to the ability of an element to combine with other elements and form chemical compounds.

Electronic symbols for the first 20 elements are given in Table 5-1. In an electronic symbol the letters represent the kernel of the atom. The *kernel* is defined as that part of the atom that includes the nucleus and all the electrons except those in the outermost shell. For example, the kernel of a chlorine atom consists of a nucleus whose charge is 17+, a *K* shell containing two electrons, and an *L* shell with eight electrons. The charge of the kernel is therefore 7+, matching the charge (7−) of the valence electrons in the *M* shell.

Electronic symbols may also be written for negative ions. For example, the electronic symbol for the chloride ion is [:C̈l̤:]$^{-}$ and the electronic symbol for the sulfide ion is [:S̤̈:]$^{--}$.

5-3 Ionization Energy

The smaller the number of electrons in the outermost shell of a neutral atom, the greater the tendency of the atom to form a positive ion by permitting electrons to be taken away from it. The energy required to remove an electron from a neutral atom can be measured experimentally

TABLE 5-1
Electronic symbols for the first 20 elements

H·							He :
Li·	Ḃe·	Ḅ̇·	·Ċ̣·	:Ṇ̇·	:Ȯ̤·	:F̤̈·	:N̤̈e:
Na·	Ṁg·	Ȧḷ·	·Ṩi·	:Ṗ̣·	:Ṡ̤·	:C̈l̤·	:Ä̤r:
K·	Ċa·						

and with considerable accuracy. The conversion of neutral particles into positive or negative ions is called *ionization.* The energy expended in removing an electron from an atom to form a positive ion is called the *ionization energy.* It is most convenient to refer to the amount of energy required to remove one electron from each atom in a *mole* of atoms and to express this in kilocalories (kcal).

The energy required to remove one electron from an atom is called the *first ionization energy* of the element; the energy required to remove a second electron is called the *second ionization energy*; and so on. The first ionization energies of the first 20 elements are listed in Table 5-2. Note that there is a general increase in ionization energy as the atomic number increases from lithium (atomic number 3) to neon (atomic number 10) and from sodium (atomic number 11) to argon (atomic number 18). This is due to the increase in nuclear charge, because the higher the nuclear charge (atomic number), the greater the attractive force exerted by the nucleus (positive) upon the electron (negative), thus making it more difficult to remove the electron from the atom. We have already stated that Li and Na have only one electron in the outermost shell; these elements, of the first 20, are the ones from whose atoms an electron is most easily removed. Conversely, fluorine and chlorine have seven electrons in the outermost shell; it is more difficult to remove an electron from one of their atoms than from any of the other elements except the noble gas elements. The extra stability of half-filled subshells is indicated by the ionization energies of nitrogen and phosphorus, which are higher than those of the elements that come immediately before and after them.

When we compare any one of the first 20 elements with the elements

TABLE 5-2

First ionization energies of the first 20 elements (kcal/mole)

H 313							He 566
Li 124	Be 215	B 191	C 261	N 335	O 314	F 402	Ne 497
Na 118	Mg 176	Al 137	Si 188	P 254	S 239	Cl 300	Ar 363
K 100	Ca 140						

FIGURE 5-3

First ionization energies of metals of Group IIA

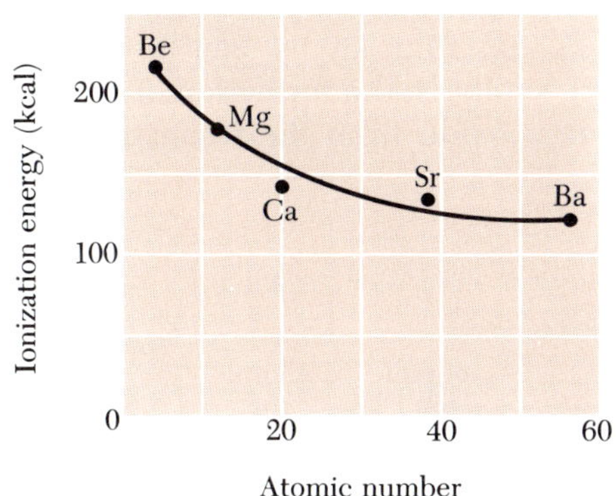

listed beneath it, we find in every instance a decrease in ionization energy with increasing atomic number. This can be attributed to the increased distance between the nucleus and the electron that is being removed. The first ionization energies of the metals of Group IIA, in kilocalories per mole, are $_{4}Be$, 215; $_{12}Mg$, 176; $_{20}Ca$, 140; $_{38}Sr$, 132; and $_{56}Ba$, 120 (Figure 5-3).

5-4 Electron Affinity and Ionic Bonding

The energy released when an atom acquires an electron, forming a negative ion, is called the *electron affinity.* Unfortunately, this cannot be easily determined by direct measurement, and in most instances the values estimated from indirect measurements are not truly reliable. The values for the three similar halogens, chlorine (atomic number 17), bromine (atomic number 35), and iodine (atomic number 53), are 87, 82, and 76 kcal per mole, respectively. Here again we see a trend: the further the outermost shell is from the nucleus, the less is the electron affinity.

Since the nonmetals have high electron affinities and the metals have very low electron affinities, only the nonmetals tend to form negative ions by accepting electrons from other atoms. In other words, atoms that need only one or two electrons to form an outermost shell of eight can readily acquire the necessary electrons from other atoms. The "other atoms" from which the electrons are most easily removed are metals of low ionization energy, i.e., atoms with only one or two electrons in the outermost shell. Elements such as lithium, sodium, and potassium readily react with such elements as fluorine and chlorine to form *ionic compounds*.

The formation of ions from atoms may be indicated with electronic symbols. With these symbols, the formation of sodium chloride, calcium chloride, and calcium oxide from their constituent elements is represented as follows:

$$\mathrm{Na\cdot} + \mathrm{\cdot\ddot{\underset{..}{Cl}}:} \longrightarrow \mathrm{Na^+} + \left[\mathrm{:\ddot{\underset{..}{Cl}}:}\right]^- \qquad \text{(NaCl, sodium chloride)}$$

sodium atom · chlorine atom · sodium ion · chloride ion

$$\mathrm{\dot{Ca}\cdot} + 2\,\mathrm{\cdot\ddot{\underset{..}{Cl}}:} \longrightarrow \mathrm{Ca^{++}} + 2\left[\mathrm{:\ddot{\underset{..}{Cl}}:}\right]^- \qquad \text{(CaCl}_2\text{, calcium chloride)}$$

calcium atom · two chlorine atoms · calcium ion · two chloride ions

$$\mathrm{\dot{Ca}\cdot} + \mathrm{\cdot\dot{\underset{..}{O}}:} \longrightarrow \mathrm{Ca^{++}} + \left[\mathrm{:\ddot{\underset{..}{O}}:}\right]^{--} \qquad \text{(CaO, calcium oxide)}$$

calcium atom · oxygen atom · calcium ion · oxide ion

In the ionic compound NaCl, each sodium "atom" is really an ion, because it bears a positive charge, and each chlorine "atom" is really an ion also, because it bears a negative charge. In any sample of the compound large enough to see or work with, there is a large number of ions—but the number of Na^+ ions is the same as the number of Cl^- ions. The formula NaCl (sometimes written Na^+Cl^-) is intended to convey the information that the two kinds of ions are present in equal number in any sample of the compound, large or small. The formula for any ionic compound is only an empirical formula.

You must always remember that the formula for an ionic compound does not represent a molecule because ionic compounds do not consist of molecules. In a crystal of an ionic compound, the ions are held together by virtue of their charges. Each ion is an independent particle, and when the compound is heated to its melting point, the ions are free to move about. But in the solid crystal they are lined up in an orderly pattern (in all three dimensions), and each ion is attracted by all its oppositely charged neighbors. (See Chapter 7, Figures 7-10 and 7-11.) There are no individual bonds between particular ions—there is no pairing of atoms. When we speak of *ionic bonding*, we are referring to the type of chemical bonding in which atoms are converted into oppositely charged ions rather than sharing a pair of electrons to form a molecule, as we discuss in Section 5-6.

Sodium metal, composed of sodium atoms, is a soft metal with a silvery appearance (when a freshly cut surface is observed); it corrodes rapidly in air because of the formation of sodium oxide on the surface and reacts rapidly with water, evolving hydrogen gas. Chlorine is a poisonous gas of a light greenish-yellow color and corrodes metals rapidly through the formation of chlorides. Yet their compound, sodium chloride, which contains sodium ions and chloride ions, is a white solid that is necessary in our diet and is used as table salt. This crystalline compound illustrates the profound difference between the properties of atoms and the properties of their corresponding ions. Because atoms and ions are so different, you must carefully differentiate between them —for example, a sodium ion is written Na^+, and a sodium atom is written Na (without the charge sign).

5-5 Sizes of Atoms and Ions

The sizes of atoms and ions can be deduced from interatomic and interionic spacings in crystals, as measured by X-ray analysis. Such spacings are actually the distances between nuclei of atoms or ions that are nearest neighbors. For example, in sodium metal the distance between the nucleus of a sodium atom and the nucleus of each of its nearest neighbors is 3.14 Å. Since the atoms are contiguous, the radius of a sodium atom is taken to be half of this value, or 1.57 Å. What lies outside the nucleus of an atom is an electron cloud which, like the earth's atmosphere, has no theoretical termination distance; but, as a practical matter, the outermost electron of the sodium atom is thought of as being within a sphere of 1.57 Å radius almost all the time.

In general, as we proceed from left to right in any period of the periodic table, each element has a smaller atom than the element preceding it. The noble gases are an exception: their atoms are larger than the atoms of the elements of next lower atomic number. All the atoms of the same period have the same number of electron shells. As we proceed from each element to the next, the magnitude of the electrostatic attraction between the nucleus and the surrounding electrons increases with increasing nuclear charge. All electron shells are thereby pulled in closer to the nuclei so that the radius of each atom is less than that of the atom that precedes it, as illustrated in Table 5-3.

The steady change in properties of the elements of the third period, from actively metallic sodium to actively nonmetallic chlorine, can be attributed to the smaller atomic size of each succeeding element. The outermost electrons are less readily removed as their distance from the

TABLE 5-3

Atomic sizes of elements of the third period

Element	$_{11}Na$	$_{12}Mg$	$_{13}Al$	$_{14}Si$	$_{15}P$	$_{16}S$	$_{17}Cl$	$_{18}Ar$
Radius (Å)	1.57	1.36	1.25	1.17	1.10	1.04	1.00	1.54

nucleus decreases. On the other hand, the attraction for enough extra electrons from other atoms to complete the octet increases as the distance between the nucleus and the outermost shell decreases. The ionization energy and electron affinity are measures of these tendencies.

In each succeeding period of the periodic table, one more shell is occupied by electrons. The effect of this on atomic size is best shown by comparing atoms in the same group, since they have the same number of electrons in the outermost shell. The atomic sizes of the elements in Groups IA and VIIA are given in Table 5-4. The alkali metals show a decreasing opposition to having an electron removed as their atomic numbers, and thus their atomic sizes, increase. Similarly, the electron affinity of the halogens decreases with increasing atomic number.

A positive ion is smaller than the parent atom (Figure 5-2). The formation of a sodium ion from a sodium atom, for example, involves the removal of the outermost electron shell, and the remaining shells are drawn closer to the nucleus because the nuclear charge now exceeds the number of extranuclear electrons. The positive ions of succeeding elements in the same group are larger because of the increased number of electron shells, but in each case the ion is smaller than its parent atom. This is illustrated with metals of Groups IA and IIA in Table 5-5.

Conversion of an atom to a negative ion by addition of one or more electrons results in increased interelectronic repulsion in the outermost

TABLE 5-4

Atomic sizes of alkali metals and halogens

Period	*Group IA (alkali metal)*	*Radius* (Å)	*Group VIIA (halogen)*	*Radius* (Å)
2	$_{3}Li$	1.33	$_{9}F$	0.72
3	$_{11}Na$	1.57	$_{17}Cl$	1.00
4	$_{19}K$	2.03	$_{35}Br$	1.14
5	$_{37}Rb$	2.16	$_{53}I$	1.35
6	$_{55}Cs$	2.35		

TABLE 5-5

Sizes of positive ions and their parent atoms

Atom	*Radius* (Å)	*Ion*	*Radius* (Å)	*Ion/atom ratio*
Li	1.33	Li^+	0.60	0.45
Na	1.57	Na^+	0.95	0.61
K	2.03	K^+	1.33	0.66
Rb	2.16	Rb^+	1.48	0.69
Cs	2.35	Cs^+	1.69	0.72
Be	0.89	Be^{++}	0.31	0.35
Mg	1.36	Mg^{++}	0.65	0.48
Ca	1.74	Ca^{++}	0.99	0.57
Sr	1.91	Sr^{++}	1.13	0.59
Ba	1.97	Ba^{++}	1.35	0.68

shell and also makes the number of electrons outside the nucleus exceed the nuclear charge. Both effects contribute toward making the negative ion larger than the parent atom, as illustrated in Figure 5-1 and in Table 5-6.

Comparison of the size of a noble gas atom with the sizes of positive and negative ions having the same number of electrons shows most clearly the effect of increasing nuclear charge. In Table 5-7 the sizes of two sets of *isoelectronic* atoms and ions (i.e., atoms and ions having the same number of electrons) are seen to decrease markedly with increasing nuclear charge.

TABLE 5-6

Sizes of negative ions and their parent atoms

Atom	*Radius* (Å)	*Ion*	*Radius* (Å)	*Ion/atom ratio*
F	0.72	F^-	1.36	1.89
Cl	1.00	Cl^-	1.81	1.81
Br	1.14	Br^-	1.95	1.71
I	1.35	I^-	2.16	1.60
O	0.74	O^{--}	1.40	1.89
S	1.04	S^{--}	1.84	1.77
Se	1.14	Se^{--}	1.98	1.74
Te	1.37	Te^{--}	2.21	1.61

TABLE 5-7

Sizes of isoelectronic atoms and ions

Ion or atom with 10 electrons	N^{3-}	O^{--}	F^{-}	Ne	Na^{+}	Mg^{++}	Al^{3+}	Si^{4+}
Nuclear charge	7	8	9	10	11	12	13	14
Radius (Å)	1.71	1.40	1.36	1.12	0.95	0.65	0.50	0.41
Ion or atom with 18 electrons	P^{3-}	S^{--}	Cl^{-}	Ar	K^{+}	Ca^{++}	Sc^{3+}	Ti^{4+}
Nuclear charge	15	16	17	18	19	20	21	22
Radius (Å)	2.12	1.84	1.81	1.54	1.33	0.99	0.81	0.64

To observe the effect of varying the number of electrons while keeping the nuclear charge constant, we must compare the sizes of differently charged ions of the same element. This is done for four metals in Table 5-8.

5-6 Covalent Bonds

Nonmetals do not ordinarily form positive ions, and metals do not form negative ions. Even some of the metals are reluctant to form positive ions, and some nonmetals practically never form negative ions. In order for two elements to form an ionic compound, one must have significantly greater electron affinity and much greater ionization energy than the other. This limits the number of possible ionic compounds, but there are many other compounds that do not contain ions. These *non-ionic compounds* consist of molecules composed of atoms that are usually held together by *covalent bonds.* A covalent bond consists of

TABLE 5-8

Sizes of differently charged ions of the same element

Ion	*Radius* (Å)	*Ion*	*Radius* (Å)
Fe^{++}	0.83	Fe^{3+}	0.60
Tl^{+}	1.49	Tl^{3+}	1.05
Pb^{++}	1.32	Pb^{4+}	0.84
Ce^{3+}	1.18	Ce^{4+}	1.02

a pair of electrons shared by two atoms. The simplest example of a covalent bond is the shared electron pair that holds the two hydrogen atoms together in the hydrogen molecule, H_2:

$$\mathrm{H\cdot + \cdot H \longrightarrow H:H}$$

Since each hydrogen atom has one electron in its *K* shell (the *K* shell is filled when it has two electrons in it), the electrons from two hydrogen atoms can form a pair that is shared equally by the two nuclei. With the two electrons held jointly by the nuclei, the two atoms are bonded together very strongly. A relatively large amount of energy, 104 kcal, is required to break the bonds in all the molecules of a mole (i.e., 2.016 g or 6.023×10^{23} molecules) of H_2; and, likewise, when two hydrogen atoms combine to form an H_2 molecule, 104 kcal are evolved per mole of H_2 formed.

If the two electrons of the covalent bond in the H_2 molecule are counted for each atom, each hydrogen atom has the same electronic arrangement as a helium atom. When two chlorine atoms, each with seven electrons in its outermost shell, share one pair of electrons between them

$$\mathrm{:\underset{\cdot\cdot}{\overset{\cdot\cdot}{Cl}}\cdot + \cdot\underset{\cdot\cdot}{\overset{\cdot\cdot}{Cl}}: \longrightarrow :\underset{\cdot\cdot}{\overset{\cdot\cdot}{Cl}}:\underset{\cdot\cdot}{\overset{\cdot\cdot}{Cl}}:}$$

each atom has the stable electronic structure of an argon atom. Thus, in forming molecules by sharing electrons, as in forming ions by gaining or losing electrons, atoms tend toward an outer-shell electron arrangement like that of the nearest noble gas element. Our explanation of covalent bond formation, first set forth in 1916 by the American chemist Gilbert N. Lewis, is therefore basically another application of the tendency of atoms to fill the outer *s* and *p* orbitals.

It is customary to indicate a covalent bond in a structural formula by drawing a line to represent the shared pair of electrons. For example, the H_2 molecule can be represented as being made up of two covalently bonded hydrogen atoms by the formula H—H. This type of structural formula is often referred to as a *line formula*.

Line formulas do not include unshared electrons. For example, the line formula for Cl_2 is Cl—Cl. If the unshared electrons are included, the structural formula is an *electronic formula* whether the bonding electrons are represented by a line or by a pair of dots. There is no difference in meaning between $\mathrm{:\underset{\cdot\cdot}{\overset{\cdot\cdot}{Cl}}—\underset{\cdot\cdot}{\overset{\cdot\cdot}{Cl}}:}$ and $\mathrm{:\underset{\cdot\cdot}{\overset{\cdot\cdot}{Cl}}:\underset{\cdot\cdot}{\overset{\cdot\cdot}{Cl}}:}$ because these are simply two ways of writing the electronic formula for Cl_2. In the line formula Cl—Cl the line represents the shared pair of electrons but the unshared electrons are not shown.

In an electronic formula the shared pair of electrons – whether represented by two dots or by a line – is placed between the two symbols that

FIGURE 5-4

Formation of H_2 molecule from two hydrogen atoms

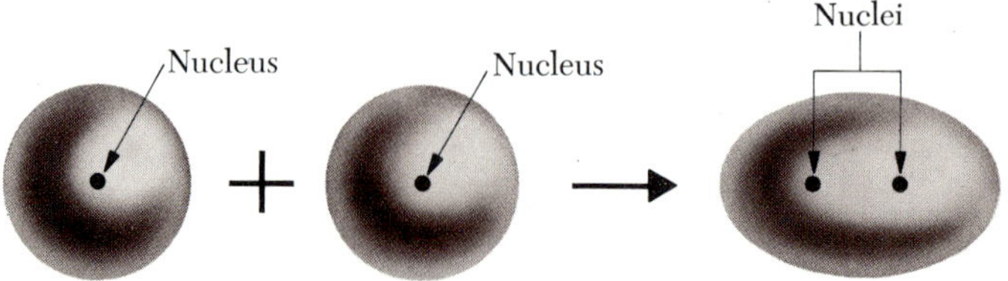

represent atomic kernels, but this must not be taken to mean that the two electrons are always situated between the two kernels. All electrons in an atom, an ion, or a molecule are in continuous motion. Those that are paired, as in a covalent bond, are nevertheless in rapid motion in the vicinity of the two kernels.

Before two hydrogen atoms form a diatomic H_2 molecule, the electron of each atom—if we may imagine each atom momentarily isolated in its ground state—occupies a 1*s* orbital, forming a spherically symmetrical cloud. We may think of bond formation as resulting from the overlap of the two spherically symmetrical *s* orbitals. The shared pair occupies an orbital in the molecule that is so situated as to make the electron density greatest along and around the line connecting the two atomic nuclei. However, this orbital cannot be visualized adequately as simply the region where the two atomic orbitals overlap, because actually it has a shape all its own (Figure 5-4).

5-7 Electronegativity

When a pair of electrons is shared by two unlike atoms, such as a hydrogen atom and a chlorine atom,

$$\mathrm{H}\cdot + \cdot\ddot{\underset{\cdot\cdot}{\mathrm{Cl}}}: \longrightarrow \mathrm{H}:\ddot{\underset{\cdot\cdot}{\mathrm{Cl}}}:$$

they are not shared equally. The electrons, on the average, spend more time in the vicinity of the chlorine nucleus than in the vicinity of the hydrogen nucleus. The cause of this unequal sharing is the greater attraction of the chlorine atom for the bonding electrons in comparison with the hydrogen atom. This tendency to attract shared electrons is called *electronegativity*. Although it is difficult to measure, electronegativity has effects that are important in determining the properties of covalent compounds. For example, HCl is considered a *polar compound* because chlorine is more electronegative than hydrogen. The two bond-

FIGURE 5-5

A polar molecule

In the HCl molecule (a), the electrons are not shared equally since chlorine is more electronegative than hydrogen. As represented by (b), the molecule is relatively positive at the hydrogen end and negative at the chlorine end.

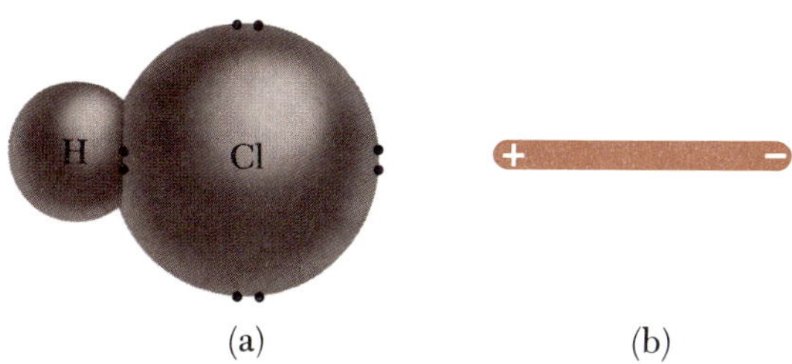

ing electrons, one from each atom, come from atoms that are initially neutral, but in the HCl molecule the covalent bond has a partial ionic character. The center of positive charge in the molecule is not at the same point as the center of negative charge; that is, the chlorine end of the molecule is somewhat negative, and the hydrogen end of the molecule is positive to the same degree (Figure 5-5). When placed in an electric field, such polar molecules tend to turn so that their negative ends are oriented toward the positive plate and their positive ends are oriented toward the negative plate (Figure 5-6). When there is no electric field the molecules are not aligned in any one direction.

The electronegativity values of the first 20 elements, as evaluated by procedures suggested by Linus Pauling, are listed in Table 5-9. Among the representative elements, electronegativity increases with atomic number in each row and decreases with atomic number in each column. Values are not yet available for the noble gas elements.

FIGURE 5-6

Polar molecules in an electric field

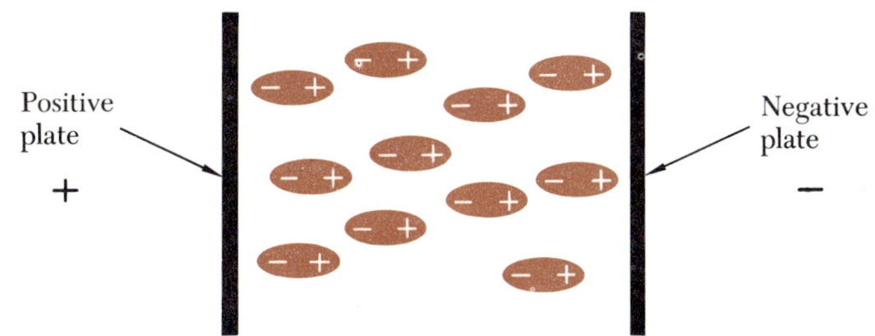

TABLE 5-9

Electronegativity values of the first 20 elements

H							He
2.1							—
Li	Be	B	C	N	O	F	Ne
1.0	1.5	2.0	2.5	3.0	3.5	4.0	—
Na	Mg	Al	Si	P	S	Cl	Ar
0.9	1.2	1.5	1.8	2.1	2.5	3.0	—
K	Ca						
0.8	1.0						

5-8 Bond Angles

The formula for water is H_2O rather than OH because an oxygen atom, with only six electrons in its outermost shell, must share a pair of electrons with each of *two* hydrogen atoms if it is to attain a noble gas structure:

$$2\mathrm{H}\cdot + \cdot\underset{\cdot}{\ddot{\mathrm{O}}}: \longrightarrow \begin{array}{l}\mathrm{H}:\underset{\cdot\cdot}{\ddot{\mathrm{O}}}:\\ \quad\ \mathrm{H}\end{array}$$

Considering the electronic configuration of an isolated oxygen atom, $1s^22s^22p^22p^12p^1$, we expect that the two incomplete p orbitals will overlap with the s orbitals of the two hydrogen atoms. Since p orbitals are perpendicular to each other, we could reasonably expect the angle between the two bonds in H_2O to be 90°. Instead, the bond angle in the H_2O molecule is about 105°. This discrepancy might be reasonably explained as a consequence of the mutual repulsion of the two hydrogen nuclei.

Since a sulfur atom is larger than an oxygen atom, the hydrogen nuclei in H_2S are farther apart than in H_2O and the repulsive force between them is weaker than in H_2O. Therefore, we would expect the bond angle in H_2S to be more nearly 90° than in H_2O. The bond angle in H_2S is 92°; in H_2Se it is 91°.

A nitrogen atom, with only five electrons in its outermost shell, shares a pair of electrons with each of *three* hydrogen atoms, thus giving it a noble gas structure; this explains the fact that the formula for ammonia is NH_3:

$$3\text{H}\cdot + \cdot\overset{\cdot}{\underset{\cdot}{\text{N}}}: \longrightarrow \text{H}:\overset{\overset{\text{H}}{\cdot\cdot}}{\underset{\underset{\text{H}}{\cdot\cdot}}{\text{N}}}:$$

Since the three incomplete p orbitals in the nitrogen atom are mutually perpendicular, we might reasonably expect the angles between the bonds in NH_3 to be 90°. Actually the angle is about 107°; the molecule can be described as a trigonal pyramid with the three hydrogen nuclei in the same plane and the nitrogen nucleus at the apex (Figure 5-7).

If we are again to explain the expanded bond angle in terms of mutual repulsion of the hydrogen nuclei, we would expect the angle in PH_3 to be smaller than in NH_3 because the phosphorus atom is larger than the nitrogen atom and in consequence of this the hydrogen nuclei are farther apart in PH_3. The bond angle in PH_3 is 93°; in AsH_3 it is 92°.

5-9 The Theory of Hybrid Bond Orbitals

Since the electronic configuration of an isolated carbon atom is $1s^22s^22p^12p^1$, we might expect carbon to form only two covalent bonds so that the formula for the simplest compound containing hydrogen and carbon alone would be CH_2. Actually, the simplest compound containing hydrogen and carbon, which is called methane, has the formula CH_4. We need an explanation for the fact that a carbon atom forms four covalent bonds instead of only two. An explanation is provided by the theory of hybrid bond orbitals, proposed by Linus Pauling in 1931.

The *ground state* of an atom is the most stable state of the uncombined atom, and the *valence state* or *bonding state* is the state in which the atom is ready to form bonds with other atoms. Ground states and

FIGURE 5-7

The NH_3 molecule

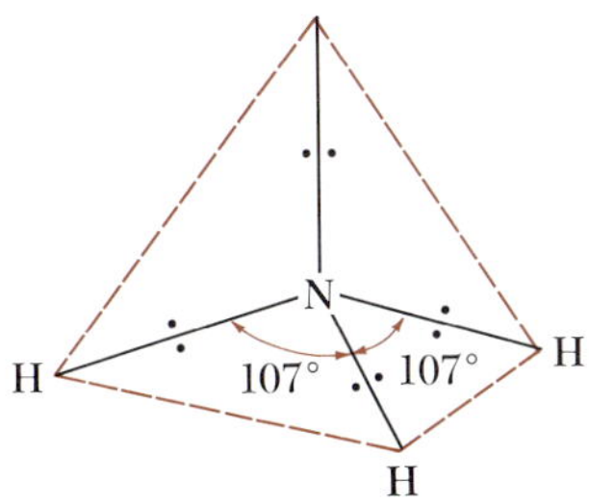

bonding states can conceivably be identical for atoms that have one, five, six, or seven electrons in the outermost shell, but they are *not* the same for atoms with two, three, or four outermost electrons. When these states are not the same, the bonding state is an excited state.

In the bonding state of a carbon atom, one of the *s* electrons is promoted to the empty *p* orbital. The energy required to do this is much less than the energy released by the formation of the two additional bonds that the promotion makes possible. In general, atoms tend to form all of the bonds of which they are capable; thus, the formula for the simplest hydrogen–carbon compound is CH_4, not CH_2.

Although *s* orbitals do not have the same shape as *p* orbitals, all the bonds in CH_4 are alike. An essential feature of the theory of hybrid bond orbitals is that new orbitals are formed that are all alike. These new orbitals partake of the nature of the original orbitals, but they become equalized. The term *hybridization* indicates in general the mixing of a number of orbitals, not all of the same type, to form the same number of hybrid orbitals that are all alike. The name given to a particular set of hybrid orbitals indicates how many orbitals of each type have mixed to form the new hybrid orbitals. For example, in methane we have sp^3 hybridization, and each orbital is an sp^3 hybrid orbital, because one *s* orbital and three *p* orbitals have mixed to form four hybrid orbitals.

The shape of a molecule can be understood by considering both hybridization and the mutual repulsion of electron pairs. With four electron pairs, maximum separation is achieved by their location at the corners of a regular tetrahedron with the nucleus of the central atom of the molecule at the geometric center of the tetrahedron. This makes the bond angle 109.5° in CH_4.

In the ground state a beryllium atom has only two electrons in the outermost shell, and both are in the 2*s* orbital. Without promotion to a higher state, it could form no covalent bonds. Actually, it forms covalent compounds such as $BeCl_2$, which, in the gaseous state, has linear molecules. The linear structure can be explained if we consider the hybridization involved. One of the *s* electrons of the beryllium atom is promoted into a *p* orbital so that in the bonding state there are two equivalent *sp* hybrid orbitals. These two *sp* hybrid orbitals, because of electronic repulsion, are situated as far apart as possible in the $BeCl_2$ molecule. In other words, the two electron pairs are located at opposite sides of the beryllium atom, forming a Cl—Be—Cl molecule in which the bond angle is 180°.

Boron, with three electrons in its valence shell, has the $1s^2 2s^2 2p$ configuration in the ground state, but it can form three bonds that are all alike. These bonds are designated as sp^2 hybrid bonds since they result from the promotion of one of the *s* electrons to a *p* orbital, making available one *s* orbital and two *p* orbitals for hybridization. This is the type of bonding that occurs in BF_3. The three electron pairs that constitute the bonds in the BF_3 molecule are located at the corners of an equilateral triangle, so as to be as far apart as possible, with the boron nucleus at the center. Thus, in the BF_3 molecule all four atoms are in the same plane and all bond angles are 120°.

In NH_3 we can explain the deviation of the bond angles from the expected 90° by assuming hybridization of the bonds. We can imagine the 2*s* orbital and the three 2*p* orbitals in the nitrogen atom to be hybridized so that in the bonding state there are four sp^3 hybrid orbitals. One of these contains a pair of unshared electrons, but each of the other three has only one electron and is available for forming a bond. Thus, the nitrogen is still only 3-covalent, the same as if no hybridization had occurred—but the hybrid orbitals are directed toward the corners of a tetrahedron, making the angle between them nearer the 109.5° tetrahedral angle than the 90° angle expected for pure *p* orbitals. The unshared electron pair, often referred to as a *lone pair*, is itself in an sp^3 hybrid orbital directed toward one of the corners of the tetrahedron. The charge concentration of the lone pair may be assumed to be closer to the nitrogen atom than any of the bond pairs. In consequence of this, the lone pair exerts a greater repulsive force on each bond pair than any bond pair exerts on either of the other bond pairs. The three N—H bonds of the ammonia molecule thus are forced slightly closer together than the four C—H bonds in methane. This accounts for the H—N—H bond angles in NH_3 being 107° instead of 109.5°.

The bond angles in the H_2O molecule also may be accounted for in terms of hybridization. We can imagine the 2*s* and 2*p* orbitals of the oxygen atom becoming hybridized to give four sp^3 hybrid orbitals with a lone pair of electrons in each of two of them but with only one electron, available for bonding, in each of the other two. Although the oxygen is still only 2-covalent, the same as if there were no hybridization, the hybrid orbitals are directed tetrahedrally. This leads us to expect the angle between the two O—H bonds to be nearer 109.5° than 90°. The fact that the bond angle is only 105° can be attributed to the presence of *two* lone pairs. It may be assumed that the two lone pairs exert an even greater repulsive force on each other than either of them exerts upon a bond pair. Since the weakest repulsive force between pairs is the force between the two bond pairs, the angle between the two O—H bonds in H_2O is even smaller than the angle between N—H bonds in NH_3.

5-10 Double and Triple Bonds

In some molecules, two atoms are chemically bonded by sharing *two* pairs of electrons. For example, in the carbon dioxide molecule there is a *double bond* between the carbon atom and each oxygen atom. The electronic formula for carbon dioxide may be written as $\underset{\cdot\cdot}{\ddot{O}}::C::\underset{\cdot\cdot}{\ddot{O}}$ or as $\underset{\cdot\cdot}{\ddot{O}}{=}C{=}\underset{\cdot\cdot}{\ddot{O}}$, and the line formula is written as O═C═O, the unshared electrons not being shown.

In a triatomic molecule such as the carbon dioxide molecule, the arrangement of the three atoms is linear, i.e., the bond angle is 180°, if the central atom has no unshared electron pairs. The arrangement is angular if the central atom has one or more unshared pairs, as in the water molecule.

There are instances in which two atoms are bonded by sharing *three* pairs of electrons. In the nitrogen molecule, N_2, there is such a *triple bond* between the two nitrogen atoms. The electronic formula for N_2 is written either as :N:::N: or as :N≡N: and the line formula is written as N≡N.

5-11 Coordinate Bonds

There are many compounds, called *coordination compounds*, that contain covalent bonds in which *both* of the shared electrons are furnished by the same atom. These bonds are called *coordinate covalent bonds* or, more briefly, *coordinate bonds*. They are also known as *dative bonds*. Such a bond accounts for the formation of the compound H_3NBF_3 by an "addition reaction" occurring between ammonia and boron trifluoride.

$$\begin{array}{ccccccc} & \text{H} & & & :\underset{\cdot\cdot}{\ddot{\text{F}}}: & & \\ \text{H}: & \underset{\cdot\cdot}{\ddot{\text{N}}}: & + & & \underset{\cdot\cdot}{\ddot{\text{B}}}: & \underset{\cdot\cdot}{\ddot{\text{F}}}: & \\ & \text{H} & & & :\underset{\cdot\cdot}{\ddot{\text{F}}}: & & \end{array} \longrightarrow \begin{array}{cccc} & \text{H} & :\underset{\cdot\cdot}{\ddot{\text{F}}}: & \\ \text{H}: & \underset{\cdot\cdot}{\ddot{\text{N}}} : & \underset{\cdot\cdot}{\ddot{\text{B}}}: & \underset{\cdot\cdot}{\ddot{\text{F}}}: \\ & \text{H} & :\underset{\cdot\cdot}{\ddot{\text{F}}}: & \end{array}$$

The formation of this compound is due to the presence of the unshared pair in the outermost shell of the nitrogen atom in NH_3 and the vacancy in the outermost shell of the boron atom in BF_3.

Another example of a coordination compound is $POCl_3$, which can be formed by the combination of oxygen with phosphorus trichloride:

$$\begin{array}{ccc} :\underset{\cdot\cdot}{\ddot{\text{Cl}}}: & \underset{\cdot\cdot}{\ddot{\text{P}}}: & \underset{\cdot\cdot}{\ddot{\text{Cl}}}: \\ & :\underset{\cdot\cdot}{\text{Cl}}: & \end{array} + :\underset{\cdot}{\ddot{\text{O}}}\cdot \longrightarrow \begin{array}{ccc} & :\ddot{\text{O}}: & \\ :\underset{\cdot\cdot}{\ddot{\text{Cl}}}: & \underset{\cdot\cdot}{\ddot{\text{P}}}: & \underset{\cdot\cdot}{\ddot{\text{Cl}}}: \\ & :\underset{\cdot\cdot}{\text{Cl}}: & \end{array}$$

With lines representing covalent bonds, the structural formulas (line formulas) for H_3NBF_3 and $POCl_3$ can be written as follows:

```
     H   F               O
     |   |               |
  H—N—B—F           Cl—P—Cl
     |   |               |
     H   F              Cl
```

The bond between the nitrogen atom and the boron atom in H_3NBF_3 is not really different from the other covalent bonds in the molecule; therefore, there is no real need for indicating that both electrons are furnished by the nitrogen atom. This is sometimes done, however, by using an arrow pointed toward the acceptor atom, as shown below. The bond between the phosphorus atom and the oxygen atom in $POCl_3$ is not really different from the bonds between the phosphorus atom and the chlorine atoms, but the arrow can be used to indicate that the phosphorus atom furnished both electrons if it is desired to emphasize this fact.

```
     H   F               O
     |   |               ↑
  H—N→B—F           Cl—P—Cl
     |   |               |
     H   F              Cl
```

5-12 The Theory of Resonance

It is not always possible to write an electronic formula for a molecule that is in agreement with what is known about the properties and structure of the substance.

If the octet principle is followed, it is possible to write an electronic formula for sulfur dioxide in two different ways:

```
      ..                  ..
  ..  S ::  ..        ..  S   ..
 :O :    O          O ::    : O :
  ..      ..         ..       ..
```

Each of these formulas shows one pair of shared electrons for one of the bonds and two pairs of shared electrons for the other bond. In other words, each formula represents one bond as single and the other as double. If one bond were really single and the other one really double, we would expect, from our knowledge of single and double bonds in other molecules, that the bond lengths and strengths would be different.

It is known that the two oxygen atoms in the sulfur dioxide molecule are located at equal distances from the sulfur atom. The two oxygen-to-sulfur bonds appear to be completely equivalent in all respects. Although we are able to write *two* electronic formulas that are in accord with the octet principle, we cannot accept either of them as representing satisfactorily the properties of sulfur dioxide.

According to the *theory of resonance*, proposed by Linus Pauling in 1931, if two (or more) reasonable electronic structures can be written for a molecule, the "real" structure is not identical to either of them but is a *resonance hybrid* that is intermediate in character between the two. For example, the "real" structure of the sulfur dioxide molecule is not correctly represented by either of the structures shown above but is a resonance hybrid that partakes of the character of both. The two bonds are equivalent; each is more than single but less than double.

No satisfactory symbolism has been invented for representing resonance structures; that is, no one electronic formula can represent a resonance hybrid. The two formulas written above for the sulfur dioxide molecule are referred to as *contributing structures*. One method for attempting to represent the structure of sulfur dioxide consists in writing the two contributing structures with braces enclosing them and a two-headed arrow separating them:

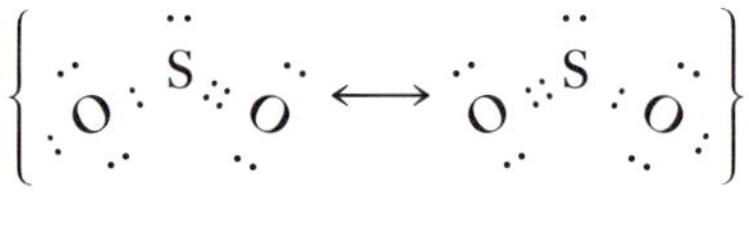

$$\left\{ :\ddot{O}:\ddot{S}::\ddot{O} \longleftrightarrow \ddot{O}::\ddot{S}:\ddot{O}: \right\}$$

This symbolism must not be taken to mean that molecules of two different structures exist in equilibrium. Only one kind of SO_2 molecule exists: a hybrid of the contributing structures.

5-13 The Concept of Oxidation State

You have now learned that certain compounds are ionic and others are covalent. Sodium and potassium form positive ions, each with a charge of 1+ because each atom loses one (and only one) electron to form an ion. A chlorine atom can gain one (but only one) electron to form a negative ion, which is a 1− ion. The high electronic affinity of chlorine and the low ionization energy of sodium have the net effect of permitting a chlorine atom to remove an electron from a sodium atom. In an H_2 molecule, neither atom has removed an electron from the other—instead, they share an electron pair equally because their electronegativities are identical. The same situation prevails in a Cl_2 molecule. In an HCl molecule, however, chlorine acquires more than an equal share of the bonding electrons because chlorine is more electronegative than hydrogen. Since the HCl molecule is polar, we say that the bond has *partial ionic character*.

In a CCl_4 molecule each bond has partial ionic character because chlorine is more electronegative than carbon. The four bonds, all alike, are distributed symmetrically in space, however, and the CCl_4 molecule is nonpolar in consequence of this. In spite of this, since each individual bond has partial ionic character, we are justified in saying that there is a *tendency* of chlorine atoms to form negative ions in both CCl_4 and HCl.

The extent to which a covalent bond has partial ionic character depends on the relative electronegativities of the two atoms involved. Therefore, the ionic character varies from one bond to another. Without concerning ourselves now with the different degrees of partial ionic character in different bonds, let us note that the ion the chlorine atom *tends* to form in HCl or in CCl_4 is the Cl^- ion, the negative ion that actually exists in the ionic compound NaCl. This is the basis for the assignment of an arbitrary *oxidation number* to each atom in a molecule.

If a compound is actually and truly ionic, the oxidation number of each ion is, by definition, the same as its charge. Thus, in NaCl the oxidation number of the sodium ion is +1 and we can say that the element sodium is in the *oxidation state* +1; the oxidation number of the chloride ion is −1 and we can say, therefore, that the element chlorine is in the oxidation state −1. In $CaCl_2$ chlorine is in the oxidation state −1, but calcium is in the oxidation state +2. In MgO the oxidation state of magnesium is +2, and the oxidation state of oxygen is −2.

If a compound is covalent, the oxidation number arbitrarily assigned to each atom is the same as if the compound were ionic. In the covalent compounds HCl, CCl_4, and PCl_3, chlorine is in the same oxidation state (−1) as in the ionic compounds NaCl and $CaCl_2$. In HCl, hydrogen is in the oxidation state +1, just as sodium is in the +1 state in NaCl. Likewise, we say that in CCl_4 carbon is in the +4 state, although carbon is not present as an ion with a 4+ charge (and never is in *any* compound). Similarly, the oxidation number of the phosphorus atom in PCl_3 is +3.

To determine which of the two atoms sharing a pair of electrons has the *positive* oxidation number, we must know the relative electronegativities of the two. Let us determine, for example, which element in $SiCl_4$ is in the positive oxidation state. Consulting Table 5-3, we note that silicon is considerably less electronegative than chlorine, so silicon is regarded as being in the +4 state and chlorine in the −1 state. Can it also be predicted whether the compound is ionic or covalent? Although compounds formed of elements having great electronegativity differences are ionic, it is difficult to know just how great the difference must be. In two covalent compounds composed of different pairs of elements, the degree of partial ionic character may differ even if the electronegativity differences are equal.

The concept of oxidation state is a highly arbitrary one, but much of its usefulness depends on this very fact. For example, you do not need to know whether $SiCl_4$ is ionic or covalent in order to deduce the correct oxidation numbers of silicon and chlorine in this compound. Conversely, oxidation numbers are useful for predicting the formulas for the products of chemical reactions.

5-14 Assignment of Oxidation Numbers

The assignment of oxidation numbers to the individual atoms in molecules is based on two arbitrary rules for counting shared electrons. The first rule provides that electrons shared by two unlike atoms are arbitrarily assigned to the more electronegative of the two atoms. According to the second rule, electrons shared by two like atoms are assigned by dividing them equally between the two atoms. In accordance with these rules, the oxidation numbers are +1 for hydrogen and −1 for chlorine in HCl; but the oxidation number of hydrogen in H_2 is zero, and the oxidation number of chlorine in Cl_2 is likewise zero (Figure 5-8). In each case the oxidation number is the difference between the kernel charge and the number of electrons in the valence shell when they are assigned by the arbitrary rules. In HCl, for example, the oxidation number −1 for chlorine corresponds to the fact that eight electrons are counted as being present in the valence shell of the chlorine atom whose kernel charge is 7+; on the other hand, no electrons are counted as present in the valence shell of the hydrogen whose kernel charge is 1+, and the oxidation number +1 corresponds to this difference. In similar fashion, the oxidation number of the oxygen atom in H_2O is seen to be −2; but in H_2O_2 the oxidation number of each oxygen atom is −1, and in OF_2 the oxidation number of the oxygen atom is +2 (Figure 5-9).

It would be inconvenient if we had to write electronic formulas for all molecules and ions in order to deduce the oxidation numbers of the atoms within them. Fortunately, this is not necessary. It is usually sufficient to keep the following three principles in mind:

1. The oxidation state of any free element is zero, regardless of the simplicity or complexity of the molecule. Sulfur in S_8, phosphorus in P_4, oxygen in O_2 and in O_3, neon in Ne, and magnesium in Mg are all in the zero oxidation state.

FIGURE 5-8

Assignment of oxidation numbers in H_2, HCl, and Cl_2

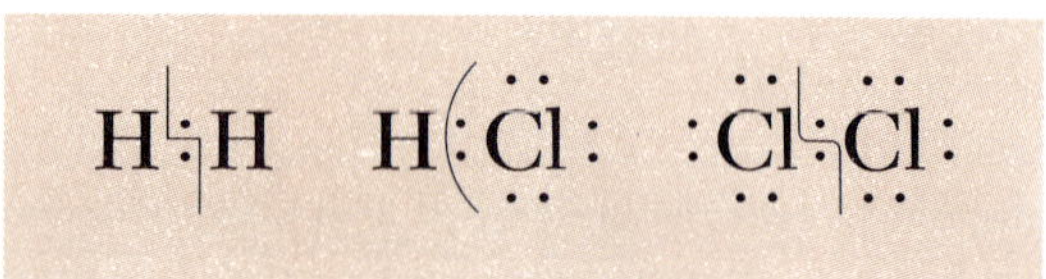

FIGURE 5-9

Assignment of oxidation numbers in H_2O, H_2O_2, and OF_2

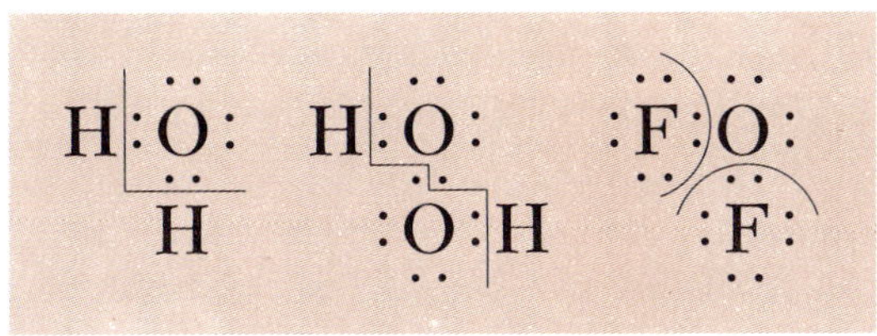

2. The sum of the oxidation numbers of all the atoms in the formula for a compound is zero because all compounds are neutral substances whether they are composed of molecules or ions. Calcium chloride, $CaCl_2$, is composed of Ca^{++} ions and Cl^- ions, but any specimen of the compound is neutral because there are twice as many singly charged Cl^- ions as there are doubly charged Ca^{++} ions. In carbon dioxide, CO_2, which is composed of neutral molecules, each oxygen atom is assigned the oxidation number -2 and the carbon atom is assigned the oxidation number $+4$ so that the sum is zero. There are no O^{--} ions in CO_2, and the assignment of $+4$ as the oxidation state of carbon does *not* mean that C^{4+} ions are present.
3. The sum of the oxidation numbers of all the atoms within an ion is equal to the charge of the ion. For example, in the hydroxide ion, OH^-, the oxidation number of the oxygen atom is -2 and the oxidation number of the hydrogen atom is $+1$ so that the sum of the oxidation numbers is -1 in agreement with the charge, 1−, of the ion. In the sulfate ion, SO_4^{--}, the oxidation number of each oxygen atom is -2; from this we deduce the oxidation number of the sulfur atom to be $+6$ so that the sum of the oxidation numbers is -2 in agreement with the charge, 2−, of the ion.

In a ternary compound, two of the three elements usually have oxidation numbers about which there is little if any uncertainty. When this is so, the oxidation state of the third element can be deduced from the values of the other two. For example, in $MgSO_4$, knowing that magnesium is in the $+2$ state and oxygen in the -2 state enables us to calculate the oxidation state of sulfur. We can represent the oxidation number of the sulfur atom by x and write the equation

$$+2 + x - 8 = 0$$

from which $x = +6$. Note that this answer is identical with the value obtained above for the oxidation number of the sulfur atom in the SO_4^{--} ion. This has to be so because $MgSO_4$ is composed of Mg^{++} and SO_4^{--} ions.

5-15 Oxidation States and the Periodic Table

One of the useful relationships associated with the periodic table is the periodicity of the oxidation states of the elements. This relationship can be summarized as follows:

1. Elements of Group IA always have the oxidation number +1 in their compounds, with the exception of hydrogen, which is in the −1 state in hydrides of Group IA and Group IIA metals, for example, NaH and CaH_2. These compounds are ionic compounds containing the hydride ion, H^-.
2. Elements of Group IIA always show the oxidation number +2 in their compounds.
3. The oxidation state +3 is the only important one for elements of Group IIIA, with the exception of thallium, which also shows the oxidation state +1.
4. The maximum oxidation number for elements of Group IVA is +4, but the +2 state is also an important one for tin and lead. The oxidation number of carbon varies from +4 as a maximum to −4 as a minimum. Thus, the oxidation number of carbon is −4 in CH_4 because carbon is more electronegative than hydrogen, but it is +4 in CCl_4 because chlorine is more electronegative than carbon. The oxidation number of carbon is +2 in $CHCl_3$, −2 in CH_3Cl, and zero in CH_2Cl_2 (Figure 5-10). The covalency of carbon is 4 in each

FIGURE 5-10

Assignment of oxidation numbers in CH_4, CCl_4, CH_3Cl, CH_2Cl_2, and $CHCl_3$

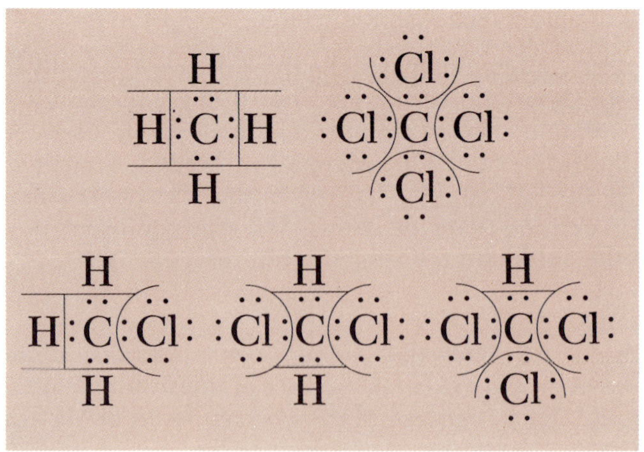

of these compounds, that is, the carbon atom shares four pairs of electrons with other atoms. In general there is no correlation between covalency and oxidation number, although the two are numerically the same in most binary compounds. Even here, however, oxidation numbers are either positive or negative, which is not true of covalency.

5. Nitrogen forms compounds in every integral oxidation state from +5 (as in HNO_3) to −3 (as in NH_3). The most important oxidation states for the other elements of Group VA are +3 and +5.
6. The most important oxidation state for oxygen is −2, but it is in the −1 state in peroxides (for example, hydrogen peroxide, H_2O_2) and in the +2 state in OF_2. The other elements of Group VIA show the oxidation numbers +4 and +6, as well as −2, in many of their most important compounds (for example, sulfur is in the +4 state in SO_2, +6 in H_2SO_4, and −2 in H_2S).
7. The highest oxidation number shown by elements of Group VIIA is +7 and the lowest is −1. Fluorine never has a positive oxidation number since it is the most electronegative of all elements.

5-16 Empirical Formulas from Oxidation Numbers

When the oxidation states of the elements in a compound are known, we can write the empirical formula for the compound. We do this by choosing the smallest numbers of the respective atoms that will make the sum of all the oxidation numbers equal to zero.

1. If the oxidation states of two elements are identical except for sign, they form compounds in which the ratio of the numbers of atoms is 1:1. *Examples:* Hydrogen (+1) and chlorine (−1) combine to form HCl; magnesium (+2) and oxygen (−2) give MgO; aluminum (+3) and nitrogen (−3) give AlN; and silicon (+4) and carbon (−4) give SiC.
2. If, neglecting signs, the oxidation state of one element is twice as great as that of the other, the atoms combine in the ratio 1:2. *Examples:* Hydrogen (+1) and oxygen (−2) combine to form H_2O; calcium (+2) and chlorine (−1) give $CaCl_2$; and silicon (+4) and oxygen (−2) give SiO_2.
3. If, neglecting signs, the oxidation states of two elements are in the ratio 1:3, the atoms combine in the ratio 3:1. *Examples:* Phosphorus (+3) and chlorine (−1) combine to form PCl_3; hydrogen (+1) and nitrogen (−3) form NH_3; and tungsten (+6) and oxygen (−2) give WO_3.

4. If, neglecting signs, the oxidation state of one element is four times that of the other, the atoms combine in the ratio 1:4. *Examples:* Silicon (+4) and chlorine (−1) give $SiCl_4$; hydrogen (+1) and carbon (−4) give CH_4; and osmium (+8) and oxygen (−2) give OsO_4.
5. If one element is in the +2 state and the other is in the −3 state, three atoms of the first combine with two atoms of the second; if one is in the +3 state and the other in the −2 state, two atoms of the first combine with three atoms of the other. *Examples:* Magnesium (+2) combines with nitrogen (−3) to form Mg_3N_2 and aluminum (+3) and oxygen (−2) give Al_2O_3.
6. If one element is in the +5 state and the other is in the −2 state, two atoms of the first combine with five atoms of the other, as in N_2O_5.
7. Other combinations of oxidation states frequently encountered include +7 and −2, as in Cl_2O_7; +5 and −1, as in PCl_5; +6 and −1, as in SF_6; +7 and −1, as in IF_7; +3 and −4, as in Al_4C_3; and +4 and −3, as in Si_3N_4.

The formulas we arrive at by this method are always empirical. It may or may not be true that a molecule contains only the minimum numbers of atoms sufficient to make the sum of the oxidation numbers equal to zero. By the oxidation number method we obtain P_2O_5 as the empirical formula for the compound containing phosphorus (+5) and oxygen (−2); actually, a single molecule of this compound contains four phosphorus atoms and ten oxygen atoms, i.e., the true or molecular formula is P_4O_{10}. It is not possible to know the molecular formula for a substance until its molecular weight has been determined by experiment.

Although it is useful to be able to write empirical formulas for compounds containing elements in specified oxidation states, it is important to remember that the empirical formula for every chemical compound is determined by chemical analysis. Oxidation numbers are inferred, by arbitrary rules, from empirical formulas, rather than conversely.

5-17 Chemical Nomenclature and Oxidation States

Several systems, both old and new, are in use for naming chemical compounds. The International Union of Pure and Applied Chemistry (IUPAC) has developed a systematic nomenclature that gives each compound a name that is based on its composition. The objective in assigning a name to a compound is to make it possible to write the formula of the compound without having to obtain additional information other than that contained in, or implied by, the name itself.

A binary compound—a compound containing only two elements—is usually designated by the suffix *-ide*. This ending is added to the name of the more electronegative element; the other element is named first. Thus, NaCl is sodium chloride and MgO is magnesium oxide. If a given pair of elements form more than one compound, the different compounds may be distinguished by adding a Roman numeral in parentheses to the name of the less electronegative element to indicate its oxidation state. Thus, $FeCl_2$ is iron(II) chloride and $FeCl_3$ is iron(III) chloride; FeO is iron(II) oxide and Fe_2O_3 is iron(III) oxide. It is not necessary to indicate the oxidation state if the two elements form only one compound. Thus, there is no need to indicate that the oxidation state of calcium is +2 in $CaCl_2$ because calcium is in the +2 state in all its compounds.

An older system of nomenclature employed the suffixes *-ous* and *-ic* to designate the lower and the higher of two oxidation states, respectively. By this system of naming compounds, $FeCl_2$ is ferrous chloride and $FeCl_3$ is ferric chloride; $SnCl_2$ is stannous chloride and $SnCl_4$ is stannic chloride. These names have the disadvantage that they do not indicate explicitly what the oxidation states are. More explicit in indicating the formula is another system, still frequently used, in which prefixes are added to the name of each element to indicate the number of atoms of that element in the formula. By this system, N_2O_3 is called dinitrogen trioxide; NO_2 is nitrogen dioxide; N_2O_4 is dinitrogen tetroxide; and so on.

Many chemical compounds were already known, and many of them were familiar substances, even before chemistry became established as a science. An example is H_2O; even chemists call this substance water. Another example is NH_3, which is usually called ammonia rather than hydrogen nitride. Fortunately, not all chemical compounds have such *common* (or *trivial*) *names*, which give no clue as to the composition of the substances.

The formulas for binary compounds are usually written with the symbol for the more electronegative element appearing last, as in NaCl, H_2O, ClF_3, and OF_2. It is unfortunate that custom dictates the writing of the formulas for ammonia and methane as NH_3 (rather than H_3N) and CH_4 (rather than H_4C), because both nitrogen and carbon are more electronegative than hydrogen. The more electronegative element may usually be taken to be the one that is either above the other or to the right of it in the periodic table.

An unfortunate source of confusion in chemical nomenclature is the use of names ending in *-ide* for several types of compounds that contain more than two elements. We have to accept as special cases, for example, the hydroxides, amides, and cyanides. Thus, NaOH is sodium hydroxide, $NaNH_2$ is sodium amide, and NaCN is sodium cyanide.

The naming of acids and salts, and other more complicated compounds, will be described as they are encountered in subsequent chapters.

NEW TERMS

Chemical bonding: a general term referring to the mechanism whereby atoms form either molecules or ions. The union of elements to form compounds and the union of like atoms to form molecules of an elementary substance are both included.

Coordinate bond: a covalent bond in which both electrons of the shared pair are furnished by the same atom.

Covalent bonding: the type of chemical bonding in which electron pairs are shared by the atoms that are thus bonded together.

Electron affinity: the energy released when an electron is added to a neutral, isolated atom, forming a negative ion.

Electronegativity: the tendency of a covalently bonded atom to attract the bonding electron pair.

Electronic formula: the type of structural formula in which all the electrons of the outermost shells of all atoms are included.

Electronic symbol: the chemical symbol for an element with dots to indicate the number of electrons in the outermost shell; for example, $:\underset{\cdot\cdot}{\overset{\cdot\cdot}{Cl}}\cdot$ for a chlorine atom and $[:\underset{\cdot\cdot}{\overset{\cdot\cdot}{Cl}}:]^-$ for a chloride ion.

Hybridization of orbitals: the mixing of a number of orbitals, not all of the same type, to form the same number of hybrid orbitals that are all alike.

Ion: an atom, or a group of atoms, in which the number of electrons and the number of protons are not the same.

Ionic bonding: the type of chemical bonding in which electrons have been transferred from atoms of low ionization energy to atoms of high electron affinity, forming positive and negative ions.

Ionization: the conversion of neutral particles into ions.

Ionization energy: the energy required to remove an electron from an atom.

Isoelectronic: having the same number of electrons.

Kernel: that part of an atom that includes the nucleus and all the electrons except those in the outermost shell.

Line formula: the type of structural formula in which a line is employed to represent a covalent bond, that is, a shared pair of electrons.

Octet rule: the tendency of atoms, when they combine, to form complete octets of electrons in their outermost shells. There are numerous exceptions to this "rule."

Oxidation number: an arbitrary number assigned to each atom in an ion or molecule so that the sum of the oxidation numbers of all the atoms is the charge of the ion or is zero for a molecule. In an ionic compound the oxidation number of each ion is the same as its charge.

Oxidation state: a term used to indicate the oxidation number of each atom of an element. For example, iron is in the same oxidation state (+3) in Fe_2O_3 as in $FeCl_3$.

Polar compound: a compound consisting of polar molecules. In a polar molecule, which contains one or more polar bonds, the positive and negative centers of charge do not coincide. A polar bond is a covalent bond between two atoms that differ in electronegativity.

Resonance hybrid: a molecule or ion to which no single electronic formula can be assigned that conforms to the observed properties of the substance. Instead, the structure is a hybrid between two (or more) contributing structures.

Valence electrons: the electrons in the outermost shell of an atom.

EXERCISES

5-1 Of what ions are the following compounds composed: LiI, KBr, Na_2S, CaO, Al_2S_3, CaF_2?

5-2 Using electronic symbols, indicate the formation of the following ionic compounds: KF, MgS, Ca_3P_2, Na_2S, Li_3P, CaF_2.

5-3 Describe the chemical bonding involved in the formation of the following molecules: F_2, H_2, HF, H_2O, NH_3, CH_4, CCl_4.

5-4 Using the periodic table and the table of electronic configurations, predict whether stable compounds with the following formulas will be found to exist: Mg_3O_2, Mg_2O_3, PO, KCl_2, RbBr, CoNe, BaSe, NaI.

5-5 Why do the ionization energies of the alkali metals decrease with increasing atomic number?

5-6 Why does nitrogen have an ionization energy greater than oxygen?

5-7 Explain why a calcium ion has a charge equal to two protons and a sodium ion has a charge equal to one proton.

5-8 State, define, describe, or explain
(a) ionic bonding, covalent bonding, chemical bonding
(b) oxidation number, oxidation state

5-9 Write line formulas for NH_3, BF_3, and NH_3BF_3.

5-10 Compare metals and nonmetals with regard to (a) electron affinity, (b) ionization energy, (c) electronegativity, and (d) tendency to form positive and negative ions.

5-11 Describe single bonds, double bonds, and triple bonds in terms of electrons.

5-12 How many sodium atoms will react with (a) 1000 chlorine atoms, (b) 1000 sulfur atoms, (c) 6.023×10^{23} bromine atoms?

5-13 Given that two elements, A and B, are to form a compound, indicate how you would attempt to predict the formula of the compound and whether it would be ionic or covalent.

5-14 How is the oxidation state of an element related to the number of electrons in the outermost shell of a neutral, isolated atom of the element?

5-15 Write acceptable names for CO and CO_2; $FeCl_2$ and $FeCl_3$; $SnCl_2$ and $SnCl_4$; P_4O_6 and P_4O_{10}.

5-16 Explain why positive ions are smaller and negative ions are larger than the parent atoms.

REFERENCES

Barrow, G. M., Kenney, M. E., Lassila, J. D., Litle, R. L., and Thompson, W. E., *Chemical Bonding*, Benjamin, New York, 1967 (paperbound). Programmed instruction.

Companion, A. L., *Chemical Bonding*, McGraw-Hill, New York, 1964 (paperbound). Covalent, ionic, and metallic bonding are described.

Ehret, W. F., "The Role of Electrons in Interatomic Relations," *Journal of Chemical Education*, **25**(5), 291 (1948). A brief treatment of the formation of ionic and covalent bonds.

Kieffer, W. F., "The Use of Electron Structure in Interpreting Chemical Reactions," *Journal of Chemical Education*, **25**(10), 537 (1948). A discussion of the arrangements of electrons in atomic orbitals and the combinations of orbitals in bonding.

Lewis, G. N., *Valence and the Structure of Atoms and Molecules*, Dover, New York, 1966 (paperbound). Reprint of the 1923 classic explaining the Lewis concepts of the covalent bond and acid–base reactions.

Pauling, L., *The Chemical Bond*, Cornell University Press, Ithaca, N.Y., 1967 (paperbound). A brief introduction to modern structural chemistry, intended especially for the use of students.

Sanderson, R. T., "Principles of Chemical Bonding," *Journal of Chemical Education*, **38**(8), 382 (1961). Twenty-five generalizations relating chemical bonding to atomic structure.

6

Gas Laws and the Kinetic Theory

6-1 The Concept of Atmospheric Pressure

Since gases are less tangible than liquids and solids, it is not quite so obvious that they have weight. The melting of a certain weight of ice produces liquid water having the same weight; likewise, the evaporation of the water produces water vapor, i.e., gaseous water, having still the same weight.

The most familiar of all gases is the air we breathe. In 1643 Evangelista Torricelli, an Italian physicist and an associate of Galileo, obtained the first evidence that air has weight. Torricelli sealed a long glass tube at one end, poured in mercury from the other end until the tube was filled, covered the open end, inverted the tube into a dish of mercury, and then uncovered the end of the tube. The mercury fell a short distance in the tube but remained about 76 cm (approximately 30 in.) higher than the surface of the mercury in the dish. When Torricelli repeated the experiment with other tubes of various lengths and

FIGURE 6-1

The barometer

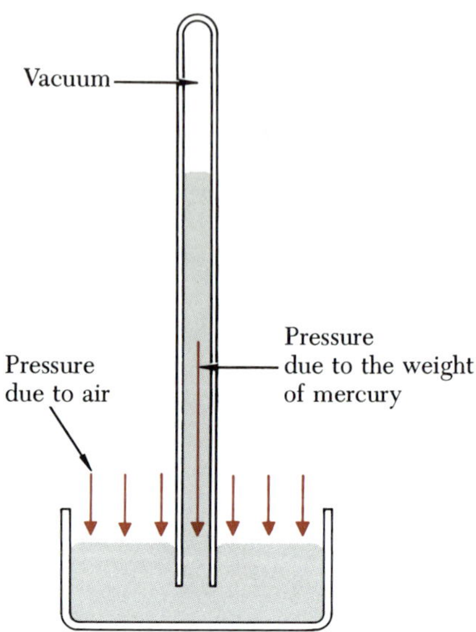

diameters the results were always the same (provided the tube was at least 76 cm long)—the mercury descended in the tube until it was 76 cm higher than the surface of the mercury in the dish.

At that time the phenomenon of suction was explained, in the Aristotelian tradition, by the phrase "nature abhors a vacuum." The Aristotelian explanation for sucking a liquid up a straw would be that the liquid rises to fill the vacuum that nature abhors. A similar explanation would be given for the mercury remaining higher in Torricelli's tube than in the dish outside. But the quantitative measurements that were made raised the critical questions: Why does the mercury in the tube stand 76 cm higher than the outside surface? Does nature's abhorrence of a vacuum not exceed 76 cm? (It must be noted that the height at which the liquid stands is greater than 76 cm if a liquid of lesser density than mercury is used.)

Galileo reasoned that the mercury was held up by the push of the outside air rather than by the pull of the vacuum inside the tube. The weight of all the air directly above the dish—a layer approximately 500 miles thick, we now realize—is a force acting downward upon the surface of the mercury in the dish. This air pressing down tends to cause the mercury to rise in the tube. The equilibrium position of the mercury level is such that the weight of the mercury in the tube above the level of the mercury in the dish is the same as the weight of a column of air of the same cross-sectional area extending from the surface of the mercury to the outer limits of the atmosphere.

Torricelli thought that the mercury column should drop on being carried up a mountain, because the amount of air above the surface of the mercury would then be less. To test Galileo's explanation he took the apparatus up a mountain and found that the height of the mercury column became progressively less with increasing altitude.

Torricelli's apparatus is the mercury *barometer*, the instrument used for measuring atmospheric pressure in scientific laboratories, weather stations, and wherever accuracy is important (Figure 6-1). The space above the mercury in the closed tube is empty space, except for a small quantity of mercury vapor, and is known as a "Torricellian vacuum."

6-2 Units of Pressure

If the column of mercury in a barometer is 76.00 cm high and 1.000 cm^2 (square centimeter) in cross section, the volume of mercury is 76.00 cm^3 (cubic centimeters). Since the density of mercury is 13.55 g per cm^3, the weight of the column is 76.00 $cm^3 \times 13.55$ g per cm^3, or 1030 g. In other words, the *pressure*, i.e., force per unit area, of the atmosphere is 1030 g per cm^2. If a larger tube is used, the weight of mercury in the tube is greater, but this force is distributed over a larger area, so the pressure is the same as before. For example, if the cross section of the tube is doubled, the weight of the mercury is twice as great. This doubled force is acting on an area of 2 cm^2 instead of only 1 cm^2, so the force per unit area is unchanged.

A pressure of 1030 g per cm^2 corresponds to a value of 14.7 psi (pounds per square inch). Atmospheric pressure varies not only with altitude but with changing weather conditions, chiefly because of the varying amounts of water vapor in the air, water molecules not being as heavy as O_2 or N_2 molecules. Fluctuating temperature also changes the density of the air.

When comparing barometer readings taken at different times or places, we do not have to convert the readings to actual pressure units. Even without knowing that a 76-cm reading means a pressure of 1030 g per cm^2, we realize that a 57-cm reading means a pressure $\frac{57}{76}$ or three-fourths as great. It has been customary to express atmospheric pressure in terms of cm Hg (centimeters of mercury) or mm Hg (millimeters of mercury). There is increasing use of the term *torr* (in honor of Torricelli) to designate the pressure exerted by a column of mercury 1 mm high.

The *standard atmospheric pressure* is defined as exactly 76.00 cm Hg (760.0 torr) at sea level and at 0°C and is referred to as *1 atmosphere*. The pressures exerted by confined gases, especially those confined under high pressures, are usually expressed in atmospheres.

6-3 Gas Density and Avogadro's Principle

With the use of modern chemical balances and "vacuum pumps," we can make accurate direct determinations of the weights of known volumes of gases confined under known pressures. To do this we evacuate a flask, weigh it, fill it with gas, and weigh it again. The weight of the gas divided by the volume is the density of the gas. Gas densities are usually expressed in grams per liter.

The density of a gas is so dependent on pressure and temperature that density figures are of no value unless accompanied by pressure and temperature information. At ordinary pressures and temperatures, gas densities vary from a few tenths of a gram to several grams per liter. Solids and liquids are generally about a thousand times as dense as gases. This strongly suggests that the molecules of a gas must be relatively far apart.

In 1811, Amedeo Avogadro, in Italy, stated as a hypothesis that *equal volumes of gases, under identical conditions of temperature and pressure, contain equal numbers of molecules.* This hypothesis is now accepted universally and is known as *Avogadro's Principle.*

The importance of Avogadro's Principle is due to the fact that it enables us to compare the masses of individual molecules of different gases by comparing the weights (masses) of equal volumes of the gases under identical conditions. For example, when we find by experimental measurement that the weight of 1 liter of the gas called sulfur dioxide is (approximately) four times the weight of 1 liter of methane gas and (approximately) twice the weight of 1 liter of oxygen gas at the same temperature and pressure, we conclude that the mass of a sulfur dioxide molecule is (approximately) four times the mass of a methane molecule and twice the mass of an oxygen molecule. The molecular weight of oxygen gas is (approximately) 32 amu. We conclude that the molecular weight of methane is approximately 16 amu and that the molecular weight of sulfur dioxide is about 64.

6-4 Boyle's Law

We sometimes speak of the pressure *of* a gas and at other times of the pressure *on* a gas. Actually, the two are the same. In some situations it is more natural to speak of the pressure the gas exerts against the walls of its container, and in other situations we focus on the pressure used to confine the gas. This difference in terminology is illustrated by the following two problems, which initially might appear to be completely different in nature.

EXAMPLE 1

If a gas exerts a pressure of 740 torr (740 mm Hg) when confined within a 5-liter container, how much pressure will it exert when confined to a 4-liter container?

EXAMPLE 2

If the volume of a sample of gas is 850 ml when confined under a pressure of 1000 g/cm², what volume will it occupy under a pressure of 800 g/cm²?

Both these questions can be answered by applying *Boyle's Law*, which states that at constant temperature the volume (V) occupied by a gas varies inversely with the confining pressure (P), or that the pressure exerted by a gas varies inversely with the volume within which it is confined. Mathematically, at constant temperature,

$$PV = \text{constant} \quad \text{or} \quad \frac{P_1}{P_2} = \frac{V_2}{V_1}$$

where P_1 and V_1 denote the original pressure and volume of the gas respectively, and P_2 and V_2 are the new pressure and volume, respectively. Boyle's Law is represented graphically in Figure 6-2. According to Boyle's Law, all gases behave alike. However, at low temperatures or at high pressures, deviations may occur. When the law is obeyed, the gas is said to exhibit *ideal behavior.*

FIGURE 6-2

Pressure–volume relationship at constant temperature, as stated by Boyle's Law

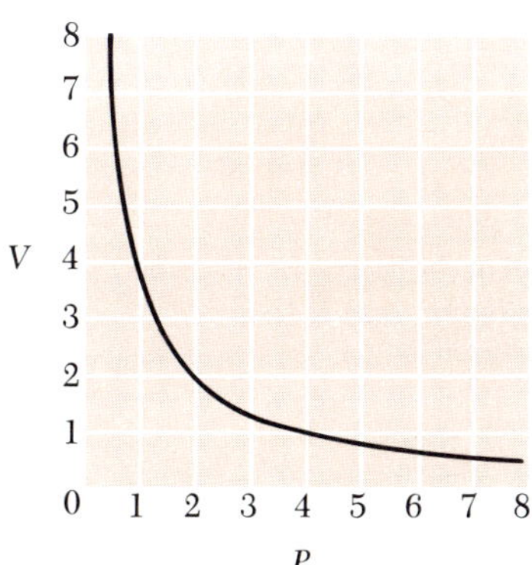

SOLUTION TO EXAMPLE 1

The volume changes to $\frac{4}{5}$ of the original volume. The pressure varies inversely; i.e., the decrease in volume causes the pressure to increase to $\frac{5}{4}$ of the original pressure. Therefore,

$$\begin{aligned}\text{New pressure} &= \frac{5}{4} \times \text{old pressure}\\ &= \frac{5}{4} \times 740 \text{ torr}\\ &= 925 \text{ torr}\end{aligned}$$

SOLUTION TO EXAMPLE 2

The pressure is reduced to $\frac{800}{1000}$ of the original pressure. Since the volume varies inversely, it increases to $\frac{1000}{800}$ of the original volume. Therefore,

$$\begin{aligned}\text{New volume} &= \frac{1000}{800} \times \text{old volume}\\ &= \frac{1000}{800} \times 850 \text{ ml}\\ &= 1064 \text{ ml}\end{aligned}$$

Only three digits are significant—a fact that is best expressed by writing the answer as 1.06×10^3 ml (or 1.06 liters).

6-5 Heat and Temperature

All too often, people confuse the concepts of heat and temperature. *Temperature* is a measure of hotness, i.e., the degree or intensity of heat, whereas *heat* itself is a form of energy. In order for a body to become hotter, i.e., to reach a higher temperature, energy in the form of heat must be allowed to enter it. If one body has a higher temperature than another and the two are placed in contact, heat will flow from the hotter body to the cooler one. Thus, temperature indicates the direction of heat flow. If the two bodies are allowed to remain in contact for a sufficiently long time, the hotter body will "cool off" and the cooler body will "warm up" until they both have the same temperature—the intensity of the heat possessed by the two bodies is the same. Note that this does *not* mean that the *amount* of heat possessed by the two bodies is the same. Temperature is not a measure of the amount of heat possessed by an object; it is a measure of heat *intensity*.

Energy can be defined as capacity for performing work—*work* meaning motion of the point of application of a force, as when a body is moved against an opposing force or is given acceleration. Energy can be classified as either kinetic or potential. *Kinetic energy* is the energy that a moving body possesses because of its motion. *Potential energy* is the energy that a body possesses because of its position with respect to other objects. Kinetic and potential energies are encountered in such familiar forms as mechanical energy, electrical energy, chemical energy, heat, light, and sound. One form of energy can be converted into another. It is especially easy for other forms of energy to be converted into heat.

The unit of heat is the *calorie* (cal), which is defined as the amount of heat required to raise the temperature of 1 g of water 1° on the Celsius scale. It is sometimes more convenient to use the *kilocalorie* (kcal, 1000 cal), which is the "big calorie," or Calorie, used in the field of nutrition.

An instrument used to measure temperature is called a *thermometer.* To ascertain the temperature of a substance, we bring it into contact with a thermometer and allow the substance and the thermometer to remain in contact until they have reached *thermal equilibrium*, that is, until they have attained the same temperature; then we read the temperature on the thermometer scale.

The most common type of thermometer consists of a piece of small-bore glass tubing attached to a reservoir or bulb (Figure 6-3). The bulb

FIGURE 6-3

The thermometer

Comparison of Celsius and Fahrenheit scales.

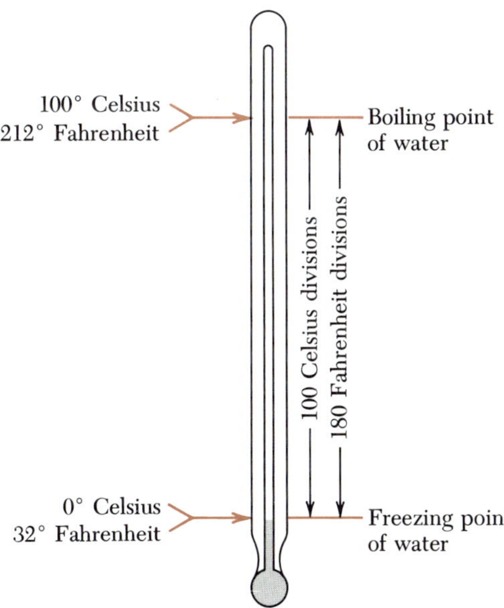

and part of the tubing are filled with a liquid (usually mercury), which expands when heated and contracts when cooled. As the temperature of the liquid in the bulb increases or decreases, the visible column of liquid rises or falls alongside the thermometer scale.

In order to translate the position of the liquid column into a temperature reading, we must use a standard reference as a basis of comparison. The two reference standards are (1) the freezing point of water and (2) the boiling point of water, at normal atmospheric pressure. The two temperature scales currently in use are those introduced by a German physicist, Gabriel Fahrenheit (1686–1736), and by a Swedish astronomer, Anders Celsius (1701–1744). Arbitrarily, the freezing point of water has been assigned the numerical value of 32° on the Fahrenheit (F) scale and 0° on the Celsius (C) scale. The boiling point of water at normal atmospheric pressure has been arbitrarily fixed as 212° on the Fahrenheit scale and 100° on the Celsius (centigrade) scale.

To provide a thermometer with a temperature scale, we first place the thermometer bulb in a mixture of ice and water. When the liquid in the bulb is at the same temperature as the ice–water mixture, we note the position of the column in the stem of the thermometer and mark the column either 32°F or 0°C at this point. When the liquid in the bulb is at the temperature at which water boils at normal atmospheric pressure, we mark the column 212°F or 100°C at the position of the column. Finally, we divide the distance between the two marks into 180 equal divisions (degrees), if we are constructing a Fahrenheit thermometer, or into 100 equal divisions (degrees), if we are making a Celsius thermometer.

Because 180 Fahrenheit degrees equal 100 Celsius degrees, we see that a Fahrenheit degree is $\frac{5}{9}$ the size of a Celsius degree and a Celsius degree is $\frac{9}{5}$ as large as a Fahrenheit degree. Noting also that a temperature reading of 0°C would be 32°F, we may convert a temperature from one scale to the other according to the equations

$$°F = 32 + \tfrac{9}{5}°C$$

$$°C = \tfrac{5}{9}(°F - 32)$$

6-6 Charles' Law

More than a century elapsed between Boyle's discovery of the pressure–volume relationship of gases (in 1660) and Jacques Charles' investigation of the effect of temperature changes on the volumes of gases. In 1787 Charles discovered that the volume of a gas varies in a linear manner with temperature (Figure 6-4) and that the volume is doubled when the Celsius temperature is raised from 0 to 273° with the pressure held constant. Stating it another way, the volume is halved when the Celsius temperature is lowered from 273 to 0° with the pressure held constant. If a gas were to follow this relationship indefinitely upon cooling, its volume would be reduced to 0 at −273°C! Actually, all gases have been found to liquefy before this temperature is reached—and the gas laws of Charles and Boyle do not apply to liquids. Nevertheless, this seemingly lowest attainable temperature has proved to be important.

In 1848 William Thomson, a British physicist later raised to peerage as Lord Kelvin, proposed a new scale of temperature with degrees the size of Celsius degrees but with −273°C redesignated as 0° absolute, 0°C as 273° absolute, 273°C as 546° absolute, and so on. Thus, measured on this new scale, the *absolute* Celsius scale, the volume of a gas is directly proportional to the temperature. The lowest point on the scale, most recently established as −273.15°C, is called *absolute zero*. The absolute Celsius scale is now known as the Kelvin scale. Temperature on the Celsius scale is converted to temperature on the Kelvin scale by adding 273.15°:

$$°K = °C + 273.15°$$

If we convert −273.15°C to Fahrenheit, we find that absolute zero is −459.67°F. Those who work with Fahrenheit temperatures have established an absolute Fahrenheit scale that they call the Rankine scale, in honor of William J. M. Rankine, a Scottish engineer and physicist. Temperature on the Fahrenheit scale is converted to temperature on the Rankine scale by adding 459.67°:

$$°R = °F + 459.67°$$

Since most scientific work is done with Celsius temperatures, the expression *absolute temperature* usually refers to the Kelvin scale. In fact,

FIGURE 6-4
Charles' Law

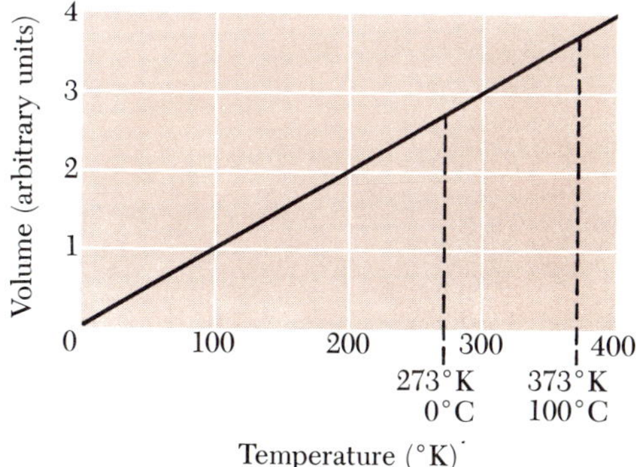

the abbreviation A instead of K is sometimes used to indicate Kelvin temperatures. It is better practice to use K and to refer to the corresponding temperature scale as the Kelvin scale—because both the Rankine and Kelvin scales are absolute scales of temperature.

The four scales of temperature are compared in Table 6-1.

TABLE 6-1

Four scales of temperature

	°*Celsius*	°*Kelvin*	°*Fahrenheit*	°*Rankine*
Boiling point of water	100	373.15	212	671.67
Freezing point of water	0	273.15	32	491.67
Absolute zero	−273.15	0	−459.67	0

According to *Charles' Law*, the volume occupied by a fixed quantity of gas confined under constant pressure is directly proportional to the absolute temperature. Mathematically, at constant pressure,

$$V = \text{constant} \times T \qquad \text{(where } T \text{ is in °K or °R)}$$

or

$$\frac{V_1}{T_1} = \frac{V_2}{T_2}$$

At high pressures and at temperatures near the liquefaction point, the behavior of real gases deviates from Charles' Law. When the law is obeyed, the gas is said to exhibit *ideal behavior.* In solving problems, we shall assume ideal behavior.

EXAMPLE

The volume of a gas is 453 ml at 27°C. What will be its volume at 127°C if the pressure is not permitted to change?

From Charles' Law, we know that

$$\text{New volume} = \text{temperature factor} \times \text{old volume}$$

The temperature *increases* from 300°K to 400°K; i.e., it increases to $\frac{400}{300}$ of its original temperature. Hence, the volume will also increase by the factor $\frac{400}{300}$:

$$\text{New volume} = \frac{400}{300} \times 453 \text{ ml} = 604 \text{ ml}$$

6-7 Amontons' Law

If the volume of a gas were held constant, you would expect the pressure exerted by the gas to be directly proportional to the absolute temperature. This corollary of Charles' Law and Boyle's Law has been verified by experiment. It is known as *Amontons' Law* because Guillaume Amontons, French physicist, discovered in 1699 that if air is confined and heated the pressure increases by a definite amount for a given rise in temperature. Mathematically, at constant volume,

$$P = \text{constant} \times T \qquad \text{(where } T \text{ is in °K or °R)}$$

or

$$\frac{P_1}{T_1} = \frac{P_2}{T_2}$$

EXAMPLE

If the pressure inside a tank of gas is 1.500×10^3 g/cm² at 77°C, what pressure will the gas exert if the tank is cooled to 20°C?

By Amontons' Law,

$$\text{New pressure} = \text{temperature factor} \times \text{old pressure}$$

The temperature *decreases* from 350°K to 293°K, that is, to $\frac{293}{350}$ of its original temperature. Hence, the pressure also decreases by the factor $\frac{293}{350}$:

$$\text{New pressure} = \frac{293}{350} \times 1.500 \times 10^3 \text{ g/cm}^2$$
$$= 1.26 \times 10^3 \text{ g/cm}^2$$

6-8 The Combined Gas Law Equation

According to Boyle's Law, the product PV is constant as long as T is fixed; according to Charles' Law, the ratio V/T is constant as long as P is fixed; and according to Amontons' Law, the ratio P/T is constant as long as V is fixed. These relationships may be combined into one expression, the *combined gas law equation*:

$$\frac{PV}{T} = \text{constant}$$

or

$$\frac{P_1V_1}{T_1} = \frac{P_2V_2}{T_2}$$

This equations holds true for any gas showing ideal behavior as long as we are dealing with a fixed mass, i.e., a fixed number of molecules, of the gas.

Conditions (i.e., temperature and pressure) under which experiments are performed vary considerably, but the combined gas law equation permits easy standardization of results. Measured gas volumes can be compared directly with one another only if they are measured at, or recalculated to, the same temperature and pressure. One atmosphere, i.e., 76.00 cm Hg (760.0 torr), has been chosen as the *standard pressure* for the measurement of gas volumes. The melting point of ice, 273.15°K (0°C), has been selected as the *standard temperature.* Unless stated otherwise, any specified gas volume is the volume that the gas would occupy under these *standard conditions* of temperature and pressure. Standard conditions are abbreviated as STP (for "standard temperature and pressure").

EXAMPLE

If the volume of a gas is 362 ml at 25°C and 745 torr, what would its volume be at standard conditions?

According to the combined gas law equation,

New volume = old volume × temperature factor × pressure factor

The temperature *decreases* by the factor $\frac{273}{298}$, which tends to cause the volume to *decrease* also by the factor $\frac{273}{298}$. The pressure *increases* by the factor $\frac{760}{745}$, and this tends to cause the volume to *decrease* by the factor $\frac{745}{760}$:

$$\begin{aligned} \text{New volume} &= 362 \text{ ml} \times \frac{273}{298} \times \frac{745}{760} \\ &= 362 \text{ ml} \times 0.916 \times 0.980 \\ &= 362 \text{ ml} \times 0.898 \\ &= 325 \text{ ml} \end{aligned}$$

6-9 Molar Volume of a Gas

The volume occupied by 1 mole of a substance is called its *molar volume.* From Avogadro's Principle it follows that the volume of 1 mole of any gas should be approximately the same as the volume of 1 mole of any other gas at the same temperature and pressure. The molar volumes of several gases at standard conditions are listed in Table 6-2. The *ideal gas* (the last item in Table 6-2) is an imaginary gas that behaves in perfect accord with Boyle's Law and Charles' Law. Most real gases behave like an ideal gas under certain conditions, but no real gas behaves ideally under all conditions. The value given in Table 6-2 for the molar volume of an ideal gas is obtained from measurements made on real gases at temperatures high enough and pressures low enough for ideal behavior to be approached closely, the measured volumes being recalculated to standard conditions by use of Boyle's Law and Charles' Law.

For the first three gases listed in Table 6-2, agreement with ideal behavior is quite close. As a practical matter, we can assume that Avogadro's Principle applies to all gases and that the molar volume of any gas at standard conditions is 22.4 liters (Figure 6-5). This information can be used profitably in several types of calculation. For example, it can be used in calculating the density of a gas if its molecular formula is known; it can be used in determining the molecular weight of a gas if its density has been measured; and it can be used to find the STP volume of a gaseous reactant or product in a chemical reaction.

EXAMPLE 1

What is the density, i.e., the mass of 1 liter, of carbon monoxide at STP?

The mass of 1 mole of CO is 12.0 g + 16.0 g = 28.0 g. The density expected at standard temperature and pressure is

$$\frac{28.0 \text{ g/mole}}{22.4 \text{ liters/mole}} = 1.25 \text{ g/liter}$$

(The value obtained experimentally is 1.2504 g/liter.)

EXAMPLE 2

The mass of 1 liter of chlorine gas at STP is 3.214 g. What is the molecular weight of chlorine?

TABLE 6-2

Volume of 1 mole of gas at standard conditions

Gas	*Molar volume (liters)*
Hydrogen	22.432
Nitrogen	22.403
Oxygen	22.392
Hydrogen chloride	22.246
Sulfur dioxide	21.889
Ideal gas	22.414

FIGURE 6-5

Molar volume of a gas

A mole of any gas occupies a volume of approximately 22.4 liters at standard conditions.

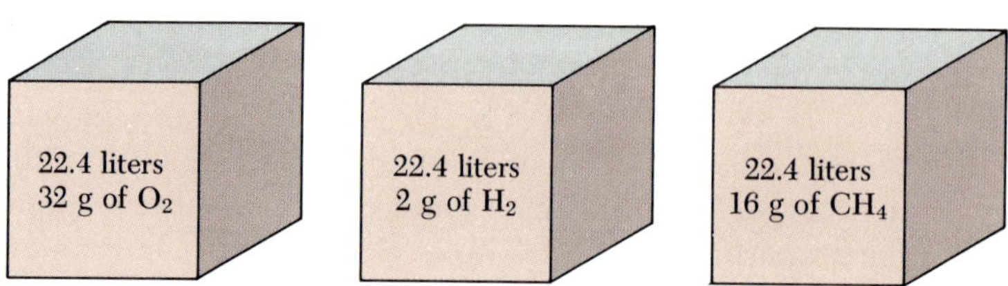

Assuming that the volume of 1 mole of chlorine gas at STP is 22.4 liters, the mass of 1 mole is 22.4 liters/mole × 3.214 g/liter = 72.0 g/mole; i.e., the molecular weight of chlorine is 72.0.

The atomic weight of chlorine is 35.453, which is approximately half the value we have calculated for the molecular weight. This indicates beyond any reasonable doubt that the chlorine molecule consists of two chlorine atoms. The correct molecular weight therefore is exactly twice the atomic weight, i.e., 2 × 35.453, or 70.906, instead of 72.0. The molar volume principle does not give exactly the correct value for the molecular weight in this example; nevertheless, we have been able to arrive at the correct value by following through with the other information at our disposal. We have deliberately chosen, for this example, a gas that does not obey the gas laws well at STP.

EXAMPLE 3

What volume of O_2, measured at STP, will be formed by the reaction of water with 10.0 g of sodium peroxide?

From the balanced equation

$$2Na_2O_2 + 2HOH \longrightarrow 4Na^+ + 4OH^- + O_2\uparrow$$

it is apparent that 2 moles of Na_2O_2 produce 1 mole of O_2, or 1 mole of Na_2O_2 produces 0.5 mole of O_2. The weight of 1 mole of Na_2O_2 is (2 × 23.0 g) + (2 × 16.0 g) = 78.0 g.

$$\frac{10.0 \text{ g}}{78.0 \text{ g/mole}} = 0.128 \text{ mole of } Na_2O_2$$

Since 1 mole of Na_2O_2 produces 0.5 mole of O_2, 0.128 mole of Na_2O_2 produces 0.0640 mole of O_2. The STP volume of the O_2 formed is

$$0.0640 \text{ mole} \times 22.4 \text{ liters/mole} = 1.43 \text{ liters}$$

6-10 The General Gas Law Equation

We have seen that for any gas sample in general, PV/T is a constant as long as Boyle's Law and Charles' Law are obeyed. The numerical value of the constant depends upon the quantity of gas in the sample, i.e., the number of moles. Under any specified conditions of temperature and pressure, the volume occupied by 2 moles of a gas is exactly twice the volume occupied by 1 mole of the same gas. If we let n represent the number of moles, we can say that PV/nT is constant for all gases under all conditions as long as the gases show ideal behavior. This universal constant is known as the *molar gas constant* and is represented by R. The equation

$$PV = nRT$$

is known as the *general gas law equation.*

The numerical value of the gas constant R can be calculated by substituting the proper known values into the general gas law equation. Substituting $n = 1$ mole, $P = 1$ atm, $V = 22.414$ liters, and $T = 273.15°K$,

$$R = \frac{PV}{nT} = \frac{(1 \text{ atm})(22.414 \text{ liters})}{(1 \text{ mole})(273.15°\text{K})} = 0.082054 \text{ liter-atm/mole-deg}$$

The numerical value of R depends on the units in which P, V, n, and T are expressed.

The general gas law equation can be used to calculate, for example, the volume of a specified quantity of gas at a given pressure and temperature, the pressure exerted by a given amount of gas, the molecular weight of a gas, or the volume of a gaseous reactant or product in a chemical reaction.

EXAMPLE 1

Calculate the volume occupied by 5.0 g of carbon monoxide gas at 25°C and 740 torr pressure.

$$T = (273 + 25)°\text{K} = 298°\text{K}$$

$$P = \frac{740 \text{ torr}}{760 \text{ torr/atm}} = 0.974 \text{ atm}$$

$$n = \frac{5.0 \text{ g}}{28.0 \text{ g/mole}} = 0.18 \text{ mole}$$

$$R = 0.082 \text{ liter-atm/mole-deg}$$

$$V = \frac{nRT}{P} = \frac{0.18 \text{ mole} \times 0.082 \text{ liter-atm/mole-deg} \times 298°\text{K}}{0.974 \text{ atm}}$$

$$= 4.5 \text{ liters}$$

EXAMPLE 2

If 5.20 g of a gas occupies a volume of 1140 ml at 78°C and 780 torr pressure, what is the molecular weight of the gas?

Let $x =$ molecular weight (so that 1 mole weighs x g). Then

$$n = \frac{5.20 \text{ g}}{x \text{ g/mole}} = \frac{5.20}{x} \text{ moles}$$

$$P = \frac{780 \text{ torr}}{760 \text{ torr/atm}} = 1.026 \text{ atm}$$

$$T = (273 + 78)°\text{K} = 351°\text{K}$$

$$V = \frac{1140 \text{ ml}}{1000 \text{ ml/liter}} = 1.140 \text{ liters}$$

$$R = 0.0821 \text{ liter-atm/mole-deg}$$

By the general gas law equation,

$$PV = \frac{5.20}{x} RT$$

Solving for x:

$$x = \frac{5.20RT}{PV} = \frac{5.20 \text{ g} \times 0.0821 \text{ liter-atm/mole-deg} \times 351°\text{K}}{1.026 \text{ atm} \times 1.140 \text{ liters}}$$
$$= 128 \text{ g/mole}$$

EXAMPLE 3

What volume of H_2, measured at 30°C and 750 torr, will be obtained from the reaction of 15.0 g of sodium metal with water?

From the balanced equation

$$2\text{Na} + 2\text{HOH} \longrightarrow 2(\text{Na}^+ + \text{OH}^-) + \text{H}_2\uparrow$$

we see that 2 moles of sodium atoms produce 1 mole of H_2 molecules, or 1 mole of sodium atoms produces 0.5 mole of H_2 molecules.

$$\frac{15.0 \text{ g}}{23.0 \text{ g/mole}} = 0.652 \text{ mole of Na atoms}$$

Since 1 mole of Na atoms produces 0.5 mole of H_2 molecules, 0.652 mole of Na atoms produces 0.326 mole of H_2 molecules. The volume of 0.326 mole of H_2 gas under the stated conditions can be obtained directly from the general gas law equation.

$$V = \frac{nRT}{P}$$

$$P = \frac{750 \text{ torr}}{760 \text{ torr/atm}} = 0.987 \text{ atm}$$

$$T = (273 + 30)°\text{K} = 303°\text{K}$$

$$R = 0.0821 \text{ liter-atm/mole-deg}$$

$$n = 0.326 \text{ mole}$$

$$V = \frac{0.326 \text{ mole} \times 0.0821 \text{ liter-atm/mole-deg} \times 303°\text{K}}{0.987 \text{ atm}}$$

$$= 8.22 \text{ liters}$$

6-11 Dalton's Law of Partial Pressures

In a paper published in 1803, John Dalton announced his discovery of the *Law of Partial Pressures*. According to this law, in a mixture of gases the molecules of each species exert the same pressure as they would exert if they were present alone, so that the total pressure is the sum of these "partial pressures" exerted by the different gases in the mixture (Figure 6-6):

$$P_{\text{total}} = P_1 + P_2 + P_3 + \cdots$$

The significance of this law is best understood in terms of moles and mole fractions. In any mixture the number of moles of each component is the number of grams of that component divided by the weight of 1 mole. The *mole fraction* of a component of a mixture is the number of moles of that component in the mixture divided by the total number of moles of all components in the mixture. Thus, if a mixture contains n_A moles of component A and n_B moles of component B, the mole fraction X_A of component A is

$$X_A = \frac{n_A}{n_A + n_B}$$

FIGURE 6-6

An example of Dalton's Law of Partial Pressures

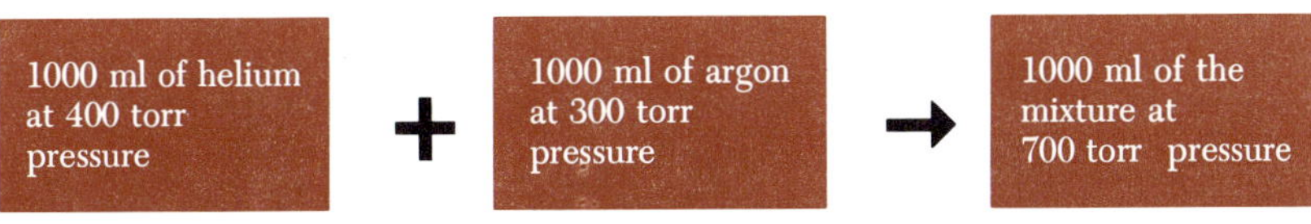

Suppose we have a mixture of 16 g of hydrogen gas (H_2) and 56 g of carbon monoxide (CO). The molecular weights of these two gases being 2 amu and 28 amu, respectively, we have

$$\frac{16 \text{ g}}{2 \text{ g/mole}} \text{ or 8 moles } H_2 \quad \text{and} \quad \frac{56 \text{ g}}{28 \text{ g/mole}} \text{ or 2 moles CO}$$

We have, then, a total of 10 moles, of which 8 are H_2 and 2 are CO. The mole fractions are $\frac{8}{10}$ or 0.8 for H_2 and $\frac{2}{10}$ or 0.2 for CO.

In the mixture for which the mole fraction of hydrogen is 0.8, it is important to realize that 0.8, i.e., 80 percent, of the molecules are H_2 molecules; similarly, since the mole fraction of carbon monoxide is 0.2, 20 percent of the molecules are CO molecules. Dalton discovered that in such a mixture 0.8 of the pressure is due to H_2 molecules and the remaining 0.2 of the pressure is exerted by CO molecules. Thus, in any gaseous mixture of A and B, the partial pressure, P_A, of component A is

$$P_A = X_A P_{total}$$

and similarly the partial pressure P_B of component B is

$$P_B = X_B P_{total}$$

A corollary to Dalton's Law was verified experimentally by E. H. Amagat, a French physicist, in 1880. *Amagat's Law* states that the total volume of any gaseous mixture is the sum of the partial volumes of the components of the mixture:

$$V_{total} = V_A + V_B + V_C + \cdots$$

The partial volume of a component is the volume that component would occupy if it were present alone at the total pressure of the mixture and at the same temperature. Insofar as the ideal gas laws are applicable, $V_A = X_A V_{total}$, $V_B = X_B V_{total}$, and so on.

The chemical composition of gas mixtures is frequently stated in terms of percentage by volume. Thus, dry air is stated to be 78 percent N_2, 21 percent O_2, and 1 percent Ar by volume. Mixing 78 cm^3 of N_2, 21 cm^3 of O_2, and 1 cm^3 of Ar will produce 100 cm^3 of mixture. In such a mixture 78 percent of the molecules are N_2 molecules, 21 percent are O_2 molecules, and 1 percent are Ar molecules. Therefore, 78 percent of the pressure is exerted by N_2 molecules, 21 percent by O_2 molecules, and 1 percent by Ar molecules.

6-12 Gay-Lussac's Law of Combining Volumes of Gases

In the discussion of Dalton's Law of Partial Pressures, we considered gases which, when mixed, do not react with each other. Of course, there are cases in which mixed gases do react. Sometimes gases react as soon as they are mixed, as when colorless NO gas comes into contact with air and reacts with the oxygen to form reddish-brown NO_2 gas:

$$2NO + O_2 \longrightarrow 2NO_2$$

More commonly, a mixture of gases requires some input of energy to get the reaction started. A mixture of H_2 and Cl_2 does not react in the dark, but light of a certain frequency (in the ultraviolet range) initiates a violent reaction:

$$H_2 + Cl_2 \longrightarrow 2HCl$$

A mixture of H_2 and O_2 does not react until a flame or an electric spark comes into contact with the mixture; then the two gases react with explosive violence:

$$2H_2 + O_2 \longrightarrow 2H_2O$$

In 1809 Joseph Louis Gay-Lussac, a French chemist and physicist, completed a series of measurements on the volumes of gases that react with each other and proposed a law now known as *Gay-Lussac's Law of Combining Volumes of Gases.* This law states that gases react in a simple numerical ratio by volume, and the volume of each gaseous product bears a simple numerical ratio to the volume of each gaseous reactant and to the volume of each other gaseous product. Thus, when Cl_2 and H_2 react, the volume of Cl_2 that reacts is the same as the volume of H_2 with which it reacts, i.e., the volume ratio is 1:1; furthermore, the volume of HCl produced is twice the volume of Cl_2 or H_2 reacting. When O_2 and H_2 react, the volume of H_2 reacting is twice the volume of O_2 with which it reacts, i.e., the volume ratio is 2:1; the volume of gaseous H_2O produced is the same as the volume of H_2 reacting.

In terms of Avogadro's Principle, H_2 and Cl_2 react in the volume ratio 1:1 because each molecule of H_2 reacts with one molecule of Cl_2. If a volume of H_2 contains, say, 1 trillion H_2 molecules, then the volume of Cl_2 with which it reacts is the volume that contains 1 trillion Cl_2 molecules. According to Avogadro's Principle, 1 trillion Cl_2 molecules occupy the same volume as 1 trillion H_2 molecules. Avogadro proposed his hypothesis in 1811 to provide an explanation for the Law of Combining Volumes, which Gay-Lussac had discovered two years earlier.

The relationship stated in the Law of Combining Volumes makes it easy to calculate the volume of a gas required to react with a given volume of another gas.

EXAMPLE

What volume of hydrogen reacts with 50 liters of carbon monoxide in the formation of methanol?

According to the equation for the reaction,

$$CO + 2H_2 \longrightarrow CH_3OH$$

2 moles of H_2 react with 1 mole of CO. If the volumes of the gases are measured at the same temperature and pressure, the volume of 2 moles of H_2 is twice the volume of 1 mole of CO. Therefore, the volume of H_2 reacting is twice the volume of the CO with which it reacts.

$$50 \text{ liters of CO} \times 2 \text{ liters of } H_2/\text{liter of CO} = 100 \text{ liters of } H_2$$

6-13 Graham's Law of Diffusion Rates

In 1829 Thomas Graham, a Scottish chemist, completed an investigation of the diffusion of gases. *Diffusion* is the spontaneous spreading of a substance throughout the space available to it. The diffusibility of gases is one of their important properties. If a gas is placed in an empty container, it almost instantly pervades the entire space inside the container. If the container is already occupied by another gas, diffusion of the added gas nevertheless occurs—although more slowly, since no pressure difference exists to hasten the process. For example, if ammonia gas is liberated in a room, its odor soon pervades the room. The presence of air does not prevent the diffusion of the ammonia.

Graham compared the rates at which different gases diffuse and found that the rate depends upon the density of the gas. In a series of gases the lightest gas diffuses most rapidly, and the gas with the greatest density diffuses most slowly. For example, the density of sulfur dioxide gas is four times that of methane; methane diffuses more rapidly than sulfur dioxide—but only *twice* as fast. Oxygen has a density 16 times that of hydrogen; hydrogen diffuses *four* times as rapidly as oxygen. Thus, we have a square-root relationship. Mathematically, *Graham's Law of Diffusion Rates* states that

$$\frac{R_1}{R_2} = \frac{\sqrt{d_2}}{\sqrt{d_1}} \quad \text{or} \quad \frac{R_1}{R_2} = \sqrt{\frac{d_2}{d_1}}$$

that is, the relative rates of diffusion of any two gases under the same conditions are inversely proportional to the square roots of their densities.

EXAMPLE

Gas A has a density of 1.24 g/liter. The density of gas B is 4.96 g/liter. Which gas will diffuse more rapidly, and in what ratio?

The ratio of the two densities is 4.96:1.24, or 4:1; i.e., gas B is four times as dense as gas A. Therefore, gas A will diffuse more rapidly than gas B by the factor $\sqrt{4}$, or 2. Gas A will diffuse twice as rapidly as gas B.

About 30 years after the Law of Diffusion Rates was announced, Avogadro's Principle attained universal acceptance. It was then realized that the density of a gas is directly proportional to the weights of the individual molecules. Thus, Graham's Law can be stated in terms of molecular weights:

$$\frac{R_1}{R_2} = \frac{\sqrt{M_2}}{\sqrt{M_1}} \quad \text{or} \quad \frac{R_1}{R_2} = \sqrt{\frac{M_2}{M_1}}$$

EXAMPLE

Which gas will diffuse more rapidly, methane (CH_4) or sulfur dioxide (SO_2), and in what ratio?

The molecular weights are 16 amu for methane (12 amu + 4 amu) and 64 amu for sulfur dioxide (32 amu + 32 amu). Thus, the ratio of the molecular weights is 64:16, or 4:1; i.e., the molecular weight of sulfur dioxide is four times that of methane. Thus, methane will diffuse more rapidly than sulfur dioxide by the factor $\sqrt{4}$, or 2—twice as fast.

6-14 The Kinetic Theory of Gases

The concept of molecules was originally based on experimental studies of the physical behavior of gases. From the work of Robert Boyle and others, there appeared to be no limit to the ability of a gas to expand when the confining pressure was reduced more and more. It did not seem reasonable that a substance could be present in every part of a given space and also be present in every part of a space a thousand times as great. The ability of a gas to fill all the space available to it is readily understandable, however, if it is assumed that the gas is composed of separate particles that are very small in comparison with the distances between them and if it is also assumed that the particles are continually in motion.

The *kinetic-molecular theory* assumes that all gases consist of very small, discrete particles that are in constant, rapid motion. These particles are the molecules of the gas, which are assumed to be so small that the average distance between them is many times the diameter of one of them. The word "kinetic" is from a Greek word referring to motion. The rapidly moving molecules collide with one another, and with the walls of the container, without loss of kinetic energy. A collision between a fast molecule and a slow molecule usually results in speeding up the slow one and slowing down the fast one (although this depends upon the angle of impact), but the sum of the kinetic energies of the two molecules after the collision will be the same as the sum of their kinetic energies prior to the collision.

The kinetic-molecular theory attributes the pressure that a gas exerts to the collisions of its molecules with the walls of the container. The average velocity of the molecules is assumed to increase as the temperature rises. Since the motion of the molecules is such an important part of the total concept, the theory is usually referred to simply as *the kinetic theory*.

The kinetic energy of a moving molecule, like that of any moving object, is the energy associated with its motion. The quantity of kinetic energy that such a molecule possesses is equal to the work required to bring it to rest and is given by the expression

$$E_k = \frac{1}{2}mu^2$$

where m is its mass and u is its velocity. The kinetic theory of gases assumes that the average kinetic energy of the molecules of a gas is directly proportional to the absolute temperature. The molecules of different gases at the same temperature have the same average kinetic energy.

The kinetic theory ignores any forces of attraction operating between the molecules of a gas. Under certain conditions of temperature and pressure, the intermolecular attractive forces are of sufficient strength that they significantly affect the behavior of a gas. As long as very low temperatures and high pressures are avoided, however, the assumptions of the kinetic theory are adequate for explaining the physical behavior of a gas.

The *ideal gas* (which has no real existence) is an imaginary gas in which there are no attractive forces between the molecules. The molecules of the ideal gas are infinitesimally small so that their own total volume is insignificant in relation to the volume occupied by the gas. The ideal gas would show ideal behavior at any temperature and pressure.

6-15 Applications of the Kinetic Theory

The assumptions of the kinetic theory account for the observed properties of gases and for the laws that express them:

1. *Compressibility.* If a gas consists of particles that are separated from one another by large spaces, then the reason that the gas is easily compressible is that the molecules are merely crowded closer together when the volume of the container is reduced (Figure 6-7).
2. *Diffusibility.* If the molecules of two different gases are in rapid motion and are relatively far apart, the spontaneous diffusion of each into the other is the result of molecules of each kind moving into the spaces between the molecules of the other kind.
3. *Boyle's Law.* When the volume of a gas is reduced, the molecules have less space in which to move; they collide with the walls of the container more frequently. Since pressure depends upon the number of impacts per second, the pressure exerted by a gas is inversely proportional to the volume it occupies.
4. *Dalton's Law.* If there is no attraction between gas molecules, each kind of molecule in a mixture of gases strikes the walls of the container the same number of times per second, and with the same force, as if it were the only kind of molecule present. Thus, the partial pressure of a gas is not changed by the presence of other gases. Each partial pressure is proportional to the number of molecules of that species present, and the total pressure exerted by the mixture of gases is the sum of the pressures exerted by the individual components.

FIGURE 6-7

The molecules of a gas

When a gas is compressed, the molecules remain the same size but there is less space between them.

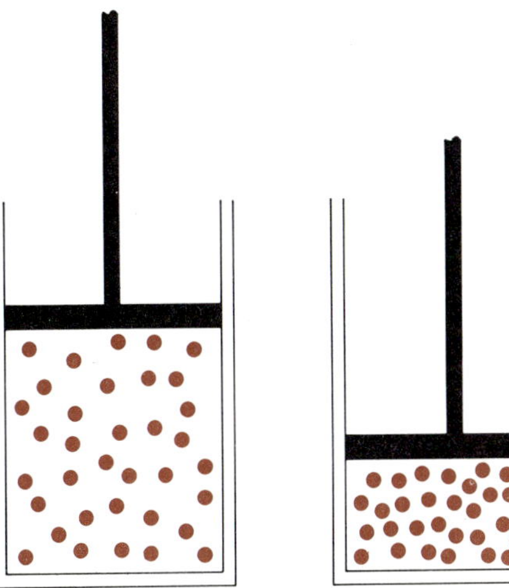

5. *Graham's Law*. If, on the average, molecules of different gases at the same temperature have the same kinetic energy, then molecules of one kind, A, must have velocities that are different from the velocities of the molecules of another gas, B, unless they have the same masses. Mathematically, if the kinetic energy is the same for the two gases,

$$\tfrac{1}{2}m_A u_A^2 = \tfrac{1}{2}m_B u_B^2$$

and

$$\frac{u_A^2}{u_B^2} = \frac{m_B}{m_A}$$

so that

$$\frac{u_A}{u_B} = \sqrt{\frac{m_B}{m_A}}$$

For example, a methane molecule (molecular weight, 16 amu), A, has a mass one-fourth as great as the mass of a sulfur dioxide molecule (molecular weight, 64 amu), B. If

$$\frac{m_B}{m_A} = 4$$

then

$$\frac{u_A}{u_B} = \sqrt{4} = 2$$

The methane molecules have an average velocity twice that of the sulfur dioxide molecules. Thus, the kinetic theory shows why methane diffuses twice as fast as sulfur dioxide and why, in general, different gases diffuse at rates that vary inversely with the square roots of their molecular weights or densities.

6. *Amontons' Law and Charles' Law*. The kinetic theory postulates that the absolute temperature is a manifestation of the average kinetic energy of the molecules and that doubling the absolute temperature means doubling the average kinetic energy. (Thus, absolute zero is the temperature of no kinetic energy.) The pressure exerted by a gas confined within a fixed volume is proportional to nE_k, i.e., the number of molecules per unit volume times their average kinetic energy. Doubling the absolute temperature doubles E_k and therefore doubles the pressure the gas exerts, provided the volume is kept constant. Alternatively, the volume must be permitted to double if the pressure is to remain the same in spite of doubled temperature.
7. *Avogadro's Principle*. If two gases are at the same temperature, then E_k (the average kinetic energy per molecule) is the same for the one gas as for the other. If the two gases are also exerting the same pressure, then nE_k (the number of molecules per unit volume multiplied by their average kinetic energy) is also the same

for the two gases. Therefore, if two gases are at the same temperature and pressure, n must be the same for both—that is, equal volumes of the two contain the same number of molecules.

When Avogadro proposed his hypothesis, he was simply making a shrewd guess that seemed reasonable. He could not deduce his hypothesis from the kinetic theory of gases, as we have done, because the kinetic theory was not well developed at that time. The beginnings of the mathematical development of the kinetic theory were made by Daniel Bernoulli in 1738 in Switzerland, but interest in the theory did not develop until a century later. Then a detailed kinetic theory of gases developed through the efforts of many physical scientists, including Rudolf Clausius in Germany, J. J. Waterston and James Clerk Maxwell in England, Ludwig Boltzmann in Austria, and J. Willard Gibbs in America.

From the equations of the kinetic theory it is possible to calculate the average velocity of gas molecules, the average distance that a molecule travels between successive collisions with other molecules, the number of collisions that a molecule has per second, etc. For example, the average velocity of H_2 molecules at 0°C is 1.84×10^5 cm per sec (equivalent to 3780 mph). At a pressure of 1 atm, the average distance that a hydrogen molecule travels between collisions at 0°C is only 1.12×10^{-5} cm; the number of collisions experienced by one molecule in 1 sec is 1.6×10^{10} (i.e., between 10 and 20 billion).

6-16 Deviations from Ideal Behavior

Boyle's Law, Charles' Law, and the other gas laws describe the behavior of real gases satisfactorily at ordinary pressures and temperatures, but at low temperatures and high pressures the behavior of any real gas deviates markedly from ideal behavior.

At high pressures gas molecules are relatively close together, and the actual volume of the molecules themselves becomes significant compared with the free space between them. As we would expect, a gas becomes less compressible at high pressures; thereafter, doubling the pressure does not reduce the volume by half. Under extremely high pressures, the space occupied by the molecules becomes a substantial fraction of the space within which the gas is confined, and the gas resists further decrease in volume, a resistance characteristic of liquids.

At low temperatures all gases become more compressible than would be predicted by the gas laws. This phenomenon is due to attractive forces between molecules that cause them to crowd closer together. The attractive forces between nonpolar molecules are called *van der Waals forces* after Johannes van der Waals, the Dutch physicist who postulated their existence in 1873. These forces, although important, are effective only over very short distances.

Because the two factors that cause nonideal behavior in gases act in opposition, the net deviation at a particular temperature and pressure depends upon the predominant factor. This varies with different gases.

Except under extreme conditions, deviation from ideal behavior is not serious enough to discredit the kinetic theory of gases. Indeed, we should expect molecules to have a volume of their own, and the very existence of liquids is evidence that there is an attraction between molecules. We should expect that the behavior of a real gas would be affected by the actual volume of the molecules and by the attractive forces between them.

The van der Waals forces of attraction between nonpolar molecules are a result of the electrical attraction between the nuclei of one molecule and the electrons of another. This is so nearly compensated for by the repulsion of nuclei by nuclei and of electrons by electrons that the net forces of attraction become significant only when the molecules are crowded close together—at temperatures low enough for the disordering effect of molecular motion to be overcome. Large molecules attract one another more strongly than small ones because they contain more electrons.

The van der Waals forces of attraction exist because of the motion of electrons in atoms and molecules. Consider two argon atoms at a moment when they happen to be close to each other, as represented in Figure 6-8. Although the electron distribution in an argon atom is highly symmetrical when averaged over even a short period of time, we can imagine that at a particular instant there is a lack of perfect symmetry with a slight preponderance of electrons in one direction (arising through the laws of chance in the motions of electrons). At this instant the left side of atom A is slightly negative, and the right side is slightly positive. The positive side of A attracts the electrons in its neighbor, atom B, leaving it similarly unsymmetrical. Thus, each atom is momentarily polar, with a consequent attraction between them. Even though the

FIGURE 6-8

Van der Waals forces of attraction between two argon atoms

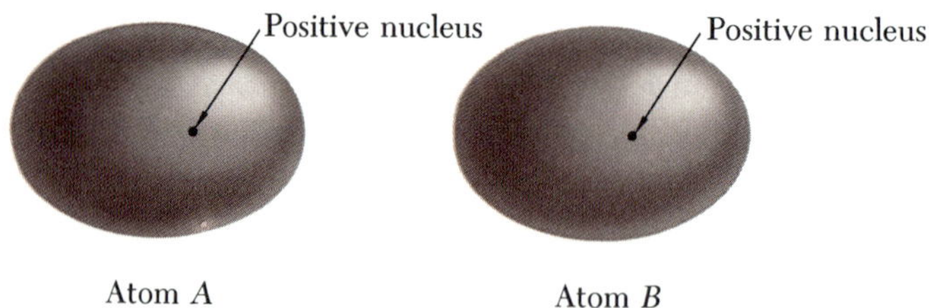

attraction between these particular atoms will persist for only a very brief period of time, any other pair of atoms has the same opportunity to develop such momentary attractions between them. Consequently, there never is an instant when van der Waals attractive forces cease to exist.

A type of attractive force that is stronger than van der Waals forces also can exist between molecules. Polar substances deviate substantially from ideal behavior because of the permanent polarity of their molecules. The positive end of each polar molecule attracts the negative end of each of its close neighbors with sufficient effectiveness to keep the substance in the liquid state at temperatures considerably above the boiling points of substances whose only intermolecular attractive forces are of the van der Waals type. Thus, HCl, mentioned in Section 5-7 as an example of a polar compound, has a boiling point 101° higher than that of argon, even though its molecular weight is several units less than the molecular weight of argon. Even more striking is the case of H_2O, which has a boiling point 346° higher than that of neon, although its molecular weight is less than the molecular weight of neon.

NEW TERMS

Absolute temperature: temperature measured from −273.15°C (−459.67°F), which is absolute zero. In scientific work the measurement is made in Celsius degrees; i.e., absolute temperature refers to the Kelvin scale, which is the absolute Celsius scale. The Rankine scale is the absolute Fahrenheit scale.

Absolute zero: the lowest possible temperature (−273.15°C; −459.67°F; 0°K). Temperatures within a few thousandths of a degree of 0°K have been attained. In the kinetic theory of gases, absolute zero is the temperature at which the kinetic energy of the molecules of an ideal gas is zero.

Barometer: an instrument for measuring atmospheric pressure.

Calorie: the quantity of heat required to raise the temperature of 1 g of H_2O one centigrade degree (specifically, from 14.5 to 15.5°C). The kilocalorie (kcal, 1000 cal) is the "big calorie," or Calorie, used in the field of nutrition.

Combined gas law equation: the equation that combines the Boyle's Law equation, the Charles' Law equation, and the Amontons' Law equation into one expression. It can be written $\frac{PV}{T} = \text{constant}$ or $\frac{P_1V_1}{T_1} = \frac{P_2V_2}{T_2}$.

Diffusion: the spontaneous spreading of a substance throughout the space available to it.

Energy: for most studies, may be defined as the capacity for performing work.

General gas law equation: the equation that states the relationship of the volume of a gas to the pressure, the temperature, and the number of moles: $PV = nRT$.

Heat: the form of energy that passes from a body of higher temperature to a body of lower temperature.

Ideal behavior of a gas: perfect conformity with Boyle's Law, Charles' Law, and Amontons' Law.

Ideal gas: an imaginary gas that behaves in perfect accord with Boyle's Law, Charles' Law, and Amontons' Law.

Kinetic energy: the energy that a moving object possesses by virtue of its motion. The quantity of kinetic energy such an object possesses is equal to the work required to bring it to rest. This is expressed by the equation $E_k = \frac{1}{2}mu^2$, where m is the mass of the object and u is its velocity.

Molar gas constant: the constant R in the general gas law equation.

Molar volume: the volume of 1 mole. Unless stated otherwise, the molar volume of a gas is the volume of 1 mole of the gas at standard conditions. The molar volume of an ideal gas is 22.414 liters at STP.

Mole fraction (of a component of a mixture): the number of moles of a component of a mixture divided by the total number of moles of all components in the mixture.

Potential energy: the energy that a body possesses by virtue of its position; for example, if you lift this book above the desk, you do work and thereby impart potential energy to the book. This energy will be released as kinetic energy if you drop the book.

Pressure: force per unit area.

Standard conditions: atmospheric pressure and 0°C.

Standard pressure: the pressure exerted by a column of mercury 760 mm (29.9 in.) high, designated 1 atmosphere (1 atm) because it is the pressure exerted by the earth's atmosphere, or 760 torr.

Standard temperature: 0° Celsius (32° Fahrenheit).

Temperature: degree of hotness, measured on a suitable scale. The temperature of an object is not to be confused with the heat that it contains.

Thermometer: an instrument used for measuring temperature.

Torr: the pressure exerted by a column of mercury 1 mm high.

Torricellian vacuum: the space above the mercury inside a mercury barometer, devoid of all matter except a small concentration of mercury vapor.

Van der Waals forces: the weak forces of attraction that exist between nonpolar molecules, due to the effect of the atomic nuclei within one molecule upon the electrons of another. The van der Waals forces are significant only when molecules are very close together. At small distances they are balanced by forces of repulsion between outer electron shells of the molecules.

Work: motion of the point of application of a force, as when a body is given acceleration or is moved against an opposing force.

The Gas Laws

Amagat's Law: The total volume of any gaseous mixture is the sum of the partial volumes of the components of the mixture. It can be written $V_{total} = V_1 + V_2 + V_3 + \cdots$.

Amontons' Law: The pressure exerted by a fixed quantity of gas confined under conditions of constant volume is directly proportional to the absolute temperature. It can be written

$$P = \text{constant} \times T \quad \text{or} \quad \frac{P_1}{T_1} = \frac{P_2}{T_2}, \qquad \text{where } T \text{ is in °K or °R}$$

Boyle's Law: At constant temperature, the volume occupied by a fixed quantity of gas varies inversely with the confining pressure. It can be written $PV = $ constant or $\frac{P_1}{P_2} = \frac{V_2}{V_1}$.

Charles' Law: The volume occupied by a fixed quantity of gas confined under constant pressure is directly proportional to the absolute temperature. It can be written

$$V = \text{constant} \times T \quad \text{or} \quad \frac{V_1}{T_1} = \frac{V_2}{T_2}, \qquad \text{where } T \text{ is in °K or °R}$$

Dalton's Law of Partial Pressures: In a mixture of gases the molecules of each species exert the same pressure they would exert if they were present alone, so that the total pressure is the sum of these partial pressures exerted by the different gases in the mixture. It can be written $P_{total} = P_1 + P_2 + P_3 + \cdots$.

Gay-Lussac's Law of Combining Volumes of Gases: Gases react in a simple numerical ratio by volume, and the volume of each gaseous product bears a simple numerical ratio to the volume of each gaseous reactant and to the volume of each other gaseous product.

Graham's Law of Diffusion Rates: The relative rates of diffusion of any two gases under the same conditions are inversely proportional to the square roots of their densities and also are inversely proportional to the square roots of their molecular weights:

$$\frac{R_1}{R_2} = \sqrt{\frac{d_2}{d_1}} = \sqrt{\frac{M_2}{M_1}}$$

EXERCISES

6-1 A liter of nitrogen weighs approximately the same as 7 liters of helium. How does the mass of a molecule of helium compare with the mass of a molecule of nitrogen?

6-2 Air is a mixture that contains approximately four molecules of nitrogen for each molecule of oxygen. A nitrogen molecule weighs seven-eighths as much as an oxygen molecule. The molecular weight of oxygen is 32 amu. Find the average molecular weight of the gases in the air.

6-3 On the basis of what you have learned in Exercise 6-2, is oxygen heavier or lighter than air? Is nitrogen heavier or lighter than air?

6-4 How would the density of a mixture of carbon monoxide and nitrogen compare with the density of either of the pure gases?

6-5 At what temperature does a Celsius thermometer give the same reading as a Fahrenheit thermometer?

6-6 The normal temperature of the human body is 98.6°F. What is this on the Celsius scale? On the Rankine scale? On the Kelvin scale?

6-7 Mercury freezes at −38.9°C, and cesium melts at 28.7°C. Convert these temperatures to the Kelvin and Fahrenheit scales.

6-8 Prove that Fahrenheit–Celsius conversions can be made by using the following formulas:

$$°F = \tfrac{9}{5}(°C + 40) - 40$$
$$°C = \tfrac{5}{9}(°F + 40) - 40$$

6-9 If we have 350 cm^3 of oxygen gas at 738 torr pressure, what volume will we have at 842 torr (temperature remaining constant)?

6-10 If a gas has a volume of 493 ml at 30°C, what volume will it occupy at 60°C if the pressure is not permitted to change?

6-11 If the volume of a gas is 1.776 liters at 22°C and 741 torr, what will its volume be at standard conditions?

6-12 What is the weight of 1348 ml of carbon monoxide at STP?

6-13 What is the weight of 1348 ml of carbon dioxide at STP?

6-14 If the weight of 376 ml of a gas at STP is 0.74 g, what is its molecular weight?

6-15 What is the molecular weight of a gas if 1.00 g has a volume of 395 ml at 19°C and 747 torr?

6-16 A gaseous mixture is 42% H_2, 36% He, and 22% CH_4 by volume. How many molecules of each kind are contained in every 100 molecules of the mixture? What is the mole fraction of each component? What is the partial pressure of each gas if the total pressure is 756 torr?

6-17 How can you tell from its molecular formula that hydrogen gas is the lightest substance known? What is the second lightest substance known?

6-18 If acetylene were to be represented by its empirical formula instead of its molecular formula, C_2H_2, what false conclusion would be reached regarding its density?

6-19 In a mixture of 3.27 g of nitrogen and 3.27 g of hydrogen, what is the partial pressure of each gas if the total pressure is 1 atm?

6-20 Arrange the following gases in order of increasing diffusion rates: CO, CO_2, O_2, He, Ne, H_2, NH_3, CH_4.

6-21 If 10 ml of a certain gas pass through a porous wall in the same length of time as 30 ml of helium under the same conditions, how does the density of the gas compare with that of helium? What is its molecular weight?

6-22 What volume of O_2 is needed to burn 13 liters of propane, C_3H_8? (The products are water and carbon dioxide.)

6-23 If a mixture of 172 ml of H_2 and 344 ml of O_2 is ignited with an electric spark, what volume of which gas will remain unreacted?

REFERENCES

Brown, E. H., "Some Early Thermometers," *Journal of Chemical Education*, **11**(8), 448 (1934). A collection of drawings of early thermometers from the time of Galileo to the time of Fahrenheit.

Day, J. H., *Kinetic Theory of Gases*, Fearon, San Francisco, 1963. Two booklets of programmed instruction.

Drago, R. S., *Prerequisites for College Chemistry*, Harcourt, Brace & World, New York, 1966 (paperbound). Chapter 5, "Gases and the Kinetic Theory," includes a selection of solved problems and self-testing exercises.

Miller, Jr., F., *College Physics*, 2nd ed., Harcourt, Brace & World, New York, 1967. Chapters 13–16 deal with the concepts of heat and temperature and with the thermal behavior of gases.

Ruckstuhl, A., "Thomas Graham's Study of the Diffusion of Gases," *Journal of Chemical Education*, **28**(11), 594 (1951). An account of the experimental work that led to the discovery of the Law of Diffusion Rates.

Stimson, H. F., "Celsius versus Centigrade: The Nomenclature of the Temperature Scale of Science," *Science*, **136**, 254 (1962). The history of the Celsius scale, including the official 1960 definition that makes the adjective "centigrade" inexact.

Wilson, R. E., "Standards of Temperature," *American Scientist*, **41**, 271 (1953). An account of the practical problem of establishing fixed reference points for thermometry.

7

The Liquid and Solid States

7-1 Properties of Liquids and Solids

In the liquid and solid states a substance lacks the compressibility that is characteristic of gases. Most of the space in liquids and solids is occupied by the molecules or ions themselves; consequently, pressures of many atmospheres have very little effect on their volumes. The densities of liquids and solids are of the order of a thousand times as great as the density of a gas under atmospheric pressure. For this reason the densities of liquids and solids are usually expressed in grams per milliliter, while the densities of gases are expressed in grams per liter. All liquids and solids show changes of volume when the temperature changes, but the changes are small compared with those shown by gases. Usually, but not always, an increase in temperature results in an increase in volume. The expansion is sometimes quite uniform, as in the case of mercury (Figure 7-1), which is widely used in thermometers.

Both liquids and gases are *fluids*; that is, they readily flow under the influence of gravity. Thus, either a liquid or a gas will assume the shape

FIGURE 7-1

Uniform expansion of mercury

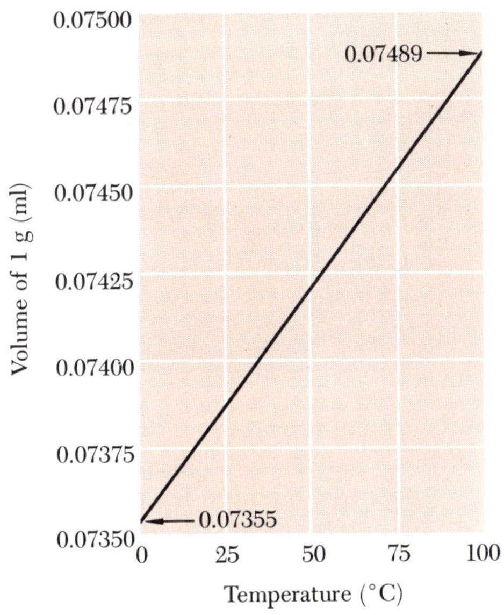

of the container in which it is placed. The volume of a gas is the same as that of its container, but a liquid specimen has its own characteristic volume and occupies only the bottom portion of the container. Consequently, a liquid has a *surface*, which a gas does not have.

A solid has a definite shape and lacks the fluidity of liquids because the molecules or ions have fixed positions—they cannot move over and beyond one another to assume new positions.

The existence of a liquid is sufficient evidence of attractive forces between its molecules, which hold the liquid together; if they did not exist, the substance would be in the gaseous state. In substances that are liquid at room temperature, such as water, and in substances that are easily liquefied by moderate cooling, such as ammonia and sulfur dioxide, strong attractive forces exist that are due to the polar character of the molecules. A molecule in the interior of a liquid is acted upon equally from all directions by the intermolecular attractive forces, which are effective only at short distances. A molecule near the surface, however, is acted upon by attractive forces beneath it that are not balanced from above (Figure 7-2). The net result is an inward pull, called *surface tension*, that causes every liquid surface to become as small as possible. Thus, a free liquid drop, such as an oil drop in water, tends to be spherical because a sphere has a minimum surface area. The spherical shapes of raindrops and teardrops are slightly distorted by the act of falling;

FIGURE 7-2

Directions of forces on a molecule in a liquid

A molecule in the interior of a liquid is attracted equally in all directions; a molecule near the surface is attracted inward.

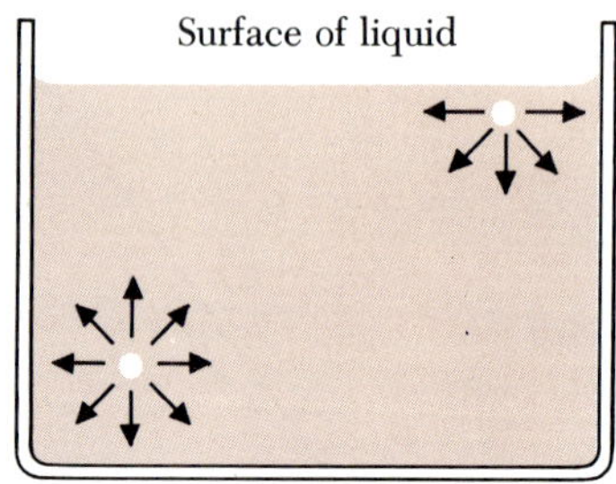

drops at rest on a solid surface, especially if they are composed of a dense liquid such as mercury, are somewhat flattened by the action of gravity.

When a liquid is present in a container of small cross section, such as a buret, a pipet, or a barometer tube, the surface of the liquid is curved and is called a *meniscus*. The meniscus is curved either upward or downward, depending on whether the containing walls are wetted by the liquid (Figure 7-3). *Wetting* is the adherence of a liquid to the surface of a solid; it is most likely to occur when there is some degree of chemical similarity between the liquid and the solid. Water wets glass, but mercury does not.

Water will rise in a glass tube such as a pipet because the wetting of the glass walls exposes a greater surface area of water and the surface

FIGURE 7-3

Capillary action

(a) Rise of a liquid that wets the wall. (b) Depression of a nonwetting liquid.

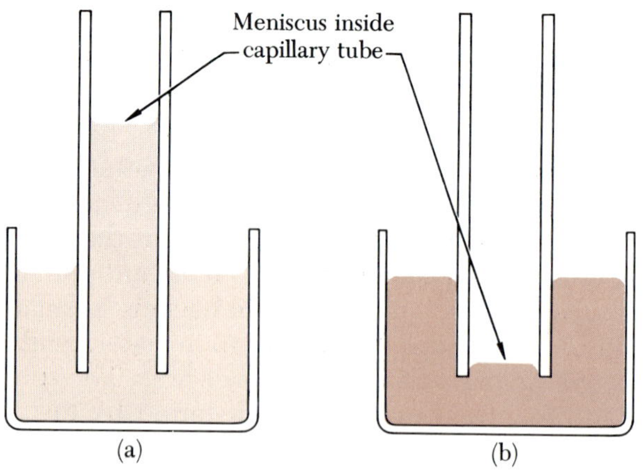

tension of the water acts to reduce this area, thereby causing the entire column of water to move upward. Equilibrium is attained when the forces acting upward balance the downward pull of gravity. The meniscus will be concave (curved inward) due to the wetting action at the glass walls.

The depression of mercury in a glass tube is due to the fact that mercury does not wet glass. The fall of the mercury inside the tube reduces the area of contact between mercury and glass. When the mercury has fallen to the equilibrium position there is a balance between the surface tension pulling downward so as to reduce the amount of surface area exposed by the mercury and the gravitational force that is acting to equalize the level of the mercury inside and outside the tube. With the meniscus curved downward the area of contact between mercury and glass is reduced to a minimum.

Tubes of very small inner diameter are known as capillary tubes, and the rise or fall of a liquid within them, just described, is called *capillary action.*

Although, by definition, fluids can flow, each individual fluid has a characteristic amount of internal friction or resistance to flow that is known as *viscosity.* The viscosities of any two liquids can be compared by measuring the time required for a specified volume of each to be forced through a tube of small diameter by a fixed pressure. Gravity causes a liquid to flow downward at a measurable rate through a narrow tube held vertically. Therefore, the simplest apparatus used to measure the viscosities of liquids consists of a pipet and a stopwatch. Viscosity varies greatly from materials having long pour times, such as heavy lubricating oils, to liquids of low viscosity, such as ether or benzene. There is roughly an inverse relation between viscosity and volatility. Viscosity is one of the frequently measured properties of blood and other animal fluids. It is also an important factor in the design of chemical engineering equipment where the cost of pumping depends greatly on the viscosity of the fluid.

The quantity of heat required to raise the temperature of 1 g of a substance 1°C is called its *specific heat.* In comparison with other liquids, water has an unusually large specific heat: 1 cal per g per °C. In nature this property has a moderating effect on the climate in regions near large bodies of water. Use is made of this property of water in such applications as hot-water heating systems and automobile cooling systems.

7-2 Volatility and Vapor Pressure

One of the familiar properties of liquids is their ability to *evaporate*, that is, to change from the liquid to the gaseous state. This ability of a liquid to evaporate is referred to as *volatility.* Two questions that concern us are why some liquids are more volatile than others and why the volatility of a liquid increases with rising temperature.

The kinetic theory, which we have found to be highly useful in explaining the behavior of gases, is also useful in understanding the properties of liquids. The molecules of a liquid, like those of a gas, move in a random fashion. The average kinetic energy of the molecules is constant at any particular temperature, but individual molecules do not all have the same amount of kinetic energy. The molecules that happen to be in the surface of a liquid at a particular moment are the ones that can evaporate, provided they have the necessary kinetic energy to overcome the attractive forces exerted upon them by their neighbors. Only some of the molecules have this much kinetic energy. Thus, evaporation involves the escape of those molecules having the greatest velocities.

Since continuing evaporation involves the continuing selective loss of high-energy molecules, the average energy of the molecules remaining in the liquid continually decreases. In other words, the temperature of the liquid drops—unless heat is supplied from an outside source. The regulation of the body temperature of many mammals is based in part on the cooling produced by evaporation of the water from the aqueous solution secreted by the sweat glands.

All liquids show an increase in volatility with rising temperature because at higher temperatures greater numbers of molecules have the necessary kinetic energy to escape from the surface. The rate of evaporation also can be increased by making the surface area greater, since evaporation occurs only at the surface.

Volatility is a property that varies from substance to substance because the attractive forces between molecules of some liquids are greater than between molecules of other liquids. The most polar substances are the least volatile. Among the nonpolar substances, the most volatile ones are those with the smallest molecular weights, because their molecules contain the fewest electrons and the van der Waals forces are therefore the weakest.

If a liquid is placed in a closed container, evaporation will reduce the volume of the liquid to a certain point but not beyond this point. This is best understood in terms of the kinetic theory. As the liquid evaporates, the molecular population of the vapor phase increases. Some of the rapidly moving molecules of vapor collide with the walls of the container and others collide with the surface of the liquid. Most of those that collide with the liquid will be captured by, and again become a part of, the liquid. The rate at which molecules return to the liquid is proportional to the molecular population of the vapor phase. The volume of liquid has been reduced to its final value when the rate at which molecules are returning to the surface of the liquid is equal to the rate at which molecules are evaporating from the surface—in other words, when a state of *dynamic equilibrium* prevails (Figure 7-4). The relative amounts of liquid and vapor remain unchanged thereafter as long as the temperature is kept constant. At equilibrium the concentration of the vapor is constant, that is, the number of molecules in any particular portion of the space occupied by the vapor remains unchanged. Therefore, the molecules of vapor exert a definite pressure that depends on their number and the temperature. The pressure that a vapor exerts when it is in equilibrium with its liquid at a particular temperature is called the *vapor pressure* of the liquid at that temperature.

The vapor pressure of a liquid is a measure of the escaping tendency of its molecules. Vapor pressures increase with rising temperatures, but the vapor pressures of different liquids at the same temperature differ greatly (Figure 7-5). For example, the vapor pressure of water at 25°C is only 23.6 torr but the vapor pressure of ether at the same temperature is 537 torr.

Certain solids, such as the moth repellents naphthalene and dichlorobenzene, give off odors that are easily detected. The sensation of odor is an effect of the molecules of the substance upon the olfactory nerve endings. The existence of an odor is evidence that the substance has vaporized. Only certain solids vaporize to a significant extent. Like a liquid, a solid has a vapor pressure, which is the pressure exerted by the vapor that is in equilibrium with the solid. Such an equilibrium can exist only within a closed space.

The vapor pressure of a solid increases with rising temperature in much the same manner as the vapor pressure of a liquid. For example,

FIGURE 7-4

Dynamic equilibrium

Evaporation in a closed vessel leads to a state of dynamic equilibrium.

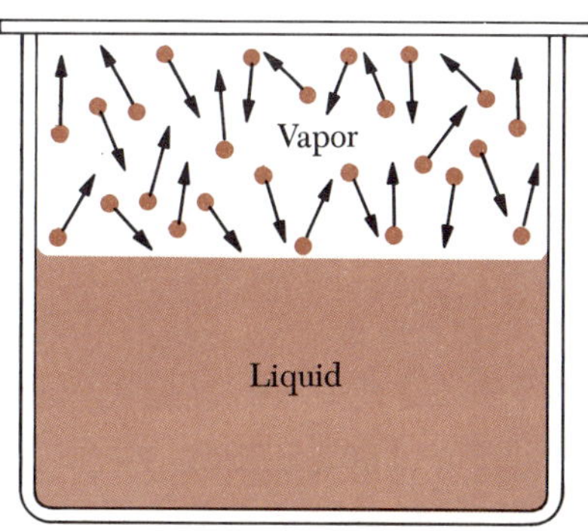

FIGURE 7-5

Change of vapor pressure with temperature

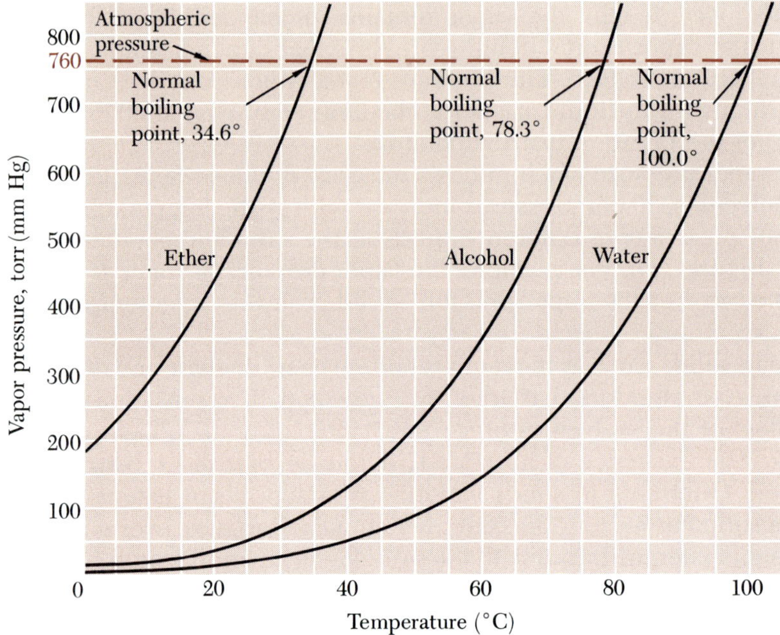

the vapor pressure of iodine is 0.202 torr at 0°C and 45.5 torr at 100°C. Solids that have measurable vapor pressures at room temperature consist of molecules that are held together in the crystal by rather feeble forces. In an ionic crystal the attractive forces are greater and the vapor pressure is low. Many solids exhibit no measurable vapor pressure, but it is a theoretical convenience to assume that every solid has a vapor pressure that increases with rising temperature.

The passage of a substance directly from the solid state to the gaseous state is called *sublimation*, and the vapor pressure of a solid is sometimes referred to as its sublimation pressure. The disappearance of snow on days when the temperature is below 0°C is due to sublimation. Mothballs slowly disappear on standing in the open because of sublimation. Most familiar is the sublimation of solid carbon dioxide, which is well known by the trade name Dry Ice.

7-3 Boiling Point and Critical Temperature

Ordinary evaporation of a liquid occurs only at its surface, but when a liquid is heated to a sufficiently high temperature, bubbles of vapor form beneath the surface. The formation of these bubbles inside the liquid and their rise to the surface is called *boiling* or *ebullition*. The temperature at which the boiling occurs is called the *boiling point*, but this temperature varies with the atmospheric pressure.

In order for a bubble of vapor to form and continue to exist beneath the surface of a liquid, the pressure exerted outward against the surrounding liquid by the vapor molecules must be sufficient to offset the crushing effect of the atmosphere overlying the liquid. Thus, boiling occurs at the temperature at which the vapor pressure of the liquid is equal to the external pressure acting downward upon the surface of the liquid. The boiling point of a liquid is not always the same temperature, because barometric pressure varies from time to time and from place to place. The *normal boiling point* of a liquid is the temperature at which its vapor pressure is 760 torr (1 atm).

We can cause a liquid to boil either above or below its normal boiling point by changing the pressure above the surface of the liquid. Thus, the vapor pressure curve for a liquid is also a boiling point curve; from the curve we can read the temperature at which the liquid will boil under any specified pressure (Figure 7-5). A liquid can be heated to a temperature above its normal boiling point without boiling if the external pressure is greater than 1 atm. As the vapor pressure curve indicates, higher external pressures necessitate higher temperatures to produce boiling. This principle receives practical application in pressure cookers.

The vapor pressure curve for any liquid terminates at a certain temperature, called the *critical temperature*, above which it is impossible for that substance to exist in the liquid state, however great the pressure applied to it. For example, the critical temperature of water is 374°C. At this temperature the vapor pressure of water is 165,467 torr.

The vapor pressures, molecular weights, boiling points, and critical temperatures of five common liquids are compared in Table 7-1. If such a comparison is limited to compounds that are not highly polar, the effect of molecular weight becomes discernible: the greater the molecular weight the less the vapor pressure at a given temperature. Consequently, the greater the molecular weight the higher the boiling point (Table 7-2) and the higher the critical temperature. This is not a gravitational effect; it is a consequence of the increased number of electrons and protons in the more massive molecules, which results in stronger intermolecular attraction.

In Table 7-3 water is compared with other, similar compounds. If water were omitted from the table, it would be true that the greater the molecular weight the higher the freezing point and the higher the boiling point. Actually, water has the highest freezing point and the highest boiling point although it has the smallest molecular weight. This and most of the other unusual properties of water are due to the highly polar character of the water molecules; this causes them to attract one another much more strongly than if they were less polar.

TABLE 7-1

Properties of selected liquids

Liquid	*Molecular weight*	*Vapor pressure at 20°C* (torr)	*Boiling point* (°K)	*Critical temperature* (°K)
Water	18	18	373	647
Alcohol	46	44	352	576
Acetone	58	185	330	508
Ether	74	442	308	467
Carbon tetrachloride	154	91	350	556

TABLE 7-2

Effect of molecular weight on freezing point and boiling point

Formula	*Molecular weight*	*Freezing point* (°C)	*Boiling point* (°C)
CCl_4	154	−23	77
$CBrCl_3$	198	−21	104
CBr_2Cl_2	243	22	135
CBr_3Cl	287	55	160
CBr_4	332	90	190

TABLE 7-3

Freezing points and boiling points of similar compounds

Formula	*Formula weight*	*Freezing point* (°C)	*Boiling point* (°C)
H_2O	18	0	100
H_2S	34	−86	−61
H_2Se	81	−65	−41
H_2Te	130	−51	−2

7-4 Liquefaction of Gases

Systematic experiments on the liquefaction of gases were initiated by Michael Faraday early in the nineteenth century. At that time the lowest temperatures attainable were those obtained with mixtures of ice and salt. Later, with mixtures of solid carbon dioxide and ether, it became possible to obtain temperatures as low as −110°C. Faraday liquefied a number of substances previously known only in the gaseous state, but he was unable to liquefy hydrogen, oxygen, nitrogen, or carbon monoxide. Later attempts to liquefy these gases by using pressures exceeding 3000 atm were likewise unsuccessful. The opinion prevailed that certain gases could not be liquefied under any circumstances; these became known as "permanent gases."

We now know that any gas can be liquefied if the temperature is reduced sufficiently. The temperature to which a gas must be cooled in order for it to be liquefied is called the *critical temperature* of the substance. We have already stated (Section 7-3) that above its critical temperature a substance cannot exist in the liquid state. The amount of pressure required to liquefy a gas at its critical temperature is called its *critical pressure*. Above its critical temperature a gas cannot be liquefied by any amount of pressure, however great.

The amount of pressure required to liquefy a gas is the same as the vapor pressure of the liquid at the same temperature. Therefore, at the normal boiling point of a substance the pressure that must be applied to convert the gaseous form to the liquid form is 760 torr. The critical pressure, the pressure required to liquefy a gas at its critical temperature, is the same as the vapor pressure of the liquid at the critical temperature. For example, the critical pressure of water is 165,467 torr. (This is the value given in Section 7-3 for the vapor pressure of water at its critical temperature.)

A *vapor* is the gaseous form of a substance below its critical temperature. The term is never used to refer to a gas above its critical temperature.

The ability of a gas to become a liquid is due to the attractive forces between its molecules. The magnitude of these attractive forces varies from one substance to another according to differences in molecular structure. Consequently, each substance has its characteristic critical temperature, just as it has its characteristic boiling point, both being related to vapor pressure. The intermolecular forces of attraction are not dependent on temperature, but the kinetic energy of the molecules is directly proportional to absolute temperature. Liquefaction results when a gas is cooled to a temperature sufficiently low that the kinetic energy of its molecules is no longer adequate to overcome the forces of attraction.

When a gas is subjected to high pressure its molecules are crowded together so that the intermolecular forces of attraction are more effective. Therefore, a gas that is under high pressure can be liquefied with less cooling than would otherwise be required. Above the critical temperature the kinetic energy of the molecules is too great to be overcome by the intermolecular forces of attraction, even when the molecules are crowded close together by the use of high pressure. Since the molecules of all gases have the same average kinetic energy at a given temperature, the critical temperature values of different substances depend on the strengths of the attractive forces and are, in a sense, a measure of these strengths. Thus, the attractive forces between water molecules are sufficiently strong that water exists as a liquid at any temperature below 374°C if sufficient pressure is applied; but the forces between helium molecules are so weak that liquid helium cannot exist at any temperature above 5.25°K (−267.9°C).

7-5 Distillation

One of the operations most commonly associated with chemistry is *distillation*, a process that involves the vaporization of a liquid by heating and the condensation of its vapor by cooling. A typical distillation apparatus for laboratory use is shown in Figure 7-6. A thermometer placed

FIGURE 7-6
Distillation

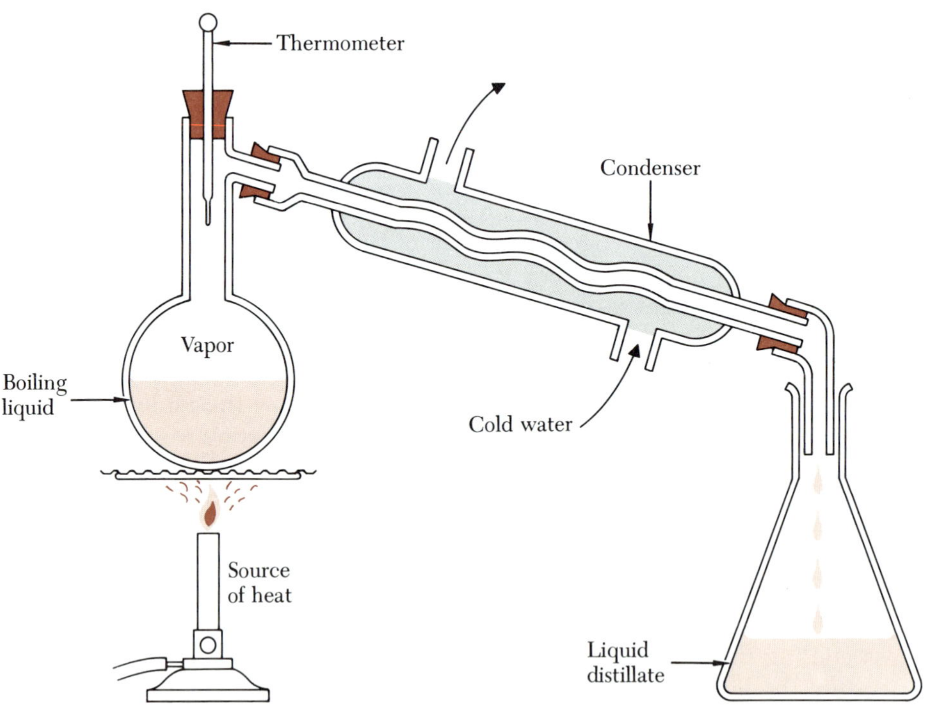

in the neck of the distilling flask registers the temperature of the vapor, i.e., the boiling point of the liquid. The vapor passes from the flask into the inner tube of the condenser where it is cooled and condensed to the liquid state at the temperature established by cold water circulating through the condenser jacket. The liquid distillate flows out of the condenser into the receiving vessel.

Simple distillation is effective in separating a pure liquid from impurities that do not vaporize at the boiling point of the liquid. Water is often purified by distillation. The nonvolatile mineral matter, which is objectionable in water that is to be used in steam irons, in storage batteries, and in certain types of laboratory work, is left behind in the distilling flask and is not present in distilled water.

If the impurities in a liquid are themselves volatile, purification by distillation requires complicated apparatus. The vapor produced by boiling a mixture of two volatile liquids usually has a composition that is different from that of the liquid mixture, but the distillate from such a mixture is never a pure liquid unless the process known as *fractional distillation* is employed. If simple distillation apparatus is employed for the distillation of a liquid mixture, it will be necessary to condense the vapor in successive portions and to subject each of these fractions to repeated redistillation. By this laborious process it is often possible to obtain each component in an essentially pure state. In practice, fractional distillation is accomplished by the use of *fractionating columns* in which there is constant contact at all levels between ascending vapor and descending liquid. Modern distillation plants are capable of separating the components of complex liquid mixtures, even those in which the components have boiling points that are close together. Such plants are used in the refining of petroleum and in separating the components of liquid air.

7-6 Heat of Vaporization

We have seen that evaporation causes the temperature of a liquid to drop unless heat is supplied from an outside source. The quantity of heat that must be supplied to vaporize a stated quantity (such as 1 g or 1 mole) of a liquid without change of temperature is known as the *heat of vaporization* and is characteristic of the liquid. The quantity of heat required to vaporize 1 mole is called the *molar heat of vaporization*. For water, a highly polar substance, the heat of vaporization is 539 cal per g (or 9700 cal per mole) at 100°C, the normal boiling point. For carbon tetrachloride, a nonpolar compound, the heat of vaporization is only 46.4 cal per g (or 7140 cal per mole) at 77°C, the boiling point.

The heat of vaporization of a liquid varies with the temperature at which the vaporization occurs; it decreases with rising temperature until it becomes zero at the critical temperature. The heat of vaporization of water decreases from 595 cal per g at 0°C to 446 cal per g at 220°C (see Table 7-4 and Figure 7-7); it becomes zero at the critical temperature, 374°C.

According to the empirical rule proposed by Frederick Trouton in 1884, dividing the molar heat of vaporization of a liquid (in calories) by the normal boiling point (in °K) gives approximately 22 cal per mole per deg. The equation for *Trouton's Rule* is, therefore,

$$\frac{\text{Molar heat of vaporization (cal/mole)}}{\text{Normal boiling point (°K)}} = 22 \text{ cal/mole per °K}$$

Trouton's Rule is followed fairly well by nonpolar liquids but not so well by polar compounds. Along with its other unusual properties, water does not conform to this rule: its molar heat of vaporization divided by its absolute boiling point is well in excess of 22 cal per mole per deg:

$$\frac{540 \text{ cal/g} \times 18 \text{ g/mole}}{373\text{°K}} = 26 \text{ cal/mole per °K}$$

When a gas is condensed to the liquid state, a quantity of heat known as the *heat of condensation* is released; this quantity of heat is equal to

FIGURE 7-7

Heat of vaporization of water

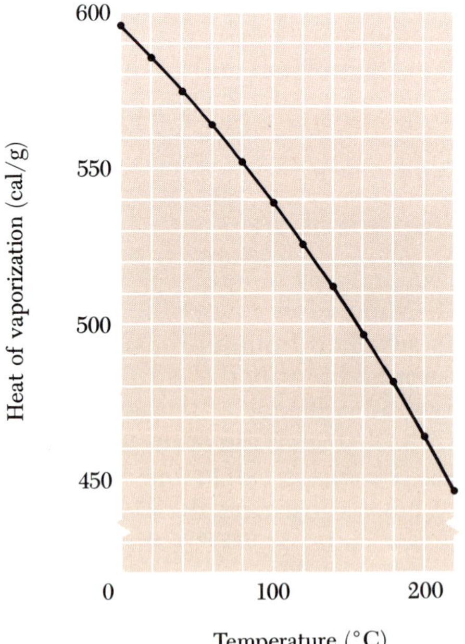

TABLE 7-4

Heat of vaporization of water

Temperature (°C)	*Heat of vaporization* (cal/g)
0	595.4
20	584.9
40	574.2
60	562.8
80	551.1
100	538.7
120	525.6
140	511.5
160	496.5
180	480.6
200	463.8
220	446.0

the heat of vaporization at the same temperature. Thus, steam produces more serious burns than does boiling water because of its large heat of condensation.

The operation of mechanical refrigerators is based on the heat of vaporization of liquids. Heat energy is removed from the interior of the refrigerator by evaporation of a liquid, and the vapor is transported to the outside of the refrigerator, where it is liquefied by compression and cooling. A satisfactory refrigerant must be a substance that is readily condensed from the gaseous state to the liquid state at the operating temperature; it must also have a high heat of vaporization to produce an adequate cooling effect inside the refrigerator. One of the first refrigerants used in mechanical domestic refrigerators was ammonia, which had been used previously as the refrigerant in commercial ice plants. Sulfur dioxide was also used extensively in refrigeration until the development, since 1930, of the nontoxic Freons, which have also found wide use in aerosol sprays and other devices. The name Freon is applied to several relatively simple organic compounds containing fluorine, such as CCl_2F_2.

7-7 Collection of Gases over Water

A commonly used procedure for collecting a gas and measuring its volume involves the physical displacement of a liquid (Figure 7-8). A bottle or cylinder is filled with the liquid and inverted in a trough that also contains the liquid. The gas that is to be collected is led from a tank or a generator through a tube beneath the surface of the liquid in the trough and upward into the inverted collection vessel. In this way the gas is prevented from escaping into or coming in contact with the air. Before the volume of collected gas is measured, the cylinder is raised or lowered until the surface of the liquid inside the inverted collection vessel is at the same level as the surface of the liquid in the trough. In this manner the pressure exerted by the collected gas is equalized with the pressure of the atmosphere, which is ascertained by reading the laboratory barometer.

Water is the liquid over which gases are usually collected, unless the gas is highly soluble in water. A sample of gas collected over water is always mixed with water vapor. It is usually safe to assume that the gas space above the water in the inverted collection vessel is saturated with water vapor. In other words, it is assumed that the water vapor within the confined space is in equilibrium with the liquid water, so that the partial pressure of the water vapor is identical with the vapor pressure of water at the prevailing temperature. The pressure exerted by the collected gas is then obtained by subtracting the vapor pressure of water from the atmospheric pressure, as indicated by Dalton's Law of Partial Pressures. The atmospheric pressure is obtained from the laboratory barometer; the vapor pressure of water at various temperatures is listed in tables (for example, in the Appendix of this book).

FIGURE 7-8

A gas collected over water

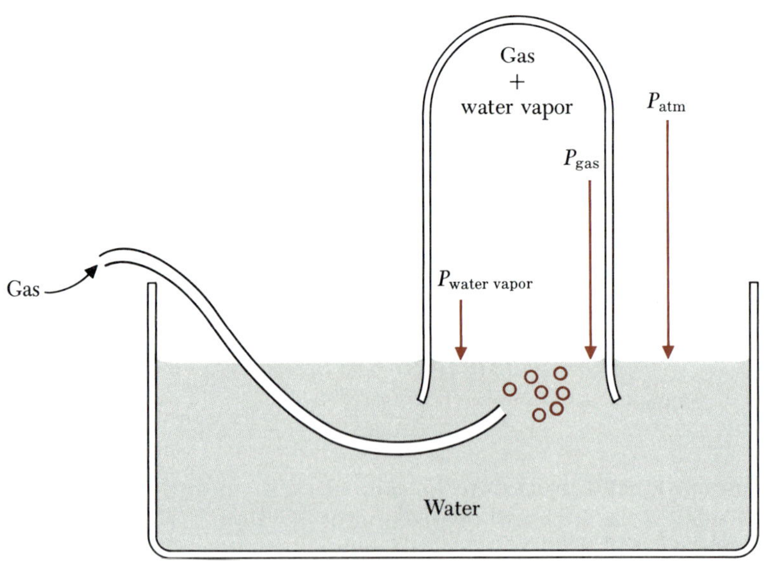

EXAMPLE 1

A gas is collected over water at 23°C. The barometric pressure is 753 torr. What is the partial pressure of the gas?

The vapor pressure of water at 23°C is 21 torr (see the Appendix). The pressure exerted by the gas is 753 torr minus the vapor pressure of water at 23°C; i.e., 753 torr − 21 torr = 732 torr.

EXAMPLE 2

A gas is collected over water at a pressure of 743 torr and a temperature of 19°C. The measured volume is 247 ml. What will be the volume of the dry gas at standard conditions?

The pressure exerted by the gas itself is 743 torr minus the vapor pressure of water at 19°C; i.e., 743 torr − 16 torr = 727 torr. At standard conditions the pressure is greater than 727 torr, so the pressure factor by which the measured volume should be multiplied is a fraction less than 1, i.e., $\frac{727}{760}$. The temperature at STP is less than 19°C, so the correct temperature factor is also a fraction less than 1, i.e., $\frac{273}{292}$. Therefore,

$$\begin{aligned} V_{\mathrm{STP}} &= 247\ \mathrm{ml} \times \frac{727}{760} \times \frac{273}{292} \\ &= 247\ \mathrm{ml} \times 0.957 \times 0.935 \\ &= 247\ \mathrm{ml} \times 0.895 \\ &= 221\ \mathrm{ml} \end{aligned}$$

7-8 Freezing and Melting

All liquids become solids when they are cooled to sufficiently low temperatures. Some substances, such as helium, do not solidify until they have almost reached absolute zero. Other substances, for example, carbon, remain solids at such high temperatures that little is known about them in either the liquid or gaseous state.

As a liquid is cooled its molecules lose kinetic energy, but the attractive forces between molecules are independent of temperature. At a sufficiently low temperature, which depends on the particular substance, the molecules cease the random motion characteristic of molecules in the liquid and gaseous states and fit themselves into an orderly three-dimensional arrangement. This is what happens when a liquid *freezes* to become a crystalline solid.

When the temperature of a cooling liquid is observed over a period of time and the readings are plotted on a graph, a curve like that in Figure 7-9 is obtained, provided that the liquid is a pure substance and that heat is removed from it at a uniform rate. As each molecule takes up its position to become part of the crystal structure, its potential energy drops. This decrease of potential energy is the source of the heat that is liberated during freezing. During the crystallization process, the average kinetic energy of the molecules in the crystalline phase is the same as in the liquid phase. The kinetic activity of the molecules in the solid phase is restricted to vibration: unlike the molecules in the liquid phase, they are not free to move about. The horizontal portion of the cooling curve indicates the temperature at which freezing occurs, which is called the *freezing point.* The temperature remains constant during this time because both the liquid and solid phases are present and the average kinetic energy of the molecules is the same in one phase as in the other. The temperature does not undergo any further reduction until crystallization is complete. The heat liberated during freezing is called the *heat of solidification* or the *heat of crystallization.* Further removal of heat after crystallization is complete results in a reduction of the temperature of the solid.

When a solid is heated to a sufficiently high temperature, it becomes liquid. This process is called *melting* or *fusion.* The *melting point* of a substance is defined as the temperature at which the solid and liquid phases can exist in equilibrium. The melting point of the pure solid is the same as the freezing point of the pure liquid. Melting points are frequently used in the identification of substances since every pure substance melts and freezes at a characteristic temperature.

The heat absorbed during melting is called the *heat of fusion.* It is, of course, the same in quantity as the heat of crystallization. The quantity of heat required to melt 1 mole of a solid is called the *molar heat of fusion.* The heat of fusion of ice is 1440 cal per mole (80 cal per g) and that of sodium chloride is 7250 cal per mole. The great difference between these two molar heats of fusion indicates that the attractive forces between the ions in crystalline NaCl are considerably greater than those between the molecules in crystalline H_2O.

The melting point of a substance is affected slightly by pressure. For example, ice skating is possible because the high pressure of the blade of the skate causes the ice to melt. This produces a thin film of water that provides lubrication, allowing the skate to move with little friction. After the skate has moved on and the pressure has been released, the water immediately freezes.

Whether the melting point of a substance is raised or lowered by the application of pressure depends on whether the substance expands or contracts when it freezes. Most liquids contract upon freezing; water is an exception. (The unusual properties of water are discussed and explained in Section 8-7.) Pressure raises the melting points of most solids, but it lowers the melting point of ice. The density of liquid water at 0°C, the freezing point, is 0.9998 g per ml; but the density of ice at the same temperature is only 0.917 g per ml. In relation to natural phenomena it is important that water expands on freezing and that ice floats on water. If ice were heavier than water, it would sink to the bottom and expose the overlying liquid to further cooling and freezing so that rivers and lakes would eventually become completely frozen from bottom to top. The weathering of rocks is partly due to the pressure exerted by water freezing in crevices.

Most cooling curves are complicated in the manner indicated by the broken line in Figure 7-9. Instead of crystallizing when the freezing point is first reached, the liquid may *supercool* (as indicated by the broken line). This happens when the molecules fail to line up properly for crystallization to begin. Removal of heat causes the temperature to fall below the freezing point, and the molecules continue to move about until by chance they fall into the correct pattern. When a "nucleus" of sufficient size has been built up, additional molecules rapidly attach themselves to it, and the crystal "grows." A large number of molecules are likely to crystallize out simultaneously once the process has been initiated in a supercooled liquid. The heat thus liberated causes the whole system to warm up to the freezing point, and thereafter the temperature remains fixed until freezing is complete.

It is instructive to consider the vapor pressures of the solid and liquid forms of a substance in the neighborhood of the melting point. In Table 7-5 the vapor pressure of ice and of supercooled liquid water are given

FIGURE 7-9

Cooling curve for a pure liquid

The broken line represents supercooling.

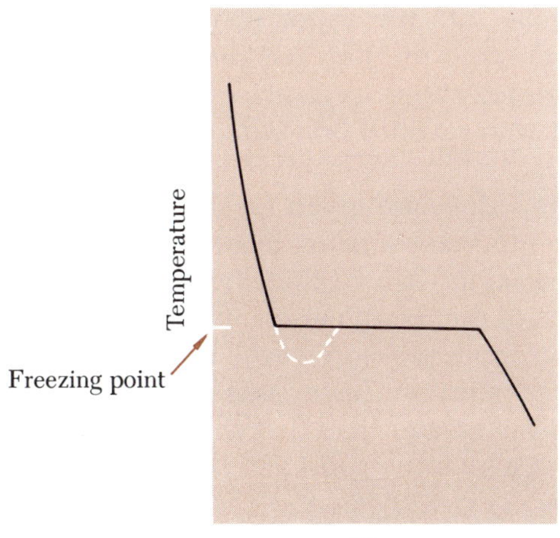

TABLE 7-5

Vapor pressure of ice and water

	Vapor pressure (torr)		
Temperature (°C)	*Ice*	*Liquid water*	*Difference*
−1.0	4.217	4.258	0.041
−0.8	4.287	4.320	0.033
−0.6	4.359	4.385	0.026
−0.4	4.431	4.448	0.017
−0.2	4.504	4.513	0.009
0.0	4.579	4.579	0
+0.2		4.647	
+0.4		4.715	
+0.6		4.785	
+0.8		4.855	
+1.0		4.926	

at intervals of 0.2° from −1°C to 0°C, along with the vapor pressure of liquid water from 0°C to +1°C. Note that the vapor pressure of ice is the same as that of liquid water at 0°C. *The vapor pressures of the solid and liquid phases of a substance are identical at the melting point.*

The vapor pressure of a solid or a liquid is a measure of its tendency to escape into the gaseous state. Below 0°C liquid water and ice cannot exist in equilibrium because the supercooled water has a higher vapor pressure than ice at these temperatures. If the supercooled liquid were in a beaker within a closed vessel that also contained ice cubes, the liquid would escape into the vapor phase, from where it would condense in solid form onto the ice, and the beaker would eventually become empty. If the ice cubes were placed in the beaker with the supercooled water, the water would immediately freeze there and the result would be the same.

In Table 7-5 no values are given for the vapor pressure of ice above 0°C because this information is not available. If we attempt to heat ice above 0°C, the ice melts. Above the true melting point of a substance, the vapor pressure of the (imaginary) superheated solid would certainly be greater than the vapor pressure of the liquid at the same temperature. Below the melting point we have seen that the vapor pressure of the supercooled liquid is greater than that of the solid. The melting point is the temperature at which the solid and the liquid have the same vapor pressure.

7-9 Internal Structure of Crystals

The first milestone on the road to understanding the nature of the solid state was the realization that most solids are composed of crystals. A *crystal* is a homogeneous piece of solid matter that has the natural shape of a polyhedron. This means that it is bounded by plane surfaces, or faces, symmetrically arranged. *Crystallography* is the science of describing precisely the external forms of crystals and the arrangement of the atoms, ions, or molecules of which they are composed. Crystallography may be said to have begun in 1669, when Nicolaus Steno noted that different crystals of quartz had identical angles between corresponding faces.

Sometimes each particle of a solid substance is a single crystal, but quite commonly, especially with metals, a single piece of the solid is an aggregate of crystals. Often a material is a collection of small solid particles, such as sand, which takes on the shape of its container just as a liquid does, but each particle may still be a single crystal. Although a crystal may have had its corners and edges rounded off sufficiently to appear noncrystalline to the eye, its crystalline structure can be observed with the aid of a microscope.

In 1784 the French mineralogist René Just Haüy was examining a large calcite crystal when he accidentally dropped it and noted that it broke along definite planes, producing small pieces having the shape of the original crystal. He surmised that the external regularity of a crystal is due to a regularity of its internal structure. Haüy's brilliant deduction, made before Dalton's atomic theory was formulated, brought increasing attention to crystallography during the early part of the nineteenth century.

Nineteenth century crystallographers were able to obtain a surprising amount of information about the arrangement of particles inside crystalline solids, but their only methods for doing so were indirect, based on logical deductions from external appearances. Much more exact and detailed information can be obtained by X-ray diffraction.

By mathematical analysis of X-ray diffraction patterns, X-ray crystallographers can "map" the relative positions occupied by the particles inside crystals and can determine whether the particles are atoms, ions, or molecules. If we let a point represent the position of each particle, the pattern of points that represents the arrangement of the particles in a crystal is known as a *space lattice*. In 1848 a French crystallographer, Auguste Bravais, showed that all possible space lattices can be assigned to one of 14 types, now known as the 14 Bravais lattices.

The space lattice is thought of as extending throughout the entire crystal, but when comparing one crystal structure with another it is necessary to consider only enough of it to represent the arrangement. This smallest part of a space lattice that is the pattern for describing the whole lattice is called a *unit cell.* The unit cell of sodium chloride is shown in Figure 7-10, where the colored spheres represent the position of the Na^+ ions and the gray spheres indicate the position of the Cl^- ions. The Na^+ and Cl^- ions have quite different sizes, as represented in the model shown in Figure 7-11.

7-10 Chemical Bonding in the Solid State

In structural chemistry, solids are classified according to the nature of the particles that compose them and the type of bonding that exists be-

FIGURE 7-10

Unit cell of sodium chloride

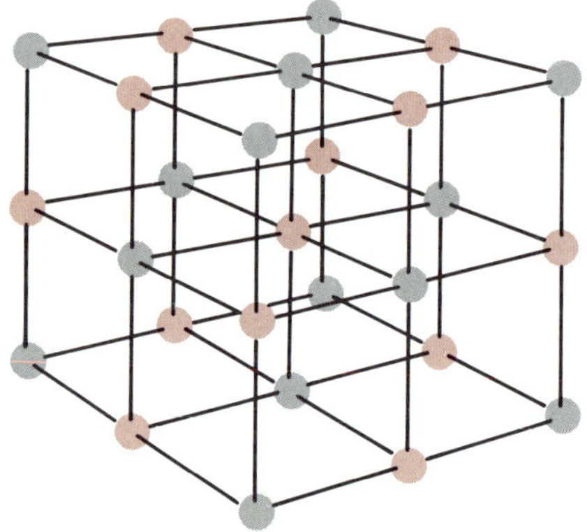

FIGURE 7-11

Structure of NaCl

The large spheres represent Cl^- ions; the small spheres, Na^+ ions.

tween the particles. From this point of view there are four types of solids: ionic, molecular, network, and metallic.

Ionic solids melt, when heated to sufficiently high temperatures, to form liquids that are good conductors of electricity because they consist of ions. The crystal lattice of an ionic solid, such as Na^+Cl^-, consists of a regular array of positive and negative ions. The relative sizes of the ions, and their relative numbers as expressed in the chemical formula of the substance, determine the arrangement of the ions in an ionic crystal. The forces that hold an ionic crystal together are the strong electrostatic forces that operate between positive and negative charges. For this reason ionic solids are usually rather hard and have high melting points (and still higher boiling points). They are soluble only in polar solvents, such as water. The solutions are good conductors of electric current because they contain ions.

Molecular solids are crystalline substances consisting of molecules. The atoms within each molecule are held together by covalent bonds, but the forces holding the crystal together are the much weaker forces of intermolecular attraction. If the molecules are nonpolar, as in the case of sulfur, S_8, or carbon tetrachloride, CCl_4, or carbon dioxide, CO_2, the only attractive forces between molecules are van der Waals forces. This type of intermolecular attraction is so weak that crystals of nonpolar substances are soft and have low melting points. In fact, many nonpolar substances are liquids or even gases at ordinary temperatures; many of those that are solid have substantial vapor pressures unless they have high molecular weights, as in the case of sulfur.

The attractive forces operating between polar molecules are appreciably stronger than those operating between nonpolar molecules, because the negative end of one molecule attracts the positive end of another. In general, substances consisting of polar molecules form harder crystals than nonpolar substances, have higher melting and boiling points, and are less volatile. Thus, their properties are intermediate between those of the ionic substances and those of the nonpolar molecular substances. Examples include water, ammonia, sulfuric acid, and the alcohols.

The *network solids* consist of atoms united by covalent bonds into a network extending throughout the crystal. For example, in diamond (Figure 7-12), each carbon atom is located at the center of an imaginary tetrahedron and is covalently bonded to the carbon atom at each of the four corners. Each carbon atom is therefore both at the center of one tetrahedron and at a corner that is common to four other tetrahedra. Thus, every carbon atom in the interior of a diamond is joined by single covalent bonds to four other carbon atoms. Diamond is a hard substance with a high melting point because the bonds are quite strong, with all the bonds of the same length and strength in an interlocking arrangement. The C—C distance is 1.54 Å.

FIGURE 7-12

Arrangement of carbon atoms in diamond

Each atom is bonded to four other atoms, all equally distant.

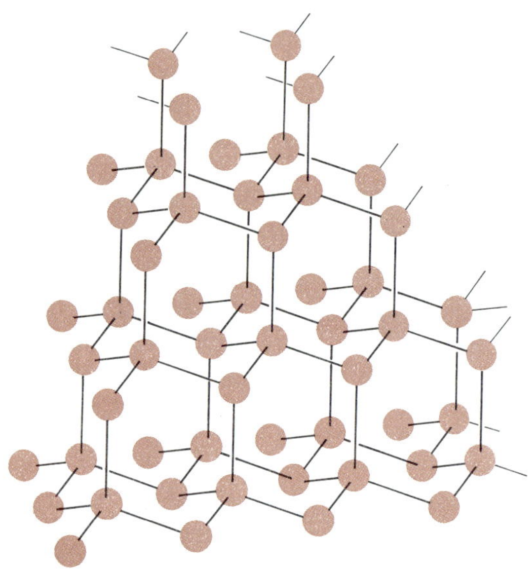

Network solids are also known as *covalent solids* or *atomic crystals*. Any solid of this type is very hard, has a high melting point, and is a nonconductor of electricity. In order for a crystal of this type to be broken, melted, or vaporized, a great number of strong bonds must be broken. Such substances are insoluble in all solvents except those that react chemically with them. Besides diamond, examples include silicon carbide, SiC (the industrial abrasive known as Carborundum), and aluminum nitride, AlN.

In a *metallic solid* the lattice points are occupied by positive ions that are held together by more or less mobile electrons. Most metal atoms in the isolated state have only one, two, or three electrons in the outermost shell, but in a metallic crystal each atom is surrounded usually by either 8 or 12 nearest neighbors. Thus, there are not enough electrons for each atom to be covalently bonded to each of its neighbors as are the carbon atoms in diamonds. Instead, according to the best evidence available, the outer-shell electrons of all the atoms are free to pass from one atom to another so that they are shared by many atoms. This special type of modified covalent bonding is known as *metallic bonding*.

The characteristic properties of metals include high conductivity for heat and electricity and the luster they have in both the solid and liquid states. All these properties are attributed to the mobility of the electrons. Most metals have high melting and boiling points, indicating that the bonding between atoms is strong. In general, they can be easily stretched or pressed into new shapes; apparently, the mobility of the electrons makes it possible for the bonds to be readily broken and replaced by new ones.

7-11 Specific Heats of Elementary Solids

In 1819 two French chemists, Pierre Dulong and Alexis Petit, discovered that the specific heats of elements in the solid state are inversely proportional to their atomic weights. In Table 7-6, where present-day values of atomic weights and specific heats have been rounded off to three digits and their products to two digits, the molar heat capacities of the solid elements listed average 6.2 cal per mole per °C. This is the average for most solid elements. This relationship between the atomic weight and specific heat of a solid element is known as the *Law of Dulong and Petit.* It is expressed in the following equation:

$$\underset{(\text{cal/g-°C})}{\text{Specific heat}} \times \underset{(\text{g/mole})}{\text{atomic weight}} = 6.2 \text{ cal/mole-°C}$$

Historically, this law has been useful for determining approximate values for atomic weights, because merely by determining the specific heat of a solid element one can estimate its atomic weight.

EXAMPLE

The specific heat of iron is determined to be 0.108 cal/g-°C. Calculate an approximate value for its atomic weight.

$$\text{Approximate atomic weight} = \frac{6.2 \text{ cal/mole-°C}}{0.108 \text{ cal/g-°C}} = 57 \text{ g/mole}$$

To appreciate the value of the Law of Dulong and Petit, let us consider the fact that the weight of silver combining with 15.9994 g of oxygen (1 mole of oxygen atoms) is 215.736 g. What is the atomic weight of silver? If the formula for silver oxide is AgO, the answer is 215.736 amu; if the formula for silver oxide is Ag_2O, the answer is 215.736 amu ÷ 2 = 107.868 amu; and so on. Until the approximate atomic weight of silver is known, the formula for silver oxide cannot be known. However, the specific heat of silver is 0.0565 cal per g, so an approximate value for

TABLE 7-6

Specific heats and molar heat capacities of selected solid elements (Molar heat capacity = atomic weight × specific heat)

Element	*Atomic weight* (g/mole)	*Specific heat* (cal/g-°C)	*Molar heat capacity* (cal/mole-°C)
Nickel	58.7	0.105	6.2
Copper	63.5	0.0921	5.9
Zinc	65.4	0.0925	6.0
Tin	119	0.0542	6.4
Antimony	122	0.0504	6.1
Iodine	127	0.0523	6.6
Lead	207	0.0306	6.3
Bismuth	209	0.0294	6.1

the atomic weight can be obtained by applying the Law of Dulong and Petit:

$$\frac{6.2 \text{ cal/mole-°C}}{0.0565 \text{ cal/g-°C}} = 110 \text{ g/mole}$$

This result indicates that the correct value to choose for the atomic weight of silver is 107.868 amu and that the formula for silver oxide is Ag_2O.

The theoretical significance of this law lies in the fact that one mole of atoms of any element is 6.023×10^{23} atoms of that element. Thus, a mole of atoms of one element has about the same heat capacity as a mole of atoms of another element because approximately the same quantity of heat is needed to increase the kinetic energy of one group of atoms as to increase the kinetic energy of an equal number of atoms of some other kind by the same amount.

A few of the elements, especially those of atomic weights below 35, do not obey the law at ordinary temperatures; but as higher temperatures are approached, all elements obey the law. This behavior is predicted in modern theoretical work based on quantum mechanics.

NEW TERMS

Boiling point: the temperature at which vapor bubbles form beneath the surface of a liquid (usually referring to *normal boiling point.*)

Capillary action: the rise or fall of liquid in a tube of very small inner diameter, due to wetting of the tube walls by the liquid and to the surface tension of the liquid.

Critical pressure: the pressure required to liquefy a gas at its critical temperature; the vapor pressure of a liquid at the critical temperature.

Critical temperature: the highest temperature at which it is possible for a substance to exist in the liquid state, however great the pressure applied to it.

Crystal: a homogeneous piece of solid matter with the natural shape of a polyhedron. It is bounded by plane surfaces, which are symmetrically arranged.

Crystallography: the science that deals with the external forms of crystals and their internal structures (the arrangements of the constituent atoms, ions, or molecules).

Distillation: the process of vaporizing a liquid by the application of heat and condensing its vapor by cooling.

Dynamic equilibrium: a state of balance between opposing actions taking place at the same rate; e.g., when the rate of recondensation equals the rate of evaporation, a state of dynamic equilibrium prevails.

Ebullition (boiling): the formation of bubbles of vapor beneath the surface of a liquid.

Fractional distillation: a process by which the components of a liquid mixture are vaporized in the order of their boiling points and the liquid condensates are collected as separate fractions.

Freezing point: the temperature at which a liquid becomes a crystalline solid (same temperature as melting point).

Heat of condensation: the quantity of heat released during condensation of a gas to the liquid state. It is equal to the heat of vaporization of the substance at the same temperature.

Heat of crystallization (heat of solidification): the quantity of heat released during freezing.

Heat of fusion: the quantity of heat absorbed during melting of a stated quantity (e.g., 1 g or 1 mole) of a solid without changing the temperature. The quantity of heat required to melt 1 mole of a solid is called *molar heat of fusion.*

Heat of vaporization: the quantity of heat that must be supplied to vaporize a stated quantity (e.g., 1 g or 1 mole) of a liquid without changing its temperature.

Ionic solid: a solid in which the lattice points are occupied by positive and negative ions. The electrostatic forces operating between the positive and negative charges hold an ionic crystal together.

Law of Dulong and Petit: The specific heats of elements in the solid state are inversely proportional to their atomic weights. It is expressed by the equation

$$\text{Specific heat} \times \text{Atomic weight} = 6.2$$

Melting point (freezing point): the temperature at which the solid and liquid phases of a substance can exist in equilibrium. At this point the solid and liquid phases have the same vapor pressure.

Meniscus: the curved upper surface of a liquid column in a container of small cross section (e.g., a buret, pipet, or capillary tube). The meniscus may be curved inward or outward, depending on whether the containing walls are wetted by the liquid.

Metallic solid: a crystalline substance in which the lattice points are occupied by positive ions held together by more or less mobile electrons.

Molecular solid: a crystalline substance consisting of molecules. The atoms within each molecule are held together by covalent bonds. If the molecules are nonpolar, the attractions between molecules are due to van der Waals forces. If the molecules are polar, the intermolecular binding is due to the attraction between the negative end of one molecule and the positive end of another.

Network solid (covalent solid or atomic crystal): a solid that consists of atoms united by covalent bonds into a network that extends throughout the crystal.

Normal boiling point: the temperature at which a liquid boils when the opposing pressure is 1 atm; the temperature at which the vapor pressure of a liquid is 760 torr.

Space lattice: the pattern of points which represents the arrangement of the constituent particles in a crystal.

Specific heat: the quantity of heat (number of calories) that must be supplied to raise the temperature of 1 g of a substance 1°C.

Sublimation: passage of a substance directly from the solid to the gaseous state.

Supercooled liquid: a liquid that has been cooled below its freezing point without freezing.

Surface tension: the inward pull, due to intermolecular attraction, that causes a liquid surface to become as small as possible.

Trouton's Rule: Dividing the molar heat of vaporization of a liquid (in calories) by the normal boiling point (in °K) gives approximately 22 cal per mole per degree.

Unit cell: the smallest part of a space lattice that is the pattern for describing the whole lattice.

Vapor: the gaseous form of a substance below its critical temperature.

Vapor pressure (of a liquid or a solid): the pressure exerted by a vapor in equilibrium with its liquid or solid.

Viscosity: the resistance of a fluid to flow.

Volatility: the tendency of a liquid to evaporate.

Wetting: the adherence of a liquid to the surface of a solid due to intermolecular attraction between the liquid and the solid.

EXERCISES

7-1 At what temperature would water boil if the barometer reading were 735 mm? 770 mm? 760 mm?

7-2 How many calories are required to evaporate 3.000 kg of water at its boiling point? To evaporate 3.000 kg of carbon tetrachloride at its boiling point?

7-3 How well do water and carbon tetrachloride follow Trouton's Rule? How do you interpret this difference?

7-4 A sample of oxygen gas generated by heating potassium chlorate is found to have a volume of 386 ml when collected over water at 22°C and 756 torr. What volume would the gas have if it were dry at standard conditions?

7-5 Why do we say that the equilibrium between a liquid and its vapor is a dynamic equilibrium?

7-6 Summarize all the desirable properties of an ideal refrigerant. Write the chemical formulas for three refrigerants.

7-7 The specific heat of iron is 0.113 cal/g-°C. What weight of steam at 100°C must be condensed to release sufficient heat to raise the temperature of 4.000 kg of iron from 85°C to 100°C?

7-8 The molar heat of vaporization of ammonia at refrigeration temperatures is about 5270 cal. When a mole of water freezes, the heat liberated is 1440 cal. How many grams of ammonia must vaporize in a referigerator that is 100% efficient to remove the heat that is released upon freezing 1.00 lb of water?

7-9 Explain how snow from a light snowfall may disappear even when the temperature is as low as −5°C or −10°C.

7-10 Suggest a procedure for separating pure iodine from a mixture of iodine and potassium iodide.

7-11 The specific heat of ice is 0.50 cal/g-°C and that of steam is 0.48 cal/g-°C. Calculate the amount of heat required to convert 50.0 lb of ice at −4°C to steam at 105°C.

7-12 Compare and explain the relative electrical conductivities of melted iodine, melted sodium, and melted sodium iodide.

7-13 Why does diamond have a small vapor pressure?

7-14 The specific heat of an unknown metal was measured and found to be 0.0238 cal/g-°C. Estimate its atomic weight. Which metal might this be?

REFERENCES

Bernal, J. D., "The Structure of Liquids," *Scientific American*, **203**(2), 124 (1960). An account of the author's attempt to construct a general theory of the liquid state.

Chemical Education Material Study, *Chemistry—An Experimental Science*, Freeman, San Francisco, 1963, Chapter 5, "Liquids and Solids." This chapter in the CHEMS text is concerned with all the condensed phases of matter, including solutions.

Lapp, R. E., and the Editors of Life, *Matter*, Time, Inc., New York, 1963, Chapter 4, "The Restless Surge of the Liquid State," and Chapter 5, "A Deceptive Facade of Solidity." Drawings and color photographs add to the reader's enjoyment.

Paul, M. A., King, E. J., and Farinholt, L. H., *General Chemistry*, Harcourt, Brace & World, New York, 1967. Chapter 9 deals with the solid and liquid states.

Sisler, H. H., *Electronic Structure, Properties, and the Periodic Law*, Reinhold, New York, 1963 (paperbound). A volume in the "Selected Topics in Modern Chemistry" series. Chapter 3 is concerned with the relation of properties to bond type.

Verhoek, F. H., "What is a Metal?" *Chemistry*, **37**(11), 6 (1964). A discussion of metallic properties and the nature of the binding holding the atoms together within a metal.

8

Oxygen, Hydrogen, and Water

8-1 Occurrence of Oxygen and Hydrogen in Nature

Oxygen is by far the most abundant element in the earth's crust, by which is meant that portion of the earth that is accessible to chemical analysis, including the atmosphere, the land, the lakes, and the seas. As can be computed from the formula H_2O, oxygen constitutes 89 percent by weight of water, which is the most abundant chemical compound on earth. Oxygen accounts for about half the mass of the minerals that make up the earth's solid surface, such as sandstone, limestone, granite, and clay. Some of the most important ores are oxides of metals. These include hematite, Fe_2O_3; bauxite, Al_2O_3; cuprite, Cu_2O; cassiterite, SnO_2; and pyrolusite, MnO_2. (Mineralogical names cannot be expected to have much chemical significance.)

Elementary oxygen, O_2, constitutes about one-fifth of the earth's atmosphere, the other major constituent being elementary nitrogen, N_2. Since air is a mixture, its density is greater than that of pure N_2 and less than that of pure O_2.

FIGURE 8-1

Hydrogen, the lightest gas

Since hydrogen gas is lighter than air, it can be poured upward.

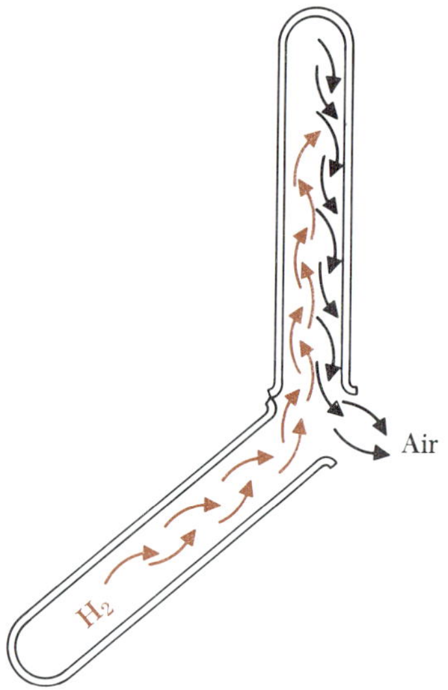

Most terrestrial hydrogen is combined with oxygen as water. Large quantities of hydrogen also occur in chemical combination with carbon in petroleum and natural gas. All plant and animal tissues are composed of compounds that contain carbon, hydrogen, and oxygen. About 90 percent of the mass of the sun is hydrogen, and the spectral analysis of starlight indicates that hydrogen is in all likelihood the most abundant element in the universe, even though the earth's crust is only about 1 percent hydrogen (Figure 1-4).

The ease with which H_2 and O_2 react to form water, coupled with the abundance of O_2 in the atmosphere and the frequent occurrence of storms accompanied by flashes of lightning, explains why there is very little elementary hydrogen on earth. In addition, the earth's gravitational force is too small to prevent the escape of molecules with such small mass. Elementary hydrogen, H_2, is the lightest substance known (Figure 8-1) and is used for inflating weather balloons. The first human flight occurred in 1783 with a hot-air balloon. Six weeks later, Jacques Charles, the French physicist whose name is attached to one of the gas laws, made the first flight with a hydrogen-filled balloon. Because of the hazard associated with highly flammable hydrogen, helium is now used for passenger balloons.

8-2 Properties of Oxygen and Hydrogen

Oxygen and hydrogen are colorless, odorless, and tasteless gases. Oxygen liquefies when cooled to −119°C (its critical temperature) under a pressure of 50 atm. The normal boiling point of the pale blue liquid oxygen is −183°C; at −219°C, liquid oxygen freezes to form a bluish-white solid. The critical temperature of hydrogen is −240°C. Liquid hydrogen is the lightest liquid known, its density being only 0.07 g per ml at the normal boiling point, −253°C. Hydrogen was first liquefied in 1898 by Sir James Dewar, who found in 1899 that liquid hydrogen freezes to form a transparent solid at −259°C.

Oxygen, especially in the liquid and solid states, is attracted into a magnetic field. Substances having this property are said to be *paramagnetic* and are believed to contain one or more unpaired electrons. This indicates that the electronic formula for oxygen may be :Ö:Ö: with two unpaired electrons.

Ordinary diatomic oxygen is partially converted into triatomic oxygen, O_3, called ozone, by the action of electrical energy or high-energy radiation. Ozone is produced by the action of lightning bolts, and its pungent odor can often be detected near sparking electric motors. It also exists in the upper atmosphere, where its presence is due to ultraviolet radiation from the sun. There it serves the useful purpose of absorbing most of the ultraviolet solar radiation, thus shielding living beings on the surface of the earth from fatal ultraviolet damage. Ozone is obtained by passing oxygen or air between layers of tinfoil that are connected to the terminals of an induction coil (Figure 8-2); the apparatus is called

FIGURE 8-2

The ozonizer

(a) End view of concentric tubes. (b) Side view, showing connections to high-voltage current.

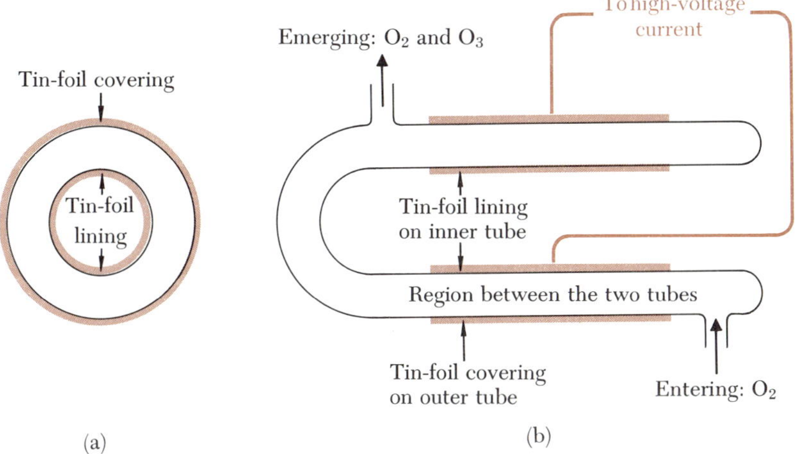

an *ozonizer.* Only a small percentage of the oxygen is converted to ozone as it passes through the ozonizer.

Because of its higher energy content, O_3 generally reacts with oxidizable substances more rapidly than O_2. Ozone is used for bleaching oils, waxes, flour, and paper pulp and for deodorizing air and sterilizing drinking water. On standing, O_3 slowly reverts to O_2.

The existence of an elementary substance in two or more different forms in the same physical state is called *allotropy.* Thus, O_2 and O_3 are the allotropes or allotropic forms of elementary oxygen. Similarly, diamond and graphite are allotropic forms of elementary carbon.

8-3 Oxidation and Combustion

At ordinary temperatures oxygen is only moderately active chemically, but its activity increases markedly at higher temperatures. Substances that undergo slow oxidation at ordinary temperatures, such as yellow phosphorus or oily rags, react at an ever-increasing rate as the temperature rises. Heat is usually liberated when oxygen unites with an element or a compound. Any reaction that is accompanied by the liberation of heat is said to be *exothermic.* Once a highly exothermic reaction has begun, it usually continues until one of the reactants has been completely consumed.

The term *combustion* is used to denote any reaction in which both heat and light are produced. In ordinary language, it refers to *burning* or *flaming.* Ordinarily, combustion involves combination with atmospheric oxygen. A temperature sufficiently high to maintain a flame is produced by the heat set free in the reaction.

The temperature at which a combustible substance bursts into flame is called its *kindling temperature* or *ignition point.* At temperatures slightly above room temperature, yellow phosphorus bursts into flame. Sodium and iron undergo slow oxidation at room temperature but undergo combustion at sufficiently high temperatures. Carbon and sulfur "take fire" when heated, even though they do not undergo oxidation at a detectable rate at room temperature. The kindling temperature of a particular substance is affected by the degree to which it is physically subdivided; for example, wood chips catch on fire at a lower temperature than large boards.

If the heat liberated by the slow oxidation of such combustible materials as oily rags is not permitted to escape as rapidly as it is produced, the kindling temperature may be reached and the material will burst into flame. Such an incident is called *spontaneous combustion.*

The kindling temperature of a substance depends on the concentration of oxygen in the gaseous mixture (the "atmosphere") in which it is to be burned. If, after being ignited in air, a burning substance is lowered into a vessel containing pure oxygen, the burning becomes much more rapid. A common test for oxygen is made with a glowing wood splint. If a gaseous mixture contains oxygen in high concentration, the oxygen will cause the glowing splint to burst into flame. The hazards of a pure oxygen atmosphere were dramatically and tragically demonstrated by the disastrous fire in the Apollo spacecraft that took the lives of three astronauts in January 1967.

The energy required to maintain body temperature in an animal and to enable the animal to perform muscular work is obtained from the oxidation of certain materials in the body. The oxygen needed for this continuous oxidation is normally acquired from the air. In the process of respiration, oxygen passes into the blood in the lungs and combines with hemoglobin to form the red oxyhemoglobin. In this form the oxygen reaches the various tissues throughout the body, where it is consumed in reactions that produce water and carbon dioxide. Patients suffering from impaired respiration are given enriched air (about 50 percent oxygen) by means of apparatus attached to storage cylinders containing pure oxygen under pressure.

Atmospheric oxygen plays an important role in the decay of wood, and dissolved oxygen is involved in the disposal of sewage. Power and heat for many industrial operations are obtained by the burning of coal, oil, or gas with atmospheric oxygen. Chemical industries are using pure oxygen on an increasing scale. Enormous quantities of oxygen are now used in the steel industry to remove carbon from iron in the production of steel in open-hearth furnaces. Large quantities of oxygen are used in the cutting and welding of metals with oxyacetylene and oxyhydrogen torches. Liquid oxygen (LOX) is used as the oxidizing agent with many rocket fuels.

In oxides, the oxidation state of oxygen is -2 (except in peroxides). Elements that are characterized by more than one oxidation state may form more than one oxide. For example, depending upon the adequacy of the oxygen supply, carbon forms either CO or CO_2. Certain oxides in which an element is not already in its highest possible oxidation state will combine with additional oxygen. Thus, carbon monoxide can be used as a fuel gas:

$$2CO + O_2 \longrightarrow 2CO_2$$

Such oxides as CO_2 and MgO do not react with oxygen because in these compounds the elements are already in the highest oxidation states possible for them.

Combustion ordinarily involves union with oxygen, but this is not always the case. For example, hydrogen will burn in a chlorine atmosphere equally as well as in an oxygen atmosphere. The following equations represent the two similar events:

$$H_2 + Cl_2 \longrightarrow 2HCl$$

$$2H_2 + O_2 \longrightarrow 2H_2O$$

Originally, the term *oxidation* was used to denote only union with oxygen, but its meaning has been extended to include any instance of an increase in oxidation number. Thus, hydrogen undergoes oxidation when it combines with chlorine as well as when it combines with oxygen, because in both cases the hydrogen goes from the 0 state to the +1 state.

In any exothermic reaction there is a definite relationship between the quantity of heat liberated and the quantities of materials reacting. When 1 mole of carbon atoms reacts with 1 mole of oxygen molecules, 1 mole of carbon dioxide and 94.4 kcal of heat are produced. This can be indicated in the equation for the reaction:

$$C(s) + O_2(g) \longrightarrow CO_2(g) \qquad \Delta H = -94.4 \text{ kcal}$$

(A parenthetical g indicates that a substance is in the gaseous state; s, the solid state; and l, the liquid state.) By definition, *heat of reaction* is the heat *absorbed* in a reaction. In the above example the heat of reaction is −94.4 kcal. The heat of reaction is negative for any exothermic reaction.

Heat of reaction is represented by the symbol ΔH (read "delta H") because every substance is assumed to have a certain *heat content* or *enthalpy*, H, so that the heat of a reaction is simply the difference between the heat content of the products and the heat content of the reactants:

$$\Delta H_{\text{reaction}} = H_{\text{products}} - H_{\text{reactants}}$$

In an exothermic reaction the heat content of the products is less than the heat content of the reactants; the difference is the amount of heat liberated.

Any reaction that is accompanied by the absorption of heat is said to be *endothermic*. The heat of reaction is just as definite for an endothermic reaction as for an exothermic reaction, and it can be indicated in the equation for the reaction:

$$N_2 + O_2 \longrightarrow 2NO \qquad \Delta H = +43.2 \text{ kcal}$$

In this example the heat of reaction is +43.2 kcal. The heat of reaction is positive for any endothermic reaction.

If a reaction can be made to proceed in either direction, the heat set free when it proceeds in one direction is the same as the heat absorbed when it goes in the other direction. This is simply an application of the Law of Conservation of Energy.

8-4 Chemical Properties of Hydrogen

At suitable temperatures, elementary hydrogen combines directly with many elements. Its reaction with oxygen (combustion) is highly exothermic:

$$2H_2(g) + O_2(g) \longrightarrow 2H_2O(g) \qquad \Delta H = -116 \text{ kcal}$$

The equation indicates that the *heat of combustion* of H_2 gas is −58 kcal per mole (the heat of the reaction divided by the number of moles of H_2). Alternatively, we may say that the *heat of formation* of gaseous H_2O from its elements is −58 kcal per mole. Heat of formation is defined as the quantity of heat absorbed during the formation of 1 mole of a substance from its elements.

The flammability of H_2 gas is one of its most dramatic characteristics. The burning of the dirigible *Hindenburg* at Lakehurst, New Jersey, May 6, 1937, was a horrifying demonstration of the burning of pure hydrogen in air. Mixtures of hydrogen with air or oxygen explode violently; therefore, hydrogen must be handled with caution. The use of liquid H_2 as a rocket fuel is a hazardous operation that is undertaken with every possible precaution.

Hydrogen gas is utilized in the oxyhydrogen torch (Figure 8-3) to produce temperatures approximating 3000°C. Temperatures approaching 5000°C are obtained with the atomic hydrogen torch (Figure 8-4), in which H_2 molecules are dissociated by an electric arc, giving monatomic molecules ("atomic" hydrogen):

$$H_2 \longrightarrow 2H \qquad \Delta H = +104 \text{ kcal}$$

The resulting H atoms impinge upon the surface of the metal to be cut or welded, where they react with oxygen. According to the principle of energy conservation, the absorption of 104 kcal in the dissociation of

FIGURE 8-3
The oxyhydrogen torch

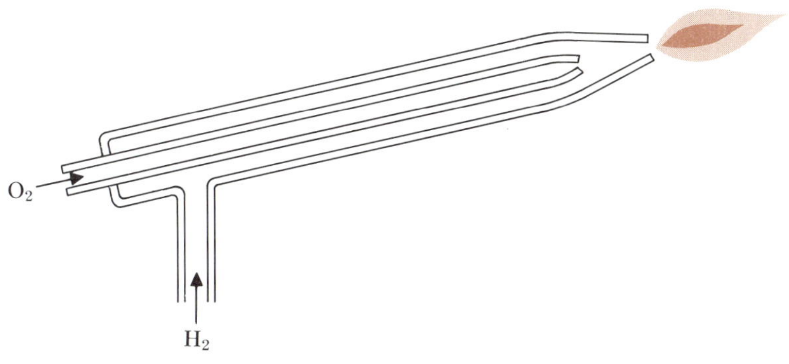

FIGURE 8-4

The atomic hydrogen torch

Molecular hydrogen is changed to atomic hydrogen in the electric arc. When the hydrogen burns, the heat of formation of molecular hydrogen from atomic hydrogen is added to the heat of combustion.

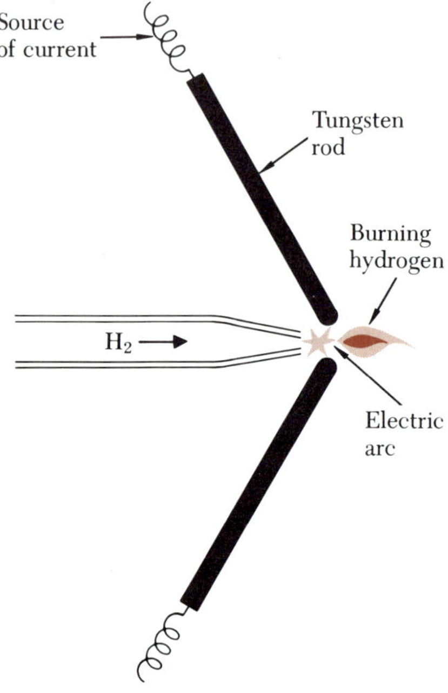

1 mole of H_2 molecules implies that 104 kcal are released when 2 moles of H atoms unite to form 1 mole of H_2 molecules. This large amount of energy, in addition to the heat of combustion of H_2 gas, is liberated when monatomic hydrogen undergoes combustion in the atomic hydrogen torch. Since the reaction of H atoms with oxygen is a surface-catalyzed process, the release of large quantities of energy occurs precisely where it is most effective in melting the metal.

A chlorine atmosphere supports the combustion of hydrogen, forming hydrogen chloride gas, HCl:

$$H_2(g) + Cl_2(g) \longrightarrow 2HCl(g) \qquad \Delta H = -44 \text{ kcal}$$

Mixtures of hydrogen gas and chlorine gas are stable in the dark but explode with considerable violence in sunlight. The reaction of fluorine with hydrogen also occurs with explosive violence, even at liquid hydrogen temperatures:

$$H_2(g) + F_2(g) \longrightarrow 2HF(g) \qquad \Delta H = -130 \text{ kcal}$$

The direct union of hydrogen with nitrogen to form ammonia, NH_3, is one of the most important reactions in the chemical industry, but it is an inherently slow reaction that requires moderate heating, high pressure, and a suitable catalyst.

Carbon does not combine directly with hydrogen, but a great number of compounds containing only carbon and hydrogen occur in petroleum and natural gas. These are known as *hydrocarbons* (not to be confused with carbohydrates, which contain carbon, hydrogen, and oxygen); the simplest is methane, CH_4. Hydrocarbons will be discussed in Chapter 14, and carbohydrates in Chapter 15.

The reactions of H_2 with compounds are of two types. Frequently the reaction involves simply the addition of hydrogen to the compound, a type of reaction called *hydrogenation*. An example is the reaction of carbon monoxide with hydrogen to form methanol, CH_3OH:

$$CO + 2H_2 \xrightarrow{\text{catalyst}} CH_3OH$$

Large amounts of hydrogen are used in the manufacture of synthetic methanol (methyl alcohol) by this reaction. Hydrogen is also used extensively in the hydrogenation of oils, such as cottonseed oil, to form solid fats.

When H_2 gas is passed over the hot oxides of some of the less active metals, the hydrogen removes oxygen from the metal and liberates the metal in the free state. For example, passing H_2 gas over black copper(II) oxide liberates reddish copper metal (Figure 8-5):

$$CuO + H_2 \xrightarrow{\text{heat}} Cu + H_2O$$

In metallurgical practice, metals are commonly released from their oxide ores by heating with coke or charcoal (essentially carbon). In the metallurgy of tungsten, the use of carbon is impractical because of the formation of tungsten carbide, but H_2 gas is quite satisfactory:

$$WO_3 + 3H_2 \longrightarrow W + 3H_2O$$

FIGURE 8-5

Reduction of a hot metal oxide by hydrogen

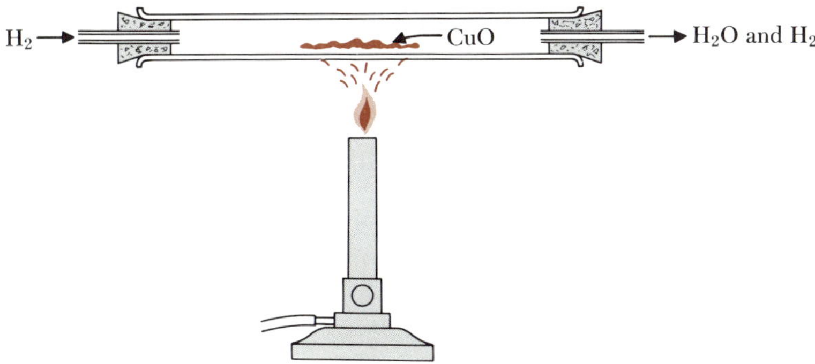

Historically, the term *reduction* has been used to denote the liberation of a metal from its ore, and the agent used to effect the liberation of the metal has been called the *reducing agent*. The release of the metal involves a change of its oxidation state from some positive value to 0. For example, when WO_3 is reduced to tungsten metal by hydrogen gas, the oxidation number of tungsten changes from +6 to 0. Simultaneously, the oxidation state of hydrogen changes from 0 to +1. Reduction is currently defined by chemists as a downward change in oxidation state and is thus the opposite of oxidation, which is an upward change in oxidation state.

Hydrogen is in the +1 state in such well-known compounds as H_2O, H_2S, NH_3, the hydrogen halides, and the hydrocarbons. *Hydrides* are binary compounds in which hydrogen is combined with an element less electronegative than itself. In these compounds hydrogen is in oxidation state −1. Binary compounds of hydrogen with the active metals of Groups IA and IIA are crystalline solids containing the *hydride ion*, H^-. An example is sodium hydride, which is obtained by bubbling H_2 gas through molten sodium metal:

$$2Na + H_2 \longrightarrow 2Na^+H^-$$

The ionic hydrides are hard and have relatively high melting points. In the liquid state these compounds are good conductors of electric current; H_2 gas is liberated at the anode (positive electrode):

$$2H^- \longrightarrow H_2\uparrow + 2e^-$$

Ionic hydrides cannot exist in the presence of water, because a reaction occurs in which the hydride ion is oxidized by water, liberating H_2 gas:

$$H^- + HOH \longrightarrow H_2\uparrow + OH^-$$

The ionic hydrides, therefore, are good reducing agents. Complex hydrides, such as $LiAlH_4$, are extremely useful reducing agents.

8-5 Preparation of Oxygen and Hydrogen

The industrial source of oxygen is air. Since oxygen is not chemically combined in air, its preparation requires only a physical separation from the other components, chiefly nitrogen. Air is first liquefied by compression and cooling; then the components are separated by fractional distillation (Section 7-3). The separation is based on the fact that the boiling point of liquid oxygen (−183°C) is 13° higher than the boiling point of liquid nitrogen. The oxygen and nitrogen are often stored and shipped in steel cylinders under pressures exceeding 100 atm.

Laboratory methods of preparing oxygen are often chosen on the basis of convenience rather than cost of the starting materials. The most commonly used method consists of the thermal decomposition of potassium chlorate, $KClO_3$, a man-made compound consisting of potassium ions, K^+, and chlorate ions, ClO_3^-. The three oxygen atoms are covalently bonded to the chlorine atom in the chlorate ion. The white crystalline solid melts when heated to 368°C; if heated to 400°C, the chlorate decomposes, releasing oxygen and producing the chloride:

$$2KClO_3 \longrightarrow 2KCl + 3O_2 \uparrow$$

The same reaction occurs rapidly when a mixture of potassium chlorate and a small amount of the black powder manganese dioxide (MnO_2) is heated to only 250°C. The weight of the manganese dioxide recoverable after the decomposition of the potassium chlorate is the same as the original weight. Although manganese dioxide does not itself undergo decomposition at this temperature, it causes the decomposition of the chlorate to occur rapidly at a much lower temperature. Here the manganese dioxide acts as a *catalyst*; its effect in causing the reaction to occur more rapidly than it otherwise would is called *catalysis*. Many industrial processes, as well as numerous laboratory operations, are successful because of the catalysts used to speed up certain reactions.

Certain other compounds of the salt type liberate at least part of their oxygen upon being heated; for example, sodium nitrate loses one-third of its oxygen:

$$2NaNO_3 \longrightarrow 2NaNO_2 + O_2 \uparrow$$

Oxides of a few of the least active metals undergo complete decomposition when heated:

$$2HgO \longrightarrow 2Hg + O_2$$

$$2Ag_2O \longrightarrow 4Ag + O_2$$

Oxides of some metals (metals that can have more than one oxidation state) liberate only part of their oxygen when they are heated. Such reactions usually require very high temperatures:

$$\underset{\text{lead(IV) oxide}}{2PbO_2} \longrightarrow O_2 + \underset{\text{lead(II) oxide}}{2PbO}$$

Whenever oxygen is prepared by thermal decomposition of a solid, the solid is placed in a test tube and heated. A delivery tube leads the oxygen into a collecting bottle, which is originally filled with water and inverted in a trough of water (Section 7-7).

Another reaction that can be employed for the generation of oxygen in laboratories is the reaction between water and sodium peroxide:

$$2Na_2O_2(s) + 2H_2O \longrightarrow 4Na^+ + 4OH^- + O_2 \uparrow$$

The electrolysis of water is a process that is used to generate both oxygen and hydrogen. Water is essentially a nonconductor of electric current, but current passes through it readily if a small amount of sulfuric acid or sodium hydroxide is dissolved in it. With the passage of direct current, H_2 gas is liberated at the cathode (negative electrode) and O_2 gas at the anode (positive electrode):

$$2H_2O \longrightarrow 2H_2\uparrow + O_2\uparrow$$

The volume of H_2 gas obtained is twice the volume of O_2 gas (Figure 8-6) because the number of H_2 molecules produced is twice the number of O_2 molecules.

Electrolysis of water as a method for manufacturing oxygen and hydrogen can be used economically only where electric power is cheap. However, hydrogen is obtained as a by-product of the electrolytic manufacture of sodium hydroxide and chlorine from sodium chloride brine:

$$(2Na^+ + 2Cl^-) + 2H_2O \longrightarrow (2Na^+ + 2OH^-) + H_2\uparrow + Cl_2\uparrow$$

The *pyrolysis* (thermal decomposition) of methane is now one of the important methods for manufacturing hydrogen gas:

$$CH_4(g) \xrightarrow[\text{catalyst}]{\text{heat}} C(s) + 2H_2(g)$$

FIGURE 8-6

Electrolysis of water

The volume of hydrogen obtained at the cathode is twice the volume of oxygen obtained at the anode.

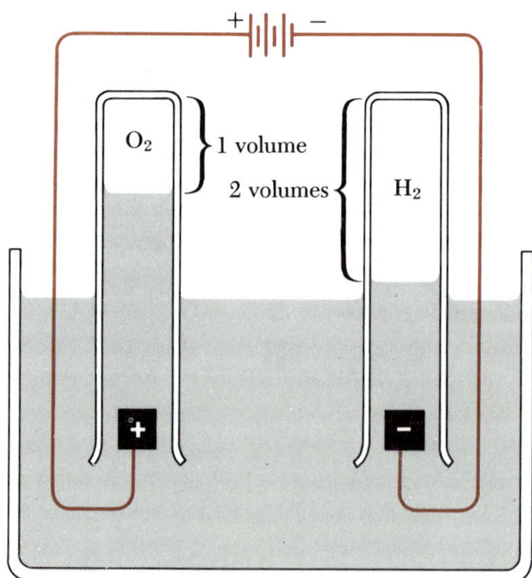

Catalyzed thermal decompositions such as this are referred to by the term "cracking." A 1:3 mixture of CO and H_2 is obtained by the reaction of steam with methane:

$$CH_4(g) + H_2O(g) \longrightarrow CO(g) + 3H_2(g)$$

Hydrogen gas can be obtained economically, mixed with carbon monoxide, by passing steam over hot coke:

$$C(s) + H_2O(g) \xrightarrow{1000°C} H_2(g) + CO(g)$$

The 1:1 mixture of CO and H_2 is used as a domestic and industrial fuel under the name "water gas." Hydrogen is also a by-product of the "coking" of coal, which involves heating coal in the absence of air to cause its "destructive distillation." The process yields coke, coal tar, and coal gas; the coal gas is about 50 percent H_2.

Hydrogen is stored and transported in thick steel cylinders at a pressure of 1000 atm; the weight of the H_2 contained in one of the cylinders is less than 1 percent of the total weight. Therefore, to save transportation costs, hydrogen is used at the plant where it is produced, whenever this is possible. In spite of the cost of manufacturing calcium hydride, it is used to produce H_2 on a small scale (by its reaction with water) since it is conveniently transported in lightweight containers.

Most of the methods used for preparing H_2 in laboratories involve the use of materials that are too expensive for industrial use. The laboratory preparations usually involve reactions of moderately active metals with dilute acids:

$$Zn + (2H^+ + SO_4^{--}) \longrightarrow (Zn^{++} + SO_4^{--}) + H_2\uparrow$$

$$2Al + 6(H^+ + Cl^-) \longrightarrow 2(Al^{3+} + 3Cl^-) + 3H_2\uparrow$$

8-6 Activity Series of Metals

Certain metals, such as sodium, react vigorously with cold water, liberating H_2 gas and forming metal hydroxides:

$$2Na + 2HOH \longrightarrow 2(Na^+ + OH^-) + H_2\uparrow$$

Certain other metals, such as magnesium, react with steam but not with cold water:

$$Mg(s) + H_2O(\text{steam}) \longrightarrow MgO(s) + H_2(g)$$

Less active metals displace hydrogen from aqueous solutions of acids—some more vigorously than others—although they do not appear to react with water itself. Still other metals, such as silver, do not displace hydrogen from acids at all.

On the basis of these and other observations, the metals have been arranged in order of their chemical activity in the presence of water. The *activity series* given in Table 8-1 is an abbreviated list, including only the most familiar metals. Each metal is more active than any of the metals below it in the series. Only those at the top of the series are able to liberate hydrogen from cold water. They are followed by metals that can displace hydrogen from steam but cannot react with cold water. Any of the metals listed above hydrogen in the series are capable of displacing it from acids, but those listed below hydrogen are unable to do this.

Hydrogen gas can reduce the oxides of metals listed below chromium but cannot reduce the oxides of metals above chromium. Oxides of metals listed near the bottom of the series, such as HgO, undergo decomposition by heat alone; that is, the metal is reduced to the free state even without the use of hydrogen.

Any metal in the series will displace any metal below it from aqueous solutions of its salts. For example, iron displaces copper from aqueous solutions of copper salts, and copper displaces silver from solutions of silver salts:

$$Fe(s) + (Cu^{++} + SO_4^{--}) \longrightarrow (Fe^{++} + SO_4^{--}) + Cu(s)$$

$$Cu(s) + 2(Ag^{+} + NO_3^{-}) \longrightarrow (Cu^{++} + 2NO_3^{-}) + 2Ag(s)$$

Since the negative ions take no part in these reactions, they may be omitted from the equations:

$$Fe(s) + Cu^{++} \longrightarrow Fe^{++} + Cu(s)$$

$$Cu(s) + 2Ag^{+} \longrightarrow Cu^{++} + 2Ag(s)$$

8-7 Hydrogen Bonding and the Properties of Water

In the H_2O molecule the oxygen atom shares a pair of electrons with each of two hydrogen atoms. Since oxygen is more electronegative than hydrogen (i.e., an oxygen atom attracts electrons more strongly than a hydrogen atom), the shared electrons spend more time near the oxygen atom. Thus, each hydrogen atom has a partial positive charge, δ^+, and the oxygen atom has a partial negative charge, δ^- (Figure 8-7b). In other words, each O—H bond is polar.

It must not be thought that the partial charge of a hydrogen atom in a water molecule is as great as the charge of a proton. The symbols δ^+ and δ^- are used to indicate only that partial charges are present on the atoms. In contrast, the positive ion in an ionic compound such as Na^+Cl^- or $Ca^{++}O^{--}$ has had one or more electrons almost completely removed from it, and the negative ion has almost complete possession of these

TABLE 8-1

Activity series of metals

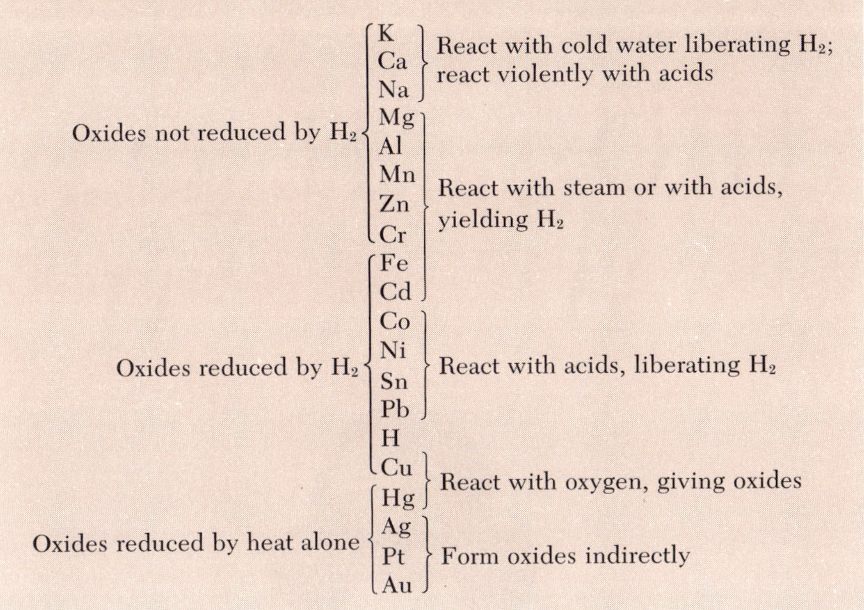

Oxides	Metal	Reactions
Oxides not reduced by H_2	K	React with cold water liberating H_2; react violently with acids
	Ca	
	Na	
	Mg	React with steam or with acids, yielding H_2
	Al	
	Mn	
	Zn	
	Cr	
Oxides reduced by H_2	Fe	
	Cd	
	Co	React with acids, liberating H_2
	Ni	
	Sn	
	Pb	
	H	
	Cu	React with oxygen, giving oxides
Oxides reduced by heat alone	Hg	
	Ag	Form oxides indirectly
	Pt	
	Au	

electrons. Thus, the positive ions in ionic compounds have charges equal to one or more protons, and the negative ions have charges equal to one or more electrons; but in polar molecules, such as HCl or H_2O, the covalently bonded atoms have only partial charges.

The structure of the water molecule is illustrated in three different ways in Figure 8-7. The relative positions of the three atomic nuclei are indicated in all three drawings. As a matter of convenience, a water molecule is commonly symbolized as

$$\begin{array}{c} \ddot{} \\ \mathrm{H} : \mathrm{O} : \\ \ddot{} \\ \mathrm{H} \end{array}$$

This representation is not intended to imply a bond angle of 90° but only that the molecule is not linear. If no electrons are shown, the formula is written as HOH or H_2O (either is to be regarded as structurally non-committal).

A water molecule acts as if it had a negative end and two positive ends. The positive ends of such a molecule attract the negative end of each of its neighbor molecules. This electrical attraction constitutes a

FIGURE 8-7

The H_2O molecule

(a) The water molecule is V-shaped, and there are two unshared pairs of electrons. (b) The angle between the O—H bonds is 105°, and each bond is polar, with each H atom carrying a partial positive charge. (c) A scale model showing the two H atoms attached to the larger O atom.

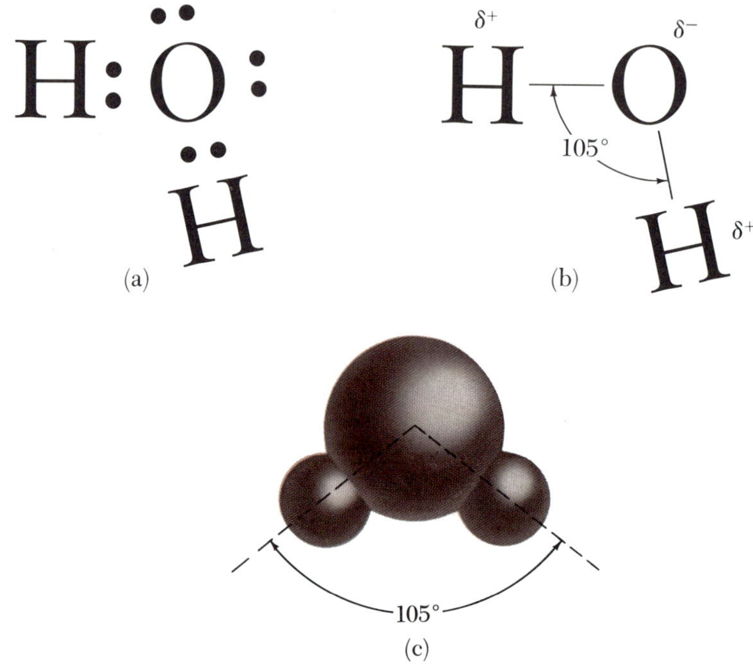

linkage between polar molecules that is important in determining the properties of a substance. Although such a linkage is far weaker than the covalent bonds that link atoms together within molecules, or the ionic bonds that hold the ions in fixed positions in a crystal of an ionic compound, it is a much stronger force than the van der Waals attraction between nonpolar molecules.

The electrostatic interaction between molecules containing an F—H, O—H, or N—H bond is substantially greater than in other cases. Molecules containing such bonds have a tendency to form linkages with other molecules containing a fluorine atom, an oxygen atom, or a nitrogen atom. This linkage is called a *hydrogen bond*, since a hydrogen atom serves as the bridge between two atoms having strong electron-attracting power.

In the vapor state, water consists of simple H_2O molecules, but in the liquid state it consists of aggregates formed by simple water molecules linking together through hydrogen bonds. In other words, water is an *associated liquid.* Liquids in which the molecules do not associate are *normal liquids.* This aggregation of water molecules can be represented by the general formula $(H_2O)_n$, where n is an integer greater than 1. If $n = 2$, the aggregate is a *dimer*, $(H_2O)_2$, with the following formula:

```
H        H
 \        \
  O—H----O—H
```

In this formula a solid line represents the covalent bond between a hydrogen and an oxygen atom within a water molecule, and the broken line indicates a hydrogen bond between atoms in different water molecules.

If $n = 3$, the aggregate is a *trimer*, $(H_2O)_3$. In general, if many molecules are associated, the resulting aggregate is called a *polymer.* In liquid water, n is variable and small.

A hydrogen bond is considerably stronger than the electrostatic attraction that exists between polar molecules with no hydrogen linkage, but a hydrogen bond between two molecules is very much weaker than a covalent bond within a molecule. For example, the energy of the covalent O—H bond within a water molecule is about 100 kcal per mole, but the energy of the hydrogen bond between two water molecules is about 3 kcal per mole. The distance between an H and an O atom that are joined by a hydrogen bond is much greater than the distance between an H and an O atom that are covalently bonded within a water molecule.

Investigations of the crystal structure of ice have revealed an arrangement of the water molecules in which each oxygen atom is surrounded symmetrically, in three dimensions, by a tetrahedral arrangement of four other oxygen atoms. Infrared spectroscopy indicates that between each two oxygen atoms is a hydrogen atom that is nearer one oxygen atom than the other. A covalent bond exists between the hydrogen atom and the nearer oxygen atom, and a hydrogen bond exists between the same hydrogen atom and the more distant oxygen atom. This arrangement results in an open structure, that is, there are holes in it (Figure 8-8), which accounts for ice being less dense than liquid water. When ice melts, some of the hydrogen bonds are broken, the open structure collapses, and the water molecules pack more closely together. Therefore, the liquid occupies a smaller volume than the ice from which it was formed.

As the temperature of liquid water is raised, more hydrogen bonds break and the volume tends to decrease. However, the increasing kinetic activity of the molecules that accompanies increasing temperatures tends to spread them farther apart. These two effects work toward opposite results. Above 4°C the kinetic effect predominates, and the density

FIGURE 8-8

Structure of ice

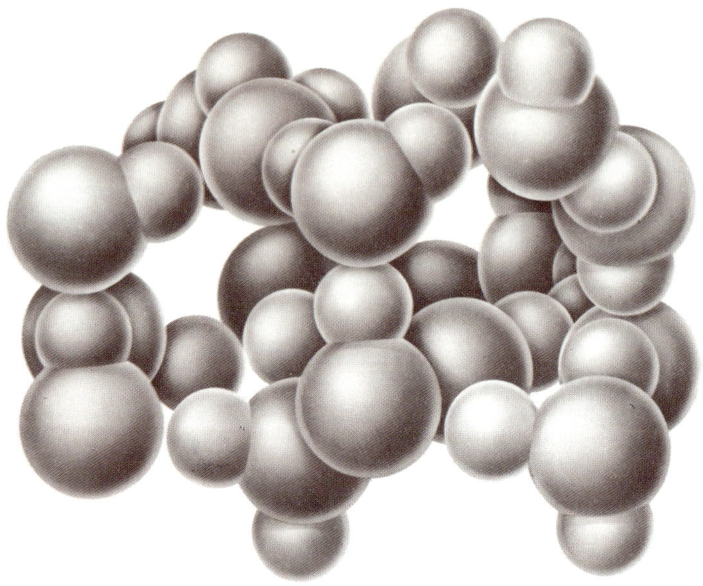

of water decreases with rising temperature. Below 4°C the structure collapse is more important, and the density increases with rising temperature between 0 and 4°C. Consequently, water has its maximum density at 4°C.

In a lake covered with ice, the water that is nearest 0°C is at the top, next to the ice, and any water that is somewhat warmer than this—provided its temperature is between 0°C and 4°C—will be in the lower layers. Thus, the warmest water is at the bottom. With the coldest water near the surface, the ice sheet thickens from the top down. The rate at which heat is lost from the warmest water is substantially reduced by the overlying layers of ice and water. Therefore, a body of water freezes only gradually over a period of time.

An input of energy is needed for the breaking of hydrogen bonds that occurs when ice melts. This energy requirement accounts for the unusually high melting point of ice and the large amount of heat required to melt it relative to other compounds of comparable structure, such as H_2S, that do not form hydrogen bonds. The high specific heat (Section 7-1) of water is partially explained by the continued breaking of hydrogen bonds with increasing temperature. Vaporization requires breaking of all remaining hydrogen bonds, so that the boiling point and heat of vaporization of water are also high.

The ability of water to dissolve large quantities of certain hydrogen compounds is related to its tendency to form hydrogen bonds with them. Such highly soluble compounds include methyl alcohol, CH_3OH, and ammonia, NH_3, both of which form hydrogen bonds themselves. In order for any substance to dissolve in water, the water molecules must become unbonded from one another and form bonds instead with the particles of the dissolving substance. Consequently, nonpolar substances such as methane are insoluble in water.

The ability of water to dissolve ionic compounds is also due to the nature of the water molecules. When an ionic compound is placed in contact with water, the polar water molecules attract the ions of the solid crystal and reduce the strength of the ionic bonds within the crystal. The negative (oxygen) ends of the water molecules attract the positive ions of the crystal, and the positive (hydrogen) ends attract the negative ions. The result is that the ions are separated from one another and are dispersed through the water. Ionic compounds do not dissolve in nonpolar solvents such as carbon tetrachloride.

8-8 Treatment of Natural Water

Rainwater is relatively pure except for dust particles and dissolved gases. When it falls to the ground and travels over and through the surface of the earth, many different types of material become dissolved or suspended in it. Appreciable quantities of soluble salts are present in water that has percolated through soil, and the suspended materials present in streams flowing through populous areas include organic material and bacteria.

Water for human consumption should be free of harmful bacteria. Water that passes through layers of soil and porous rock undergoes a natural filtration that removes most bacteria; therefore, water from deep wells is often safe for drinking. This water is likely, however, to contain calcium, magnesium, and iron(II) compounds which prevent the effective cleansing action of soap. Such water is known as *hard water.*

Harmful bacteria can be removed from water that is to be used for drinking by filtration through beds of sand and by chlorination. Water to be used for chemical purposes may need to be purified by distillation, but this is an expensive operation. The removal of much of the hardness of a city water supply (to render it fit for use in laundries and steam boilers) is accomplished by a process called *water softening.* For example, in a large-scale water-softening operation, sodium carbonate ("soda ash") is used to bring about the removal of Ca^{++} ions by precipitation as calcium carbonate. If the compound responsible for the hardness of the water is calcium sulfate, the following reaction occurs:

$$(Ca^{++} + SO_4^{--}) + (2Na^+ + CO_3^{--}) \longrightarrow CaCO_3 \downarrow + (2Na^+ + SO_4^{--})$$

The softened water contains the same number of moles of sodium sulfate as the moles of calcium sulfate originally present; but the sodium salt formed does not interfere with the cleansing action of soap, nor does it form boiler scale.

In recent years processes have been developed for softening water by ion exchange. Two major types of porous solids are used as *ion exchangers*: (1) the zeolites, which are naturally occurring aluminosilicate minerals, such as $NaAl(SiO_3)_2$ or $NaAlSi_2O_6$, and (2) the ion-exchange resins, which are synthetic organic polymers.

In the process of ion exchange, ions previously held at electrically charged sites within the porous solid have been exchanged for ions in a solution that surrounds and invades the solid. For example, when water containing Ca^{++} or Mg^{++} ions is allowed to filter through thick layers of zeolite, the Na^+ ions in the zeolite are replaced by Ca^{++} or Mg^{++} ions and the Na^+ ions replace the Ca^{++} or Mg^{++} ions in the solution:

$$2NaAlSi_2O_6 + Ca^{++} \longrightarrow Ca(AlSi_2O_6)_2 + 2Na^+$$

Thus, the effect of the ion exchange in the zeolite is to soften the water. This is the principle upon which domestic water softeners operate.

Ion-exchange reactions are reversible. The direction the reaction takes is determined by the ion concentrations. If a concentrated NaCl solution is allowed to stand in contact with a zeolite that has been used for softening water, the calcium aluminosilicate is reconverted to sodium aluminosilicate:

$$Ca(AlSi_2O_6)_2 + 2Na^+ \longrightarrow 2NaAlSi_2O_6 + Ca^{++}$$

This reaction, the reverse of the preceding one, leaves sodium zeolite ready to serve as a water softener again. When, through use, it is again converted to calcium zeolite, sodium zeolite can be regenerated again by the same procedure as before. Thus, the only material needed to keep an ion-exchange water softener operating over a period of many years is common salt.

Both cation exchangers and anion exchangers are available. Water can be completely demineralized by being passed through an exchanger that replaces all metal ions with H^+ ions and then through another exchanger that exchanges OH^- ions for any other anions. The effluent water is as pure as carefully distilled water.

In the United States the amount of water used each year averages about 2500 tons per capita. One of the critical problems attending the population explosion now under way may very well be the problem of obtaining an adequate supply of fresh water. Recent advances in developing desalinization (desalting) processes that are economical and practical give promise that the oceans may become an important source of water for uses that cannot be served by untreated saline water.

8-9 Hydrates

When an aqueous solution of a soluble salt is concentrated by evaporation, the crystals that separate frequently contain the salt and water combined in definite proportions. Such compounds are known as *hydrates*. A familiar example is $CuSO_4 \cdot 5H_2O$, commonly known as blue vitriol. Its chemical name is copper(II) sulfate 5-hydrate, or cupric sulfate pentahydrate. The dot (·) used in the formula is structurally noncommittal; its only purpose is to indicate that the hydrate is formed by the chemical combination of 5 moles of H_2O with 1 mole of $CuSO_4$ and also that the 5 moles of H_2O can in turn be removed from the hydrate. The formula for the hydrate is sometimes written $CuSO_4(H_2O)_5$.

When the blue crystals of copper(II) sulfate 5-hydrate are heated, water is lost and the crystalline structure of the hydrated salt appears also to be lost. The salt that remains is a white powder, $CuSO_4$, which may appropriately be referred to as anhydrous copper(II) sulfate:

$$CuSO_4 \cdot 5H_2O \rightleftharpoons CuSO_4 + 5H_2O$$

This equation is written with a double arrow to indicate that the reaction is reversible, because addition of water to the anhydrous salt produces the hydrated salt again.

The solution obtained by dissolving copper(II) sulfate 5-hydrate in water cannot be distinguished from the solution that is formed by dissolving anhydrous copper(II) sulfate. Certainly, most salts become hydrated in aqueous solution; that is, all ions, positive and negative, are probably hydrated in the presence of water. It is impossible to be sure how many water molecules are attached to an ion in solution, but crystal structure investigations of hydrates outside of solution have produced some information. For example, in copper(II) sulfate 5-hydrate, the copper(II) ion, Cu^{++}, is combined with four water molecules, $[Cu(H_2O)_4]^{++}$, and the sulfate ion is combined with one water molecule, $[SO_4(H_2O)]^{--}$.

Water molecules combine with metal ions by coordinate bonds, that is, covalent bonds in which both electrons of the shared pair are furnished by the same atom (oxygen atoms in this case):

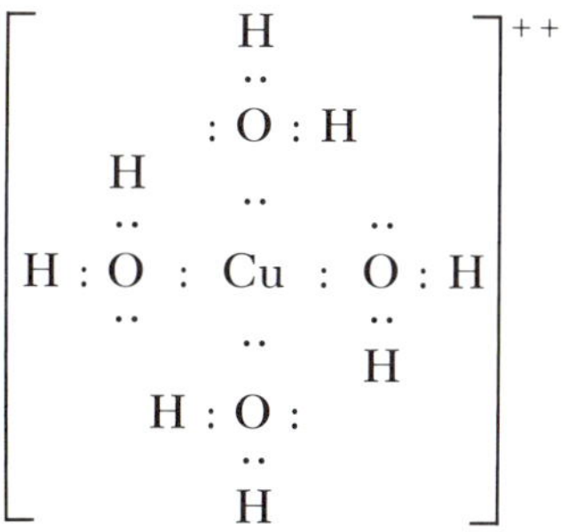

```
┌                          ┐++
             H
             ..
           : O : H
     H
             ..
     ..                ..
 H : O  :   Cu   :  O : H
     ..                ..
             ..
                        H
         H : O :
             ..
└            H             ┘
```

Water molecules are attached to negative ions by hydrogen bonds.

In some hydrates water molecules occupy certain positions in the crystal structure without being attached by chemical bonds to either the positive or the negative ions. Such water is called *lattice water*. An example is the hydrate $KAl(SO_4)_2 \cdot 12H_2O$, in which six H_2O molecules are lattice water molecules and the other six are attached by coordinate bonds to the aluminum ion.

Water of hydration should not be confused with the water taken up by an oxide in becoming a hydroxide (or given up by a hydroxide in becoming an oxide). For example, when aluminum hydroxide, $Al(OH)_3$, loses water it becomes aluminum oxide, Al_2O_3; but the crystal structure of Al_2O_3 is entirely different from that of $Al(OH)_3$, and $Al(OH)_3$ does not "contain" water in the sense that a hydrate contains water of hydration.

If the crystals of certain hydrates are placed in the open atmosphere they will lose part or all of their water of hydration. This spontaneous loss of water of hydration to the atmosphere is known as *efflorescence*. Whether a hydrate is stable against efflorescence when exposed to air depends upon the humidity of the air, the nature of the hydrate, and the temperature.

A *hygroscopic* substance absorbs moisture from the air. If a hygroscopic substance takes up so much water vapor that an aqueous solution is formed, the substance is *deliquescent*. Any substance that deliquesces is very soluble in water, but finely divided substances, such as flour, may be hygroscopic even though they are not soluble in water and are thus incapable of deliquescence.

Hydrate formation is sometimes, but not always, involved in hygroscopic or deliquescent behavior. For example, sodium hydroxide forms no crystalline hydrate, but pellets of NaOH exposed to the air take up so much water that they form a solution. Both anhydrous calcium chloride and $CaCl_2 \cdot 6H_2O$ deliquesce. When the anhydrous salt takes up water vapor, all of it is converted to $CaCl_2 \cdot 6H_2O$ before the first drop of solution forms. The use of calcium chloride to keep down the dust on dirt roads is an application of its deliquescent property.

8-10 Heavy Hydrogen and Heavy Water

In 1931 Harold C. Urey, an American chemist, made one of the important discoveries of this century when he found that one atom in every 5000 hydrogen atoms has a mass of 2 amu.

A difference of one or more units in atomic mass from one isotope to another in the heavy elements causes only minor variations in the properties of these isotopes (because the percentage difference in their masses is so small), but the same difference in atomic mass of the very light elements can result in an appreciable variation in properties. Since a *heavy hydrogen* atom has twice the mass of an ordinary hydrogen atom, the properties of heavy hydrogen and ordinary hydrogen differ considerably.

Usually, the different isotopes of an element are not given special names. However, the properties of heavy hydrogen are so different from those of ordinary hydrogen that Urey named it *deuterium* (from the Greek root meaning "two"). It is represented by the symbol D. Deuterium is not regarded as a new chemical element, but its discovery was more significant than the discovery of some of the new elements.

There are three isotopic varieties of the diatomic hydrogen molecule: H_2, HD, and D_2. Their melting points and boiling points are compared in Table 8-2. (The boiling point of HD is missing from the table because it cannot be determined: at temperatures approaching the boiling point, HD undergoes partial conversion to H_2 and D_2.) The density of D_2 gas is, of course, twice as great as that of H_2 gas because the molecular weight of D_2 is twice that of H_2.

The composition of a deuterium nucleus accounts for its greater mass. The nucleus of a light hydrogen atom consists only of a proton, but the nucleus of a deuterium atom consists of a proton and a neutron. Although the nucleus of a deuterium atom is not one of the "fundamental" particles, it has been given the special name *deuteron*.

A third isotope of hydrogen with a mass of 3, *tritium* (T), has also been discovered. Only very small amounts of tritium occur naturally because of its radioactive instability. The fact that it exists at all is due to the action of cosmic rays. The tritium nucleus consists of a proton and two neutrons.

Since the two heavy hydrogen isotopes have been named deuterium and tritium, the most abundant isotope, whose atomic nucleus consists of a lone proton, has been named *protium*. No special symbol, other than H, is used to represent protium.

Since hydrogen, as it occurs in nature, is a mixture of the two isotopes protium (1.007825 amu) and deuterium (2.01410 amu), all hydrogen compounds occurring in nature are mixtures of molecules containing protium atoms and molecules containing deuterium atoms. Water that

TABLE 8-2

Isotopic varieties of the hydrogen molecule

	H_2	HD	D_2
Melting point (°K)	13.95	16.60	18.65
Boiling point (°K)	20.38		23.6

is composed of deuterium and oxygen (D_2O) is called *deuterium oxide* or, less precisely, *heavy water*. Since protium atoms are 5000 times as abundant as deuterium atoms, the properties of ordinary water are approximately the same as those of protium oxide.

Heavy water was first obtained by the electrolysis of ordinary water. Since protium oxide undergoes electrolytic decomposition at a somewhat faster rate than deuterium oxide, the heavy water tends to remain behind in the electrolytic cell. If the cell contents are continuously replenished by the addition of ordinary water, the residue becomes enriched in deuterium oxide.

Although heavy water and ordinary water appear identical, many of their physical properties are different, as indicated in Table 8-3. Since heavy water is about 10 percent heavier than ordinary water, it can be detected by precision methods for determining density when there is only 1 part of it in 100,000 parts of an H_2O–D_2O mixture.

Heavy water can be used as a moderator in nuclear reactors because it reduces the speed of neutrons to within the desired range without absorbing them to any great extent. After the fall of Norway in World War II, the Germans attempted to increase the production of heavy water in a plant at Rjukan. Fear that the Germans might be planning to use heavy water as a moderator to produce plutonium for the manufacture of atomic bombs led to a famous British commando raid on the plant in March 1943.

Protium oxide and deuterium oxide are identical chemically except for differences in reaction rates. This property, also true of other hydrogen compounds, makes it possible to study chemical reactions and biochemical processes such as digestion and metabolism by replacing protium with deuterium in molecules of chemical reagents or in the food given to an organism. The subsequent whereabouts of the deuterium atoms can be ascertained by density determinations. Thus, deuterium

TABLE 8-3

Physical properties of ordinary water and deuterium oxide

	Ordinary water (H_2O)	*Deuterium oxide* (D_2O)
Density at 20°C (g/ml)	0.998	1.105
Maximum density (g/ml)	1.000	1.107
Temperature of maximum density (°C)	4.0	11.6
Melting point (°C)	0.0	3.8
Boiling point (°C)	100.0	101.42
Heat of vaporization (cal/mole)	9720	9970
Heat of fusion (cal/mole)	1440	1500

can be used as a "tracer" element without recourse to radioactive isotopes.

A deuterium atom and a protium atom can both be present in the same water molecule. As a matter of fact, the heavy water present in small amounts in ordinary water is mostly HOD. The presence of two molecules of HOD in 10,000 molecules of H_2O is much more probable than the presence of one molecule of D_2O. The properties of pure HOD are intermediate between those of protium oxide and those of deuterium oxide.

Like ordinary water, D_2O and HOD show strong tendencies toward hydrogen bond formation.

8-11 Peroxides

Compounds in which the oxidation number of oxygen is -1 are called *peroxides*. Examples include hydrogen peroxide, H_2O_2; sodium peroxide, Na_2O_2; and barium peroxide, BaO_2. The metal peroxides contain the peroxide ion, O_2^{--}, which has the electronic structure $[:\ddot{\underset{..}{O}}:\ddot{\underset{..}{O}}:]^{--}$. Hydrogen peroxide is a covalent compound,

$$\begin{array}{ccc} & & \text{H} \\ & \cdot\cdot & \cdot\cdot \\ : & \text{O} : & \text{O} : \\ & \cdot\cdot & \cdot\cdot \\ & \text{H} & \end{array}$$

characterized by an oxygen–oxygen bond. (One of the hydrogen atoms is in a plane approximately perpendicular to the plane of the other atoms.)

The metals that form peroxides are zinc, cadmium, mercury, and most of the members of Groups IA and IIA. Sodium peroxide is formed by the direct union of sodium with oxygen. Barium peroxide is formed by heating barium oxide in air. Hydrogen peroxide must be obtained by indirect procedures; for example, a metal peroxide may be added to an acidic solution. When barium peroxide is added to a solution of sulfuric acid, barium sulfate precipitates, leaving a solution of hydrogen peroxide:

$$BaO_2(s) + H_2SO_4 \longrightarrow BaSO_4(s) + H_2O_2$$

Hydrogen peroxide is unstable because of its tendency to decompose, forming water and elementary oxygen:

$$2H_2O_2 \longrightarrow 2H_2O + O_2\uparrow$$

The decomposition is accelerated by light and is catalyzed by many materials, including dust and blood.

The -1 oxidation state of oxygen in H_2O_2 is intermediate between the 0 state of elementary oxygen and the -2 state in water and most other oxygen compounds. Consequently, H_2O_2 can be reduced by strong reducing agents or oxidized by strong oxidizing agents. For example, it oxidizes hydrogen iodide to elementary iodine,

$$2HI + H_2O_2 \longrightarrow I_2 + 2H_2O$$

and it reduces silver oxide to elementary silver,

$$Ag_2O + H_2O_2 \longrightarrow 2Ag + H_2O + O_2 \uparrow$$

Whenever H_2O_2 acts as an oxidizing agent, the product of its own reduction is H_2O. Whenever H_2O_2 acts as a reducing agent, the product of its own oxidation is elementary oxygen.

NEW TERMS

Activity series: a listing of chemical elements in the order of their chemical activity in the presence of water.

Allotropy: the existence of a free element in two or more forms in the same physical state. O_2 and O_3 are the allotropic forms of free oxygen. Graphite and diamond, both crystalline solids, are allotropes of carbon.

Anode: the positive electrode in an electrolytic cell.

Associated liquid: a liquid whose molecules are linked together to form aggregates. An aggregate consisting of two molecules is a *dimer*. An aggregate consisting of three molecules is a *trimer*. If a large number of molecules are linked together, the aggregate is a *polymer*. (In *normal liquids* the molecules do not associate.)

Catalysis: the increase in the rate of a reaction resulting from the presence of a substance, called the catalyst, which is still present in the same quantity at the end of the reaction.

Cathode: the negative electrode in an electrolytic cell.

Combustion: any chemical process accompanied by the evolution of light and heat. Commonly, combustion involves union with oxygen.

Deliquesce: to become wet and dissolve gradually in moisture absorbed from the air.

Deuterium: the hydrogen isotope of mass number 2; its symbol is D. Also known as heavy hydrogen.

Deuteron: the nucleus of a deuterium atom.

Effloresce: to lose water of hydration to the atmosphere spontaneously.

Electrode: one of the terminals at which an electric current enters or leaves an electrolytic solution.

Electrolysis: chemical change brought about by an electric current.

Endothermic process: a process of any kind in which heat is absorbed from the surroundings, as in the melting of ice, the evaporation of a liquid, or the union of carbon and sulfur to form CS_2. Any chemical reaction in which heat is absorbed is an endothermic reaction.

Exothermic process: a process of any kind in which heat is evolved, as in the liquefaction of a gas, the freezing of a liquid, or the burning of a candle. Any chemical reaction in which heat is evolved is an exothermic reaction.

Hard water: water in which it is not easy to obtain the cleansing action of soap. Hardness of water is due to Ca^{++}, Mg^{++}, and Fe^{++} ions.

Heat of combustion: the quantity of heat absorbed during the combustion of a specified quantity of an element or compound. (This definition is formulated to be in technical agreement with the definition of *heat of reaction* given below. The quantity of heat absorbed during combustion is, of course, negative; i.e., a combustion reaction is always exothermic.)

Heat of formation: the quantity of heat absorbed during the formation of a specified quantity (for example, 1 mole) of a substance from its constituent elements. (If the formation of a substance from its elements is an exothermic reaction, the heat of formation of the substance is negative.)

Heat of reaction: the quantity of heat absorbed when a reaction occurs. The heat of reaction is positive for an endothermic reaction and negative for an exothermic reaction.

Heavy water (deuterium oxide): water that is composed of deuterium and oxygen. Its formula is D_2O.

Hydrate: a crystalline substance that contains chemically combined water in a definite proportion.

Hydride: a binary compound in which hydrogen is combined with an atom less electronegative than itself. In all hydrides, hydrogen is in the oxidation state -1.

Hydrocarbon: a binary compound consisting of hydrogen and carbon.

Hydrogenation: any reaction in which hydrogen is simply added onto a compound, as in the formation of methanol, CH_3OH, from CO and H_2.

Hydrogen bond: the bond between two molecules linked together through a hydrogen atom which serves as the bridge between two atoms having strong electron-attracting power. Hydrogen bonding is an especially strong electrostatic attraction between polar molecules containing F—H, O—H, or N—H bonds. Hydrogen bonds are much weaker than ionic or covalent bonds but much stronger than van der Waals forces.

Hygroscopic: having a tendency to absorb moisture from the air.

Ion exchange: the process by which ions held at electrically charged sites within a porous solid are exchanged for ions in a solution in contact with the solid.

Kindling temperature: the temperature at which a substance bursts into flame.

Lattice water: water that is incorporated into the crystalline structure of a solid hydrate without being attached by chemical bonds to any other component units of the crystal.

Oxidation: an upward change in oxidation state. (A downward change in oxidation state is *reduction*.)

Paramagnetic: subject to being attracted into a magnetic field. (Substances subject to repulsion by a magnetic field are *diamagnetic*.)

Peroxides: binary compounds that contain oxygen in the oxidation state -1.

Protium: the hydrogen isotope of mass number 1; its symbol is H.

Pyrolysis: decomposition brought about by heat.

Reduction: a downward change in oxidation state. (Reduction is the opposite of *oxidation*.)

Spontaneous combustion: combustion resulting from self-ignition, the heat evolved during slow oxidation being responsible for raising the material to its kindling temperature.

Tritium: the hydrogen isotope of mass number 3; its symbol is T.

EXERCISES

8-1 Many gases, for example, H_2, N_2, and O_2, are purchased in tanks. Why can O_3 not be purchased in tanks?

8-2 Define exothermic reaction, endothermic reaction, combustion, heat of combustion, spontaneous combustion, allotropy, oxide, hydride, oxidation, reduction, hydrogenation.

8-3 List some compounds that will not react with oxygen, and explain why they will not.

8-4 Compute the amount of heat liberated by (a) the burning of 100.0 g of H_2 and (b) the burning of 100.0 oz of H_2.

8-5 In the reaction $CH_4 + 2O_2 \rightarrow CO_2 + 2H_2O$, which element is oxidized? Which element is reduced? What changes in oxidation state have these two elements undergone?

8-6 Describe in some detail a laboratory method for the preparation of oxygen, and explain why it is not used commercially.

8-7 How many grams of oxygen could be obtained by the complete thermal decomposition of 10.0 moles of silver oxide?

8-8 If $KClO_3$, Na_2O_2, $NaNO_3$, and HgO all cost the same amount per kilogram (which is certainly not the case), which would be the most economical source of oxygen?

8-9 Describe the commercial preparation of oxygen, and explain why it is not used as a laboratory method.

8-10 Compare the reactions of sodium oxide and sodium peroxide with water.

8-11 Which would give the higher temperature, a gas–air torch or a gas–oxygen torch? Why?

8-12 Write equations for the following reactions:
(a) sodium and water
(b) potassium and "heavy water" (D_2O)
(c) zinc and dilute hydrochloric acid
(d) magnesium and dilute sulfuric acid
(e) lithium hydride and water

8-13 On the basis of the activity series of the metals, why is gold a precious metal?

8-14 How do you explain the fact that water molecules are polar? That water molecules are associated in the liquid state? That water pipes burst when water freezes inside them?

8-15 Methane, ammonia, and water have molecular weights that are nearly the same: 16, 17, and 18, respectively. The boiling points of the three compounds are 112°K, 240°K, and 373°K, respectively. Why are their boiling points so different?

8-16 Calculate the empirical formula for the compound that is 18.3% lithium, 71.1% aluminum, and 10.6% hydrogen.

8-17 By means of suitable chemical equations, show that water sometimes acts as an oxidizing agent and sometimes as a reducing agent and sometimes takes part in reactions in which it is neither.

8-18 Can water always be used to extinguish fires? Why or why not?

8-19 Calculate the percentage of water in Epsom salts, $MgSO_4 \cdot 7H_2O$.

8-20 When borax, a hydrated salt, is heated it loses water, and the weight is reduced to 52.77% of the original value. The formula of the anhydrous salt that remains is $Na_2B_4O_7$. Calculate the formula of the hydrate.

8-21 Calculate how many moles of each of the following hydrates would have to be dehydrated to obtain 5.00 kg of the anhydrous salt: $CuSO_4 \cdot 5H_2O$, $CuSO_4 \cdot H_2O$, $Na_2CO_3 \cdot 10H_2O$.

8-22 People who wear eyeglasses are plagued by having their glasses coated with a film of water when they enter a warm room on a cold day. Why does this happen?

8-23 Suppose 1.000 kg of ice (at 0°C) is placed in 1.000 kg of water at 40°C in a vessel that permits no heat to leave or enter. Certain physical changes will occur. When these changes have ceased and a condition of equilibrium has been reached, what will the vessel contain? What temperature will prevail?

8-24 A student who knew that deuterium gas has twice the density of protium gas had difficulty reconciling this fact with the statement that heavy water is only 10% heavier than ordinary water. How would you point out the fallacy in his reasoning?

8-25 Why has there been unusual interest in the isotopes of hydrogen?

8-26 Explain the fact that heavy water does not have its maximum density at the same temperature as ordinary water.

8-27 Calculate the percentage of water of hydration in $KAl(SO_4)_2 \cdot 12D_2O$.

REFERENCES

Asimov, I., "The Composition of the Atmosphere," *Journal of Chemical Education*, **32**(12), 633 (1955). A discussion of the relative amounts of the different isotopic species of oxygen and nitrogen in the atmosphere.

Chalmers, B., "How Water Freezes," *Scientific American*, **200**(2), 114 (1959). A detailed consideration of the mechanism of formation of ice crystals, including the various shapes of snowflakes.

Chemical Bond Approach Project, *Chemical Systems*, McGraw-Hill, New York, 1964, Chapter 18, "Water." This last chapter in the CBA text is devoted entirely to the structure and chemistry of water.

Davis, K. S., and Day, J. A., *Water, the Mirror of Science*, Anchor Books (Doubleday), Garden City, N.Y., 1961 (paperbound). The peculiarities of water are emphasized.

Dole, M., "History of Oxygen," *Science*, **109**, 77 (1949). A discussion of the origin of oxygen as an element and of oxygen as a free element in the earth's atmosphere.

Farber, E., "Oxygen, the Element with Two Faces," *Chemistry*, **39**(5), 17 (1966). The life-giving and life-taking aspects of our most abundant element.

Gillam, W. S., and McCutchan, J. W., "Demineralization of Saline Waters," *Science*, **134**, 1041 (1961). An account of desalination processes that may solve our impending water crisis.

Katz, J. J., "The Biology of Heavy Water," *Scientific American*, **203**(1), 106 (1960). The effects on living organisms of replacing H_2O with D_2O.

Keirstead, R. E., "Water, a Basic Natural Resource," *Journal of Chemical Education*, **32**(2), 99 (1955). A discussion of man's freshwater needs and the adequacy of the supply.

Leopold, L. B., Davis, K. S., and the Editors of Life, *Water*, Time, Inc., New York, 1966. An account of the many aspects of water with numerous color photographs.

Miller, L. E., "Chemistry in the Stratosphere and Upper Atmosphere," *Journal of Chemical Education*, **31**(3), 112 (1954). The variation in the chemical nature of the atmosphere with altitude.

Partington, J. R., "The Discovery of Oxygen," *Journal of Chemical Education*, **39**(3), 123 (1962). A consideration of the independent discoveries of Priestley and Scheele by an eminent chemical historian.

Urey, H. C., "Significance of Hydrogen Isotopes," *Industrial and Engineering Chemistry*, **26**, 803 (1934). By the American chemist who discovered deuterium.

Watt, G. W., Hatch, L. F., and Lagowski, J. J., *Chemistry*, Norton, New York, 1964, Chapter 7, "The Atmosphere." Normal and accidental components of the atmosphere are discussed.

9
Acids and Bases

9-1 Acids

Water has always been the most widely used solvent. As a natural consequence, other substances have been described and classified according to their behavior toward water and the properties that they show in the presence of water. For example, whenever we speak of a substance as soluble or insoluble, it may be assumed that we are referring to the ability or inability of the substance to dissolve in water. Whenever the reference is to another solvent, the solvent will be named.

The early studies of acids emphasized behavior in water. Those compounds that were observed to have the properties of this class of compounds were given names that include the word *acid.* The following are examples:

1. Hydrochloric acid, HCl, known commercially as muriatic acid.
2. Nitric acid, HNO_3, known to the alchemists as *aqua fortis.*
3. Sulfuric acid, H_2SO_4, once called oil of vitriol.

4. Acetic acid, $HC_2H_3O_2$, present in vinegar.
5. Carbonic acid, H_2CO_3, present in "carbonated" beverages.
6. Acetylsalicylic acid, $HC_9H_7O_4$, better known as aspirin.

The English word *acid* is derived from the Latin word *acidus*, meaning "sour." Use of this word dates from the seventeenth century. Thus, according to the origin of the name, the acids are the "sour substances." As usually happens in the development of a scientific discipline, the meaning of the word has changed considerably since it first came into use.

Some of the properties that characterize acids were listed by Robert Boyle in 1680, and the list has been extended with the growth of chemical knowledge. Substances commonly regarded as acids have the following properties, some to a greater degree than others:

1. They have a sour taste. This is actually a property of the aqueous solution of an acid, since our taste buds are bathed with an aqueous fluid.
2. Their aqueous solutions impart characteristic colors to complex substances known as *indicators*. For example, they cause the vegetable dyestuff known as litmus to develop a red color and the synthetic indicator called bromphenol blue to turn yellow.
3. They react with the class of compounds known as bases to form salts and water. For example,

$$\underset{\text{base}}{NaOH} + \underset{\text{acid}}{HCl} \longrightarrow \underset{\text{salt}}{NaCl} + \underset{\text{water}}{H_2O}$$

4. They contain hydrogen, and hydrogen gas is produced when active metals, such as zinc and magnesium, react with them. For example,

$$\underset{\substack{\text{active}\\\text{metal}}}{Zn} + \underset{\text{acid}}{2HCl} \longrightarrow \underset{\text{salt}}{ZnCl_2} + \underset{\substack{\text{hydrogen}\\\text{gas}}}{H_2\uparrow}$$

5. In aqueous solutions they give rise to the *hydronium ion*, H_3O^+. For example,

$$\underset{\text{water}}{H_2O} + \underset{\substack{\text{hydrogen}\\\text{chloride}}}{HCl} \longrightarrow \underset{\substack{\text{hydronium}\\\text{ion}}}{H_3O^+} + \underset{\substack{\text{chloride}\\\text{ion}}}{Cl^-}$$

$$\underset{\text{water}}{H_2O} + \underset{\substack{\text{acetic}\\\text{acid}}}{HC_2H_3O_2} \rightleftharpoons \underset{\substack{\text{hydronium}\\\text{ion}}}{H_3O^+} + \underset{\substack{\text{acetate}\\\text{ion}}}{C_2H_3O_2^-}$$

Aqueous solutions of acids are able to conduct electric current because they contain ions that are produced when the acids are dissolved in water. It is important to note that the compounds under discussion here are covalent compounds and that the ions present in their aqueous solutions are formed by a chemical reaction with water. Pure HCl, for example, is a gas under ordinary conditions. When HCl is liquefied, the pure liquid HCl is found to be a nonconductor of electric current because it is a covalent compound; it consists of neutral molecules, not ions.

The aqueous solution of a *strong acid*, such as HCl, conducts electric current especially well because virtually all its molecules react with water to form ions. The presence of a large number of ions enables the solution to conduct electric current very well. This is also the reason that an active metal reacts rapidly with such a solution.

If only a small percentage of the molecules of an acid react with water, the acid is called a *weak acid*. Acetic acid is a weak acid, as indicated by the double arrow in the equation for its reaction with water. This reaction is reversible, and it is as complete as it is capable of becoming when only a small fraction of the molecules of the acid have reacted with water. Consequently, an aqueous solution of acetic acid (or any other weak acid) is a relatively poor conductor of electric current and it reacts only slowly with an active metal.

The hydronium ion, H_3O^+, is responsible for the characteristic properties of an aqueous solution of an acid. This ion can be regarded as a hydrogen ion, H^+, to which a water molecule is attached. The H^+ ion is unique among ions—it contains no electrons at all. In fact, it is nothing more than a proton, and its radius is only about 10^{-13} cm, compared with 10^{-8} cm for other simple ions. This means that the H^+ ion has an extraordinarily large charge for its size. Certainly such a positive ion could not exist in aqueous solution, where it would be surrounded by polar H_2O molecules that have unshared electron pairs. A coordinate bond would be established immediately between such an ion and a water molecule:

$$H^+ + \underset{\cdot\cdot}{\overset{\overset{\displaystyle H}{\cdot\cdot}}{:O:}}H \longrightarrow \left[H\underset{\cdot\cdot}{\overset{\overset{\displaystyle H}{\cdot\cdot}}{:O:}}H \right]^+$$

The greater the concentration of hydronium ions in a solution, the more acidic is the solution, that is, the more evident are its acidic properties. Thus, a solution of a weak acid is not as sour as a solution of a strong acid and is not able to conduct electric current as well nor react with active metals as rapidly. An indicator that shows its "acid color" when a strong acid is present may fail to do so in the presence of a weak acid. The minimum concentration of H_3O^+ ions required to make the indicator show its acid color differs for different indicators.

9-2 Bases

We have noted that one of the properties of acids is their ability to react with bases to form salts and water. Historically, bases have been regarded essentially as antiacids; they are substances which, when added to a solution containing an acid, cause the solution to lose its acidic properties. There are many bases, of which the following are well known:

1. Sodium hydroxide, NaOH, formerly called caustic soda or soda lye.
2. Potassium hydroxide, KOH, formerly called caustic potash or potash lye.
3. Calcium hydroxide, $Ca(OH)_2$, commonly called slaked lime.
4. Ammonia, NH_3, a gas that is highly soluble in water.

Substances commonly regarded as bases have the following properties:

1. They have a peculiar taste that has been variously described as caustic or bitter or soapy. It is a taste that is in marked contrast to the sourness of acids.
2. They impart characteristic colors to indicators. For example, they turn litmus blue and phenolphthalein pink.
3. They react with acids to form salts and water.
4. Their aqueous solutions contain the *hydroxide ion*, OH^-.

Unlike the acids, some of the most important bases are ionic compounds; that is, in the crystalline state they consist of ions, not molecules. For example, sodium hydroxide is a crystalline solid consisting of equal numbers of Na^+ and OH^- ions. When sodium hydroxide dissolves in water, the polar water molecules attract the Na^+ and OH^- ions and break the bonds between the ions in the crystal. The resulting solution is a good conductor of electric current because the dissolved base is present entirely in the form of ions. If solid NaOH is heated until it melts, the molten base is likewise able to conduct electric current because it is a liquid that consists entirely of positive and negative ions.

Some bases are covalent compounds. When they dissolve in water, they react incompletely with the water, giving rise to OH^- ions. For example,

$$\underset{\text{ammonia}}{NH_3} + \underset{\text{water}}{H_2O} \rightleftharpoons \underset{\text{ammonium ion}}{NH_4^+} + \underset{\text{hydroxide ion}}{OH^-}$$

$$\underset{\text{methylamine}}{CH_3NH_2} + \underset{\text{water}}{H_2O} \rightleftharpoons \underset{\text{methylammonium ion}}{CH_3NH_3^+} + \underset{\text{hydroxide ion}}{OH^-}$$

Usually the covalent bases are weak bases; their aqueous solutions do not conduct electricity as well as solutions of sodium hydroxide and the other ionic bases.

Some bases that are commonly referred to as weak bases owe their apparent weakness to their slight solubility. For example, $Ba(OH)_2$ is an ionic compound but its saturated aqueous solution does not contain a high concentration of OH^- ions because the compound is not highly soluble.

A base is described as a *strong base* if it is possible to establish a high concentration of OH^- ions by dissolving it in water. The best-known strong bases are NaOH and KOH, which are called *alkalies*. A base is usually classified as a *weak base* if it is not possible to establish a high concentration of OH^- ions by dissolving it in water. An indicator that shows its "basic color" when a strong base is present may fail to do so in the presence of a weak base. The minimum concentration of OH^- ions required to affect the indicator in this manner varies with different indicators.

Any solution that has the properties characteristic of solutions of bases is said to be an *alkaline* solution. The greater the concentration of hydroxide ions in a solution, the more strongly alkaline is the solution.

9-3 Salts

Historically, salts have been thought of as compounds that can be formed by allowing appropriate acids and bases to interact. For example, sodium chloride (NaCl) can be obtained by allowing aqueous solutions of sodium hydroxide (NaOH) and hydrochloric acid (HCl) to interact. A salt consists of positive ions derived from its parent base and negative ions derived from its parent acid.

Crystalline salts were once thought to consist of molecules. Within 20 years of the discovery of X-rays, crystallographers were using them to investigate the structures of crystals. These investigations soon revealed that crystalline salts are composed of ions, not molecules. When a salt dissolves in water, the polar water molecules attract the positive ions and the negative ions and break the bonds between them in the crystal. An aqueous solution of a salt is able to conduct electric current because the salt is present in the form of separate ions. If a salt is heated until it melts, the molten salt is likewise able to conduct electricity because the melt consists of positive and negative ions that are now able to move. In this respect, salts resemble ionic bases but stand in marked contrast to acids and covalent bases. In the solid state the ions of a salt or an ionic base are firmly held in place by the electrical forces of attraction that exist between each ion and its oppositely charged neighbors.

TABLE 9-1

Relationship between formulas of salts and the charges of their constituent ions

Positive ion	*Negative ion*	*Salt*
Ba^{++}	SO_4^{--}	$BaSO_4$
Al^{3+}	SO_4^{--}	$Al_2(SO_4)_3$
Ba^{++}	NO_3^-	$Ba(NO_3)_2$
Al^{3+}	NO_3^-	$Al(NO_3)_3$
Al^{3+}	PO_4^{3-}	$AlPO_4$
Ba^{++}	PO_4^{3-}	$Ba_3(PO_4)_2$
K^+	PO_4^{3-}	K_3PO_4
Ag^+	S^{--}	Ag_2S

Since a molten salt (or ionic base) is liquid, the ions of which it consists are free to move toward the oppositely charged electrodes.

Compounds that are ionic are electrically neutral because they contain positive ions and negative ions in such numbers that their total charges are equal. The positive sodium ion, Na^+, bears the same quantity of charge as the negative chloride ion, Cl^-, and this means that sodium chloride, NaCl, is electrically neutral because it consists of equal numbers of Na^+ and Cl^- ions. On the other hand, the positive calcium ion, Ca^{++}, bears twice the amount of charge of a sodium ion or a chloride ion. This means that the electroneutrality of a calcium chloride crystal requires the presence of twice as many Cl^- ions as Ca^{++} ions, so that the formula for the salt is $CaCl_2$. Several other examples of the relationship between the formula of a salt and the charges of its ions are given in Table 9-1.

The principle of electroneutrality enables us to derive the formulas of salts from the charges of their ions and also to derive the charge of an ion when the formula for the salt and the charge of the other ion are known. For example, knowing that the formula for manganese nitrate is $Mn(NO_3)_2$ and that it contains NO_3^- ions enables us to conclude that the manganese ion in this salt is bipositive, i.e., Mn^{++}.

9-4 Neutralization

When hydronium and hydroxide ions come together, they react to form water molecules. This happens, for example, when hydrochloric acid and sodium hydroxide solutions are brought together. The equation

$$NaOH + HCl \longrightarrow NaCl + H_2O$$

does not give a faithful picture of this chemical happening. As written, the equation fails to focus attention on the H_3O^+ ions in the acid solution and the OH^- ions in the base solution. Reactions between strong acids and strong bases in aqueous solutions can more adequately be represented by the equation

$$H_3O^+ + OH^- \longrightarrow 2H_2O$$

This process, known as *neutralization*, is typical of acids and bases.

In a solution of a weak acid there are more acid molecules than hydronium ions. The reaction between a solution of a weak acid and a solution of a strong base may be represented, therefore, as a reaction between hydroxide ions and the molecules of the acid. For example,

$$\underset{\text{(a weak acid)}}{\underset{\text{acetic acid}}{HC_2H_3O_2}} + \underset{\text{(from a strong base)}}{\underset{\text{hydroxide ion}}{OH^-}} \longrightarrow \underset{\text{water}}{H_2O} + \underset{\text{ion}}{\underset{\text{acetate}}{C_2H_3O_2^-}}$$

In a solution of a weak base there are more molecules of the base than hydroxide ions. The reaction between a solution of a weak base and a solution of a strong acid can properly be represented, therefore, as a reaction between hydronium ions and the molecules of the base. For example,

$$\underset{\text{(from a strong acid)}}{\underset{\text{hydronium ion}}{H_3O^+}} + \underset{\text{(a weak base)}}{\underset{\text{ammonia}}{NH_3}} \longrightarrow \underset{\text{water}}{H_2O} + \underset{\text{ion}}{\underset{\text{ammonium}}{NH_4^+}}$$

When our primary concern is with the formula of the salt that is present after an acid has reacted with a base, we may prefer to write an equation in which we use the molecular formulas for both the acid and the base, regardless of whether they are weak or strong. The formula for the salt can readily be inferred from the charges of the positive ion of the base and the negative ion of the acid.

EXAMPLE

Complete and balance the equation for the neutralization of sulfuric acid with sodium hydroxide, given that their formulas are H_2SO_4 and NaOH, respectively.

Since this is a neutralization reaction, the products are water and a salt. The positive ion of the base is Na^+, and the negative ion of the acid is SO_4^{--}. Therefore, the formula for the salt that is formed when sodium hydroxide neutralizes sulfuric acid is Na_2SO_4. Thus, before balancing, we have

$$NaOH + H_2SO_4 \longrightarrow Na_2SO_4 + H_2O \qquad \text{(not balanced)}$$

If you have learned how to balance equations as described in Chapter 2, you will have no difficulty in arriving at the final result:

$$2NaOH + H_2SO_4 \longrightarrow Na_2SO_4 + 2H_2O$$

The formula for any salt can be predicted from the formulas of its parent acid and its parent base by mere inspection. All that is required is a familiarity with the implications of the formulas as to the charges of the ions. The relationships between formulas of salts and the formulas of the corresponding acids and bases are illustrated in Table 9-2.

The formula for an acid may show that there is more than one replaceable hydrogen atom in each molecule. For example, the formula for sulfuric acid, H_2SO_4, shows that there are two replaceable hydrogen atoms. The balanced equation written above for the neutralization of sulfuric acid by sodium hydroxide shows that 2 moles of NaOH are required to neutralize 1 mole of H_2SO_4. However, when equal numbers of moles of NaOH and H_2SO_4 are brought together, the sulfuric acid is half-neutralized; the salt that is formed has the formula $NaHSO_4$. This compound is an example of an *acid salt*. It is a salt, but it is also an acid because it is capable of reacting with another mole of sodium hydroxide:

$$\underset{\text{base}}{NaOH} + \underset{\text{acid}}{H_2SO_4} \longrightarrow \underset{\text{salt}}{NaHSO_4} + \underset{\text{water}}{H_2O}$$

$$\underset{\text{base}}{NaOH} + \underset{\text{acid}}{NaHSO_4} \longrightarrow \underset{\text{salt}}{Na_2SO_4} + \underset{\text{water}}{H_2O}$$

Many acids contain hydrogen that is not replaceable in addition to that which is. For example, the molecular formula for acetic acid is $C_2H_4O_2$, but one of the hydrogen atoms is acidic and the others are not. This

TABLE 9-2

Relationship between formulas of salts and formulas of the corresponding bases and acids

Base	*Acid*	*Salt*
$NaOH$	HCl	$NaCl$
KOH	HNO_3	KNO_3
$Ca(OH)_2$	H_2SO_4	$CaSO_4$
NH_4OH	H_3AsO_4	$(NH_4)_3AsO_4$
$NaOH$	H_2CO_3	Na_2CO_3
$Ca(OH)_2$	$HC_2H_3O_2$	$Ca(C_2H_3O_2)_2$
$Ca(OH)_2$	H_3PO_4	$Ca_3(PO_4)_2$

information is implied by writing the formula $HC_2H_3O_2$. The salt $NaC_2H_3O_2$ that is formed by the reaction of 1 mole of sodium hydroxide with 1 mole of acetic acid is not an acid salt, because the hydrogen atoms within the acetate ion are not acidic.

9-5 Oxides of Metals and Nonmetals

Many elements combine with oxygen to form binary compounds called oxides. Oxides of virtually all the elements are known, although some must be prepared by indirect procedures because they cannot be obtained by direct union.

In the oxides of the various elements, oxygen is in the oxidation state −2. The oxidation states of the elements combined with oxygen in the oxides vary from +1, as in Cu_2O and H_2O, to +7, as in Cl_2O_7 and Mn_2O_7.

Practically all oxides of nonmetals are acidic; if they dissolve in water, the solutions turn litmus red and show the other characteristics of acid solutions because these oxides react with water to form acids at the time they dissolve:

$$CO_2 + H_2O \longrightarrow \underset{\text{carbonic acid}}{H_2CO_3}$$

$$P_4O_{10} + 6H_2O \longrightarrow \underset{\text{phosphoric acid}}{4H_3PO_4}$$

Oxides of metals in which the metals are in their lowest oxidation states are basic; if they dissolve in water, the solutions turn litmus blue and show the other characteristics of alkaline solutions because these oxides react with water to form ionic bases at the time they dissolve:

$$\underset{\text{sodium oxide}}{Na_2O} + H_2O \longrightarrow \underset{\text{sodium hydroxide}}{2NaOH}$$

$$\underset{\text{calcium oxide}}{CaO} + H_2O \longrightarrow \underset{\text{calcium hydroxide}}{Ca(OH)_2}$$

In their highest oxidation states, many of the metals that are characterized by variable oxidation states form acidic oxides:

$$Mn_2O_7 + H_2O \longrightarrow \underset{\text{permanganic acid}}{2HMnO_4}$$

$$CrO_3 + H_2O \longrightarrow \underset{\text{chromic acid}}{H_2CrO_4}$$

All compounds whose names include the word *acid* contain hydrogen; those that also contain oxygen are known collectively as *oxyacids*. Nitric acid, HNO_3, and sulfuric acid, H_2SO_4, are examples of oxyacids. Hydrochloric acid, HCl, is an example of an acid that is not an oxyacid; it may be called a *binary acid* since it contains only two elements.

9-6 Anhydrides of Oxyacids and Basic Hydroxides

The *anhydride* of an oxyacid is the oxide formed when the acid loses the elements of water. Most such oxides will react with water to form the corresponding oxyacid. Thus, CO_2 is the anhydride of carbonic acid, H_2CO_3, because carbonic acid loses the elements of water to form CO_2 and CO_2 reacts with water to form carbonic acid.

The oxidation state of each element in an oxyacid is the same as in its anhydride. For example, carbon is in the +4 state in both CO_2 and H_2CO_3. This principle makes it possible to write the empirical formula for an acidic anhydride whenever the formula for the oxyacid is known.

EXAMPLE 1

Write the formula for the anhydride of silicic acid, H_2SiO_3.

The oxidation number of silicon in H_2SiO_3 is +4. Therefore, the empirical formula for the acidic anhydride corresponding to H_2SiO_3 is SiO_2, because this is the simplest formula for an oxide of silicon that shows silicon and oxygen in oxidation states +4 and −2, respectively.

EXAMPLE 2

What is the formula for the acidic oxide that reacts with water to form phosphoric acid, H_3PO_4?

In H_3PO_4 phosphorus is in the oxidation state +5. The simplest formula for an oxide of phosphorus with phosphorus in the +5 state and oxygen in the −2 state is P_2O_5. Therefore, the empirical formula for the anhydride of phosphoric acid is P_2O_5. (Molecular weight determinations show that the molecular formula for this oxide is P_4O_{10}.)

Oxides that react with water to form basic hydroxides are called *basic oxides* or *basic anhydrides*. The oxidation state of each element is the same in the oxide as in the hydroxide. We can use this relationship to deduce the formula of a basic oxide from the formula of the corresponding basic hydroxide.

EXAMPLE 1

What is the formula for the oxide that reacts with water to form calcium hydroxide, $Ca(OH)_2$?

In $Ca(OH)_2$ the oxidation state of calcium is +2. The oxide corresponding to $Ca(OH)_2$ contains calcium in the +2 state and oxygen in the −2 state. Therefore, its formula is CaO.

EXAMPLE 2

What is the formula for the anhydride of $Fe(OH)_3$?

In $Fe(OH)_3$ the oxidation state of iron is +3. The corresponding basic oxide contains iron in the +3 state and oxygen in the −2 state. Therefore, the formula for this basic anhydride is Fe_2O_3.

The formula for the anhydride of an oxyacid or a basic hydroxide may easily be derived without reference to oxidation numbers by mentally removing all "H" as water. For example, removing both H atoms as H_2O from H_2SiO_3 leaves SiO_2; removing both H atoms as H_2O from $Ca(OH)_2$ leaves CaO. This method is slightly more complicated in such cases as $Fe(OH)_3$ or H_3PO_4, because after removal of one water molecule, two molecules must be mentally combined to remove the remaining hydrogen atoms.

Writing chemical equations for the decomposition of basic hydroxides or oxyacids to produce water and the corresponding anhydride is easy if we recall how the formula of the anhydride is inferred from the formula of the base or acid.

EXAMPLE 1

Complete the equation

$$Al(OH)_3 \xrightarrow{\text{heat}}$$

Since the oxidation state of aluminum in the hydroxide (base) is +3, the formula for the anhydride must be Al_2O_3. The first step in completing the equation is to write the formulas for the products:

$$Al(OH)_3 \longrightarrow Al_2O_3 + H_2O \qquad \text{(not balanced)}$$

Then we must balance the equation by inserting the necessary coefficients:

$$2Al(OH)_3 \longrightarrow Al_2O_3 + 3H_2O$$

EXAMPLE 2

Complete the equation

$$HIO_3 \xrightarrow{\text{heat}}$$

Since the oxidation state of iodine in the acid is +5, the formula for the anhydride is I_2O_5, and the equation is completed as follows:

$$2HIO_3 \longrightarrow H_2O + I_2O_5$$

9-7 Predicting Formulas for Oxyacids and Basic Hydroxides

When an acidic oxide or a basic oxide reacts with water, the resulting oxyacid or basic hydroxide contains all the elements in their original oxidation states. For example,

$$CaO + H_2O \longrightarrow Ca(OH)_2$$
$$SO_3 + H_2O \longrightarrow H_2SO_4$$

On this basis we can often predict the formula for the oxyacid or basic hydroxide corresponding to the anhydride. Thus, the formula for the hydroxide corresponding to the oxide K_2O is KOH, and the formula for the hydroxide that corresponds to the oxide MgO is $Mg(OH)_2$. For the oxide Al_2O_3 we would predict the formula for the corresponding hydroxide to be $Al(OH)_3$. There is a compound with the formula $Al(OH)_3$, but there is also a compound with the formula AlO(OH), and Al_2O_3 is the anhydride of both of them.

It is prudent to be cautious in writing the equations for the reactions of oxides with water. For example, it could not be predicted with certainty that SO_3 reacts with water to form H_2SO_4; the formula could conceivably be H_4SO_5 or H_6SO_6, because the sulfur would be in the +6 state in either case. It is a fact, however, that sulfuric acid is H_2SO_4 rather than H_4SO_5 or H_6SO_6. Again, the oxide N_2O_5 might conceivably react with water to form H_5NO_5 or H_3NO_4 or HNO_3; actually, nitric acid is HNO_3 rather than H_3NO_4 or H_5NO_5. On the other hand, there are three phosphoric acids, with the formulas HPO_3, H_3PO_4, and $H_4P_2O_7$; they all have the same anhydride, P_4O_{10}, because phosphorus is in the +5 state in each of them.

9-8 Reactions of Basic and Acidic Oxides

When a basic oxide reacts with an acid, the products are water and the same salt that would result from the reaction of the corresponding basic hydroxide with the same acid. For example, if we are given the equation for the reaction of calcium hydroxide with nitric acid,

$$Ca(OH)_2 + 2HNO_3 \longrightarrow Ca(NO_3)_2 + 2H_2O$$

we can infer that the corresponding basic oxide, CaO, reacts with nitric acid according to the equation

$$CaO + 2HNO_3 \longrightarrow Ca(NO_3)_2 + H_2O$$

The reaction of an acidic oxide with a basic hydroxide produces water and the same salt that would be obtained if the basic hydroxide were reacting with the corresponding oxyacid. Thus, from the equation for the reaction of potassium hydroxide with nitric acid,

$$KOH + HNO_3 \longrightarrow KNO_3 + H_2O$$

we are able to write the equation for the reaction of potassium hydroxide with the anhydride of nitric acid:

$$2KOH + N_2O_5 \longrightarrow 2KNO_3 + H_2O$$

The reaction between a basic oxide and an acidic oxide produces the same salt that would be obtained if the corresponding basic hydroxide were reacting with the oxyacid of which the acidic oxide is the anhydride. From the equation

$$2KOH + H_2SO_3 \longrightarrow K_2SO_3 + 2H_2O$$

we can infer that the reaction that occurs when gaseous sulfur dioxide comes into contact with solid potassium oxide will proceed according to the equation

$$K_2O(s) + SO_2(g) \longrightarrow K_2SO_3(s)$$

The classification of an oxide as acidic or basic is sometimes complicated by the fact that the oxide does not dissolve in water and react with it to a sufficient degree to give a solution that is either acidic or alkaline. Magnesium oxide, for example, dissolves to only a small extent, and it also dissolves slowly; a period of time must elapse before it is possible to observe the pink color it eventually gives with phenolphthalein. However, the oxide "dissolves" readily in strong acids, such as hydrochloric acid, and this quickly establishes that it is a basic oxide:

$$MgO(s) + 2HCl \longrightarrow MgCl_2 + H_2O$$

On the other hand, tin(IV) oxide is not noticeably soluble in water, but it goes into solution by chemical reaction when it is treated with strong bases, such as sodium hydroxide:

$$SnO_2(s) + 2NaOH \longrightarrow Na_2SnO_3 + H_2O$$

This behavior indicates that tin(IV) oxide is an acidic oxide.

Certain oxides that are insoluble in water will go into solution if treated with either a strong acid or a strong base. An example is tin(II) oxide:

$$SnO(s) + 2HCl \longrightarrow SnCl_2 + H_2O$$

$$SnO(s) + 2NaOH \longrightarrow Na_2SnO_2 + H_2O$$

In reacting with HCl, tin(II) oxide behaves as a basic oxide; in reacting with NaOH, it behaves as an acidic oxide. Such oxides are said to be *amphoteric* (Greek: *amphoteros*, both). Amphoteric behavior also characterizes other types of compounds, including the hydroxides corresponding to the amphoteric oxides, such as $Sn(OH)_2$.

9-9 The Brønsted-Lowry Concept of Acids and Bases

The definitions of acids and bases as substances that produce H_3O^+ ions and OH^- ions, respectively, in their aqueous solutions seem to imply that acid–base reactions are limited to aqueous solutions. In 1923 J. N. Brønsted in Denmark and T. Martin Lowry in England independently proposed broadened definitions of acids and bases that are not limited to aqueous solutions. They proposed that an *acid* be defined as any molecule or ion that donates a proton to any other molecule or ion and that a *base* be defined as any molecule or ion that accepts a proton from any other molecule or ion. In simplest terms, an acid is a proton donor and a base is a proton acceptor, according to the definitions suggested by Brønsted and Lowry. Inherent in these definitions are the implications that a molecule or an ion can act as an acid only in the presence of a base and that a molecule or an ion that is capable of acting as a base can do so only in the presence of an acid. To avoid confusion between old and new definitions, we may designate a proton donor as a *Brønsted-Lowry acid* and a proton acceptor as a *Brønsted-Lowry base.*

An aqueous solution of a strong acid contains a high concentration of H_3O^+ ions, and an aqueous solution of a strong base contains a high concentration of OH^- ions. The neutralization that occurs when an aqueous solution of a strong acid is mixed with an aqueous solution of a strong base involves a reaction in which a proton is transferred from a hydronium ion to a hydroxide ion:

$$\underset{\text{Brønsted-Lowry acid}}{H_3O^+} + \underset{\text{Brønsted-Lowry base}}{OH^-} \rightleftharpoons H_2O + H_2O$$

The H_3O^+ ion is acting as a Brønsted-Lowry acid in this reaction. When a proton is transferred from the H_3O^+ ion, a molecule of water remains.

Another molecule of water is formed upon acceptance of the proton by the hydroxide ion, which is acting as a Brønsted-Lowry base.

In an aqueous solution of a weak acid there are more acid molecules than hydronium ions, and the reaction that occurs between such a solution and a solution of a strong base may involve, at least in part, the direct transfer of a proton from an acid molecule to a hydroxide ion:

$$\underset{\text{Brønsted-Lowry acid}}{HC_2H_3O_2} + \underset{\text{Brønsted-Lowry base}}{OH^-} \rightleftharpoons H_2O + C_2H_3O_2^-$$

When a proton is transferred from an acetic acid molecule, an acetate ion remains; the acceptance of the proton by the hydroxide ion produces a molecule of water.

The reaction that occurs when a solution of a strong acid is added to a solution of a weak base may involve, at least in part, the direct transfer of a proton from a hydronium ion to a molecule of the base:

$$\underset{\text{Brønsted-Lowry acid}}{H_3O^+} + \underset{\text{Brønsted-Lowry base}}{NH_3} \rightleftharpoons NH_4^+ + H_2O$$

When a proton is transferred to an ammonia molecule, an ammonium ion is produced; the removal of the proton from a hydronium ion produces a molecule of water.

Any reaction in which a proton is transferred from a Brønsted-Lowry acid to a Brønsted-Lowry base is called *protolysis*. If the reaction occurs in aqueous solution, and if one of the products is water, the reaction is an example of *neutralization in aqueous solution*. If protolysis in aqueous solution does not form water as one of the products, the term neutralization ought not be applied to it.

Protolysis occurs when an acid dissolves in water:

$$\underset{\text{Brønsted-Lowry acid}}{HCl} + \underset{\text{Brønsted-Lowry base}}{H_2O} \rightleftharpoons H_3O^+ + Cl^-$$

By definition, the water is a Brønsted-Lowry base in this reaction, because it accepts a proton from the HCl. Protolysis also occurs when a covalent base dissolves in water:

$$\underset{\text{Brønsted-Lowry acid}}{H_2O} + \underset{\text{Brønsted-Lowry base}}{NH_3} \rightleftharpoons NH_4^+ + OH^-$$

In this reaction, the water is a Brønsted-Lowry acid, because it donates a proton to the NH_3. A molecule or ion that can act as a proton donor in the presence of a base or as a proton acceptor in the presence of an acid is said to be *amphiprotic*. Water is an example of an amphiprotic substance. The HSO_4^- ion is an example of an amphiprotic ion: it can accept a proton from a very strong acid, and it can donate a proton to a strong base:

$$H_3O^+ + HSO_4^- \rightleftharpoons H_2SO_4 + H_2O$$

$$HSO_4^- + OH^- \rightleftharpoons H_2O + SO_4^{--}$$

The term "amphoteric" has long been used to refer to substances that can neutralize either acids or bases. "Amphiprotic" is a more recent term that is used to refer to the ability to act either as a Brønsted-Lowry acid or as a Brønsted-Lowry base.

9-10 Conjugate Acid-Base Pairs

According to the Brønsted-Lowry definitions, there must be a base corresponding to every acid and an acid corresponding to every base:

base	+	proton	$\rightleftharpoons$	acid
NH_3	+	H^+	$\rightleftharpoons$	NH_4^+
$C_2H_3O_2^-$	+	H^+	$\rightleftharpoons$	$HC_2H_3O_2$
O^{--}	+	H^+	$\rightleftharpoons$	OH^-
OH^-	+	H^+	$\rightleftharpoons$	H_2O
H_2O	+	H^+	$\rightleftharpoons$	H_3O^+
SO_4^{--}	+	H^+	$\rightleftharpoons$	HSO_4^-

An acid and a base that are related in this manner are called a *conjugate acid–base pair.* For example, NH_3 is the base that is conjugate to the acid NH_4^+; NH_4^+ is the acid that is conjugate to the base NH_3; and NH_4^+ and NH_3 are a conjugate pair because they differ only in the presence or the absence of a proton.

Since protons under ordinary conditions are incapable of independent existence, the reaction of a Brønsted-Lowry acid with a Brønsted-Lowry base is a proton transfer reaction in which one of the products is a Brønsted-Lowry base and the other product is a Brønsted-Lowry acid:

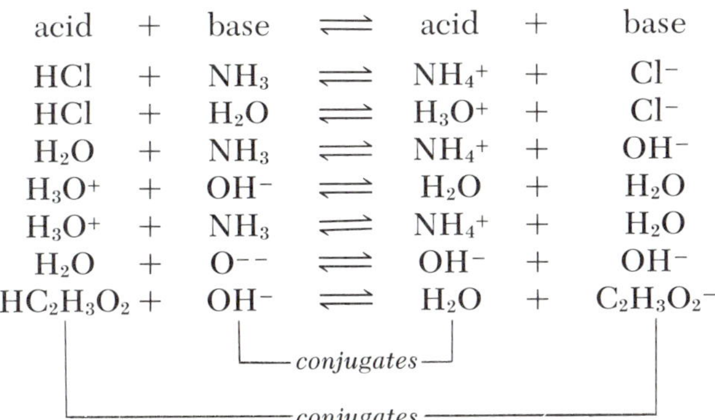

acid	+	base	$\rightleftharpoons$	acid	+	base
HCl	+	NH_3	$\rightleftharpoons$	NH_4^+	+	Cl^-
HCl	+	H_2O	$\rightleftharpoons$	H_3O^+	+	Cl^-
H_2O	+	NH_3	$\rightleftharpoons$	NH_4^+	+	OH^-
H_3O^+	+	OH^-	$\rightleftharpoons$	H_2O	+	H_2O
H_3O^+	+	NH_3	$\rightleftharpoons$	NH_4^+	+	H_2O
H_2O	+	O^{--}	$\rightleftharpoons$	OH^-	+	OH^-
$HC_2H_3O_2$	+	OH^-	$\rightleftharpoons$	H_2O	+	$C_2H_3O_2^-$

In each case the new acid is the old base with a proton added to it; thus, the new acid and the old base are a conjugate pair. The new base is the old acid from which a proton has been removed; therefore, the new base and the old acid are also a conjugate pair.

Since the products of an acid–base reaction are themselves an acid and a base, it should be anticipated that there will be a tendency for the new acid to donate a proton to the new base, forming the old base and the old acid. In brief, most acid–base reactions, especially in aqueous solution, are incomplete; they are reversible and proceed only to the point of equilibrium. At this point the rate of the reverse reaction is equal to the rate of the forward reaction.

The strength of a Brønsted-Lowry acid is measured by its tendency to donate a proton, and the strength of a Brønsted-Lowry base is measured by its tendency to accept a proton. Whether an acid–base reaction reaches equilibrium when only a little or a great deal of the starting acid and the starting base have reacted depends on the relative strengths of the new acid and the new base as compared to the old acid and the old base. Hydrogen chloride is a strong acid; the HCl molecule gives up its proton readily. From this it follows that the chloride ion is necessarily a weak base; its tendency to accept a proton, forming an HCl molecule, is correspondingly slight. On the other hand, OH^- ion is a strong base; it readily acquires a proton, forming an H_2O molecule. From this it follows that H_2O is a weak acid. *The stronger an acid, the weaker is its conjugate base; the stronger a base, the weaker is its conjugate acid.* Therefore, the reaction between a strong acid and a strong base goes virtually to completion before reaching equilibrium, because the new acid and the new base are correspondingly weak and the reverse reaction has only a slight tendency to occur as compared to the forward reaction.

When hydrogen chloride is dissolved in water, the forward reaction proceeds almost to completion:

$$\underset{\text{acid}}{HCl} + \underset{\text{base}}{H_2O} \rightleftharpoons \underset{\text{acid}}{H_3O^+} + \underset{\text{base}}{Cl^-}$$

In this equation we have used arrows of unequal lengths to emphasize that in the equilibrium solution the concentration of H_3O^+ ions is much greater than the concentration of HCl molecules. In other words, the forward reaction is much more complete than the reverse reaction: The tendency for a proton to be transferred from an HCl molecule to an H_2O molecule is far greater than the tendency for such a transfer to occur between an H_3O^+ ion and a Cl^- ion. This is the inevitable consequence of the H_3O^+ ion being a weaker acid than HCl and the Cl^- ion being a weaker base than H_2O.

In an aqueous solution of acetic acid, the concentration of acetic acid molecules is several times greater than the concentration of hydronium ions:

$$\underset{\text{acid}}{HC_2H_3O_2} + \underset{\text{base}}{H_2O} \rightleftharpoons \underset{\text{acid}}{H_3O^+} + \underset{\text{base}}{C_2H_3O_2^-}$$

As indicated again by arrows of unequal length, equilibrium is attained when one of the reactions is far more complete than its opposite. Here, equilibrium is reached as soon as only a small fraction of the acetic acid molecules have transferred protons to H_2O molecules. This is because the tendency for a proton to be transferred from an H_3O^+ ion to a $C_2H_3O_2^-$ ion is considerably greater than the tendency for a proton to be transferred from an $HC_2H_3O_2$ molecule to an H_2O molecule. It is the inevitable consequence of the H_3O^+ ion being a stronger acid than $HC_2H_3O_2$ and the $C_2H_3O_2^-$ ion being a stronger base than H_2O.

Every acid–base reaction favors the formation of the weaker acid and the weaker base. When the stronger acid donates a proton, the conjugate base that is formed is weaker than the base that accepted it. Likewise, the acceptance of the proton by the stronger base produces an acid that is weaker than the one that donated it. At the conclusion of a protolytic reaction there is established an equilibrium that is characterized by the presence of relatively greater amounts of the weaker acid and the weaker base. If the weaker acid and the weaker base are the starting acid and the starting base, equilibrium will be reached when the reaction between them is very incomplete. If the starting species are the stronger acid and the stronger base, equilibrium will not be reached until the reaction between them has proceeded much further toward completion. The stronger the starting acid and base and the weaker their conjugates, the more nearly complete the reaction will be.

9-11 The Leveling Effect of Water

The following acids have traditionally been classified as the strong acids:

Perchloric acid	$HClO_4$
Hydriodic acid	HI
Hydrobromic acid	HBr
Sulfuric acid	H_2SO_4
Hydrochloric acid	HCl
Nitric acid	HNO_3

In general, no two acids have exactly the same acid strength. In aqueous solution the acids named above ionize virtually completely;

this makes it impossible to distinguish their acid strengths in aqueous solution. In other solvents, properly chosen, they do not appear to be equally strong; in fact, perchloric acid is a stronger acid than hydriodic acid, and so on through the list.

In accordance with the general rule that an acid–base reaction proceeds well toward completion if the products are a weaker acid and a weaker base, each of these acids reacts virtually completely with water; the acid that is formed is the hydronium ion, H_3O^+. Each of them is a stronger acid than H_3O^+—so much so that the reaction is virtually complete in each case. They appear to have the same acid strength in aqueous solution because the same acid, H_3O^+ ion, is responsible for the acidity in each instance. This phenomenon, which occurs because of the extent of the reaction with water, is known as the *leveling effect*. Acids stronger than H_3O^+ are leveled to the strength of H_3O^+ when placed in aqueous solution. Acids that are weaker than H_3O^+ do not react completely with water; instead, each one reacts to an extent that is proportional to, and is a measure of, its acid strength.

The intrinsic differences in the proton-donating tendencies of the individual substances in our list of traditional strong acids will be evident only in a solvent that has relatively weak proton-accepting tendencies. In other words, the solvent must be a weaker base than water in order, for example, for hydrochloric acid to react with the solvent to a lesser degree than perchloric acid. An example of such a differentiating solvent is anhydrous acetic acid; it is a weak acid in aqueous solution, and as a solvent it is only a very weak base.

9-12 Hydrolysis of Ions

Aqueous solutions of the salt sodium chloride, NaCl, are found to be neither acidic nor alkaline when they are tested with acid–base indicators. Unlike acids, which form H_3O^+ ions in aqueous solutions, and bases, whose aqueous solutions contain OH^- ions, sodium chloride forms neither H_3O^+ ions nor OH^- ions when it dissolves in water. Not only is the solution electrically neutral, which is true of all solutions, but it is also neutral in the sense that it is neither acidic nor alkaline.

Aqueous solutions of the salt sodium acetate, $NaC_2H_3O_2$, are alkaline because hydroxide ions are formed in the reaction of acetate ions with water:

$$\underset{\text{acid}}{H_2O} + \underset{\text{base}}{C_2H_3O_2^-} \rightleftharpoons \underset{\text{acid}}{HC_2H_3O_2} + \underset{\text{base}}{OH^-}$$

As indicated by the arrows of unequal length, the reaction reaches equilibrium as soon as only a fraction of the acetate ions have reacted with water. This is to be expected, since the OH^- ion is a stronger base than the $C_2H_3O_2^-$ ion and $HC_2H_3O_2$ is a stronger acid than H_2O. Regardless of the fact that the forward reaction occurs only to a slight extent, it is of great importance that it occurs at all because it forms OH^- ions. The formation of even a relatively small number of OH^- ions in an aqueous solution is of real significance because it makes the solution alkaline instead of neutral.

The reaction of acetate ion with water is an example of *hydrolysis*. In Brønsted-Lowry terminology, hydrolysis is defined as a proton transfer reaction in which H_2O is a reactant. In the hydrolysis of acetate ion, H_2O is the proton donor. Recalling that H_2O is a product when neutralization occurs in aqueous solution, we see that hydrolysis and neutralization are defined so as to make the one the opposite of the other.

Aqueous solutions of the salt ammonium chloride, NH_4Cl, are acidic in consequence of the hydrolysis of ammonium ions, which produces hydronium ions:

$$\underset{\text{acid}}{NH_4^+} + \underset{\text{base}}{H_2O} \rightleftharpoons \underset{\text{acid}}{H_3O^+} + \underset{\text{base}}{NH_3}$$

The reaction is very incomplete because H_3O^+ ion is a stronger acid than NH_4^+ ion and NH_3 is a stronger base than H_2O. The fact that the reaction occurs even to a slight degree is, however, of great importance because it makes the solution acidic instead of neutral.

Aqueous solutions of sodium chloride are neutral because neither of its ions reacts with water to produce detectable quantities of H_3O^+ or OH^- ions. An aqueous solution of a salt is alkaline only if its negative ion acts as a Brønsted-Lowry base, which requires that it be derived from a weak acid, for example, $HC_2H_3O_2$. An aqueous solution of a salt is acidic only if its positive ion acts as a Brønsted-Lowry acid, which requires that it be derived from a weak base, for example, NH_3.

We can make correct predictions regarding the hydrolysis of salts by using the following simple rules:

1. If a salt derived from a strong acid and a weak base is soluble in water, its aqueous solution is acidic. *Examples:* NH_4Cl, $(NH_4)_2SO_4$, and NH_4NO_3 are derived from strong acids and a weak base; all of them dissolve in water to give solutions that are acidic because a small but significant fraction of the NH_4^+ ions hydrolyze, forming H_3O^+ ions.
2. If a salt derived from a weak acid and a strong base is soluble in water, its aqueous solution is alkaline. *Examples:* $NaC_2H_3O_2$ and KCN are derived from strong bases and weak acids; they dissolve in water to give solutions that are alkaline because a small but significant fraction of the negative ions hydrolyze, forming OH^- ions.

3. If a salt of a strong acid and a strong base is soluble in water, its aqueous solution is neutral. *Examples:* NaCl and KNO_3 are derived from strong acids and strong bases; they dissolve in water to give solutions that are neutral because neither ion hydrolyzes in either case.
4. If a salt of a weak acid and a weak base is soluble in water, its aqueous solution will be acidic, neutral, or alkaline, depending on the relative weakness of the parent acid and the parent base. In solutions of these salts, both ions tend to hydrolyze and the outcome depends on the relative abilities of the positive ions to donate protons to water and of the negative ions to accept protons from water. If the abilities are equally matched, as in the case of ammonium acetate, $NH_4C_2H_3O_2$, the solution remains neutral in spite of hydrolysis. Otherwise the solution is acidic, as in the case of ammonium formate, NH_4CHO_2, or alkaline, as in the case of ammonium cyanide, NH_4CN. We cannot make a prediction for a specific case unless we have available specific information as to the relative weakness of the parent acid and the parent base.

9-13 The Lewis Concept of Acids and Bases

Another definition for acids and bases, which is even more general than the Brønsted-Lowry definition, was proposed by G. N. Lewis in 1923. Lewis defined an acid as an electron-pair acceptor and a base as an electron-pair donor. For example, in the reaction

$$H^+ + :\overset{\displaystyle H}{\underset{\displaystyle H}{\overset{..}{\underset{..}{N}}}}:H \longrightarrow \left[H:\overset{\displaystyle H}{\underset{\displaystyle H}{\overset{..}{\underset{..}{N}}}}:H \right]^+$$

the proton is a Lewis acid and the ammonia molecule is a Lewis base.

In any reaction in which a coordinate bond is formed (Section 5-11), a molecule or ion with an incomplete group of electrons around one of its atoms acts as a Lewis acid in accepting a pair of electrons from another molecule or ion. The molecule or ion that furnishes these electrons is acting as a Lewis base. Note that in a Lewis acid–base reaction the base does not give up a pair of electrons—it only allows the acid to share them.

In every Brønsted-Lowry acid–base reaction the base donates a pair of electrons to a proton. This means that every Brønsted-Lowry base may also be classed as a Lewis base. Conversely, every Lewis base satisfies the definition of a Brønsted-Lowry base when it donates a pair of electrons to a proton, thereby being a proton acceptor. The difference in the two acid–base concepts lies in the definition of an acid. Whereas the reaction

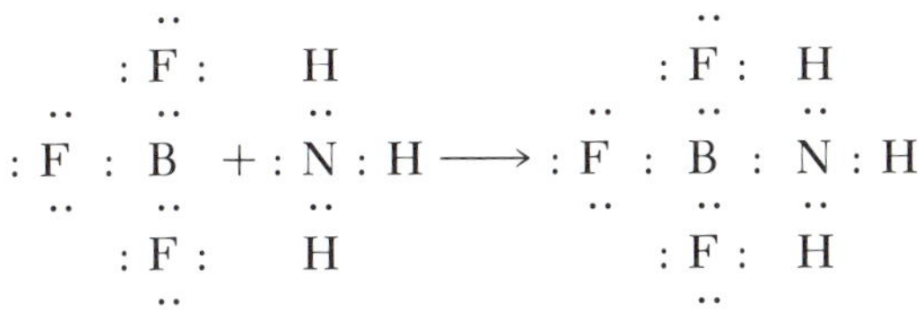

```
            ..                       ..
           : F :    H               : F :  H
      ..    ..      ..          ..    ..    ..
    : F  :  B  + : N : H ⟶    : F  :  B  :  N : H
      ..    ..      ..          ..    ..    ..
           : F :    H               : F :  H
            ..                       ..
```

involves the formation of a coordinate bond and is therefore a Lewis acid–base reaction, no proton is involved, so it does not satisfy the Brønsted-Lowry definition for an acid–base reaction.

The Brønsted-Lowry definition limits acids to those substances that contain one or more protons available for sharing electrons made available by any base. The Lewis definition of acids includes all substances that contain any kind of atom that can share electrons with any base. The Lewis extension of the traditional concepts has made possible the correlation of a much larger amount of information.

NEW TERMS

Acid (Brønsted-Lowry definition): a proton donor. In aqueous solution, acids give rise to hydronium (H_3O^+) ions.

Acid salt: a compound that is both an acid and a salt; for example, $NaHSO_4$.

Acidic anhydride: an acidic oxide, designated as the anhydride of a specified acid; for example, we say that SO_2 is an acidic oxide or that SO_2 is the anhydride of sulfurous acid.

Alkali: a highly soluble strong base; for example, NaOH and KOH.

Alkaline solution: a solution that has properties characteristic of solutions of bases.

Amphiprotic substance: a substance capable of acting as a proton donor in the presence of a proton acceptor and as a proton acceptor in the presence of a proton donor.

Amphoterism: the ability to act as an acid and also as a base.

Anhydride: the oxide formed when an oxyacid or a basic hydroxide loses the elements of water. The oxidation state of each element is the same in the oxide as in the acid or base of which it is the anhydride.

Base (Brønsted-Lowry definition): a proton acceptor. In aqueous solution, bases give rise to hydroxide (OH^-) ions.

Basic anhydride: a basic oxide, designated as the anhydride of a specified base; for example, we say that CaO is a basic oxide, or that CaO is the anhydride of the base calcium hydroxide.

Conjugate acid–base pair: two species so related that one is formed from the other by transfer of a proton. For example, NH_3 is the base conjugate to the acid NH_4^+, NH_4^+ is the acid conjugate to the base NH_3, and NH_4^+ and NH_3 are a conjugate acid–base pair.

Hydrolysis (Brønsted-Lowry definition): any protolytic reaction in which water is a reactant.

Indicator: a substance, natural or synthetic, that exhibits different colors in aqueous solutions, depending on the degree of acidity or alkalinity of the solutions.

Leveling effect (of water on an acid): the effect of water (the solvent) in reducing the acid strength of a solute to the level of the acid strength of the hydronium ion. Thus, H_3O^+ is the strongest acid that can exist in aqueous solution.

Lewis acid: an electron-pair acceptor.

Lewis base: an electron-pair donor.

Neutral solution: a solution that is neither acidic nor alkaline.

Neutralization: the reaction between an acid and a base. In Brønsted-Lowry terminology, neutralization is any protolytic reaction in which the solvent is a product. Neutralization in aqueous solution involves the reaction between H_3O^+ and OH^- ions, forming H_2O molecules, with the positive ion of the base and the negative ion of the acid remaining in solution or precipitating as an insoluble salt.

Oxyacid: an acid that contains both hydrogen and oxygen combined with a third element.

Protolysis: any proton-transfer reaction (Brønsted-Lowry acid–base reaction).

Salt: an ionic compound consisting of the positive ion of a base and the negative ion of an acid.

Strong acid: an acid with a strong tendency to donate protons. In aqueous solutions, strong acids give rise to a high concentration of ions, enabling the solutions to serve as good conductors of electricity. An aqueous solution of a strong acid is said to be highly acidic because of the large concentration of hydronium ions.

Strong base: a base with a strong tendency to accept protons. In aqueous solutions, strong bases give rise to a high concentration of ions, enabling the solutions to serve as good conductors of electricity. An aqueous solution of a strong base is said to be highly alkaline because of the large concentration of hydroxide ions.

Weak acid: an acid with a relatively small tendency to donate protons. In aqueous solutions, weak acids give rise to a relatively small concentration of ions, so that the solutions are only slightly acidic and only fair or poor conductors of electric current.

Weak base: a base with a relatively small tendency to accept protons. In aqueous solutions, weak bases give rise to a relatively small concentration of ions, so that the solutions are only slightly alkaline and only fair or poor conductors of electric current.

EXERCISES

9-1 List some of the properties of compounds whose names include the word *acid*. Why do aqueous solutions of some of these compounds conduct electric current much better than others?

9-2 List some of the properties of compounds that have been regarded traditionally as bases. Why do aqueous solutions of some of these compounds conduct electricity better than others?

9-3 Write the formulas of the salts that are composed of the following ions: Ca^{++} and PO_4^{3-}, Mg^{++} and Cl^-, Na^+ and Se^{--}, Ba^{++} and Te^{--}, Ag^+ and AsO_4^{3-}.

9-4 Write the formulas of the acid and the base that would be required to form the following salts: KI, $Ba(C_2H_3O_2)_2$, K_2SO_4, $CaCO_3$, $AlPO_4$, $RaSO_4$.

9-5 Write the formulas of all the salts, including the acid salts, that could be formed by reactions between the following acids and bases: RbOH and H_2SO_4, NaOH and H_3AsO_4, $Ca(OH)_2$ and H_2SO_4, KOH and H_3PO_4.

9-6 Illustrate the reactions of a nonmetallic oxide, a basic metallic oxide, and an acidic metallic oxide with water.

9-7 Write the formula for the oxide that would react with water to give H_3PO_4, HNO_3, H_2SeO_4, $HClO_4$, $HClO_3$.

9-8 *Neutralization.* Write and balance the following neutralization equations:
(a) $KOH + HBr \longrightarrow$
(b) $CsOH + H_2SeO_4 \longrightarrow$
(c) $NH_4OH + H_3PO_4 \longrightarrow$
(d) $Sr(OH)_2 + HNO_3 \longrightarrow$
(e) $Fe(OH)_2(s) + H_2SO_4 \longrightarrow$
(f) $Cu(OH)_2(s) + H_3PO_4 \longrightarrow$
(g) $Fe(OH)_3(s) + HBrO_2 \longrightarrow$
(h) $Ga(OH)_3(s) + H_3AsO_4 \longrightarrow$

9-9 *Decomposition of acids and bases.* Complete and balance the following equations (in each case, water is one of the two products and the anhydride of the acid or base is the other):
(a) $Ba(OH)_2 \longrightarrow$
(b) $Fe(OH)_3 \longrightarrow$
(c) $Cu(OH)_2 \longrightarrow$

(d) $HNO_3 \longrightarrow$
(e) $H_3PO_4 \longrightarrow$
(f) $H_2CO_3 \longrightarrow$
(g) $Zn(OH)_2 \longrightarrow$
(h) $H_2ZnO_2 \longrightarrow$
(i) $H_2SiO_3 \longrightarrow$
(j) $H_4SiO_4 \longrightarrow$
(k) $HMnO_4 \longrightarrow$
(l) $Ga(OH)_3 \longrightarrow$
(m) $Th(OH)_4 \longrightarrow$

9-10 *Reactions of basic oxides with acids.* Complete and balance:
(a) $Na_2O + HCl \longrightarrow$
(b) $K_2O + H_2SO_4 \longrightarrow$
(c) $CaO + HNO_3 \longrightarrow$
(d) $BaO + H_2SO_4 \longrightarrow$
(e) $Al_2O_3 + HCl \longrightarrow$
(f) $Al_2O_3 + H_2SO_4 \longrightarrow$
(g) $Na_2O + H_3PO_4 \longrightarrow$
(h) $CaO + H_3PO_4 \longrightarrow$
(i) $Al_2O_3 + H_3PO_4 \longrightarrow$
(j) $ThO_2 + H_3PO_4 \longrightarrow$

9-11 *Reactions of acidic oxides with bases.* Complete and balance:
(a) $NaOH + SO_3 \longrightarrow$
(b) $Ca(OH)_2 + CO_2 \longrightarrow$
(c) $Al(OH)_3 + P_4O_{10} \longrightarrow$
(d) $NaOH + CO_2 \longrightarrow$
(e) $Ca(OH)_2 + N_2O_5 \longrightarrow$
(f) $Ga(OH)_3 + SO_2 \longrightarrow$
(g) $Th(OH)_4 + Mn_2O_7 \longrightarrow$

9-12 *Reactions of acidic oxides with basic oxides.* Complete and balance:
(a) $CaO + CO_2 \longrightarrow$
(b) $Na_2O + SO_3 \longrightarrow$
(c) $Al_2O_3 + P_4O_{10} \longrightarrow$
(d) $Ga_2O_3 + SO_2 \longrightarrow$
(e) $BaO + SO_3 \longrightarrow$
(f) $ThO_2 + P_4O_{10} \longrightarrow$
(g) $CaO + Mn_2O_7 \longrightarrow$
(h) $CuO + CO_2 \longrightarrow$
(i) $Fe_2O_3 + CrO_3 \longrightarrow$

9-13 Solutions of some salts are neutral, but others are acidic and still others are alkaline. Explain.

9-14 State and illustrate the Brønsted-Lowry concept of hydrolysis.

9-15 List several strong acids and their conjugate bases. What can be said of the strength of these bases?

9-16 List several strong bases and their conjugate acids. Discuss the strength of these conjugate acids.

9-17 Write equations for the reaction of the following acids and bases with water: HCl, $HClO_4$, NH_4^+, NH_3, CN^-. Identify each of these as a Brønsted-Lowry acid, a Brønsted-Lowry base, a Lewis acid, or a Lewis base. Identify the role played by the water in each reaction.

REFERENCES

Hamm, D. I., *Fundamental Concepts of Chemistry*, Appleton-Century-Crofts, New York, 1969, Chapter 10, "Acids and Bases. Proton-Transfer Reactions." An extensive treatment of classical and Brønsted-Lowry concepts of acids and bases.

Sisler, H. H., *Chemistry in Non-Aqueous Solvents*, Reinhold, New York, 1961 (paperbound). An account of the relation between the properties of solvents and the processes that occur in them, including acid–base reactions.

Sisler, H. H., VanderWerf, C. A., and Davidson, A. W., *College Chemistry*, 3rd ed., Macmillan, New York, 1967, Chapter 27, "Chemistry in Nonaqueous Solvents." Acid–base reactions in a variety of solvents.

VanderWerf, C. A., *Acids, Bases, and the Chemistry of the Covalent Bond*, Reinhold, New York, 1961 (paperbound). An organic chemist's interpretation of the Brønsted-Lowry and Lewis concepts.

10

Solutions and Colloids

10-1 The Nature of Solutions

When sugar crystals are stirred with water, a clear mixture of sugar and water is obtained. Such a homogeneous mixture, in which the particles of the individual substances cannot be seen even with a microscope, is called a *solution*.

Solutions may be classified as gaseous, liquid, or solid. Any mixture of gases is a gaseous solution; air is a familiar example. Ideally, the molecules in a gaseous mixture move independently of one another and the mixture conforms to Dalton's Law of Partial Pressures (Section 6-11). Liquid solutions may be formed by mixing certain liquids or by dissolving a gas or a solid in a liquid. Solid solutions are less common, but certain alloys belong to this class; for example, brass is a solid solution of zinc in copper. Solid-state electronic devices, such as transistors, employ ultrapure germanium or silicon into which exceedingly small quantities of properly selected impurities have been introduced, forming a solid solution.

When two liquids come into the presence of each other, they may form a homogeneous mixture (i.e., a solution), or one liquid may persist in floating atop the other with an easily observed interface between the two phases (Figure 10-1). Two or more liquids that mix with one another to form solutions are said to be *miscible*; for example, water and alcohol are miscible. Liquids that do not intermingle to form solutions, such as water and gasoline, are said to be *immiscible*. If two liquids, A and B, are only partially miscible, a limited amount of each will dissolve in the other. If liquid A is present in excess (i.e., if more is present than can dissolve in liquid B), two layers form and each layer is a solution. In one layer the solvent is B, and it contains as much A as can dissolve in it; in the other layer the solvent is A, and it contains as much B as can dissolve in it. Briefly stated, one layer is a saturated solution of A in B and the other layer is a saturated solution of B in A.

Liquids that are miscible consist of molecules that are of similar character. Nonpolar liquids, such as carbon disulfide, CS_2, and carbon tetrachloride, CCl_4, are readily miscible with one another, but they will not dissolve in water because of its highly polar character. On the other hand, polar compounds, such as methyl alcohol, CH_3OH, and ethyl alcohol, C_2H_5OH, are miscible with one another and with water.

When a solution is formed by dissolving a solid or a gas in a liquid, the liquid is called the *solvent* and the solid or gas is called the *solute*. These terms may be used wherever one substance can be regarded as having dissolved *in* another; the medium is the solvent, and the dissolved substance is the solute.

The feature that distinguishes solutions from heterogeneous mixtures is the fineness of dispersion. It is impossible to distinguish individual particles of any component of a solution by visual methods because the

FIGURE 10-1

Miscible and immiscible liquids

(a) When two immiscible liquids are brought together, the less dense liquid forms a layer overlying the more dense one. (b) When two miscible liquids are brought together, they mix to form a single homogeneous layer (solution).

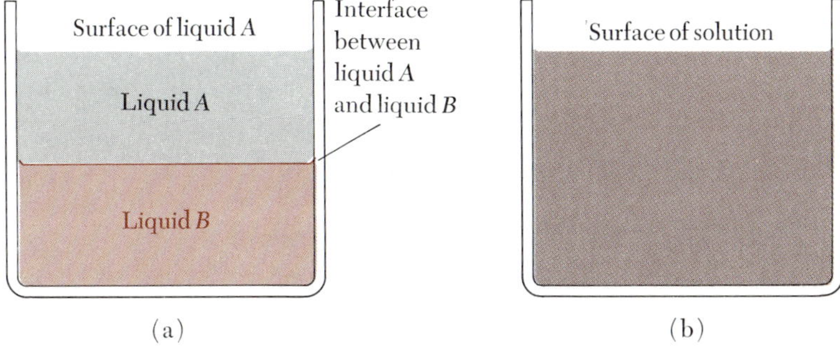

particles are too small to be visible. The process of dissolution occurs because the solvent molecules have an attraction for the solute molecules that is comparable with the attraction that the solute molecules have for one another. In the dissolved state the solute is molecularly dispersed, that is, each particle of solute is an individual molecule, except that in solutions of electrolytes (Section 10-9) some or all of the solute particles are ions.

The effect of temperature on solubility is generally to make solids more soluble in liquids and to make liquids more soluble in other liquids. There are exceptions, however, because some solids become less soluble in water with rising temperature and some liquid pairs become less miscible at higher temperatures. With increasing temperature, gases become less soluble in liquids without exception.

The effect of pressure on the solubility of a solid in a liquid is negligible. Its effect on the miscibility of liquids is also very slight. On the other hand, the effect of pressure on the solubility of a gas in a liquid is quite pronounced. William Henry, who was John Dalton's closest friend, noted in 1803 that the mass of a gas that dissolves in a given volume of a liquid at a specified temperature is directly proportional to the pressure of the gas. This relationship is known as *Henry's Law.* If there is a mixture of gases, the solubility of each individual gas is directly proportional to its partial pressure (and, therefore, to its mole fraction) in the mixture (Figure 10-2).

A familiar example of Henry's Law is afforded by the behavior of carbonated beverages. As long as the cap is on the bottle, there is a fairly large pressure of CO_2 gas above the surface of the liquid. Removal of the cap allows CO_2 gas to escape from the bottle, and this in turn allows dissolved CO_2 to escape as bubbles from the liquid.

FIGURE 10-2

Henry's Law

The solubility of a gas is directly proportional to the pressure.

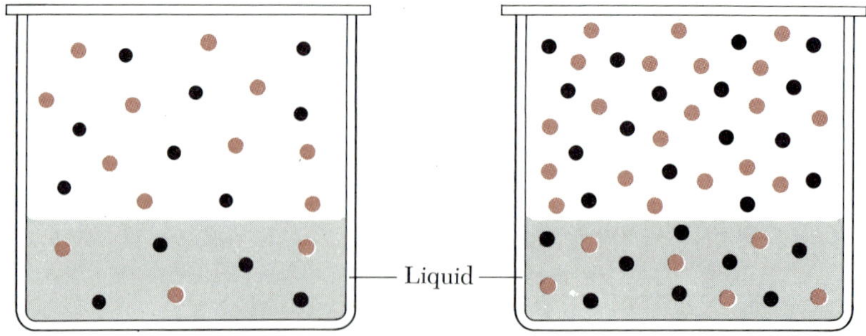

Another application of Henry's Law is encountered by divers and by workmen operating under high pressure. Breathing compressed air causes considerably more than the normal quantity of nitrogen to dissolve in the blood. When a diver is brought up rapidly and begins to breathe air at normal pressure again, much of the nitrogen that dissolved under high pressure comes out of solution, forming bubbles that block the circulation of the blood. This painful and dangerous condition is known as "the bends."

Henry's Law does not apply to cases in which large proportions of the dissolved gases react chemically with the solvent, for example, HCl in water.

10-2 Concentration

If only a small quantity of the solute is present in comparison with the amount of solvent, a solution is said to be *dilute*. The solution is described as *concentrated* if it contains a relatively large amount of solute. Since the terms *dilute* and *concentrated* are too vague to be generally useful, chemists usually employ the term *concentration* when they have reference to the quantity of solute in some specified volume of solution.

A chemically significant method of expressing the concentration of a solution is to state the number of moles of the solute contained in 1 liter of the solution. The *molarity* of a solution is defined as this number. A molar (M) solution, i.e., a $1M$ (pronounced "one-molar") solution, of sugar in water contains 1 mole (342 g) of sugar, $C_{12}H_{22}O_{11}$, in 1 liter of solution. Note that a molar solution is *not* prepared by dissolving 1 mole of the solute in 1 liter of the solvent. Instead, the 1 mole of solute is dissolved in a quantity of solvent (usually water) substantially less than 1 liter in volume, and then the solution is diluted with more solvent until the volume becomes exactly 1 liter. This is usually done in a volumetric flask, which is a flask with a mark on its neck to indicate the point to which the flask must be filled in order to contain the specified volume.

Unless otherwise specified, we assume the solvent is water, so such an expression as $3M$ HCl refers to a solution in which the solvent is water, with HCl (the solute) present in a concentration of 3 moles per liter. If there is any likelihood of ambiguity, we should refer to the solution specifically as an aqueous solution to indicate that the solvent is water.

The number of moles of solute present in a solution is directly proportional to its volume and to its concentration (expressed in moles per liter). Thus,

$$\text{Number of moles} = \text{molarity} \times \text{volume (in liters)}$$

This relationship enables us to calculate the molarity of a solution if we know the volume and the moles present. We can also use the relationship to calculate the volume of solution required to contain a designated number of moles.

EXAMPLE 1

How many grams of sugar are needed to prepare 500 ml (0.500 liter) of 0.250*M* solution?

One liter of 0.250*M* solution contains 0.250 mole; 0.500 liter of 0.250*M* solution contains

$$0.250 \text{ mole/liter} \times 0.500 \text{ liter} = 0.125 \text{ mole}$$

Since the formula for sugar is $C_{12}H_{22}O_{11}$, the formula weight is

$$(12 \times 12 \text{ amu}) + (22 \times 1.0 \text{ amu}) + (11 \times 16 \text{ amu}) = 342 \text{ amu}$$

Therefore, 1 mole of sugar weighs 342 g and 0.125 mole of sugar weighs

$$0.125 \text{ mole} \times 342 \text{ g/mole} = 42.8 \text{ g}$$

EXAMPLE 2

What is the molarity of a solution prepared by dissolving 13.2 g of table salt, NaCl, in sufficient water to give 2.000 liters?

Since the formula for the salt is NaCl, the formula weight is

$$22.99 \text{ amu} + 35.45 \text{ amu} = 58.44 \text{ amu}$$

Therefore, 1 mole of NaCl weighs 58.44 g.

$$\frac{13.2 \text{ g}}{58.44 \text{ g/mole}} = 0.226 \text{ mole of NaCl}$$

$$\frac{0.226 \text{ mole}}{2.000 \text{ liters}} = 0.113M$$

EXAMPLE 3

What volume of 2.00*M* NaOH would be needed to furnish 12.0 g of NaOH for a reaction?

One mole of NaOH weighs 40.00 g.

$$\frac{12.0 \text{ g}}{40.00 \text{ g/mole}} = 0.300 \text{ mole}$$

$$\frac{0.300 \text{ mole}}{2.00 \text{ moles/liter}} = 0.150 \text{ liter}$$

EXAMPLE 4

How would you prepare 200 ml of 3.0*M* HCl from 12*M* HCl solution?

In 200 ml of 3.0*M* HCl solution there are 3.0 moles/liter × 0.200 liter = 0.600 mole.

$$\frac{0.600 \text{ mole}}{12 \text{ moles/liter}} = 0.050 \text{ liter} = 50 \text{ ml of } 12M \text{ HCl}$$

Place 50 ml of 12*M* HCl in a 200-ml volumetric flask and add enough water to bring the level of the solution up to the mark.

10-3 Titration

The simplest method for determining the concentration of a base solution is to measure the volume of an acid solution of known concentration required to neutralize it. This determination is made by a process called *titration* (Figure 10-3). A measured volume of the base solution is placed in a flask, together with a few drops of an indicator solution, and the acid solution is added from a buret until the base solution is neutralized. This point is signaled by the indicator, which changes color. The volume of acid solution added is then read from the buret. From the information thus obtained you can calculate the concentration of the base solution, provided you know the equation for the reaction for which the indicator has signaled the end point.

EXAMPLE

Neutralization of 25.00 ml of NaOH solution required 23.78 ml of 0.0492*M* H_2SO_4 solution. What is the concentration of the base solution?

Write a balanced equation for the reaction:

$$2NaOH + H_2SO_4 \longrightarrow Na_2SO_4 + 2H_2O$$

The equation shows that the number of moles of NaOH in the 25.00 ml of base solution is twice the number of moles of H_2SO_4 in the acid solution used in the titration. The number of moles of H_2SO_4 used in the titration is

$$0.0492 \text{ mole/liter} \times 0.02378 \text{ liter} = 0.001170 \text{ mole}$$

The number of moles of NaOH is twice 0.001170, or 0.002340 mole. The concentration of the NaOH solution is

$$\frac{0.002340 \text{ mole}}{0.02500 \text{ liter}} = 0.0936M$$

FIGURE 10-3

Titration of base with acid

The buret contains an acid solution of known concentration; the flask contains a base solution with an indicator.

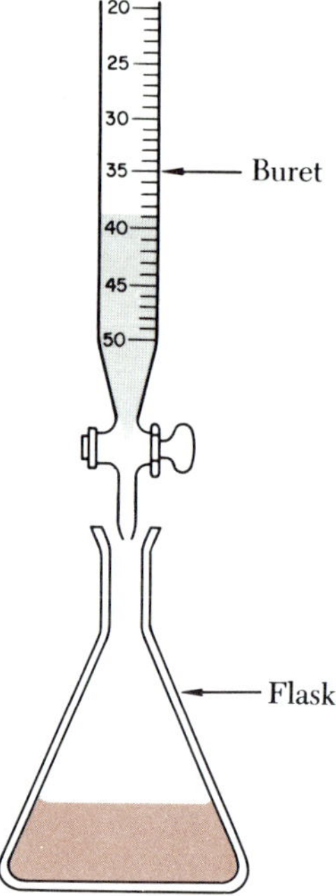

The concentration of an acid solution can likewise be determined by titrating with a base solution of known concentration. Precipitation reactions, suitably chosen, can also serve as the means for determining the concentrations of solutions by titration. In fact, any reaction is suitable for use in titration, provided the equation for the reaction is known and some means is available for indicating the end point.

As an example of a precipitation titration, the concentration of chloride ion, Cl^-, in a solution may be determined by adding from a buret a solution of a silver salt (usually $AgNO_3$) of known concentration. Before the titration is begun, a small amount of Na_2CrO_4 or K_2CrO_4 is added to the chloride solution. When sufficient Ag^+ ions have been added to precipitate virtually all the Cl^- ions as white AgCl, reddish Ag_2CrO_4 begins to precipitate, indicating the end point.

10-4 Molality

For many purposes, a useful method of stating the concentration of a solution is to give the number of moles of solute that are present in 1.000

kg (1000 g) of solvent. *Molality*, usually designated by m, is the name given to this number. To calculate the molality of a solution, we divide the number of grams of solute by the number of grams in 1 mole of solute to obtain the number of moles of solute present. We then divide this number by the number of kilograms of solvent in the solution to obtain the number of moles of solute per kilogram of solvent.

EXAMPLE 1

A solution is prepared by dissolving 1.0 mole of sugar in 2000.0 g of water. What is its molality?

The 1.0 mole of the solute is dissolved in 2000.0 g, which is 2.000 kg, of solvent. The quantity of solute per kilogram of solvent is 1.0 mole/2.000 kg or 0.50 mole/kg. The solution is, therefore, 0.50m; i.e., the molality of the solution is 0.50.

EXAMPLE 2

If 25 g of sugar, $C_{12}H_{22}O_{11}$, is dissolved in 1000.0 g of water, what is the molality of the solution?

One mole of sugar weighs 342 g.

$$\frac{25\text{ g}}{342\text{ g/mole}} = 0.073\text{ mole}$$

Since 0.073 mole is dissolved in 1.000 kg of water, the solution is 0.073m.

EXAMPLE 3

What is the molality of a solution obtained by dissolving 32.5 g of sodium chloride in 1500.0 g of water?

One mole of sodium chloride weighs 58.44 g.

$$\frac{32.5\text{ g}}{58.44\text{ g/mole}} = 0.556\text{ mole}$$

This quantity is dissolved in 1500.0 g of water, which is

$$\frac{1500.0\text{ g}}{1000\text{ g/kg}} = 1.500\text{ kg}$$

The quantity of salt present per kilogram of water is

$$\frac{0.556\text{ mole}}{1.500\text{ kg}} = 0.371\text{ mole/kg}$$

Thus, the molality of the solution is 0.371.

10-5 Effect of the Solute on the Freezing Point of the Solvent

One of the most frequently investigated aspects of solutions has been the effect the solute has on the physical properties of the solvent. In general, addition of a solute lessens the tendency of the solvent to evaporate and makes it necessary that the solution be heated to a temperature higher than the boiling point of the pure solvent in order for it to boil (unless the solute is itself a volatile substance); for the solution to freeze, it must be cooled to a temperature lower than the freezing point of the pure solvent.

It is common knowledge that solutions freeze at lower temperatures than pure liquids. Addition of various antifreezes to the water in automobile radiators is a well-known procedure for preventing the water from freezing when it is cooled to 0°C (32°F). The extent to which the freezing point is lowered is related to the molality. In concentrated solutions the situation is complicated and somewhat unpredictable, but in dilute solutions the freezing-point depression is directly proportional to the molality. The freezing point of water, for example, is lowered at the rate of 1.86°C for each mole of solute per kilogram of water (if the solute is not an electrolyte). If 0.100 mole (34.2 g) of sugar, $C_{12}H_{22}O_{11}$, is dissolved in 1.000 kg of water, the freezing point is lowered 0.100 mole × 1.86°C/mole, or 0.186°C; that is, the solution would begin to freeze at −0.186°C instead of 0°C.

This example indicates a method by which the molecular weight of a substance can be determined experimentally. If ΔT_f represents the amount that the freezing point is lowered (Δ means "the change in"), m the molality, and K_f a constant characteristic of the particular solvent known as the molal freezing-point depression constant (1.86°C/mole in the case of water), then the relationship can be stated as follows:

$$\Delta T_f = K_f m$$

From this relationship the molality of a solution can be calculated if the freezing-point lowering has been measured and the value of the constant K_f is known:

$$m = \frac{\Delta T_f}{K_f}$$

On the other hand, the molality of the solution is the number of moles of solute per kilogram of solvent. Since the number of moles of solute present is the number of grams of solute divided by the weight of 1 mole, we have the necessary information for calculating the molecular weight of the solute.

EXAMPLE

A solution prepared by dissolving 9.00 g of glucose in 150.0 g of water has a freezing point of −0.620°C. What is the molecular weight of glucose?

From the freezing-point lowering 0.620°C, we first calculate the molality of the solution (number of moles of glucose per kilogram of water):

$$m = \frac{0.620°\text{C}}{1.86°\text{C/mole}} = 0.333 \text{ mole}$$

For the weight (grams) of glucose per kilogram of water, we have

$$\frac{9.00 \text{ g glucose}}{150.0 \text{ g water}} \times 1000 \text{ g water/kg water} = 60.0 \text{ g glucose/kg water}$$

Since the solution contains 0.333 mole of glucose per kilogram of water and 60.0 g of glucose per kilogram of water, it follows that

$$0.333 \text{ mole glucose} = 60.0 \text{ g glucose}$$

and

$$1 \text{ mole glucose} = \frac{60.0 \text{ g}}{0.333} = 180 \text{ g}$$

Thus the molecular weight of glucose is 180 amu. (Check this against the formula, $C_6H_{12}O_6$. You should not expect results to be this good in all cases, because the laws of solutions are limited in their accuracy, just as the gas laws are.)

10-6 Elevation of the Boiling Point

The temperature of a liquid must be raised above its normal boiling point for the liquid to boil when a nonvolatile substance is dissolved in it. The extent to which the boiling point is raised is related to the molality. As with the lowering of the freezing point, the situation is complicated and somewhat unpredictable in concentrated solutions, but the elevation of the boiling point is directly proportional to the molality in dilute solutions. The boiling point of water, for example, is raised at the rate of 0.512°C for each mole of solute per kilogram of water (if the solute is not an electrolyte). If 0.100 mole (34.2 g) of sugar is dissolved in 1.000 kg of water, the boiling point is raised 0.100 mole × 0.512°C/mole, or 0.051°C; that is, the solution would boil at 100.051°C instead of 100.000°C.

This example indicates another method by which the molecular weight of a substance can be determined from experimental data. If ΔT_b represents the amount that the boiling point is raised, m the molality, and K_b a constant characteristic of the particular solvent known as the molal boiling-point elevation constant (0.512°C/mole in the case of water), the relationship can be stated as follows:

$$\Delta T_b = K_b m$$

From this relationship the molality of a solution can be calculated if the boiling-point elevation has been measured and the value of the constant K_b is known:

$$m = \frac{\Delta T_b}{K_b}$$

Knowing also the composition of the solution in terms of grams of solute and grams of solvent, we have the information we need to calculate the molecular weight of the solute.

EXAMPLE

A solution prepared by dissolving 3.89 g of a certain compound in 187 g of chloroform boils at 62.27°C. What is the molecular weight of the solute?

Consulting Table 10-1, we find that the boiling point of pure chloroform is 61.26°C. The boiling-point elevation is the difference between the boiling point of chloroform and the boiling point of the solution:

$$\Delta T_b = 62.27°\text{C} - 61.26°\text{C} = 1.01°\text{C}$$

Table 10-1 also shows that K_b for chloroform is 3.63°C/mole. From the boiling-point elevation we now calculate the molality (the number of moles of solute in 1 kg of solvent):

$$m = \frac{1.01°\text{C}}{3.63°\text{C/mole}} = 0.278 \text{ mole}$$

The quantity of solvent in which the 3.89 g of solute is dissolved is

$$\frac{187 \text{ g}}{1000 \text{ g/kg}} = 0.187 \text{ kg}$$

The weight of solute per kilogram of solvent is

$$\frac{3.89 \text{ g solute}}{0.187 \text{ kg solvent}} = 20.8 \text{ g solute/kg solvent}$$

TABLE 10-1

Molal freezing-point depression and boiling-point elevation constants

Solvent	T_f (°C)	K_f (°C/mole)	T_b (°C)	K_b (°C/mole)
Benzene	5.49	5.08	80.1	2.53
Camphor	176	37.7	208	5.95
Carbon tetrachloride	−22.8	31.8	76.8	5.03
Chloroform	−63.5	4.68	61.26	3.63
Water	0.00	1.86	100.00	0.512

Thus, the solution contains 0.278 *mole* of solute per kilogram of solvent and 20.8 *grams* of solute per kilogram of solvent. Therefore,

$$0.278 \text{ mole of solute} = 20.8 \text{ g solute}$$

and the weight of 1 mole of the solute is

$$\frac{20.8 \text{ g}}{0.278 \text{ mole}} = 74.8 \text{ g/mole}$$

Thus, the molecular weight of the solute is calculated to be 75 amu. (We will not claim three-digit accuracy because we do not expect the laws of solutions to be completely accurate.)

10-7 Lowering of the Vapor Pressure

In a mixture of two similar liquids, the properties of the mixture are an average of the properties of the individual liquids. Thus, if the densities of two liquids are 0.74 g per ml and 0.86 g per ml, one would expect the density of a mixture of equal volumes of the two liquids to be 0.80 g per ml. If this is found to be true, the mixture is showing one of the characteristics of an *ideal solution*, that is, the molecules of each kind are in an environment very much like the one they were in before mixing.

The vapor pressure of an ideal solution can be calculated from its composition. The vapor pressure of each component of an ideal solution is proportional to the mole fraction of that component in the solution. According to the law announced by the French chemist François Raoult in 1886 (and therefore known as *Raoult's Law*), the vapor pressure, P, of a liquid in an ideal solution is given by the expression

$$P = XP°$$

where $P°$ is the vapor pressure of the pure liquid at the same temperature and X is its mole fraction in the solution. If a solution is prepared by dissolving a nonvolatile solid in a volatile liquid, the above expression gives the *total* vapor pressure. Therefore, at all temperatures the vapor pressure of the solution is less than the vapor pressure of the pure solvent. For example, at the normal boiling point of the pure solvent, the vapor pressure of the solution is less than 1 atm. This explains why a solution containing a nonvolatile solute must be heated to a temperature higher than the boiling point of the pure solvent in order for it to boil.

The same considerations explain why a solution must be cooled to a temperature lower than the freezing point of the pure solvent in order for it to begin to freeze. At the freezing point the vapor pressure of the solid is the same as the vapor pressure of the liquid (Section 7-8). At the temperature at which the pure solvent would freeze (0°C in the case of water), the vapor pressure of the solvent component of a solution is less than the vapor pressure of the solid. It is not possible, therefore, to have frozen solvent in equilibrium with a solution at the freezing point of the pure solvent. If the temperature is reduced, both the vapor pressure of the solid and the vapor pressure of the liquid solvent in the solution become smaller. The vapor pressure of the solid falls off more rapidly than the vapor pressure of the liquid solvent, so that there is some lower temperature at which the vapor pressure of the solid and the vapor pressure of the liquid solvent are identical. This temperature is the freezing point of the solution.

The mole fraction of the solvent, X_A, and the mole fraction of the solute, X_B, are related by the expression $X_A = 1 - X_B$. Therefore, the Raoult's Law expression, $P_A = X_A P_A°$, can be written in the form

$$P_A = (1 - X_B)P_A° = P_A° - X_B P_A°$$

Solving this for X_B, we get

$$X_B = \frac{P_A° - P_A}{P_A°}$$

The numerator, $P_A° - P_A$, is the amount that the vapor pressure is lowered by having the solute dissolved in it. This lowering of the vapor pressure can be appropriately symbolized as ΔP. Therefore,

$$\Delta P = P_A° - P_A = X_B P_A°$$

This equation states that the amount by which the vapor pressure of the solvent is reduced is proportional to the mole fraction of solute. Written in the form

$$\frac{\Delta P}{P_A°} = X_B$$

it states that the *relative* lowering of the vapor pressure has the same numerical value as the mole fraction of solute. It suggests still another method for determining the molecular weight of the solute. Upon measuring the vapor pressure of the solution we can calculate the mole fraction of the solute:

$$X_B = \frac{\Delta P}{P_A^\circ}$$

Since we know the number of grams of solute and the number of grams of solvent used to prepare the solution, we can calculate the molecular weight of the solute.

EXAMPLE

At 20°C the vapor pressure of ether, $(C_2H_5)_2O$, is 442.2 torr. At the same temperature the vapor pressure of a solution prepared by dissolving 10.0 g of a nonvolatile organic compound in 100 g of ether is 426.0 torr. What is the molecular weight of the solute?

Letting M_A represent the molecular weight of the solvent, W_A the number of grams of solvent in the solution, M_B the molecular weight of the solute, and W_B the number of grams of solute in the solution, the mole fraction of the solute is

$$X_B = \frac{W_B/M_B}{W_A/M_A + W_B/M_B}$$

Therefore,

$$P_A^\circ - P_A = \left[\frac{W_B/M_B}{W_A/M_A + W_B/M_B}\right] P_A^\circ$$

Rearranged, the equation reads

$$M_B = M_A\left(\frac{W_B}{W_A}\right)\left(\frac{P_A}{P_A^\circ - P_A}\right)$$

Now we can substitute the numerical values given for the different quantities in the statement of the problem. The molecular weight of ether, obtained by adding the atomic weights of its constituent elements, is

$$\begin{aligned} M_A &= 2[(12.01 \text{ amu} \times 2) + (1.008 \text{ amu} \times 5)] + 16.00 \text{ amu} \\ &= 74.12 \text{ amu} \\ M_B &= 74.12 \text{ amu} \times \frac{10.0 \text{ g}}{100 \text{ g}} \times \frac{426.0 \text{ torr}}{16.2 \text{ torr}} \\ &= 74.12 \text{ amu} \times 0.100 \times 26.3 \\ &= 195 \text{ amu} \end{aligned}$$

Since the lowering of the vapor pressure of the solvent is proportional to the mole fraction of solute, we might expect the elevation of the boiling point and the depression of the freezing point also to be proportional to the mole fraction of the solute because these effects are due to the lowering of the vapor pressure. They are, indeed, but in dilute solutions it can be shown that molality in a stated solvent is approximately proportional to the mole fraction of solute. By working with molalities instead of mole fractions we need not know the molecular weight of the solvent. Both K_f and K_b can be evaluated empirically by determining the lowering of the freezing point and the raising of the boiling point produced by solutes of known molecular weight. The data in Table 10-1 were obtained in this way.

10-8 Osmosis

The kinetic activity of the molecules of a solute causes the concentration of a solution to approach uniformity in due course of time. Various materials, such as parchment paper or animal bladder, may serve as *semipermeable membranes*—they may allow one component of a solution, but not the other, to pass through them. Usually it is the solvent that is able to pass through. If a solution is placed on one side of such a semipermeable membrane and pure solvent is placed on the other side, the solvent molecules diffuse out of the solution at a slower rate than they diffuse into it because they are present in lower concentration on the solution side. The result is a net flow of solvent through the semipermeable membrane into the solution—a phenomenon known as *osmosis*.

The osmotic flow of solvent into a solution and the attendant increase in the volume of the solution may be prevented by the application of mechanical pressure to the surface of the solution. The amount of pressure that is just sufficient to prevent osmosis is called the *osmotic pressure* of the solution (Figure 10-4). Measurements of the osmotic pressures of dilute solutions have shown them to be directly proportional to the molality (if the solute is not an electrolyte) and to the absolute temperature.

Since the osmotic pressure, the lowering of the vapor pressure, the elevation of the boiling point, and the lowering of the freezing point are all proportional to molality, they are referred to collectively as the colligative ("tied together") properties of solutions. In general, the term *colligative properties* has been broadened to refer to all properties that depend on, or vary with, the number of molecules and not primarily on

FIGURE 10-4

Osmotic pressure

The osmotic pressure of a solution is the pressure that must be applied to the solution to prevent osmosis.

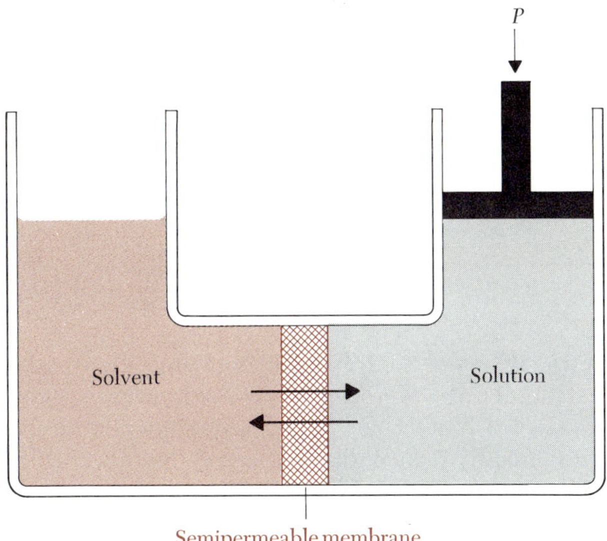

their nature; for example, the pressure exerted by a gas is a colligative property.

Osmosis plays an important role in various physiological functions of plants and animals, such as the passage of water through the semipermeable membranes of roots and blood cells and the action of the kidneys.

10-9 Solutions of Electrolytes

There is a certain class of compounds that affect the freezing point of water to a greater extent than can be predicted from the equation $\Delta T_f = K_f m$. They likewise affect the boiling point to a greater extent than can be predicted from the equation $\Delta T_b = K_b m$. Furthermore, the relative reduction of the vapor pressure exceeds the mole fraction of solute, and the osmotic pressure is proportionately magnified. Aqueous solutions of these compounds conduct an electric current. Such substances are known as *electrolytes* and include acids, bases, and salts.

An electric current is, by definition, a movement of charged particles.

In solutions of electrolytes, the charged particles are ions, i.e., atoms or groups of atoms that bear one or more positive or negative charges.

Svante Arrhenius, a Swedish chemist, proposed a theory of ionization in 1887 to explain electrolytic conduction and the abnormal colligative properties of solutions of electrolytes. Like the other scientists of his day, Arrhenius thought that all compounds consist of molecules. We have already discussed the fact that some compounds, notably salts, consist of ions instead of molecules. Since Arrhenius did not know this, he assumed that the ions present in a solution of any electrolyte originate through the dissociation of the molecules of the solute:

$$NaCl \longrightarrow Na^+ + Cl^-$$

$$HCl \longrightarrow H^+ + Cl^-$$

$$NaOH \longrightarrow Na^+ + OH^-$$

If each molecule were to split into two ions, two particles would be present instead of only one. In this way Arrhenius explained why 1 mole of NaCl or HCl or NaOH in 1 kg of water lowers the freezing point of water approximately *twice* the expected amount: 3.72°C instead of 1.86°C. By assuming that the dissociation gives charged particles (ions) instead of neutral atoms, Arrhenius explained the ability of these solutions to conduct an electric current.

Our present-day knowledge enables us to correct Arrhenius on two counts: (1) Salts and the ionic hydroxides (for example, NaCl and NaOH) consist of ions, not molecules, in the crystalline state; when they dissolve in water, the crystals dissociate, i.e., they break apart, but there is no dissociation of molecules. (2) Although acids, such as HCl, consist of molecules (as Arrhenius supposed), the formation of ions when they dissolve in water is due to an interaction with the solvent:

$$HCl + H_2O \longrightarrow H_3O^+ + Cl^-$$

Arrhenius correctly supposed that in a solution of sodium sulfate, Na_2SO_4, there are singly charged Na^+ ions and doubly charged SO_4^{--} ions, there being twice as many of the former as of the latter:

$$Na_2SO_4(s) \longrightarrow 2Na^+ + SO_4^{--}$$

In solutions of this salt the effect on the freezing point of water is three times as great as for a solution of a nonelectrolyte of the same molality. In a calcium chloride ($CaCl_2$) solution there are Ca^{++} ions and Cl^- ions:

$$CaCl_2(s) \longrightarrow Ca^{++} + 2Cl^-$$

In a solution of aluminum sulfate, $Al_2(SO_4)_3$, there are Al^{3+} ions and SO_4^{--} ions:

$$Al_2(SO_4)_3(s) \longrightarrow 2Al^{3+} + 3SO_4^{--}$$

In "mole language," if 1 mole of NaCl is dissolved in water, the solution contains 1 mole of Na^+ ions and 1 mole of Cl^- ions, a total of 2 moles of ions. When 1 mole of Na_2SO_4 is dissolved in water, 2 moles of Na^+ ions and 1 mole of SO_4^{--} ions, a total of 3 moles, are formed. Similarly, there are 3 moles of ions in 1 mole of $CaCl_2$ and 5 moles of ions in 1 mole of $Al_2(SO_4)_3$. Thus, if 0.0100 mole of each of these salts is dissolved in separate kilogram quantities of water, the expected freezing points are NaCl, −0.0372°C; Na_2SO_4, −0.0558°C; $CaCl_2$, −0.0558°C; and $Al_2(SO_4)_3$, −0.0930°C. (The actual freezing points are somewhat higher; that is, the freezing point is lowered somewhat less than expected because of interionic attraction.)

Arrhenius was aware that some electrolytes do not conduct electric current or depress the freezing point of water as effectively as others. To explain these closely correlated facts, he made the important assumption that *only a small percentage* of the molecules are dissociated into ions. He symbolized this situation by using two arrows in the expression representing the dissociation:

$$\underset{\text{hydrocyanic acid molecules}}{HCN} \rightleftharpoons \underset{\text{hydrogen ions}}{H^+} + \underset{\text{cyanide ions}}{CN^-}$$

$$\underset{\text{acetic acid molecules}}{HC_2H_3O_2} \rightleftharpoons \underset{\text{hydrogen ions}}{H^+} + \underset{\text{acetate ions}}{C_2H_3O_2^-}$$

Acids such as hydrocyanic acid and acetic acid are examples of weak acids. As with the strong acid HCl, the formation of ions actually involves reaction with water:

$$HCN + H_2O \rightleftharpoons H_3O^+ + CN^-$$

$$HC_2H_3O_2 + H_2O \rightleftharpoons H_3O^+ + C_2H_3O_2^-$$

The main points in the Arrhenius theory of ionization in solution are the following:

1. Aqueous solutions of electrolytes contain ions, which are charged atoms or groups of atoms, whose motion through the solution under the influence of an applied voltage constitutes an electric current.
2. In a solution where molecules dissociate into ions, the total number of particles is greater and the effect on the freezing point (and other related properties) of the solvent is correspondingly greater.
3. Electrolytes differ in the extent to which they form ions in solution. Those that give solutions that are good conductors of electricity because large numbers of ions are present are called *strong electrolytes*. Those that give slightly conducting solutions because relatively small numbers of ions are present are called *weak electrolytes*.

10-10 Metathetical Reactions

Metathesis is the name given to the type of reaction in which two substances react to produce two new substances of the same class or classes. There are three types of metathetical reactions:

1. Two salts react, producing two salts. For example,

$$AgNO_3 + NaCl \longrightarrow AgCl\downarrow + NaNO_3$$

This reaction occurs when an aqueous solution of silver nitrate is mixed with an aqueous solution of sodium chloride. The reaction occurs because one of the products, silver chloride, is insoluble in water (the solvent) and therefore forms a *precipitate*. This fact is indicated by a downturned arrow following the formula for silver chloride in the equation.

In general, if neither of the possible products of the reaction between two salts is insoluble, there will be no reaction between the salts. For example,

$$NaCl + KNO_3 \longrightarrow \text{no reaction}$$

The sodium chloride solution contains Na^+ and Cl^- ions; the potassium nitrate solution contains K^+ and NO_3^- ions. When the two solutions are mixed, no reaction occurs because both of the possible products, potassium chloride (KCl) and sodium nitrate ($NaNO_3$), are soluble in water. Since no precipitate forms, the solution contains all four ions, K^+, Na^+, Cl^-, and NO_3^- (Figure 10-5).

FIGURE 10-5

Two salts that do not react

When NaCl and KNO_3 solutions are mixed, no reaction occurs.

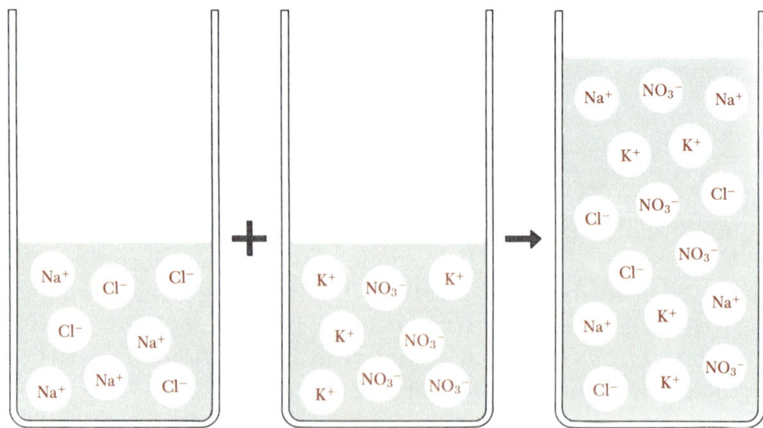

2. A salt of a weak acid reacts with a strong acid to form a salt and a weak acid. For example,

$$NaC_2H_3O_2 + HCl \longrightarrow NaCl + HC_2H_3O_2$$

The sodium acetate solution contains Na^+ and $C_2H_3O_2^-$ ions, and the hydrochloric acid solution contains H^+ (actually, H_3O^+) and Cl^- ions. When the two solutions are mixed, the H^+ and $C_2H_3O_2^-$ ions combine to form $HC_2H_3O_2$ molecules. The reaction occurs because $HC_2H_3O_2$ is a weak acid; acetic acid molecules do not dissociate readily into ions, and it is impossible for a solution to contain high concentrations of both H^+ and $C_2H_3O_2^-$ ions at the same time.

Another example of this kind of metathetical reaction is the following:

$$CaCl_2 + H_2SO_4 \xrightarrow{\text{heat}} CaSO_4 + 2HCl\uparrow$$

When "concentrated" sulfuric acid, i.e., pure hydrogen sulfate, is added to solid calcium chloride and the mixture is heated, a reaction occurs. The upturned arrow indicates that hydrogen chloride is a gas, which escapes as quickly as it is formed. (HCl gas is highly soluble in water, but in this reaction water is absent.) The escape of the HCl gas permits the reaction to continue until one of the reactants is completely consumed.

A third example is the following:

$$Na_2CO_3 + 2HCl \longrightarrow 2NaCl + H_2O + CO_2\uparrow$$

This reaction occurs when hydrochloric acid (the aqueous solution of hydrogen chloride) is added to solid Na_2CO_3 or to an aqueous solution of Na_2CO_3. Carbonate ions, CO_3^{--}, would be expected to combine with H^+ ions to form H_2CO_3 (carbonic acid) molecules. Perhaps they do; but carbonic acid is an unstable compound which cannot be isolated, and it decomposes as rapidly as it is formed:

$$H_2CO_3 \longrightarrow H_2O + CO_2\uparrow$$

The escape of the CO_2 gas assures the continuation of the reaction.

3. A salt reacts with a base, producing a salt and a base. For example,

$$MgCl_2 + 2NaOH \longrightarrow 2NaCl + Mg(OH)_2\downarrow$$

When aqueous solutions of sodium hydroxide and magnesium chloride mix, sodium chloride and magnesium hydroxide are formed. Magnesium hydroxide is of low solubility and removes itself from the solution.

A second example of this kind of metathetical reaction is

$$NH_4Cl + NaOH \longrightarrow NaCl + NH_3\uparrow + H_2O$$

The combination of the NH_4^+ ions from ammonium chloride and the OH^- ions from sodium hydroxide produces an unstable base, ammonium hydroxide, which decomposes as fast as it forms:

$$NH_4OH \longrightarrow NH_3\uparrow + H_2O$$

Since one of the products, NH_3, escapes as a gas, the reaction proceeds until it is complete.

10-11 Ionic Equations

Since the reactions described in the previous section consist primarily of the combination of ions to form the products, the reactions may be represented by equations in which only the participating ions are shown. For example, the metathetical reaction between silver nitrate and sodium chloride is actually a reaction in which only Ag^+ and Cl^- ions take part; the Na^+ and NO_3^- ions are *spectator ions* (Figure 10-6):

$$Ag^+ + NO_3^- + Na^+ + Cl^- \longrightarrow AgCl\downarrow + Na^+ + NO_3^-$$

Since the Na^+ and NO_3^- ions do not take part in the reaction they may be omitted from the equation. Thus, instead of writing the *complete ionic equation*, we may write the *net ionic equation* as follows:

$$Ag^+ + Cl^- \longrightarrow AgCl\downarrow$$

This equation represents the reaction that occurs when *any* silver salt solution is mixed with a solution of *any* ionic chloride, such as KCl, $CaCl_2$, or $MgCl_2$. We can also include HCl in the list; although HCl is a covalent substance, its aqueous solution contains a high concentration of Cl^- ions.

The reaction that occurs when sodium acetate and hydrochloric acid solutions are mixed occurs whenever any ionic acetate is mixed with any strong acid. This reaction can be represented by the net or general ionic equation

$$H^+ + C_2H_3O_2^- \rightleftharpoons HC_2H_3O_2$$

In the specific case of sodium acetate and hydrochloric acid solutions, the complete ionic equation is

$$H^+ + Cl^- + Na^+ + C_2H_3O_2^- \rightleftharpoons HC_2H_3O_2 + Na^+ + Cl^-$$

The reaction between sodium carbonate and hydrochloric acid can be represented by the complete ionic equation

$$2Na^+ + CO_3^{--} + 2(H^+ + Cl^-) \longrightarrow 2(Na^+ + Cl^-) + H_2O + CO_2\uparrow$$

FIGURE 10-6

Spectator ions

In the reaction between NaCl and $AgNO_3$, Na^+ and NO_3^- are spectator ions.

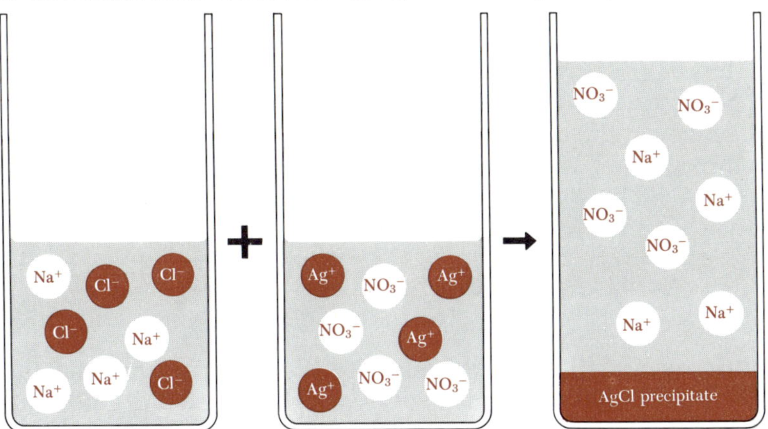

It can also be represented by the net or general ionic equation

$$2H^+ + CO_3^{--} \longrightarrow H_2CO_3 \longrightarrow H_2O + CO_2\uparrow$$

The same result would be obtained by allowing, say, potassium carbonate or calcium carbonate to react with hydrochloric acid or any other strong acid.

The precipitation of magnesium hydroxide, which occurs whenever any magnesium salt and any strong base are brought together, is represented by the general equation

$$Mg^{++} + 2OH^- \longrightarrow Mg(OH)_2\downarrow$$

This represents what happens—*all* that happens—when sodium hydroxide solution and magnesium chloride solution are mixed. It also represents what happens when *potassium* hydroxide and magnesium *sulfate* are brought together. If we consider only sodium hydroxide, potassium hydroxide, magnesium nitrate, and magnesium sulfate, we see that either hydroxide may react with either magnesium salt, but the result is the same in each case:

$$2(Na^+ + OH^-) + (Mg^{++} + 2NO_3^-) \longrightarrow Mg(OH)_2\downarrow + 2(Na^+ + NO_3^-)$$
$$2(K^+ + OH^-) + (Mg^{++} + 2NO_3^-) \longrightarrow Mg(OH)_2\downarrow + 2(K^+ + NO_3^-)$$
$$2(Na^+ + OH^-) + (Mg^{++} + SO_4^{--}) \longrightarrow Mg(OH)_2\downarrow + (2Na^+ + SO_4^{--})$$
$$2(K^+ + OH^-) + (Mg^{++} + SO_4^{--}) \longrightarrow Mg(OH)_2\downarrow + (2K^+ + SO_4^{--})$$

Since the result is the same in each case, it saves space and mental effort if we simply write the one equation

$$Mg^{++} + 2OH^- \longrightarrow Mg(OH)_2\downarrow$$

When magnesium nitrate and sodium hydroxide are used, the NO_3^- and Na^+ ions do not take part in the reaction. They are therefore spectator ions. Similarly, in the last example above, the SO_4^{--} and K^+ ions are spectator ions. Sodium nitrate, potassium sulfate, etc. are products of the above reactions only in the sense that if the solution is separated from the $Mg(OH)_2$ precipitate and the water is evaporated, crystals of $NaNO_3$, K_2SO_4, etc. will be obtained. At the time the starting solutions are mixed, however, the only chemical event is the precipitation of magnesium hydroxide.

10-12 Solubilities of Common Ionic Solids

From the discussion in the two preceding sections it is apparent that the solubility or insolubility of an ionic solid in water is one of its most important properties. Chemical analysis is based to a considerable degree on such differences. The following are general statements regarding the solubilities of some of the most common of these compounds. A knowledge of these statements will be of value to anyone contemplating the further study of chemistry.

1. The following types of ionic compounds are generally soluble in water:
   a. All acetates, for example, $NaC_2H_3O_2$, $Ca(C_2H_3O_2)_2$
   b. All nitrates, for example, $AgNO_3$, $Pb(NO_3)_2$
   c. All chlorides, bromides, and iodides [except those of silver, mercury(I), and lead], for example, $NaCl$, $CuBr_2$, CaI_2
   d. All sulfates (except those of barium, strontium, and lead), for example, $MgSO_4$, $CuSO_4$, $Al_2(SO_4)_3$
   e. All alkali metal (Group IA) hydroxides (and barium hydroxide), for example, $NaOH$, KOH
   f. All common salts of sodium, potassium, and ammonium, for example, Na_2CO_3, K_3PO_4, $(NH_4)_2S$, $NaHCO_3$, KI
   g. All alkaline-earth metal (Group IIA) hydrogen carbonates (bicarbonates), for example, $Ca(HCO_3)_2$, $Ba(HCO_3)_2$
2. The following inorganic ionic solids are insoluble or only very slightly soluble in water:
   a. All metal hydroxides (except those of the alkali metals and ammonium), for example, $Mg(OH)_2$, $Fe(OH)_3$
   b. All carbonates (except those of the alkali metals and ammonium), for example, $CaCO_3$, $PbCO_3$
   c. All phosphates (except those of the alkali metals and ammonium), for example, $Ca_3(PO_4)_2$, $AlPO_4$
   d. All sulfides (except those of the alkali metals and ammonium), for example, CuS, Bi_2S_3

10-13 Colloidal Dispersions

It is not always easy to distinguish between heterogeneous mixtures and homogeneous mixtures. Between coarse suspensions, which are heterogeneous mixtures, and ordinary true solutions, which are homogeneous mixtures, there is an intermediate class of dispersions in which the dispersed particles are said to be in the *colloidal state*. Matter in the colloidal state is in an intermediate state of subdivision with the particles ranging from about 10 Å to about 2000 Å in diameter.

In coarse suspensions the suspended particles are so large that they soon separate out upon standing—provided there is a density difference between the particles and the medium. The particles of a coarse suspension may be separated from the medium by filtration—provided the filter pores are large enough to allow the molecules of the medium to pass through but are too small to permit passage of the suspended particles. The particles of a coarse suspension are sufficiently large to be visible with a microscope—or even with the unaided eye.

In an ordinary true solution, the dispersed particles are small molecules or ions that are not much larger than the molecules of the medium itself. Consequently, they do not separate out upon standing, they cannot be separated by filtration, and they are too small to be visible.

The particles in a colloidal dispersion may be either giant molecules or aggregates of small molecules. They do not separate out upon standing, they cannot be separated by ordinary filtration, and they cannot be visually distinguished from the medium. In these respects the properties of colloidal dispersions are the properties of ordinary true solutions. Yet, colloidal dispersions differ from ordinary true solutions in certain respects. For example, when gold is colloidally dispersed in water, the particles of colloidal gold are of various sizes, each containing hundreds of thousands of gold atoms. The effect of the presence of the dispersed gold on the freezing point of the water, for example, is so small that it cannot be measured. The colligative properties of such a dispersion are determined by the number of colloidal particles present, not by the total number of gold atoms.

If the particles in a colloidal dispersion are giant molecules, the colligative properties will be determined by the number of molecules, as in ordinary true solutions. However, substances that consist of giant molecules usually cannot be dispersed in sufficient quantity to form systems of high concentrations. Consequently, their colligative properties also are usually small, although not immeasurably small.

Many colloidal dispersions may be distinguished from ordinary true solutions by examination with a beam of light. The path of such a beam through a true solution containing small molecules is invisible, but the path of a narrow beam of light through most colloidal dispersions is easily seen as a bright streak (Figure 10-7). The phenomenon is due to the scattering of light by the particles of the dispersed material, much as a ray of sunlight in a darkened room is visible because the light is scattered by the dust particles in the air. This property of colloids was discovered in 1857 by Michael Faraday. It was studied in detail by John Tyndall in 1869 and is generally known as the *Tyndall effect*.

The *ultramicroscope* is an instrument developed for the examination of a Tyndall beam under magnification. It is a microscope so constructed that the line of vision is directed toward a dark background, with the illuminating beam passing through the observation cell at right angles to the direction of observation. Although the colloidal particles are too small to be visible, even with magnification, they are observed in the ultramicroscope as luminous points that are seen to be in constant, irregular motion as they leave and enter the focal plane of the microscope. This random motion is called *Brownian motion* after Robert Brown, who first observed it. The motion is attributed to the impacts with molecules of the dispersion medium, which act to keep the colloidal particles suspended in apparent defiance of gravity. Even so, the particles presumably would settle out eventually if convection currents could be eliminated completely—which may be impossible.

Although they are larger than ordinary molecules, colloidal particles are sufficiently small that they have a great amount of surface relative to their volumes. Consequently, it is possible for large numbers of molecules or ions to become adsorbed to the surfaces of the particles. *Adsorption* is the adherence of any substance to a surface. Wetting (Section 7-1) is a special case of adsorption in which a liquid adheres to the surface of a solid. In many aqueous colloidal dispersions the particles have an electrical charge. In most cases all the particles are positive or else all of them are negative—in consequence of selective adsorption of one of the ions of water or one of the ions of an electrolyte that may be present at low concentration (either by intent or by accident). Electrical repulsion between particles of like charge is another factor favoring the dispersion of colloidal material. In an electric field, charged colloidal

FIGURE 10-7

The Tyndall effect

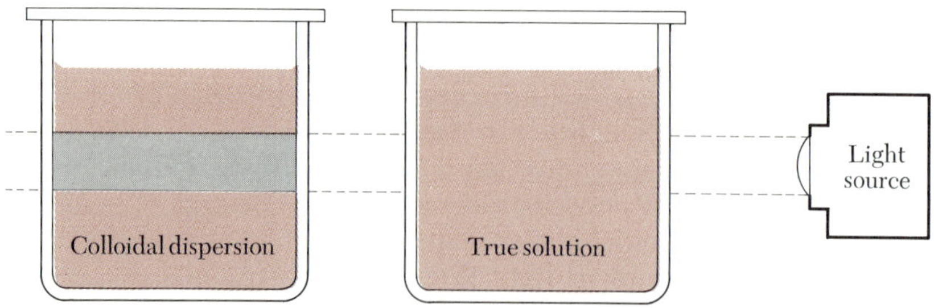

particles move toward the electrode of opposite charge; this migration is called *electrophoresis*.

The *Cottrell precipitator*, a device for the abatement of the smoke nuisance, operates on the principle that charged particles will migrate toward an electrode of opposite charge and will become deposited upon it. Smoke particles are given an electrical charge before they start up the smokestack. As they ascend the stack, the particles pass over electrodes of large area that neutralize the charges of the particles. This causes the particles to coagulate and settle out, thereby lessening the soot content of the emerging gases. Many industrial plants use Cottrell precipitators for the control of air pollution.

NEW TERMS

Brownian motion: the rapid and random motion that the particles in colloidal dispersions and coarse suspensions may be seen to display under appropriate conditions of illumination.

Colligative properties: properties that depend on how many particles are present and not, primarily, on their nature. The pressure exerted by a gas is a colligative property. The colligative properties of solutions are osmotic pressure, lowering of vapor pressure, depression of freezing point, and elevation of boiling point.

Colloid: any substance in the colloidal state.

Colloidal state: a state of subdivision intermediate between coarse suspensions, which are heterogeneous, and ordinary solutions, which are homogeneous.

Complete (total) ionic equation: an equation, representing a reaction involving ionic compounds, in which all the individual ions, including spectator ions, are represented.

Electrolyte: a compound that conducts electricity in the molten state or in the dissolved state.

Electrophoresis: the movement of charged particles in an electric field.

General ionic equation: same as net ionic equation; so called because it represents all reactions of a given type.

Henry's Law: the solubility of a gas is directly proportional to the pressure of the gas.

Ideal solution: a solution in which the molecules of each kind have an environment very much like the one they have when present only with molecules of their own kind; ideal solutions obey Raoult's Law.

Metathesis: a reaction in which two substances react to produce two new substances of the same class or classes.

Miscibility: the ability of two or more liquids to form a solution.

Molality of a solution: the number of moles of solute per kilogram of solvent, usually designated by m.

Molarity of a solution: the number of moles of solute in 1 liter of the solution, usually designated by M.

Net ionic equation: an equation, representing a reaction involving ionic compounds, from which spectator ions are omitted.

Osmosis: a net flow of solvent through a semipermeable membrane into a solution.

Osmotic pressure: the amount of pressure which, when applied to a solution, is just sufficient to prevent osmosis.

Raoult's Law: the vapor pressure of a liquid in an ideal solution is directly proportional to its mole fraction in the solution. It can be written $P = XP°$, where $P°$ is the vapor pressure of the pure liquid at the same temperature and X is its mole fraction in the solution.

Semipermeable membrane: a membrane that allows passage of one component of a solution but not others. Usually, the membrane is permeable to the solvent but not to solutes, as in osmosis.

Solute: in a solution, the dissolved substance.

Solvent: in a solution, the medium in which the solute is dissolved.

Spectator ions: the ions that are "partners" of the ions participating in a reaction but which do not take an actual part in the reaction themselves.

Strong electrolyte: any electrolyte (acid, base, or salt) that is a good conductor of electricity in the dissolved state because of the high concentration of its ions.

Titration: the process of determining the volume of a solution needed to react with a given volume of another solution (or with a given weight of some material). An important example is the addition of an acid solution to a base solution (or vice versa) to determine the concentration of the base (or acid).

Tyndall effect: the scattering of light by the particles in a coarse suspension or a colloidal dispersion, causing the path of a beam of light to be visible from the side.

Ultramicroscope: a microscope designed for viewing a colloidal dispersion while it is illuminated from the side by an intense beam of light.

Weak electrolyte: any electrolyte that is only a fair or poor conductor of electricity in the dissolved state because the substance furnishes only a few ions in solution.

EXERCISES

10-1 For each of the following substances, find the weight (in grams) required to prepare 1.000 liter of a 1.00M solution, 3.000 liters of a 4.00M solution,

5.000 liters of a $0.030M$ solution, 250 ml of a $3.0M$ solution, and 125 ml of a $0.010M$ solution:

(a) sucrose, $C_{12}H_{22}O_{11}$
(b) sodium chloride, NaCl
(c) magnesium sulfate 7-hydrate, $MgSO_4 \cdot 7H_2O$

10-2 What volume of a $0.0540M$ solution of barium hydroxide, $Ba(OH)_2$, is required to neutralize 21.8 ml of an $0.0872M$ solution of hydrochloric acid?

10-3 If 3 moles of a solid are dissolved in 4 moles of a chosen solvent, what is the molality of the solution? This question cannot be answered without additional information. What additional information is required?

10-4 If 200.0 g of sugar, $C_{12}H_{22}O_{11}$, are dissolved in 1200.0 g of water, what is the molality of the solution? To what temperature must it be cooled before ice can begin to form?

10-5 If the freezing point of an aqueous solution is $-0.345°C$, what is the molality of the solution? If the solution was prepared by dissolving 9.82 g of a solid in 200.0 g of water, what is the molecular weight of the solid?

10-6 What would be the boiling point, at 1 atm, of a solution containing 64.5 g of glucose, $C_6H_{12}O_6$, in 350.0 g of water?

10-7 The freezing point of an aqueous solution of a nonvolatile compound is $-0.876°C$. What is the boiling point of the solution?

10-8 Calculate the boiling point of a solution composed of 200.0 g of water, 0.0500 mole of NaCl, 0.1500 mole of Na_2SO_4, and 0.2000 mole of sucrose, $C_{12}H_{22}O_{11}$.

10-9 (a) A solution made by dissolving 10.80 g of glucose in 100.0 g of water is found to have a vapor pressure of 22.854 torr. At the same temperature the vapor pressure of pure water is 23.109 torr. Assuming Raoult's Law to be valid for this solution, calculate the molecular weight of glucose.
(b) The empirical formula for glucose is CH_2O. According to the result obtained in (a), what is the molecular formula of glucose?

10-10 What is meant by the statement that osmotic pressure is a colligative property?

10-11 What is the origin of the ions in an aqueous solution of Na_2SO_4? H_2SO_4?

10-12 What freezing points would you expect for aqueous solutions of the following compositions:
(a) $2.50m$ (molal) solution of sucrose, $C_{12}H_{22}O_{11}$
(b) $0.250m$ solution of Na_2SO_4
(c) $0.50m$ solution of HCl

10-13 If only 0.1% of the molecules of solute in a solution of a weak acid, HA, react with water to give H_3O^+ and A^- ions, what percentage of the HA molecules are still present as molecules?

10-14 *Metathetical reactions.* Complete and balance the following as molecular equations, as complete ionic equations, and as net ionic equations:

(a) $NH_4Cl + Ca(OH)_2 \longrightarrow$
(b) $NaNO_3 + H_2SO_4 \longrightarrow$
(c) $AgNO_3 + NaCl \longrightarrow$
(d) $CaCl_2 + H_2SO_4 \longrightarrow$
(e) $(NH_4)_2CO_3 + Sr(NO_3)_2 \longrightarrow$
(f) $Ag_2SO_4 + NaI \longrightarrow$
(g) $H_2S + CuSO_4 \longrightarrow$
(h) $NaOH + FeCl_3 \longrightarrow$
(i) $H_3PO_4 + FeCl_3 \longrightarrow$
(j) $CuSO_4 + Ca(OH)_2 \longrightarrow$

10-15 *Ionic equations.* Complete and balance the following equations, assuming that the product is a (neutral) insoluble solid or a (neutral) molecule of a covalent compound.

(a) $H^+ + OH^- \longrightarrow$
(b) $H_3O^+ + OH^- \longrightarrow$
(c) $Ag^+ + Br^- \longrightarrow$
(d) $Cu^{++} + S^{--} \longrightarrow$
(e) $Fe^{3+} + OH^- \longrightarrow$
(f) $Ca^{++} + AsO_4^{3-} \longrightarrow$
(g) $Mg^{++} + SiO_3^{--} \longrightarrow$
(h) $Ca^{++} + F^- \longrightarrow$
(i) $Mn^{++} + OH^- \longrightarrow$
(j) $Ag^+ + S^{--} \longrightarrow$
(k) $H^+ + C_2O_4^{--} \longrightarrow$
(l) $H^+ + BO_3^{3-} \longrightarrow$
(m) $H_3O^+ + NO_2^- \longrightarrow$
(n) $NH_4^+ + OH^- \longrightarrow$

10-16 Suppose that you were given a clear liquid and asked to determine whether it is a pure compound, a solution of an electrolyte, a solution of a nonelectrolyte, or a colloidal dispersion. How would you proceed?

10-17 An excess of $BaCl_2$ solution was added to 20.0 ml of a K_2SO_4 solution. The precipitated barium sulfate weighed 0.695 g. What was the molarity of the K_2SO_4 solution?

REFERENCES

Kieffer, W. F., *The Mole Concept in Chemistry*, Reinhold, New York, 1962 (paperbound). In Chapter 6, "Chemical Equivalence," the author recommends abandonment of the conventional designation of "gram-equivalent weight" and the correlated "normality" scale of concentrations.

Masterton, W. L., and Slowinski, E. J., *Chemical Principles*, 2nd ed., Saunders, Philadelphia, 1969, Chapter 11, "Solutions." Principles of solubility and colligative properties of solutions.

Mortimer, C. E., *Chemistry: A Conceptual Approach*, Reinhold, New York, 1967, Chapter 8, "Solutions." Includes consideration of interionic attraction.

Sienko, M. J., *Chemistry Problems*, Benjamin, New York, 1967 (paperbound). Chapter 11 is concerned with calculations involving molality, molarity, Raoult's Law, etc.

Vold, M. J., and Vold, R. D., *Colloid Chemistry*, Reinhold, New York, 1964 (paperbound). The science of large molecules, small particles, and surfaces.

11

Reaction Rates and Equilibrium

11-1 Chemical Kinetics

It is common knowledge that some chemical reactions occur much more rapidly than others. It is also well known that the rate of a reaction can be altered by changing the conditions, for example, the temperature. The *rate of a reaction* is expressed as the amount of reactants converted into products, or the change in concentrations of the reactants and products, in a specified length of time. The amounts of the reactants consumed or products formed are usually expressed in moles; their concentrations, and the changes in their concentrations, are expressed in moles per liter. The unit of time employed in describing reaction rates may be an hour, a minute, or a second; if the reaction is very fast, the unit of time may be a millisecond (msec), a microsecond (μsec), or a nanosecond (nsec); for a quite slow reaction the unit of time may be a day or a week.

The study of the rates of chemical reactions and the factors that affect them is called *chemical kinetics*. One of the objectives of chemical kineticists is to arrive at possible mechanisms by which a particular reaction occurs. Many chemical reactions occur in a sequence of steps. The *mechanism of a reaction* is the sequence of steps by which the reactants are converted into the products in that particular reaction.

The factors that have been found, by observation and experiment, to affect reaction rates are (1) the nature of the reactants, (2) the concentrations of the reactants, (3) the temperature, and (4) the presence of a catalyst. The usual procedure for discovering how the rate of reaction is affected by each of these factors is to keep all factors constant except the one under study.

The fact that the nature of the reactants is an important factor affecting the rates of reactions is well illustrated by comparing the behavior of copper and magnesium when they are heated in a flame. Copper forms copper oxide slowly under these conditions, but magnesium burns to form magnesium oxide very rapidly. The effect of concentration is shown when a glowing splint bursts into flame upon being inserted into a bottle containing pure oxygen. Closely related is the effect of increased surface area upon the rate at which a solid reacts with a gas or with a constituent of a solution; coal mine explosion disasters are often the result of combustible coal dust being suspended in air. The effect of temperature upon reaction rate is made quite obvious when an explosive mixture of hydrogen and oxygen, or coal dust and air, is ignited by a match. The effect of the presence of a catalyst is familiar from the use of manganese dioxide as a catalyst in the preparation of oxygen by heating potassium chlorate.

11-2 Concentrations and Reaction Rates

The procedure usually followed for learning how the rate of a reaction is affected by the concentration of a particular reactant is to perform experiments in which all other factors are kept constant—including the concentrations of the other reactants. As soon as the rate of reaction has been measured in a series of experiments in which the concentration of the one reactant is varied in a systematic fashion (doubling, tripling, etc.), it will be known how the rate is related to the concentration of that particular reactant. The rate is arrived at by observing how fast some one reactant disappears or how fast some specified product forms. In reactions involving gases, the rate of reaction can sometimes be followed by noting the change of pressure with time. For example, the reaction between nitrogen(II) oxide and hydrogen,

$$2NO(g) + 2H_2(g) \longrightarrow N_2(g) + 2H_2O(g)$$

can be studied kinetically by measuring the change in pressure as the reaction proceeds. With the passage of time, the pressure steadily decreases because only 3 moles of gas are formed when 4 moles of gas react.

In studies of the reaction between nitrogen(II) oxide and hydrogen, it has been established that doubling the initial concentration of H_2 doubles the reaction rate, tripling the concentration of H_2 triples the reaction rate, etc. In other words, the rate of reaction is found to be proportional to the concentration of H_2. On the other hand, doubling the initial concentration of NO has been found to quadruple the reaction rate, tripling the concentration of NO makes the reaction take place nine times as fast, and so on. In other words, the reaction rate is found to be proportional to the *square* of the NO concentration. These findings can be stated in the form of an equation known as the *rate expression* for the reaction:

$$\text{Rate} = k[H_2][NO]^2$$

Each formula in brackets represents the concentration of that particular reactant. The proportionality constant k is called the *specific rate constant.* It is characteristic of a particular reaction but it varies with temperature; the higher the temperature, the greater the value of k.

All rate expressions are of the form

$$\text{Rate} = k[A]^m[B]^n \cdots$$

where m is the power to which the concentration of A must be raised, n is the power to which the concentration of B must be raised, and so on (as indicated by the dots) for each of the other reactants.

It is important to note that the exponents (m, n, etc.) in the rate expression *must be determined by experiment.* In general, the exponents are *not* the same as the coefficients in the balanced net equation for the reaction. As we have just seen, the exponents for the reaction between H_2 and NO are 1 and 2, respectively, although the coefficients in the equation for the reaction are 2 and 2. This will be explained in the next section.

11-3 Reaction Mechanisms

One of the challenging aspects of chemical kinetics is the problem of accounting for the exponents in the rate expression. This is attempted by assuming that a reaction takes place in a sequence of steps and that the slowest step in the sequence is the one that determines the measured reaction rate. Since the products of the slowest step are the reactants in the subsequent steps, the slowest step must be the *rate-determining step*, because it controls the rate at which any subsequent step can occur.

Both bond forming and bond breaking are involved in a chemical reaction. For example, in the formation of water from elementary hydrogen and elementary oxygen, H—H bonds and O—O bonds are broken and H—O bonds are formed. Energy is required to break the H—H bonds and the O—O bonds, but the formation of H—O bonds releases energy. This particular reaction is exothermic, that is, the heat of formation of water is negative, because *more* energy is released in the formation of the H—O bonds than is required to break the H—H and O—O bonds.

The *collision theory* of chemical reactions assumes that each step in a reaction requires the occurrence of a collision between the reacting particles. The theory assumes, furthermore, that many colliding molecules do not undergo chemical reaction because they do not collide with sufficient energy to cause bond breaking. The kinetic theory of gases assumes that most molecules of a gas have velocities near the average but that some molecules have velocities far in excess of the average. Only the exceptional molecules with the highest velocities have sufficient kinetic energy to break bonds upon impact. The greater the amount of energy required to break a bond, the smaller the number of colliding molecules that will have sufficient kinetic energy to react upon impact. Therefore, with a reaction taking place in a sequence of steps it is almost a certainty that one of the steps will be inherently slower than the others and will consequently be the rate-determining step.

The rate expression found experimentally for the reaction between nitrogen(II) oxide and hydrogen can be accounted for by assuming that the reaction takes place in the following steps:

$$\begin{array}{lrcl}
\text{STEP 1:} & H_2(g) + NO(g) + NO(g) & \longrightarrow & N_2O(g) + H_2O(g) \\
\text{STEP 2:} & H_2(g) + N_2O(g) & \longrightarrow & N_2(g) + H_2O(g) \\
\hline
\text{NET EQUATION:} & 2H_2(g) + 2NO(g) & \longrightarrow & N_2(g) + 2H_2O(g)
\end{array}$$

If Step 1 is assumed to be the slow step, then we can explain why the reaction rate is dependent on the first power of the concentration of H_2 and on the second power of the concentration of NO. Surely Step 1 would be the slow one because it involves a three-body collision, whereas Step 2 involves a two-body collision. The simultaneous collision of three molecules is much less probable than a collision involving only two molecules.

Three-body collisions are so highly improbable that many chemical kineticists object to the mechanism proposed above in favor of an alternative mechanism that involves only two-body collisions:

$$\begin{array}{lrcl}
\text{STEP 1:} & 2NO(g) & \longrightarrow & N_2O_2(g) \\
\text{STEP 2:} & N_2O_2(g) + H_2(g) & \longrightarrow & N_2O(g) + H_2O(g) \\
\text{STEP 3:} & N_2O(g) + H_2(g) & \longrightarrow & N_2(g) + H_2O(g)
\end{array}$$

The second step is assumed to be the rate-determining step. Since the collision of two NO molecules is required to form the N_2O_2 for the second step, the concentration of N_2O_2 is proportional to the second power of the NO concentration; i.e., $[N_2O_2] = k_1[NO]^2$, where k_1 is the specific rate constant for the first step. Therefore, the rate of the second step is proportional to the square of the concentration of NO multiplied by the concentration of H_2; i.e., $k_2[N_2O_2][H_2] = k_2k_1[NO]^2[H_2]$. This mechanism accounts for the observed rate law as well as the first one, and it involves no three-body collisions.

There is no way of knowing for certain whether a proposed sequence of steps is really the mechanism for a reaction. All that we can do is to propose a mechanism that is consistent with the data. Even if no other mechanism seems plausible at the moment, we must not assume that our proposed mechanism is the only possible one.

11-4 Chain Reactions

Some chemical reactions apparently proceed by a mechanism in which a species that is consumed in one step is regenerated in a subsequent step. Such a series of steps is referred to as a *chain mechanism*, and a reaction that involves such a mechanism is called a *chain reaction*. The reaction between gaseous hydrogen and gaseous chlorine in sunlight is a reaction of this type.

At ordinary temperatures, hydrogen and chlorine form hydrogen chloride only slowly in the dark. Exposure of a mixture of hydrogen and chlorine to sunlight results in an explosion. The absorption of a photon (a quantum of light energy) by a chlorine molecule is the first step in a succession of rapid events that constitute a chain reaction. The absorption of the photon causes the Cl_2 molecule to dissociate into two *activated*, i.e., energy-rich, chlorine atoms:

$$Cl_2 + \text{energy} \longrightarrow 2Cl$$

Either of the chlorine atoms can then react with an H_2 molecule to form a molecule of hydrogen chloride and a hydrogen atom:

$$Cl + H_2 \longrightarrow HCl + H$$

The hydrogen atom can then react with a Cl_2 molecule to form another molecule of hydrogen chloride and a new chlorine atom:

$$H + Cl_2 \longrightarrow HCl + Cl$$

The new chlorine atom, in turn, can react with another H_2 molecule, and thus a *chain reaction* that can result in the formation of a very large number of HCl molecules is established. The chain will not be terminated until an atom combines with another atom. For example,

$$H + H \longrightarrow H_2$$

or

$$Cl + Cl \longrightarrow Cl_2$$

or

$$H + Cl \longrightarrow HCl$$

The reactions that terminate the chain do not occur with great frequency because fewer atoms are present at any moment than H_2 or Cl_2 molecules. The probability of a collision between an atom and an H_2 or Cl_2 molecule is much greater than the chance that two atoms will collide.

11-5 Energy of Activation

The extra energy that colliding molecules must have to break bonds upon impact and thus make possible a chemical reaction between them is called *energy of activation*. The magnitude of the energy of activation of a reaction depends on the nature of the reactants, because some bonds are more easily broken than others.

An increase in temperature means that the average kinetic energy of the molecules is greater. Therefore, at higher temperatures a greater fraction of the colliding molecules have the necessary energy to break bonds upon contact. The increase in the total number of collisions is only a minor factor in causing the rates of reactions to become greater at higher temperatures. The increase in reaction rate is principally due to the increase in the fraction of the collisions that are effective because of the increased kinetic energy of the colliding molecules. The magnitude of the temperature effect is different for different reactions because the energies of activation are different. The increase in kinetic energy due to, say, a 10° rise in temperature will obviously have the greatest effect on the reaction with the lowest energy of activation. Many reactions double or even triple their rates for each 10° rise in temperature.

Each step in a reaction mechanism is presumably characterized by its own energy of activation. The energy of activation of the slowest step must be the most important in determining the rate at which reaction products are formed. A catalyst causes a change of mechanism in some manner—the activation energy for the rate-determining step in the mechanism with a catalyst is less than for the rate-determining step in the mechanism without a catalyst.

The details of how a catalyst operates must be investigated separately in each instance. A substance that is a catalyst for one reaction is not usually a catalyst for a different reaction.

11-6 Reversible Reactions and Chemical Equilibrium

If hydrogen and iodine are present together in a closed container at, say, 600°C, both elements are in the gaseous state and a reaction occurs that is represented by the equation

$$H_2(g) + I_2(g) \longrightarrow 2HI(g)$$

If the hydrogen and iodine are originally present in equimolar quantities at 600°C, the reaction continues until 80 percent of the hydrogen and 80 percent of the iodine are converted to hydrogen iodide. If pure hydrogen iodide is heated in a closed container at 600°C, a reaction occurs that is represented by the equation

$$2HI(g) \longrightarrow H_2(g) + I_2(g)$$

and it continues until 20 percent of the hydrogen iodide has decomposed. Since the second reaction is the exact opposite of the first one, either may be referred to as a *reversible reaction.* Both reactions may be represented by a single equation with double arrows. If the equation written is

$$H_2 + I_2 \rightleftharpoons 2HI$$

the H_2 and I_2 are referred to, arbitrarily, as *reactants* because they appear on the left; the HI is referred to arbitrarily as the *product* because it appears on the right; the formation of HI is designated the *forward reaction*; and the decomposition of HI is referred to as the *reverse reaction.* Alternatively the equation written may be

$$2HI \rightleftharpoons H_2 + I_2$$

with HI designated as the reactant, H_2 and I_2 as the products, the decomposition of HI as the forward reaction, and the formation of HI as the reverse reaction. Ordinarily, a chemical equation is written so that the starting materials appear on the left.

If we start with only those materials written on the left, the rate of the forward reaction steadily declines because of the diminishing concentrations of the reactants. During this period of time the concentrations of the products steadily increase and, consequently, the rate of the reverse reaction continually increases. Eventually a condition is reached in which the rates of the opposing reactions are equal. This condition is referred to as a state of *chemical equilibrium*; the mixture is an *equilibrium mixture*; and each component of the mixture is an *equilibrant.*

When the *point of equilibrium* is reached, the equilibrium is a dynamic situation. The forward and reverse reactions continue, but at the same rate. The concentration of each equilibrant remains unchanged, and both the forward and reverse reactions *appear* to have ceased, only because each equilibrant is being consumed at the same rate it is being produced.

Deuterium has been used to prove that the forward and reverse reactions continue even after the point of equilibrium has been reached. When a small quantity of deuterium gas, D_2, is added to an equilibrium mixture of hydrogen, iodine, and hydrogen iodide, we find that the mixture almost immediately contains the following chemical species: H_2, I_2, HI, D_2, DI, and HD.

Any reaction has its own characteristic point of equilibrium under any given set of conditions. The relative amounts of the various equilibrants in an equilibrium mixture serve to define the point of equilibrium. If the temperature changes, the system adjusts to the change by establishing a new point of equilibrium at which the relative amounts of reactants and products will be different. Some reversible reactions are virtually complete, in one direction or the other, at the point of equilibrium. Other reactions attain equilibrium when they have reached some intermediate degree of completion.

11-7 The Equilibrium Constant

The relationship that prevails among the concentrations of the equilibrants in an equilibrium mixture is given by the *equilibrium constant*, K, for the reaction, which expresses the equilibrium ratio of products to reactants. All reversible reactions may be represented at once by the generalized equation

$$aA + bB \rightleftharpoons cC + dD$$

where a is the coefficient for reactant A, and so on. It has been discovered by experiment that equilibrium constants take the form of the expression

$$K = \frac{[C]^c[D]^d}{[A]^a[B]^b}$$

Each capital letter in brackets represents the equilibrium concentration of the designated equilibrant. The small letters used to represent coefficients in the chemical equation appear as exponents in the expression for the equilibrium constant; the equilibrium concentration of each equilibrant must be raised to the power indicated by its exponent.

The expression defining the equilibrium constant contains the concentrations of the products in the numerator and the concentrations of the reactants in the denominator. Therefore, the size of the equilibrium constant is large if a large fraction of the reactants is converted into products during the approach to equilibrium. A very small equilibrium constant characterizes a reaction that is very incomplete at the point of equilibrium. It is a theoretical convenience to assume that all reactions are reversible and that reactions that appear to be irreversible are simply reactions that have exceptionally large equilibrium constants.

The numerical value of K for a particular reaction is calculated by substituting into the expression for the equilibrium constant values for the concentrations of the individual equilibrants (obtained by chemical analysis of equilibrium mixtures).

EXAMPLE

At 458°C an equilibrium mixture was found to contain 5.617×10^{-3} mole of hydrogen per liter, 5.936×10^{-4} mole of iodine per liter, and 1.270×10^{-2} mole of hydrogen iodide per liter. Calculate the equilibrium constant for the reaction

$$H_2 + I_2 \rightleftharpoons 2HI$$

(Note the statement that the mixture is an *equilibrium* mixture.)

$$\begin{aligned} K &= \frac{[HI]^2}{[H_2][I_2]} \\ &= \frac{(1.270 \times 10^{-2} \text{ mole/liter})^2}{(5.617 \times 10^{-3} \text{ mole/liter})(5.936 \times 10^{-4} \text{ mole/liter})} \\ &= 48.4 \end{aligned}$$

Table 11-1 presents a set of data listing the equilibrium concentration of each component in five different equilibrium mixtures of hydrogen, iodine, and hydrogen iodide at 458°C. A value for the equilibrium constant, calculated from each set of data, is listed in the last column of the table. Although this set of data was obtained under carefully controlled conditions, there is some variation in the calculated K values. This is due partly to experimental error and partly to inherent limitations resulting from the fact that the expression we use to calculate K is for an ideal mixture, much as Boyle's Law and Charles' Law are for ideal gases.

In the above example the calculated value for K is a unitless quantity and is therefore numerically independent of the units in which the equilibrium concentrations are expressed. The concentration units in the numerator cancel those in the denominator because the sum of the exponents in the numerator is the same as in the denominator. However, for the reversible reaction

$$PCl_5(g) \rightleftharpoons PCl_3(g) + Cl_2(g)$$

the equilibrium constant expression

$$K = \frac{[PCl_3][Cl_2]}{[PCl_5]}$$

TABLE 11-1

Data on equilibrium mixtures of H_2, I_2, and HI at 458°C

Mixture	$[H_2]$ (moles/liter)	$[I_2]$ (moles/liter)	$[HI]$ (moles/liter)	$K = \frac{[HI]^2}{[H_2][I_2]}$
1	5.617×10^{-3}	5.936×10^{-4}	1.270×10^{-2}	48.4
2	4.580×10^{-3}	9.733×10^{-4}	1.486×10^{-2}	49.5
3	4.567×10^{-3}	1.058×10^{-3}	1.544×10^{-2}	49.3
4	3.842×10^{-3}	1.524×10^{-3}	1.687×10^{-2}	48.6
5	1.696×10^{-3}	1.696×10^{-3}	1.181×10^{-2}	48.5
			AVERAGE =	48.9

contains two concentrations in the numerator and only one in the denominator. Here, the calculated value for K does depend on the concentration units employed. *Equilibrium concentrations are always expressed in moles per liter.*

EXAMPLE

At a certain temperature, 1 liter of an equilibrium mixture was found to contain 0.035 mole of PCl_3, 0.147 mole of Cl_2, and 0.0083 mole of PCl_5, all gases. Calculate the equilibrium constant for the reaction

$$PCl_5(g) \rightleftharpoons PCl_3(g) + Cl_2(g)$$

$$K = \frac{(0.035 \text{ mole/liter})(0.147 \text{ mole/liter})}{(0.0083 \text{ mole/liter})}$$
$$= 0.62 \text{ mole/liter}$$

The practice of expressing equilibrium concentrations in moles per liter is so well established that K values such as the one just calculated are usually written without indicating the units.

The definition for the equilibrium constant may be said to constitute a *Law of Chemical Equilibrium*, which may be stated in words as follows: Whenever a reversible reaction is in a state of chemical equilibrium, the ratio of the product of the molar concentrations of the substances formed by the reaction to the product of the molar concentrations of the reactants, each raised to the power corresponding to its coefficient in the equation for the reaction, is equal to a constant referred to as the equilibrium constant for the reaction. The meaning of this law is that addition or removal of a quantity of one of the equilibrants (i.e., a change in its concentration) will be followed by a change in the concentration of each of the other equilibrants sufficient to preserve the original ratio of numerator to denominator in the equilibrium constant expression for the reaction, that is, to maintain the value of K for the reaction.

11-8 Shifting the Point of Equilibrium

Changing the composition of an equilibrium mixture by altering the conditions in some way is referred to as *shifting the point of equilibrium.* If the change in conditions causes an adjustment to occur so as to produce additional amounts of products, we say the point of equilibrium has been shifted to the right. If the adjustment results in an increased concentration of reactants, we say the point of equilibrium has been shifted to the left.

Two ways in which conditions can be altered to shift the point of equilibrium are:

1. The concentration of one of the equilibrants can be changed by the addition of more of the substance or by the removal of some of it.
2. The volume of the system can be reduced or increased; that is, additional pressure can be applied or the applied pressure can be reduced.

The effect of either of these changes upon an equilibrium can be predicted with substantial accuracy.

Let us consider the reversible reaction

$$H_2(g) + CO_2(g) \rightleftharpoons H_2O(g) + CO(g)$$

for which the equilibrium constant expression is

$$K = \frac{[H_2O][CO]}{[H_2][CO_2]}$$

If we add more H_2 to the equilibrium mixture without changing the volume of the container that the mixture occupies, we are establishing a higher concentration of H_2. Considering the expression for the equilibrium constant for this reaction, we see that the numerical value of the denominator has been increased. Automatically, the other concentrations change so that the ratio of numerator to denominator is restored to its previous value, i.e., the value of K. Thus, the preservation of equilibrium entails shifting the point of equilibrium to the right with a consequent reduction of CO_2 concentration and a corresponding increase in the concentrations of H_2O and CO. In general, increasing the concentration of one of the *reactants* will shift the point of equilibrium to the *right*; increasing the concentration of one of the *products* will shift the point of equilibrium to the *left*.

If we bring into contact with an equilibrium mixture some material that removes a portion of one of the equilibrants, we are reducing the concentration of that equilibrant. For example, a hygroscopic material reduces the concentration of H_2O. Considering again the above reaction, if we reduce the concentration of H_2O the concentrations of the other equilibrants will also change to such an extent as to preserve the ratio of numerator to denominator. Specifically, the point of equilibrium will shift to the right, that is, the CO concentration will increase and there will be a corresponding decrease in the concentrations of H_2 and CO_2. In general, reducing the concentration of one of the products will shift the point of equilibrium to the right; reducing the concentration of one of the reactants will shift the point of equilibrium to the left.

If the volume of the container of an equilibrium mixture is reduced, subjecting the mixture to increased pressure, the concentration of each equilibrant is increased in the same proportion. Whether this will have any effect on the point of equilibrium can be judged from the type of equilibrium constant expression that applies to the reaction. If the equilibrium constant is a unitless number, an increase in pressure will have no effect on the equilibrium. More simply stated, the sum of the coefficients of the gaseous products in the chemical equation must be equal to the sum of the coefficients of the gaseous reactants if the application of pressure is to have no effect on the equilibrium.

Suppose the reaction in question is the one represented by the equation

$$PCl_5(g) \rightleftharpoons PCl_3(g) + Cl_2(g)$$

Here the sum of the coefficients of the products is 2, but for the reactant it is only 1. Doubling the pressure doubles the concentration of each equilibrant. Consider the equilibrium constant expression

$$K = \frac{[PCl_3][Cl_2]}{[PCl_5]}$$

If each concentration were doubled, the ratio of numerator to denominator would also double, and the value of the expression would no longer be that characteristic of this particular reaction. The Law of Chemical Equilibrium tells us that, for this reaction, when the pressure is increased the point of equilibrium will shift to the left so that the concentrations of the products decrease and the concentration of the reactant correspondingly increases.

In general, the effect of reducing the volume of an equilibrium system involving gases, i.e., increasing the applied pressure, is to shift the equilibrium *in the direction of fewer molecules* of gaseous substances.

Conversely, reducing the pressure shifts the equilibrium in the direction of more molecules of gaseous substances.

11-9 Effect of Temperature on Equilibrium

The rate of almost every chemical reaction becomes greater at higher temperatures; the rate of an endothermic reaction increases more than the rate of its exothermic opposite. Consequently, virtually all reactions have different equilibrium constants at different temperatures. In 1884 the Dutch chemist Jacobus van't Hoff concluded, from experimental observations, that the equilibrium constant for a reaction becomes larger at higher temperatures if the forward reaction is endothermic and becomes smaller at higher temperatures if the reverse reaction is endothermic. This relationship is known as *van't Hoff's Law.* Bearing in mind that the reverse reaction is endothermic if the forward reaction is exothermic and that the forward reaction is endothermic if the reverse reaction is exothermic, we can predict with confidence whether a change in temperature will shift an equilibrium toward the right or toward the left.

The only kind of reaction whose equilibrium constant would not be affected by a change of temperature would be one for which the heat of reaction is zero. Strictly speaking, no such reaction is known, although the ΔH value is in some cases quite small. Thus, all equilibria are shifted somewhat by any change in temperature. The extent to which an equilibrium is shifted, and the equilibrium constant changed, by a specified change in temperature is related to the magnitude of the heat of reaction.

11-10 The Principle of Le Chatelier

The effects of changes in concentration, pressure, and temperature upon equilibrium were summarized in a single principle of broad application by the French chemist Henri Le Chatelier in 1888. The Principle of Le Chatelier (also called the Principle of Mobile Equilibrium) can be applied to all types of dynamic equilibria, chemical and physical.

Once a system has reached an equilibrium state, it *never* changes from that state spontaneously. The only changes which any system undergoes spontaneously are those in the direction toward equilibrium, and such a reaction will proceed only far enough to attain the equilibrium state. As we have already seen, this equilibrium state may be approached from either direction. *Le Chatelier's Principle* states that the change that takes place in an equilibrium system when conditions are altered is a change that tends to restore the original conditions.

Van't Hoff's Law is only a special case of the Principle of Le Chatelier. An increase in temperature favors the endothermic reaction—the reaction whose occurrence tends to prevent an increase in temperature.

An increase in the pressure applied to an equilibrium mixture favors the reaction that results in a decrease in pressure. For example, in an equilibrium mixture of gases, high pressure favors the change that reduces the number of molecules present in the system.

Again, in accord with Le Chatelier's Principle, the effect of increasing the concentration of one of the equilibrants is to shift the point of equilibrium in the direction that reduces the concentration of that particular substance.

The Principle of Le Chatelier can be applied only to systems that are already in a state of equilibrium when the change in conditions occurs. A mixture of H_2 and O_2 gases can be preserved for an indefinite length of time with no apparent reaction at room temperature. If the temperature is increased enough, a sudden reaction of explosive violence occurs. This reaction is exothermic. It appears that here an increase in temperature has favored an exothermic reaction. However, this situation does not violate Le Chatelier's Principle, because a mixture containing only H_2 and O_2 is not in a state of equilibrium—true chemical equilibrium does not exist unless all reactants and products are present, with the opposing reactions occurring at equal rates. Le Chatelier's Principle affords no basis for predicting what will happen if conditions are changed in a system that is not already in equilibrium.

When an appropriate catalyst is added to a reaction mixture, the speed of the reaction is increased. If the reaction is reversible, the catalyst increases the rates of both the forward and reverse reactions to the same extent. Therefore, since a reversible reaction can proceed only as far as the point of equilibrium, a catalyst serves only to reduce the time required for equilibrium to be reached. This is not the same as shifting the point of equilibrium. If a catalyst is added to a system that has already attained equilibrium, there are no changes in the concentrations of any of the equilibrants.

The Principle of Le Chatelier also applies to equilibria that involve no chemical reactions. For example, if a liquid and its vapor are in equilibrium within a closed container, we can predict correctly that an increase in temperature will shift the equilibrium in the direction of more vapor and less liquid because the conversion of liquid to vapor is endothermic.

Since ice has a smaller density than water, the application of pressure to ice at 0°C will cause the ice to melt. The system responds to increased pressure by forming more of the material that has less volume; that is, the higher pressure favors the process that tends to prevent an increase in pressure.

11-11 An Application of the Principle of Le Chatelier: The Haber Process

Ammonia is an economically important substance because of its many useful properties. Since elementary nitrogen is abundant in the atmosphere and elementary hydrogen can be manufactured at small cost, it has long been thought that it might be economically feasible to synthesize ammonia from its elements. When we examine the equation

$$N_2 + 3H_2 \rightleftharpoons 2NH_3 \qquad \Delta H = -22 \text{ kcal}$$

we note that the reaction is reversible, that the forward reaction is the exothermic one, and that the product consists of half as many molecules as the reactants. One more fact required to complete the picture is that the reaction is by nature a slow one.

According to the Principle of Le Chatelier, the formation of ammonia, i.e., shifting the equilibrium toward the right, is favored by the use of high pressure. On the other hand, the use of high temperature shifts the equilibrium to the left. Finally, of course, the continuous removal of ammonia, as it is formed, favors the forward reaction.

In practice, low temperatures should be avoided because the rate of any reaction is reduced by lowering the temperature, and this reaction is normally slow. However, since high temperatures have an unfavorable effect on the equilibrium, a compromise is necessary, and temperatures of 550-600°C are used. Fortunately, the use of high pressure not only shifts the point of equilibrium toward the right but increases the concentration of each reactant, thus increasing the rate at which the reaction proceeds. Even with the use of the highest feasible pressures (100–200 atm) and the optimum temperature, the reaction is still too slow to be practical unless an effective catalyst is used. Fritz Haber, a German chemist, received the Nobel prize in chemistry in 1918 for developing this process for the synthesis of ammonia with the aid of a catalyst. The *Haber process* is one of the most successful methods for the "artificial fixation" of atmospheric nitrogen, i.e., the conversion of atmospheric nitrogen into useful compounds.

11-12 Chemical Equilibrium Involving Solids and Gases

Expressions for equilibrium constants do not include terms for solid equilibrants. It has been found experimentally that the concentrations of the gaseous substances involved in a chemical equilibrium are not affected by changing the amount of any solid equilibrant. It is necessary, of course, for some of each solid equilibrant to be present in order for equilibrium to exist, but the *amount* is irrelevant.

The equilibrium involving copper oxide, copper metal, hydrogen gas, and gaseous water is represented by the following equation:

$$CuO(s) + H_2(g) \rightleftharpoons Cu(s) + H_2O(g)$$

This equilibrium involves two phases, the gas phase and the solid phase. As such, it is an example of *heterogeneous equilibrium*. We write the expression for the equilibrium constant for the reaction as follows:

$$K = \frac{[H_2O(g)]}{[H_2(g)]}$$

The ratio between the concentration of gaseous H_2O and the concentration of H_2 gas is constant as long as equilibrium prevails. Introducing more H_2 gas into the system that is already in equilibrium will shift the point of equilibrium toward the right, increasing the amount of gaseous H_2O, until the equilibrium ratio of $[H_2O(g)]$ to $[H_2(g)]$ is restored. On the other hand, introduction of more solid copper oxide into the system will have no effect on the amounts of the other equilibrants.

The equilibrium constant expression for the reaction

$$CaCO_3(s) \rightleftharpoons CaO(s) + CO_2(g)$$

involves the concentration of only one substance:

$$K = [CO_2(g)]$$

Since the concentration of a gas is proportional to its pressure, this expression means that at any given temperature there is only one pressure at which CO_2 gas can be in equilibrium with solid $CaCO_3$ and solid CaO. At 900°C this pressure is 790 torr. If solid $CaCO_3$ and solid CaO are present in the same container where the temperature is 900°C, three possibilities exist: (1) if CO_2 is present at a pressure of 790 torr, equilibrium prevails and there will be no change in the amounts of $CaCO_3$ and CaO in the system; (2) if CO_2 is present at a pressure in excess of 790 torr, the reaction will proceed to the left, producing more $CaCO_3$ and consuming CaO and CO_2 until the quantity of CO_2 consumed is sufficient to reduce the pressure of the remaining CO_2 to 790 torr; and (3) if CO_2 is present at a pressure less than 790 torr, or if no CO_2 is present, the reaction will proceed to the right, consuming $CaCO_3$ and producing CaO and CO_2 until a sufficient quantity of CO_2 is present to exert a pressure of 790 torr.

11-13 Equilibrium in Saturated Solutions of Solids

A solid can be dissolved in a liquid (provided the liquid is a solvent for the solid), a small quantity at a time, to give a series of solutions of differing concentrations; however, this cannot continue indefinitely. Eventually a point is reached beyond which no more solid dissolves; addition of more solid to the system simply results in the solid falling to the bottom of the container without dissolving. If the solid remains there indefinitely, in undiminished quantity, in spite of intermittent stirring of the solution or shaking of the vessel, the solution has reached the highest concentration possible at that particular temperature. When this situation exists, we say that the solution is *saturated.*

In a saturated solution, in the presence of excess solid, there is a state of equilibrium. The equilibrium is dynamic—the solid continues to dissolve, but the number of molecules going into solution, say, each minute, is the same as the number of molecules leaving the solution and attaching themselves to the solid each minute. After this equilibrium condition has been established, the concentration of solute in the solution undergoes no further change, and neither does the amount of undissolved solid in contact with the solution. If more solid is placed in the container, none of it dissolves. The concentration of the saturated solution is not affected by the amount of undissolved solid that happens to be present.

If all the undissolved solid is removed from contact with the saturated solution, as by filtration, we still have a solution of the same concentration. A saturated solution, therefore, may be defined as a solution that is or would be in equilibrium with the undissolved solid.

The *solubility* of a substance describes the concentration of its saturated solution. The factors determining the solubility of a solid are (1) the nature of the solid, (2) the nature of the solvent, and (3) the temperature. The act of dissolving is usually accompanied by the absorption or liberation of heat. On the basis of Le Chatelier's Principle, we can predict whether the solubility will increase or decrease with rising temperature if we know whether the solution process is endothermic or exothermic.

If we wish, we can describe the equilibrium between undissolved solid and saturated solution by means of a "chemical equation," even though the process is nonchemical. For example, iodine dissolves in water without reacting with it to any appreciable extent. The equilibrium between the saturated solution and excess undissolved solid may be expressed in the following manner:

$$I_2(s) \rightleftharpoons I_2 \text{ (saturated solution)}$$

From this point of view, the concentration of the saturated solution is the "equilibrium constant" for the process:

$$K = [I_2 \text{ (saturated solution)}]$$

If the solid is an ionic one, K takes on a special form, which is discussed in Section 11-21.

11-14 Equilibrium in Solutions of Weak Electrolytes

A weak electrolyte is a covalent compound that yields a small concentration of ions in solution by incomplete reaction with the solvent. Most acids and bases are weak electrolytes.

Pure acetic acid (hydrogen acetate) is a covalent compound composed of $HC_2H_3O_2$ molecules. When dissolved in water, this compound partially ionizes; a fraction of the $HC_2H_3O_2$ molecules react with the water in accordance with the equation

$$HC_2H_3O_2 + H_2O \rightleftharpoons H_3O^+ + C_2H_3O_2^-$$

In consequence of this incomplete reaction, an aqueous solution of acetic acid contains a small concentration of H_3O^+ ions and $C_2H_3O_2^-$ ions; but the concentration of $HC_2H_3O_2$ molecules is considerably greater than that of the ions.

There is a dynamic equilibrium between the molecular and ionized forms of the acid, as indicated by the double arrows in the equation. Thus, the proportion of molecules and ions remains constant.

Application of the Law of Chemical Equilibrium to this reaction gives the expression

$$\frac{[H_3O^+][C_2H_3O_2^-]}{[H_2O][HC_2H_3O_2]} = \text{constant}$$

Except at high concentrations of the solute, the concentration of water in aqueous solutions varies so slightly that the value of $[H_2O]$ in such expressions as this one may be regarded as constant without introducing any substantial error. Therefore, the equilibrium constant for this reaction is written in the form

$$K = \frac{[H_3O^+][C_2H_3O_2^-]}{[HC_2H_3O_2]}$$

This constant is called the *ionization constant* or *dissociation constant* of acetic acid.

The numerical value for the ionization constant of acetic acid may be calculated by substituting into the ionization constant expression the values for the individual concentrations of a particular solution. For example, by measurements of electrical conductance it is known that in a $0.1000M$ solution of acetic acid at room temperature, 1.35 percent of the molecules form ions while the remaining 98.65 percent retain their identity as molecules. Thus,

$$[H_3O^+] = [C_2H_3O_2^-] = 0.0135 \times 0.1000M = 0.00135M$$

and

$$[HC_2H_3O_2] = 0.9865 \times 0.1000M = 0.09865M$$

Therefore,

$$K = \frac{(0.00135 \text{ mole/liter})(0.00135 \text{ mole/liter})}{(0.09865 \text{ mole/liter})}$$

$$= 1.85 \times 10^{-5} \text{ mole/liter}$$

The degrees of ionization of acetic acid at various concentrations and the corresponding values for the ionization constant are given in Table 11-2. The last solution is almost a hundred times as dilute as the first, and the degree of ionization is more than nine times as great, but the value of the ionization constant is essentially the same in each case. This result is typical of those obtained for weak electrolytes and is indicative of a true equilibrium between molecules and ions.

Ammonia is one of the best-known weak bases. The equilibrium between molecules and ions in an aqueous solution of ammonia can be represented by the equation

$$NH_3 + H_2O \rightleftharpoons NH_4^+ + OH^-$$

The expression for the ionization constant is

$$K = \frac{[NH_4^+][OH^-]}{[NH_3]}$$

TABLE 11-2

Ionization of acetic acid at different concentrations

Concentration (moles/liter)	*Percent ionized*	*Ionization constant* (moles/liter)
0.10000	1.350	1.85×10^{-5}
0.05000	1.905	1.85×10^{-5}
0.02000	2.988	1.84×10^{-5}
0.009842	4.222	1.83×10^{-5}
0.005912	5.401	1.82×10^{-5}
0.001028	12.38	1.80×10^{-5}

AVERAGE = 1.83×10^{-5} mole/liter

The conductance of a $0.100M$ solution of ammonia indicates that 1.34 percent of the solute is ionized. Thus,

$$[NH_4^+] = [OH^-] = 0.0134 \times 0.100M = 0.00134 \text{ mole/liter}$$

and

$$[NH_3] = 0.9866 \times 0.100M = 0.09866 \text{ mole/liter}$$

Therefore,

$$K = \frac{(0.00134 \text{ mole/liter})(0.00134 \text{ mole/liter})}{(0.09866 \text{ mole/liter})}$$

$$= 1.82 \times 10^{-5} \text{ mole/liter}$$

The fact that the ionization constants of acetic acid and ammonia are substantially identical is an intriguing coincidence.

11-15 Use of Ionization Constants

Knowledge of the numerical value for the ionization constant of a weak electrolyte makes it possible to calculate the degree of ionization of the electrolyte, and the concentrations of its ions, in any solution containing it.

EXAMPLE

Given that the ionization constant for acetic acid is 1.8×10^{-5}, calculate the hydronium-ion concentration and the degree (percent) of ionization in a $0.010M$ solution of the acid.

Since we are interested in hydronium-ion concentration, we may begin by letting

$$x = [H_3O^+]$$

When we inspect the chemical equation

$$HC_2H_3O_2 + H_2O \rightleftharpoons H_3O^+ + C_2H_3O_2^-$$

we are reminded that acetate ions and hydronium ions are formed in equal numbers. Therefore, we can also say

$$x = [C_2H_3O_2^-]$$

The chemical equation also shows that the number of molecules of acetic acid undergoing ionization is the same as the number of hydronium ions formed. The equilibrium concentration of acetic acid molecules is the concentration that they would have in the absence of any ionization, minus the reduction in their concentration that results from the ionization:

$$0.010 - x = [HC_2H_3O_2]$$

Substituting these values into the expression for the ionization constant, we have

$$\frac{[H_3O^+][C_2H_3O_2^-]}{[HC_2H_3O_2]} = K$$

$$\frac{x^2}{0.010 - x} = 1.8 \times 10^{-5}$$

$$x^2 = 1.8 \times 10^{-7} - (1.8 \times 10^{-5})x$$

This equation can be written in the more familiar form of a quadratic equation:

$$x^2 + (1.8 \times 10^{-5})x - 1.8 \times 10^{-7} = 0$$

The solution of the generalized quadratic equation

$$ax^2 + bx + c = 0$$

is given by the formula

$$x = \frac{-b \pm \sqrt{b^2 - 4ac}}{2a}$$

Substituting $a = 1$, $b = 1.8 \times 10^{-5}$, and $c = -1.8 \times 10^{-7}$, we obtain the result

$$x = 4.16 \times 10^{-4}M$$

Since there are only two significant digits in the numbers for the ionization constant and for the concentration of the acid, we are justified in retaining only two digits of our answer. Thus, the final answer for the hydronium-ion concentration is

$$x = 4.2 \times 10^{-4}M$$

The solution to a quadratic equation is time-consuming and frequently can be avoided. Reviewing our solution to this problem, we recognize that the small value for the ionization constant, 1.8×10^{-5}, means that the ratio

$$\frac{x^2}{0.010 - x}$$

is small. Therefore, x must be small compared to the number 0.010, from which it is to be subtracted. Therefore, $0.010 - x$ is substantially equal to 0.010. This enables us to write the simplified equation

$$\frac{x^2}{0.010} = 1.8 \times 10^{-5}$$

Solution of this equation does not require the quadratic formula.

$$x^2 = 1.8 \times 10^{-7} = 18 \times 10^{-8}$$

$$x = \sqrt{18 \times 10^{-8}} = \sqrt{18} \times \sqrt{10^{-8}}$$

$$= 4.24 \times 10^{-4}M$$

Again, we are justified in having only two digits in the answer:

$$x = 4.2 \times 10^{-4}M$$

We see that we were justified in neglecting x in the expression $0.010 - x$, because we have obtained the same answer as by solving the quadratic equation.

Use of the approximate solution is justified when the degree of ionization is less than about 5 percent. In case of doubt, we can calculate the approximate solution and then ascertain whether the value for x thus calculated is in excess of 5 percent of the molarity of the solution. (Fortunately, the error introduced by the simplified solution increases the value of x.) Thus, if the resulting x is greater than 5 percent of the acid concentration, the results of the simplified solution should be discarded and the answer obtained by solution of the quadratic equation.

The degree of ionization is calculated by dividing the concentration of the acid in ionic form by the total concentration:

$$\frac{4.2 \times 10^{-4}M}{0.010M} = 0.042 = 4.2\% \text{ ionized}$$

11-16 The Common-Ion Effect

A $0.1M$ solution of acetic acid has the sour taste typical of acidic solutions and causes indicators to show their "acid colors." If sodium acetate is added to the solution, the sourness lessens and the indicator changes its hue. Other acetates, such as potassium acetate and calcium acetate, produce the same effect as sodium acetate. In general, the ionization of a weak acid is reduced, that is, the concentration of H_3O^+ ions is decreased, in the presence of one of its own salts.

A dilute solution of ammonia has the properties characteristic of weak bases; for example, it causes the indicator phenolphthalein to turn pink. If a sufficient amount of an ammonium salt, such as ammonium chloride or ammonium nitrate, is dissolved in the solution, the phenolphthalein will no longer show its "alkaline color." In general, the ionization of a weak base is reduced, that is, the concentration of OH^- ions is decreased, in the presence of one of its own salts.

Since sodium acetate and the other acetates have the acetate ion in common with acetic acid, the influence of an acetate in decreasing the ionization of acetic acid is known as the *common-ion effect*. The effect of ammonium salts in reducing the ionization of ammonia is likewise an example of the common-ion effect. In general, if we add to a weak electrolyte a substance providing an ion in common with it, ionization of the electrolyte is repressed. Thus, addition of HCl to a solution of acetic acid reduces the concentration of acetate ions, and the addition of a strong acid to a solution containing H_2S reduces the concentration of sulfide ions.

The magnitude of the common-ion effect can be calculated from the value for the ionization constant.

EXAMPLE

If a liter of $0.10M$ acetic acid solution contains 0.50 mole of sodium acetate, calculate the hydronium-ion concentration.

We may begin by letting

$$x = [H_3O^+]$$

Acetate ions are derived from both sodium acetate and acetic acid. For each H_3O^+ ion derived from acetic acid, there is one $C_2H_3O_2^-$ ion; all of the sodium acetate is ionized. Thus,

$$0.50 + x = [C_2H_3O_2^-]$$

and

$$0.10 - x = [HC_2H_3O_2]$$

From what we have learned about acetic acid in Section 11-14, we know that the concentration of H_3O^+ ions in the solution would be small, even without the common-ion effect making it smaller. Since x is small in comparison with 0.10, it is also small in comparison with 0.50. Therefore, we may say that, nearly enough,

$$0.50 = [C_2H_3O_2^-]$$

$$0.10 = [HC_2H_3O_2]$$

$$K = 1.8 \times 10^{-5} = \frac{[H_3O^+][C_2H_3O_2^-]}{[HC_2H_3O_2]} = \frac{0.50x}{0.10}$$

$$x = \frac{1.8 \times 10^{-5}}{5}$$

$$= 3.6 \times 10^{-6}M$$

In Section 11-14 we saw that in a 0.10*M* solution of acetic acid, in the absence of other solutes, the concentration of hydronium ions is 0.00135*M*. Thus, the presence of the 0.50*M* sodium acetate in the 0.10*M* acetic acid solution has reduced the hydronium-ion concentration from $135 \times 10^{-5}M$ to $0.36 \times 10^{-5}M$.

In qualitative terms, the common-ion effect is merely another special case of the Principle of Le Chatelier. The acetate ion is one of the products in the ionization of acetic acid. Increasing the concentration of one of the products of a reaction always shifts the point of equilibrium to the left. If we increase the concentration of acetate ions, we thereby decrease the concentration of hydronium ions and increase the concentration of acetic acid molecules.

11-17 Ionization of Water

Pure water is a very poor conductor of electricity, but its slight conductance is measurable. Measurements indicate that a very small fraction of the molecules, one in 555 million, are dissociated into H^+ and OH^- ions, the H^+ ions immediately becoming attached by coordinate bonds to other water molecules:

$$H_2O + H_2O \rightleftharpoons H_3O^+ + OH^-$$

The ionization of water thus gives equal numbers of the hydronium ions characteristic of aqueous solutions of acids and the hydroxide ions that characterize aqueous solutions of bases. Since the two kinds of ions are present in equal number, the properties of neither one predominate. Consequently, water does not have the sour taste often present in aqueous solutions of acids nor the bitter taste characteristic of aqueous solutions of bases, and it does not affect the color of either red or blue litmus. Water is *neutral*, and any aqueous solution is neutral if the concentrations of H_3O^+ and OH^- ions are equal. For example, a solution of sodium chloride is neutral because neither the Na^+ nor the Cl^- ions have any effect on either of the ions produced from water nor on the equilibrium in which water molecules, hydronium ions, and hydroxide ions are involved.

When an acid is dissolved in water, it forms more H_3O^+ ions. This shifts the point of equilibrium to the left. By the common-ion effect, the slight ionization of water is further suppressed and the concentration of OH^- ions is correspondingly reduced. It is impossible, however, to reduce the hydroxide-ion concentration to zero in any aqueous solution. When a base is dissolved in water, the increase in the hydroxide-ion concentration is accompanied by a corresponding decrease in the hydronium-ion concentration; but it is impossible to reduce the hydronium-ion concentration to zero in any aqueous solution.

In pure water at 25°C the concentration of H_3O^+ ions has been measured to be 1×10^{-7} mole per liter; the concentration of OH^- ions is also 1×10^{-7} mole per liter. In pure water and in all aqueous solutions the concentrations of H_3O^+, OH^-, and H_2O are such as to satisfy the equilibrium relationship

$$\frac{[H_3O^+][OH^-]}{[H_2O]^2} = K$$

Except at high concentrations of solute, $[H_2O]$ varies so little that it may be regarded as a constant. Therefore, $[H_2O]^2$ is combined with K to give a constant, K_w:

$$[H_3O^+][OH^-] = K_w$$

The constant K_w is the *ion-product constant of water*, and its value at 25°C is obtained by substituting into the K_w expression the measured values for the concentrations of H_3O^+ and OH^- in pure water at 25°C:

$$(1 \times 10^{-7})(1 \times 10^{-7}) = 1 \times 10^{-14} = K_w$$

Consequently, the hydroxide-ion concentration of any aqueous solution can be calculated if its hydronium-ion concentration is known:

$$[OH^-] = \frac{1 \times 10^{-14}}{[H_3O^+]}$$

In like manner, the hydronium-ion concentration of any aqueous solution can be calculated if its hydroxide-ion concentration is known.

EXAMPLE

If the hydroxide-ion concentration of a solution is $1 \times 10^{-3}M$, what is its hydronium-ion concentration?

$$[H_3O^+] = \frac{1 \times 10^{-14}}{1 \times 10^{-3}} = 1 \times 10^{-11}M$$

Since either the $[H_3O^+]$ or the $[OH^-]$ can be calculated if the other is known, we do not need to state both $[H_3O^+]$ and $[OH^-]$ in order to characterize a solution. Most often the acidity, alkalinity, or neutrality of a solution is expressed in terms of $[H_3O^+]$. To understand such characterization of solutions, it is necessary to remember these definitions:

1. In a neutral solution, $[H_3O^+] = 1 \times 10^{-7}M$.
2. In an acidic solution, $[H_3O^+] > 1 \times 10^{-7}M$.
3. In an alkaline solution, $[H_3O^+] < 1 \times 10^{-7}M$.

11-18 The pH Scale

The concentration of the hydronium ions in a solution is conveniently expressed in terms of a function called pH, which was suggested in 1909 by the Danish biochemist S. P. Sørensen. The pH is sometimes called the "hydrogen-ion exponent" since it refers to the negative power of 10 that represents the concentration of hydronium ions (hydrated hydrogen ions). Thus, if $[H_3O^+] = 10^{-3}$, pH = 3.

Expressed mathematically, pH is the negative logarithm of the hydronium-ion concentration, which is the logarithm of the reciprocal of the hydronium-ion concentration:

$$\text{pH} = -\log [H_3O^+] = \log \frac{1}{[H_3O^+]}$$

EXAMPLE 1

What is the pH of $0.01M$ HCl?

Since HCl is completely ionized in dilute solution, the hydronium-ion concentration is $0.01M$, i.e., $10^{-2}M$.

METHOD 1: $\text{pH} = -\log 10^{-2} = -(-2) = 2$

METHOD 2: $\text{pH} = \log \frac{1}{[H_3O^+]} = \log \frac{1}{10^{-2}} = \log 10^2 = 2$

EXAMPLE 2

If the pH of a solution is 8, what is the hydronium-ion concentration?

$$\text{pH} = -\log [H_3O^+] = 8$$

$$\log [H_3O^+] = -8$$

$$[H_3O^+] = 10^{-8}$$

We may go beyond the suggestion of Sørensen and define pOH as the negative logarithm of the hydroxide-ion concentration:

$$\text{pOH} = -\log [OH^-]$$

EXAMPLE

What is the pOH of $0.001M$ NaOH?

Since NaOH is a completely ionized compound, the hydroxide-ion concentration is $0.001M = 10^{-3}M$. Therefore,

$$pOH = -\log 10^{-3} = 3$$

We may go a step further and define pK_w as the negative logarithm of the ion-product constant of water:

$$pK_w = -\log K_w = -\log 10^{-14} = 14$$

This leads to the convenient relationship

$$pH + pOH = 14$$

so that the pH of a solution is readily calculated if its pOH is known, and the pOH can likewise be determined from the pH:

$$pH = 14 - pOH$$

$$pOH = 14 - pH$$

The relationships among the various methods for expressing the acidity or alkalinity of a solution are indicated in Table 11-3. A solution can be characterized with respect to its acidity or alkalinity by any one of the four quantities $[H_3O^+]$, $[OH^-]$, pH, or pOH. If one of these is known, the other three can be calculated.

In most instances the concentration of H_3O^+ or OH^- ions in a solution is not exactly a negative integral power of 10 but must be expressed as a negative power of 10 multiplied by some factor. For example, in $0.002M$ HCl the concentration of H_3O^+ ions is $2 \times 10^{-3}M$. In such cases a table of logarithms must be consulted in order to find the pH.

TABLE 11-3

Methods for expressing acidity or alkalinity of aqueous solutions

Solution	$[H_3O^+]$	$[OH^-]$	pH	pOH	*Nature of solution*
$0.1M$ HCl	10^{-1}	10^{-13}	1	13	Strongly acidic
$10^{-5}M$ HCl	10^{-5}	10^{-9}	5	9	Acidic
Pure water	10^{-7}	10^{-7}	7	7	Neutral
NaCl solution	10^{-7}	10^{-7}	7	7	Neutral
$10^{-5}M$ NaOH	10^{-9}	10^{-5}	9	5	Basic
$0.1M$ NaOH	10^{-13}	10^{-1}	13	1	Strongly basic

EXAMPLE

Calculate the pH of a 0.002*M* HCl solution.

$$\begin{aligned} pH &= -\log (2 \times 10^{-3}) = -(\log 2 + \log 10^{-3}) \\ &= -(0.30 - 3) = -(-2.70) \\ &= 2.70 \end{aligned}$$

In Table 11-4 the two-digit logarithms of the numbers 2 through 9 are listed. This is all the information needed for calculating pH when the hydronium-ion concentration is expressed as a power of 10 multiplied by an integral factor.

11-19 Buffer Solutions

Certain solutions, called *buffer solutions*, maintain their pH values at virtually constant levels despite the addition of small quantities of strong acid or strong base. There are many substances and mixtures that can serve as buffers, but the ones that can be most simply explained in terms of chemical equilibrium are prepared by dissolving a weak acid and one of its salts, or a weak base and one of its salts, in the same solution.

If a solution contains both acetic acid (a weak acid) and sodium acetate (a completely ionized salt), the common-ion effect suppresses the ionization of the acetic acid, that is, the point of the following equilibrium is shifted far to the left:

$$HC_2H_3O_2 + H_2O \rightleftharpoons H_3O^+ + C_2H_3O_2^-$$

If a small amount of a strong base, such as NaOH, is added to this acid buffer solution, H_3O^+ ions react with the added OH^- ions to form water molecules, but the hydronium-ion concentration is maintained at almost

TABLE 11-4

Logarithms of the integers 2-9

Number	*Logarithm*	*Number*	*Logarithm*
2	0.30	6	0.78
3	0.48	7	0.85
4	0.60	8	0.90
5	0.70	9	0.95

the same value through further reaction of $HC_2H_3O_2$ molecules with water. On the other hand, if a small amount of a strong acid, such as HCl, is added to the solution, $C_2H_3O_2^-$ ions react with the added acid so that the hydronium-ion concentration is virtually unaffected.

Solutions that contain a weak base and one of its salts are basic buffer solutions that maintain virtually constant hydroxide-ion concentrations against the addition of small quantities of strong acid or strong base. The mechanism of their buffer action is analogous to that of a solution containing a weak acid and one of its salts.

Most body fluids are protected by buffer action against appreciable change in pH. The pH of the blood is normally between 7.35 and 7.45; gastric juice has a pH of 1.6 to 1.8; and saliva is nearly neutral (pH 6.2 to 7.4). Since many of the reactions that take place in body tissues form acidic products, the blood must have a mechanism for preventing appreciable increases in its acidity. If illness causes the pH of the blood to go below 7.0 or above 7.8, death occurs.

11-20 pH Indicators

Certain organic substances occurring in nature, and others of synthetic origin, possess different colors in solutions having different pH values. These substances, called *acid–base indicators* or *pH indicators*, are often employed to determine the pH values of solutions. Such substances are themselves weak acids or weak bases.

The indicator called methyl orange is an example of a pH indicator that is a weak acid. In its aqueous solution there is an equilibrium between the ionized and the nonionized forms. If we symbolize this weak-acid indicator by the "formula" HInd, we may represent the equilibrium by the equation

$$\underset{\text{red}}{\text{HInd}} + H_2O \rightleftharpoons H_3O^+ + \underset{\text{yellow}}{\text{Ind}^-}$$

The feature that makes this particular weak acid useful as an indicator is the contrast between the red color of the molecules and the yellow color of the ions. An increase in the hydronium-ion concentration of the solution, brought about by addition of an acid, shifts the point of equilibrium to the left so that there is an increase in the concentration of the red form of the indicator. In solutions of smaller hydronium-ion concentration the point of equilibrium is shifted to the right toward the yellow form.

The color that methyl orange shows in a solution of a particular pH is determined by the relative amounts of the yellow form and the red form, i.e., by the ratio of their concentrations. From the expression for the ionization constant of the indicator,

$$\frac{[H_3O^+][Ind^-]}{[HInd]} = K$$

we can see that the ratio in question is given by the following equation:

$$\frac{[Ind^-]}{[HInd]} = \frac{K}{[H_3O^+]}$$

When the hydronium-ion concentration of the solution has the same numerical value as the ionization constant of the indicator, that is, when the ratio is equal to 1, the red form and yellow form are present in equal amounts. This causes the solution to have an orange color, that is, a color intermediate between yellow and red. When the pH is such that 90 percent of the indicator is in the red form, the eye cannot detect any change in color arising from a further increase in hydronium-ion concentration. If the pH of the solution is such that 90 percent of the indicator is in the yellow form, the eye cannot detect any change in color from a further decrease in hydronium-ion concentration. The limits between which this indicator undergoes color changes perceptible to the eye are pH 3.1 and pH 4.4. Within these limits the pH of a solution can be determined by observing the color that it imparts to methyl orange. The determination is made by comparison with color standards obtained by adding the indicator to a series of buffer solutions of known pH values.

A great number of pH indicators are known, each with its characteristic color-change interval. Determination of the pH of a particular solution requires the use of an indicator with the proper color-change interval; that is, the pH of the solution must be between the limits within which the indicator undergoes perceptible color changes. The measurement and control of pH is important in scientific work, both pure and applied. This is done most conveniently and most accurately with electrical instruments known as pH meters.

11-21 The Solubility Product Principle

The equilibrium that exists in a saturated solution of a slightly soluble ionic solid in the presence of excess solid (compare Section 11-13) may be represented by a chemical equation. For example,

$$AgCl(s) \rightleftharpoons Ag^+ + Cl^-$$

The formation of a precipitate upon mixing solutions containing the necessary ions is a chemical reaction from any point of view; the above equation is simply the reverse of the net ionic equation representing the formation of a silver chloride precipitate when solutions of, say, silver nitrate and sodium chloride are mixed.

Application of the Law of Chemical Equilibrium to the reversible process represented by the above equation leads to the following expression for the equilibrium constant:

$$K_{sp} = [Ag^+][Cl^-]$$

In this expression the equilibrium constant, K_{sp}, is the *solubility product constant* for silver chloride.

If we add a concentrated solution of NaCl, or any other soluble ionic chloride, to a saturated solution of AgCl, we increase the concentration of Cl^- ion. The product, $[Ag^+][Cl^-]$, would now be greater than the numerical value of K_{sp} for AgCl if it were not for the fact that solid AgCl separates from the solution. This is an example of the common-ion effect as it applies in the case of the heterogeneous equilibrium existing between a strong electrolyte and its ions in solution.

If we add a concentrated solution of $AgNO_3$, or any other soluble silver salt, to a saturated solution of AgCl, we increase $[Ag^+]$ so that the product, $[Ag^+][Cl^-]$, would again exceed the numerical value of K_{sp} for AgCl if solid AgCl did not separate from the solution. This example also illustrates the common-ion effect.

The fact that in a saturated solution of AgCl the product of the ion concentrations, $[Ag^+][Cl^-]$, is a constant is an example of the *solubility product principle*. Before attempting to state this principle, it would be well for us to inspect the expressions for the solubility product constants listed in Table 11-5. Note that if an ionic solid has more than one ion of one kind in its formula, the concentration of that ion must be raised to the corresponding power when the solubility product con-

TABLE 11-5

Solubility product expressions for selected compounds

Chemical formula	*Solubility product expression*
$BaSO_4$	$K_{sp} = [Ba^{++}][SO_4^{--}]$
CaF_2	$[Ca^{++}][F^-]^2$
Ag_2S	$[Ag^+]^2[S^{--}]$
$MgNH_4PO_4$	$[Mg^{++}][NH_4^+][PO_4^{3-}]$
$Fe(OH)_3$	$[Fe^{3+}][OH^-]^3$
$Ca_3(PO_4)_2$	$[Ca^{++}]^3[PO_4^{3-}]^2$
$FePO_4$	$[Fe^{3+}][PO_4^{3-}]$
Bi_2S_3	$[Bi^{3+}]^2[S^{--}]^3$

stant for the compound is calculated. This leads to the following statement of the solubility product principle: The product of the concentrations of the ions in a saturated solution of a slightly soluble strong electrolyte is a constant for a given temperature, provided each concentration is raised to the power equal to the subscript of the corresponding ion in the chemical formula.

The solubility product principle applies only to strong electrolytes that are slightly soluble, i.e., compounds that form saturated solutions in which the ion concentrations are of the order of $0.01M$ or less. The principle is useful because it applies to any solution that is in equilibrium with a slightly soluble ionic solid, regardless of how the ions got into the solution and regardless of the proportions in which they are present.

By analysis of a saturated solution of AgCl containing no other electrolytes, it has been found that the concentration of Ag^+ ions in the solution equals $1.28 \times 10^{-5}M$ at 25°C. With no other electrolytes present, the concentration of Cl^- ions must also equal $1.28 \times 10^{-5}M$. For the solubility product constant we then have

$$K_{sp} = [Ag^+][Cl^-] = (1.28 \times 10^{-5})(1.28 \times 10^{-5})$$
$$= 1.64 \times 10^{-10} \quad \text{at } 25°C$$

A simple experimental method for determining K_{sp} values is to determine the volumes of properly chosen solutions that must be mixed in order to produce the first detectable precipitation.

EXAMPLE

It was found, at 25°C, that only 0.16 ml of $0.001M$ $AgNO_3$ needed to be added to 1.000 liter of $0.001M$ NaCl to produce the first signs of cloudiness (indicating precipitation) in the stirred solution. What is the solubility product constant for AgCl?

$$[Cl^-] = 0.001M$$
$$= 1 \times 10^{-3}M$$

$$[Ag^+] = 0.001M \times \frac{0.16}{1000}$$
$$= 1.6 \times 10^{-7}M$$

$$K_{sp} = [Ag^+][Cl^-]$$
$$= (1 \times 10^{-3})(1.6 \times 10^{-7})$$
$$= 1.6 \times 10^{-10} \quad \text{at } 25°C$$

In principle, the most direct procedure is to shake the slightly soluble solid with a large volume of water until a saturated solution is obtained, filter it to separate the solution from the excess solid, measure exactly

1 liter of the saturated solution, evaporate the water from the solution, and weigh the solid residue. It is sometimes difficult, though, to be sure when saturation has been reached by this procedure.

EXAMPLE

A saturated solution of $PbCl_2$ at 20°C was found to contain 4.50 g of $PbCl_2$ per liter. Calculate the solubility product constant of lead chloride.

$$\frac{4.50\ \text{g}}{278\ \text{g/mole}} = 1.62 \times 10^{-2}\ \text{mole}\ PbCl_2$$

Therefore,

$$[Pb^{++}] = 1.62 \times 10^{-2}\ \text{mole/liter}$$

$$[Cl^-] = 3.24 \times 10^{-2}\ \text{mole/liter}$$

$$\begin{aligned} K_{sp} &= [Pb^{++}][Cl^-]^2 \\ &= (1.62 \times 10^{-2})(3.24 \times 10^{-2})^2 \\ &= (1.62 \times 10^{-2})(10.5 \times 10^{-4}) \\ &= 1.70 \times 10^{-5} \end{aligned}$$

If the numerical value for the solubility product constant for a compound is known, it can be used to determine whether a precipitate will form when a stated volume of an electrolyte solution of given concentration is added to a specified volume of a solution of another electrolyte of known concentration.

EXAMPLE

Will a precipitate form if 5.0 ml of 0.00001*M* NaCl is added to 95.0 ml of 0.0001*M* $AgNO_3$?

Each solution is diluted by the other when they are mixed. If no precipitation occurs to reduce the ion concentrations, the concentrations of the ions after mixing can be calculated from the dilution ratios:

$$\begin{aligned} [Ag^+] &= \frac{95.0}{100.0} \times 0.0001M \\ &= 9.5 \times 10^{-5}M \end{aligned}$$

$$\begin{aligned} [Cl^-] &= \frac{5.0}{100.0} \times 0.00001M \\ &= 5.0 \times 10^{-7}M \end{aligned}$$

$$\begin{aligned} [Ag^+][Cl^-] &= (9.5 \times 10^{-5})(5.0 \times 10^{-7}) \\ &= 4.75 \times 10^{-11} \end{aligned}$$

Since the product of the ion concentrations, 4.75×10^{-11}, is substantially less than K_{sp} for AgCl, 1.6×10^{-10}, precipitation is impossible in this instance. Precipitation cannot occur from a solution that is less than saturated.

The solubility product principle can likewise be employed to calculate the concentration of an ion remaining in solution after precipitation has occurred. Precipitation cannot proceed beyond the point where the ion concentrations have been reduced sufficiently so that the solution in contact with the precipitate is a saturated solution.

EXAMPLE

If 10 ml of $6.0M$ HCl is added to 90 ml of $0.20M$ $AgNO_3$, what will be the silver-ion concentration in the solution after precipitation of AgCl has ceased?

If no precipitation were to occur, the $[Ag^+]$ after mixing would be $\frac{90}{100} \times 0.20M = 0.18M$, and the $[Cl^-]$ would be $\frac{10}{100} \times 6.0M = 0.60M$. If *all* the Ag^+ ions were to precipitate out as AgCl, the $[Cl^-]$ in the solution would be reduced only to $0.60 - 0.18 = 0.42M$. Actually, of course, not all of the Ag^+ ions will precipitate out, but we may be sure that enough will do so to make the remaining $[Cl^-]$ approximately $0.42M$. At equilibrium, $[Ag^+][Cl^-] = 1.6 \times 10^{-10}$. If $[Cl^-]$ is approximately $0.42M$, then the approximate value for $[Ag^+]$ is

$$[Ag^+] = \frac{1.6 \times 10^{-10}}{0.42} \cong 3.8 \times 10^{-10}M$$

NEW TERMS

Buffer solution: a solution capable of maintaining its pH value at a virtually constant level despite the addition of a small quantity of a strong acid or strong base.

Chain reaction: a reaction that yields products that cause further reactions of the same kind and thus one which becomes self-sustaining.

Chemical equilibrium: the situation that exists when a chemical reaction is taking place at the same rate as the reverse reaction.

Chemical kinetics: the branch of chemistry that deals with reaction rates and mechanisms.

Collision theory: the theory that assumes that a reaction takes place in one or more steps involving collisions between reacting particles.

Common-ion effect: the reduction in the degree of ionization of a weak electrolyte brought about by the presence of a strong electrolyte providing an ion in common.

Energy of activation: the extra energy colliding particles must have in order to produce chemical reaction.

Equilibrant: a component of an equilibrium system. Each reactant and each product in an equilibrium mixture is an equilibrant.

Equilibrium constant: a constant characteristic of a particular reaction that is in a state of chemical equilibrium at a particular temperature, found by calculating the ratio of the equilibrium concentrations of products to reactants, each raised to the appropriate power.

Equilibrium mixture: a system, consisting of a mixture of substances, in which a state of chemical equilibrium prevails.

Forward reaction: the reaction in which the reactants are consumed; the left-to-right reaction. (The right-to-left reaction is the *reverse reaction.*)

Heterogeneous equilibrium: equilibrium involving more than one phase, for example, the equilibrium between a liquid and its vapor or between a solid and its vapor, or a chemical equilibrium involving both solid and gaseous substances.

Ionization constant (of a weak electrolyte): the equilibrium constant for the reaction in which molecules of a weak electrolyte produce ions.

Ion-product constant (of water): the constant, K_w, that represents the product $[H^+][OH^-]$ for any aqueous solution.

Law of Chemical Equilibrium: at the point of equilibrium a chemical reaction is characterized by a certain ratio of the concentrations of products to reactants, each raised to the appropriate power.

Le Chatelier's Principle: the change that takes place in an equilibrium system when conditions are altered is a change that tends to restore the original conditions.

pH: a number used to designate the degree of acidity or alkalinity of a solution, defined as the negative logarithm of the hydronium-ion concentration.

Point of equilibrium: the point at which the reverse reaction has attained the same rate as the forward reaction so that a condition of equilibrium prevails.

Rate expression: an equation that expresses the relationship of the rate of a reaction to the concentrations of the reactants. Rate expressions are of the form, $\text{Rate} = k[A]^m[B]^n$.

Rate of reaction: the quantity of reactants converted into products in a specified length of time.

Reactants: in an equation representing a reversible reaction, the substances whose formulas are written on the left. (The substances represented on the right are *products.*)

Reaction mechanism: the series of steps by which the reactants are transformed into products in a chemical reaction.

Saturated solution: a solution of such a concentration that if excess solute is present, equilibrium exists between the dissolved and undissolved phases.

Solubility: the concentration of the saturated solution of a substance.

Solubility product constant: the equilibrium constant, K_{sp}, that applies to the equilibrium between crystals of an ionic solid and its saturated solution.

Solubility product principle: the product of the concentrations of the ions in a saturated solution of a slightly soluble strong electrolyte is a constant for a given temperature, provided each concentration is raised to the power equal to the subscript of the corresponding ion in the chemical formula.

Specific rate constant: the proportionality constant k in the rate expression for a reaction.

Van't Hoff's Law: the equilibrium constant for an endothermic reaction increases with rising temperature, and for an exothermic reaction it decreases with rising temperature.

EXERCISES

11-1 "Heating a reaction mixture will increase the rate of reaction if the reaction is endothermic, but not if it is exothermic." Do you believe this? Defend your answer.

11-2 How would you prepare a "3-minute egg" on a high mountain? Explain.

11-3 Why is a suspension of flour, cellulose powder, or coal dust in air so dangerously explosive?

11-4 Write the expression for the equilibrium constant for each of the following reactions:
(a) $4NH_3(g) + 5O_2(g) \rightleftharpoons 4NO(g) + 6H_2O(g)$
(b) $COCl_2(g) \rightleftharpoons CO(g) + Cl_2(g)$
(c) $2NO(g) + O_2(g) \rightleftharpoons 2NO_2(g)$

11-5 At 400°C, 1.00 liter of an equilibrium mixture of nitrogen, hydrogen, and ammonia gases was found to contain 0.40 mole of H_2, 0.60 mole of N_2, and 0.14 mole of NH_3. Calculate the equilibrium constant.

11-6 In which direction will the equilibrium be shifted by an increase in (1) the concentration of O_2, (2) pressure, and (3) temperature in each of the following reactions:
(a) $2SO_2(g) + O_2(g) \rightleftharpoons 2SO_3(g) + \text{heat}$
(b) $N_2(g) + O_2(g) + \text{heat} \rightleftharpoons 2NO(g)$
(c) $2H_2(g) + O_2(g) \rightleftharpoons 2H_2O(g) + \text{heat}$
(d) $2CO(g) + O_2(g) \rightleftharpoons 2CO_2(g) + \text{heat}$

11-7 Write the expressions for the equilibrium constants for the following reactions:

(a) $NH_4Cl(s) \rightleftharpoons NH_3(g) + HCl(g)$
(b) $C(s) + CO_2(g) \rightleftharpoons 2CO(g)$

11-8 Hydrofluoric acid, HF, is 11% ionized in a 0.050*M* solution. Calculate its ionization constant.

11-9 Calculate the hydronium-ion concentration in a 0.090*M* solution of HCN for which the ionization constant is 4.0×10^{-10}.

11-10 Calculate the hydronium-ion concentration of a solution containing 0.090 mole of HCN and 1.25 moles of NaCN per liter.

11-11 Calculate the hydroxide-ion concentration in a 0.020*M* solution of hydrochloric acid, HCl.

11-12 Calculate the pH and pOH of the following solutions:

(a) a solution in which the hydronium-ion concentration is 1.0×10^{-3} mole/liter
(b) a solution in which the hydroxide-ion concentration is 4.0×10^{-5} mole/liter
(c) a $1.00 \times 10^{-3}M$ solution of acetic acid, 12.6% ionized
(d) a 0.020*M* solution of formic acid, for which the ionization constant is 1.8×10^{-4}

11-13 In an aqueous solution containing 0.20 mole of acetic acid in a total volume of 350 liters, the acid is 16.4% ionized. Calculate the weight (in grams) of H_3O^+ ions in the entire solution.

11-14 The solubility of silver chromate at room temperature is 0.0028 g of Ag_2CrO_4 per 100 ml of H_2O. Calculate K_{sp} for silver chromate.

11-15 If 0.010 mole of sodium chromate is dissolved in 100.0 ml of water, how many grams of silver chromate can then be dissolved?

REFERENCES

Bard, A. J., *Chemical Equilibrium*, Harper & Row, New York, 1968 (paperbound). Solves problems by using material balance and electroneutrality equations.

Barrow, G. M., Kenney, M. E., Lassila, J. D., Litle, R. L., and Thompson, M. E., *Chemical Equilibrium*, Benjamin, New York, 1967 (paperbound). Programmed instruction.

Carnell, P. H., and Reusch, R. N., *Molecular Equilibrium*, Saunders, Philadelphia, 1963 (paperbound). Programmed instruction in the basic principles of equilibrium.

Gaidis, J. M., and Rochow, E. G., "The Catalyzed Decomposition of Potassium Chlorate," *Journal of Chemical Education*, **40**(2), 78 (1963). Report on investigations of the role played by MnO_2 and the factors that determine its effectiveness.

Garrett, A. B., Lippincott, W. T., and Verhoek, F. H., *Chemistry: A Study of Matter*, Blaisdell, Waltham, Mass., 1968, Chapter 24, "Chemical Kinetics." A well-written treatment of reaction rates and mechanisms.

Guyon, J. C., and Jones, B. E., *Introduction to Solution Equilibrium*, Allyn and Bacon, Boston, 1969 (paperbound). Solubility product, acid–base, complex ion, and oxidation–reduction equilibria.

Hutchinson, E., *Chemistry: The Elements and Their Reactions*, 2nd ed., Saunders, Philadelphia, 1964, Chapters 15 and 16. Extensive lists of problems.

King, E. L., *How Chemical Reactions Occur*, Benjamin, New York, 1963 (paperbound). An introduction to kinetics and reaction mechanisms.

Sienko, M. J., and Plane, R. A., *Chemistry*, 3rd ed., McGraw-Hill, New York, 1966, Chapter 13, "Chemical Equilibrium." Concisely written; excellent questions.

12

Oxidation–Reduction and Electrochemistry

12-1 Chemical Change in Electrochemical Cells

Electrochemistry is the science that deals with the relation between electric current and chemical change. *Electrochemical cells* are devices in which chemical reactions are produced by electric current or electric current is produced by chemical reactions. The production of chemical reactions by electric current is called *electrolysis*, and the electrochemical cells in which electrolysis occurs are known as *electrolytic cells*. The production of electric current by a chemical reaction is known as *voltaic action*, and the electrochemical cells in which voltaic action occurs are called *voltaic cells*.

The reactions that take place in voltaic cells are spontaneous reactions; that is, they take place of their own accord without the action of some external agency such as heat, light, or shock. The reactions that occur in electrolytic cells are nonspontaneous; they do not occur until electric current from an outside source is passed through the cell.

The reactions that occur in electrochemical cells of either type are oxidation–reduction reactions. We have defined oxidation (Section 8-3) as an upward change in oxidation state, and we have defined reduction (Section 8-4) as a downward change in oxidation state. In an electrochemical cell, oxidation occurs at one electrode and reduction at the other. *The anode is the electrode at which oxidation occurs, and the cathode is the electrode at which reduction occurs.* This statement applies in both electrolytic and voltaic cells.

The study of the chemical changes that occur in electrochemical cells serves to deepen our understanding of oxidation–reduction reactions in general, since all electrochemical reactions are oxidation–reduction reactions.

12-2 Electrolysis and Changes in Oxidation States

An aqueous solution containing an acid, a base, or a salt may be shown to be a conductor of electric current by immersing strips of metal in the solution to serve as electrodes through which the current can leave and enter the solution and then connecting them to a source of current with an electric light bulb included in the circuit (Figure 12-1). If the solution is a good conductor (because of the presence of a large number of

FIGURE 12-1

Apparatus for testing conductivity of solutions

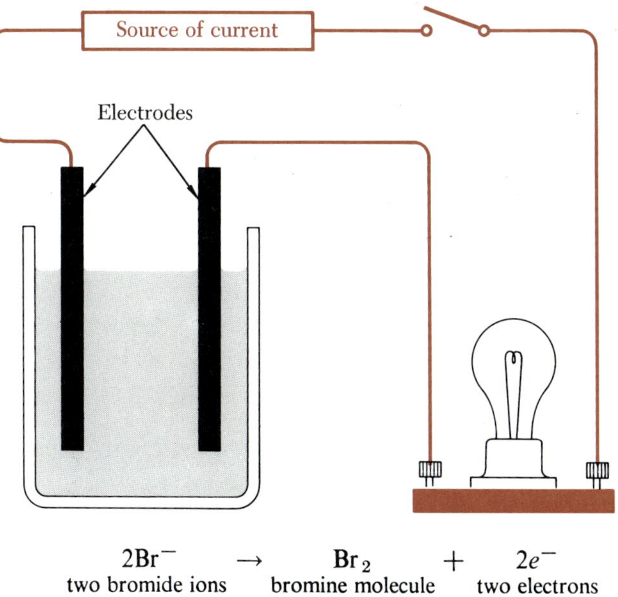

$$2Br^- \rightarrow Br_2 + 2e^-$$

two bromide ions — bromine molecule — two electrons

ions), the bulb glows brightly; if it is a poor conductor, the bulb glows feebly; and if it is a nonconductor, the light remains "off."

Either alternating current or direct current can be used to test the electrical conductivity of a solution. If direct current is used, one electrode remains positive and the other electrode remains negative during the entire time that the current is "on." Under these circumstances, chemical changes occur at both electrodes as long as the current is "on." For example, when direct current is passed through a solution containing copper(II) bromide, $CuBr_2$, copper plates out upon the negative electrode and elemental bromine forms at the positive electrode (Figure 12-2). The total cell reaction is referred to as the electrolysis of copper(II) bromide. The equation for the cell reaction indicates that the copper(II) bromide has been decomposed into its elements:

$$CuBr_2 \longrightarrow Cu + Br_2$$

It is more instructive, however, to consider separately the processes that occur at the two electrodes.

The copper metal forms at the surface of the negative electrode because this is where the positive copper ions acquire the electrons needed to convert them to neutral atoms:

$$\underset{\text{copper(II) ion}}{Cu^{++}} + \underset{\text{two electrons}}{2e^-} \longrightarrow \underset{\text{copper atom}}{Cu}$$

FIGURE 12-2

Electrolysis

Chemical changes occur at both electrodes.

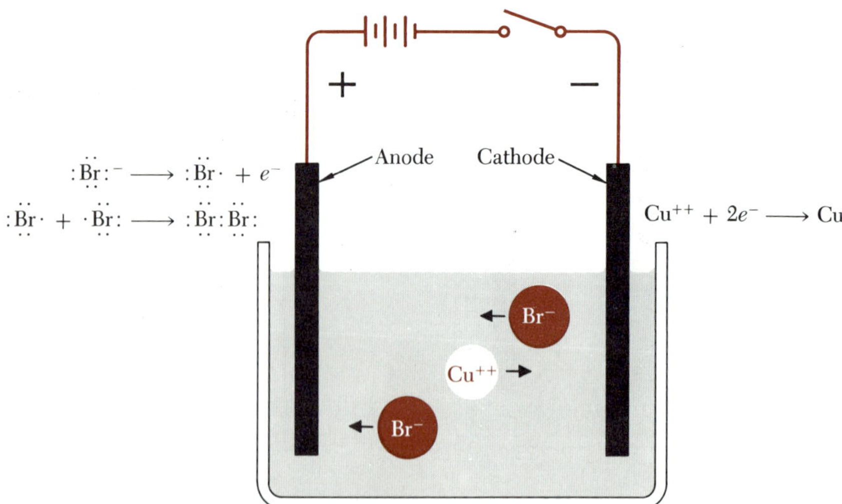

The process occurring at this electrode is reduction, because the oxidation state of the copper changes downward from +2 to 0. Therefore, this electrode is the cathode. (All positive ions are called *cations* because they go to the cathode when electric current is passed through a solution.)

At the positive electrode, the negative bromide ions lose electrons and are thereby converted to neutral atoms that combine in pairs to form molecules of elemental bromine:

$$\underset{\text{two bromide ions}}{2Br^-} \longrightarrow \underset{\text{bromine molecule}}{Br_2} + \underset{\text{two electrons}}{2e^-}$$

The process occurring at this electrode is oxidation, because the oxidation state of bromine changes upward from −1 to 0. Therefore, this electrode is the anode. (All negative ions are called *anions* because they go to the anode when electric current is passed through a solution.)

The equations we have written for the individual electrode reactions are called *ion–electron equations* because both ions and electrons appear in them. The equation for the total reaction occurring in an electrolytic cell may be obtained by adding the ion–electron equations that represent the two electrode reactions. The number of electrons shown to be gained in one ion–electron equation must be the same as the number of electrons shown to be lost in the other ion–electron equation when the two are added to give the equation for the total cell reaction:

$$\begin{array}{lrl} \text{CATHODE:} & Cu^{++} + 2e^- & \longrightarrow Cu \\ \text{ANODE:} & 2Br^- & \longrightarrow Br_2 + 2e^- \\ \hline & Cu^{++} + 2Br^- & \longrightarrow Cu + Br_2 \end{array}$$

Certain compounds conduct electric current in the pure liquid state, but such compounds are solids at room temperature, so they must be heated to their melting points to be able to conduct electric current. All these compounds are ionic compounds; examples are sodium chloride, NaCl, and sodium hydroxide, NaOH. When melted sodium chloride is present in an electrolytic cell, the sodium ions are attracted to the cathode, where they combine with electrons to form sodium atoms, i.e., sodium metal:

$$Na^+ + e^- \longrightarrow Na$$

Chloride ions are attracted to the anode, where each ion gives up an electron and becomes a chlorine atom. The chlorine atoms combine in pairs to form molecules of chlorine gas:

$$2Cl^- \longrightarrow Cl_2 + 2e^-$$

Before we add the ion–electron equations to obtain the equation for the total cell reaction, we must make the number of electrons gained in the one equation the same as the number lost in the other equation. We do this by multiplying each item in the cathode equation by 2:

$$\begin{array}{lrcl} \text{CATHODE:} & 2Na^+ + 2e^- & \longrightarrow & 2Na \\ \text{ANODE:} & 2Cl^- & \longrightarrow & Cl_2 + 2e^- \\ \hline & 2Na^+ + 2Cl^- & \longrightarrow & 2Na + Cl_2 \end{array}$$

12-3 Faraday's Law of Electrolysis

Early in the nineteenth century the passage of electric current through aqueous solutions of acids, bases, or salts was found to cause chemical changes at the two electrodes. Several active metals were discovered by Sir Humphry Davy, who obtained them by the process of electrolytic reduction (Section 1-8); he passed current from a voltaic cell through the melted hydroxides.

The idea that the charges of ions are related as small whole numbers is associated with the findings made by Michael Faraday in his investigations of electrolysis. In 1832 Faraday began investigating the relationship of the *amount* of chemical change produced by an electric current to (1) the strength of the current (i.e., the *amperage*), (2) the length of time the current was passed through the electrolytic cell, and (3) the chemical identity of the electrolyte. He found that passage of current through a specified solution produces an amount of chemical change that is directly proportional to the amperage and to the time during which the current flows. For example, a 5-A current passing through a copper sulfate solution for 2 min will liberate the same amount of copper as a 1-A current in 10 min. A *coulomb* is the amount of electricity carried by a 1-A current in 1 sec:

$$\text{Amperes} \times \text{seconds} = \text{coulombs}$$

Faraday found, then, that the amount of chemical change is directly proportional to the number of coulombs used.

Faraday investigated the relationship of the amount of electrolysis to the chemical identity of the electrolyte by passing the same current through each of several solutions contained in electrolytic cells connected in series, as shown in Figure 12-3. He found that the number of coulombs required for the liberation of 1 mole of silver atoms (107.870 g) from a silver salt will liberate only $\frac{1}{2}$ mole (31.77 g) of copper atoms from $CuCl_2$, $Cu(NO_3)_2$, $CuSO_4$, etc. This same number of coulombs (96,490) will liberate 1 mole of chlorine atoms (35.453 g) from NaCl, $CuCl_2$, etc. but only $\frac{1}{2}$ mole (16.032 g) of sulfur atoms from Na_2S, K_2S, etc. Faraday's results thus indicated that the quantity of charge carried by a chloride ion is the same as that carried by a silver ion, that a copper ion has twice

FIGURE 12-3

Electrolysis in cells in series

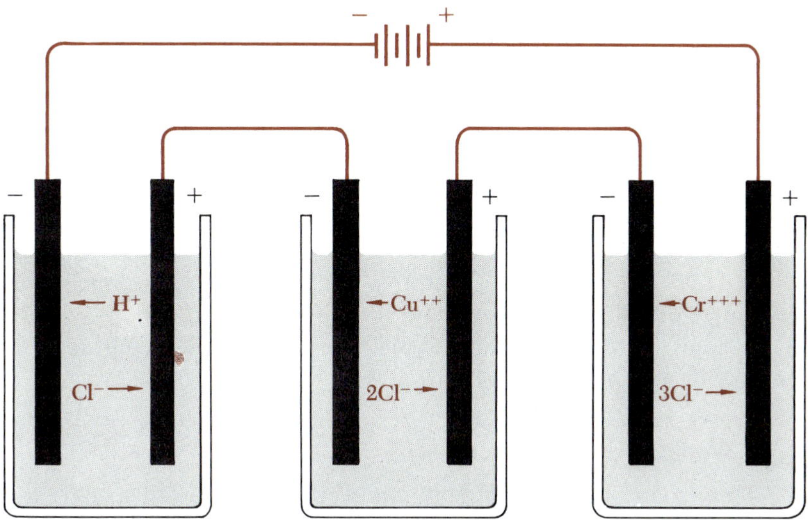

the charge of a silver ion, that a sulfide ion has twice the charge of a chloride ion, etc.

The findings of Faraday led the Irish physicist G. Johnstone Stoney to propose the electron theory (1874). This theory assumed electricity to be "atomistic" in much the same manner as matter itself. In other words, Stoney assumed that there is a fundamental particle of negative electricity, for which he proposed, in 1891, the name *electron*. Proof of the existence of the electron was subsequently obtained, as described in Section 3-5. We can now say that a silver ion, Ag^+, is a silver atom that has lost one electron; a chloride ion, Cl^-, is a chlorine atom that has gained one electron; a calcium ion, Ca^{++}, is a calcium atom that has lost two electrons; a sulfide ion, S^{--}, is a sulfur atom that has gained two electrons; etc. In electrolysis, a positive ion is converted to an atom by gaining the number of electrons required to neutralize its charge. For example,

$$\underset{\text{silver ion}}{Ag^+} + \underset{\text{one electron}}{e^-} \longrightarrow \underset{\text{silver atom}}{Ag}$$

$$\underset{\text{copper ion}}{Cu^{++}} + \underset{\text{two electrons}}{2e^-} \longrightarrow \underset{\text{copper atom}}{Cu}$$

The electrolytic liberation of atoms of a nonmetal at the positive electrode consists of the conversion of each negative ion to an atom by losing its "extra" electrons. For example,

$$2Cl^- \longrightarrow Cl_2 + 2e^-$$

$$S^{--} \longrightarrow S + 2e^-$$

Since 96,490 coulombs will liberate 1 mole of silver atoms, and since each silver atom is formed by the combination of one electron and one silver ion, it is apparent that 96,490 coulombs is the total charge of a mole of electrons. This quantity of electricity is now called a *faraday.*

Faraday's Law of Electrolysis may be stated in the following (modernized) manner: 1 faraday discharges 1 mole of singly charged ions, $\frac{1}{2}$ mole of doubly charged ions, etc. at each electrode. Thus, if 1 faraday is passed through the cells shown in Figure 12-3, there will be liberated 1 mole of H atoms, $\frac{1}{2}$ mole of Cu atoms, and $\frac{1}{3}$ mole of Cr atoms, respectively. One mole of Cl atoms will be liberated in each cell.

Knowledge of this law enables us to calculate the weights of the products obtained in electrolytic processes.

EXAMPLE 1

How many moles of Cd^{++} ions will be converted to Cd atoms by 3 faradays?

Since 1 faraday will liberate 1 mole of singly charged ions but only $\frac{1}{2}$ mole of doubly charged ions, 3 faradays will convert 3 faradays $\times$ $\frac{1}{2}$ mole/faraday = 1.5 moles of Cd^{++} ions to Cd atoms.

EXAMPLE 2

How many grams of silver will be plated out by 0.01 faraday?

One faraday will liberate 1 mole of singly charged atoms, such as Ag^+; 0.01 faraday will convert 0.01 mole of Ag^+ ions to Ag atoms. Since the weight of 1 mole of silver is 107.870 g, the weight of the silver is

$$0.01 \text{ mole} \times 107.870 \text{ g/mole} = 1.07870 \text{ g}$$

If the 0.01 mole is itself regarded as an experimentally determined quantity, it should be borne in mind that it has only one significant figure; in this case the answer is best expressed as 1 g. If the question had stated that 0.010 faraday was used (note the two significant figures), the answer would best be expressed as 1.1 g; that is, we would round off our answer to two digits. (See the Appendix for a discussion of significant figures.)

EXAMPLE 3

How many grams of copper metal will be plated out by the passage of a 1.92-A current for 30.0 min through a $CuSO_4$ solution?

Convert time to seconds:

$$30.0 \text{ min} \times 60.0 \text{ sec/min} = 1800 \text{ sec} \qquad \text{(3 significant figures)}$$

Calculate coulombs (multiply amperes by seconds):

$$1.92 \text{ A} \times 1800 \text{ sec} = 3456 \text{ coulombs} \qquad \text{(3 significant figures)}$$

Calculate faradays (1 faraday is 96,490 coulombs):

$$\frac{3456 \text{ coulombs}}{96{,}490 \text{ coulombs/faraday}} = 0.0358 \text{ faraday}$$

Calculate the weight liberated by 1 faraday: since CU^{++} ion is doubly charged, 1 faraday will liberate exactly $\frac{1}{2}$ mole.

$$0.5 \text{ mole} \times 63.54 \text{ g/mole} = 31.77 \text{ g Cu}$$

Multiply faradays by grams per faraday:

$$0.0358 \text{ faraday} \times 31.77 \text{ g Cu/faraday} = 1.14 \text{ g Cu}$$

Faraday's Law is the basis for one of the most easily comprehended methods of determining the numerical value of Avogadro's Number (Section 2-9). The total charge of 1 mole of electrons (1 faraday) is the charge of one electron multiplied by the number of electrons in 1 mole of electrons (Avogadro's Number, N). The total charge of 1 mole of electrons is 96,490 coulombs. The charge of one electron (Section 3-6) is 1.602×10^{-19} coulomb. Therefore, Avogadro's Number is

$$N = \frac{96{,}490 \text{ coulombs}}{1.602 \times 10^{-19} \text{ coulomb}}$$
$$= 6.023 \times 10^{23}$$

12-4 The Role of Electrons in Oxidation–Reduction Reactions

When magnesium burns in oxygen, electrons are transferred from magnesium atoms to oxygen atoms, forming the ionic compound MgO, which consists of Mg^{++} and O^{--} ions. In becoming oxidized, the magnesium goes from the zero oxidation state to the +2 state by losing electrons. At the same time the oxygen is reduced, going from the zero state to the −2 state by gaining electrons:

$$Mg \longrightarrow Mg^{++} + 2e^{-}$$
$$O + 2e^{-} \longrightarrow O^{--}$$

In its reaction with, say, sulfur or chlorine, magnesium undergoes the same change as in its reaction with oxygen—it loses electrons and changes from the zero state to the +2 state. Simultaneously, either sulfur or chlorine gains electrons to form negative ions and is reduced from the zero state to a lower oxidation state. Thus, the term *oxidation* applies to any instance of chemical change in which electrons are lost, whether to oxygen atoms, to sulfur atoms, to chlorine atoms, to a metallic electrode in electrolysis, or to any of a variety of substances classified as *oxidizing agents.* Likewise, the term *reduction* includes any reaction in which electrons are gained, whether from a metallic electrode in electrolysis or from any of a variety of substances classified as *reducing agents.*

In any chemical reaction involving oxidation there is also reduction and, conversely, reduction is always accompanied by oxidation. The total reaction, involving both oxidation and reduction, is called an *oxidation–reduction reaction* or a *redox reaction.*

In electrolysis the number of electrons gained by the anode is the same as the number lost by the cathode. This means that the number of electrons lost by the reducing agent is the same as the number gained by the oxidizing agent—a general relationship that holds true not only in electrolysis but in all oxidation–reduction reactions in which electron transfer takes place.

There are oxidation–reduction reactions in which there is no outright transfer of electrons from the reducing agent to the oxidizing agent. In these reactions, covalent bonds are either formed or broken. For example, the union of a hydrogen atom with a chlorine atom to form a hydrogen chloride molecule involves no actual transfer of an electron. However, a covalent bond is established between the two atoms,

$$\mathrm{H}\cdot + \cdot\overset{\cdot\cdot}{\underset{\cdot\cdot}{\mathrm{Cl}}}: \longrightarrow \mathrm{H}:\overset{\cdot\cdot}{\underset{\cdot\cdot}{\mathrm{Cl}}}:$$

and in the resulting HCl molecule the shared electrons are attracted more strongly by the chlorine atom than by the hydrogen. Thus, in HCl we assign to the hydrogen atom the oxidation state +1 and to the chlorine atom the oxidation state −1. Defining oxidation as an upward change in oxidation state and reduction as a downward change in oxidation state enables us to say that the hydrogen has been oxidized and that the chlorine has been reduced. These broad definitions for oxidation and reduction make it possible, for example, to place the reaction of hydrogen with chlorine in the same category as the reaction of magnesium with oxygen or sodium with chlorine or hydrogen with oxygen—all being examples of oxidation–reduction reactions. Reactions in which electrons are literally transferred from the reducing agent to the oxidizing agent are special cases of oxidation-reduction reactions.

If we speak of the reaction between hydrogen and chlorine, for example, as involving loss and gain of electrons, we are speaking somewhat figuratively. Our justification lies in the fact that the shared electrons in the HCl molecule "belong" more to the chlorine atom than to the hydrogen, and in this sense the formation of hydrogen chloride in this reaction involves a partial transfer of electrons.

12-5 Types of Oxidation–Reduction Reactions

The direct union of one chemical element with another is one of the simplest types of oxidation–reduction reactions. An equally simple type of oxidation-reduction reaction is the decomposition of a compound into its elements. A third simple type involves the displacement of an element by another element:

$$2NaBr + Cl_2 \longrightarrow 2NaCl + Br_2$$

$$Zn + CuSO_4 \longrightarrow Cu + ZnSO_4$$

Slightly more complicated oxidation–reduction reactions occur when an element combines with a compound or when a compound decomposes to give an element and a compound:

$$2CO + O_2 \longrightarrow 2CO_2$$

$$CO + Cl_2 \longrightarrow COCl_2$$

$$2KClO_3 \longrightarrow 2KCl + 3O_2$$

$$2KNO_3 \longrightarrow 2KNO_2 + O_2$$

Reactions of a somewhat more subtle type, known as *disproportionation reactions*, occur in some instances. In these reactions an element starts out in an intermediate oxidation state and ends up in a mixture of a higher and a lower state. For example, when chlorine dissolves in aqueous sodium hydroxide, part of the chlorine is oxidized to the +1 state and the rest of it is reduced to the −1 state:

$$Cl_2 + 2NaOH \longrightarrow NaClO + NaCl$$

Disproportionation also occurs when hydrogen peroxide decomposes,

$$2H_2O_2 \longrightarrow 2H_2O + O_2$$

and when monopositive copper ions change into dipositive ions and elemental copper:

$$2Cu^+ \longrightarrow Cu^{++} + Cu$$

In addition to these easily categorized types of oxidation–reduction reactions, there are many that are complicated to the point where classification is hardly worth the effort. The balancing of the equations for complicated oxidation–reduction reactions can be very difficult if we depend on the simple technique of inspection described in Section 2-10. In such cases, a systematic method can be very helpful.

12-6 Balancing Oxidation–Reduction Equations

The balancing of the skeleton equation for an oxidation–reduction reaction, which is already complete except for coefficients, is most readily done by consideration of the changes in oxidation states. This method of balancing oxidation–reduction equations, commonly known as the *oxidation number method*, involves selecting the smallest coefficients for the oxidizing agent and for the reducing agent that will make the total decrease in oxidation number for the one equal to the total increase in oxidation number for the other. In any oxidation–reduction reaction the oxidizing agent is reduced and the reducing agent is oxidized. This amounts to saying that the oxidizing agent acquires electrons from the reducing agent, and the number of electrons gained by the oxidizing agent must be the same as the number lost by the reducing agent. Therefore, the algebraic sum of all the changes in oxidation numbers must equal zero.

EXAMPLE 1

Consider the reaction between hydrogen sulfide and nitric acid to give sulfur, nitrogen(II) oxide, and water, according to the skeleton equation

$$H_2S + HNO_3 \longrightarrow NO + S + H_2O$$

It should be evident, after you have assigned the correct oxidation state to each element on both sides of the equation, that sulfur has been oxidized and nitrogen has been reduced:

$$\underset{-2}{H_2S} + \underset{+5}{HNO_3} \longrightarrow \underset{+2}{NO} + \underset{0}{S} + H_2O$$

decrease = 3

increase in oxidation state = 2

It takes three sulfur atoms (three H_2S molecules) to undergo an increase in oxidation number equal to the decrease in oxidation number experienced by two nitrogen atoms (two HNO_3 molecules). We indicate this by writing 3 in front of H_2S on the left and in front of S on the right and 2 in front of HNO_3 on the left and in front of NO on the right:

$$3H_2S + 2HNO_3 \longrightarrow 2NO + 3S + H_2O \qquad \text{(not balanced)}$$

We must obtain the coefficient for H_2O by inspection. If we choose to inspect hydrogen, we note that there are eight hydrogen atoms on the left; in order to have eight hydrogen atoms on the right, we place the coefficient 4 in front of H_2O:

$$3H_2S + 2HNO_3 \longrightarrow 2NO + 3S + 4H_2O \qquad \text{(balanced)}$$

We can check the equation and see that it is correctly balanced by noting that the number of oxygen atoms is six on each side of the equation.

EXAMPLE 2

Balance the equation for the reaction between copper and dilute nitric acid, which gives copper(II) nitrate, nitrogen(II) oxide, and water:

$$HNO_3 + Cu \longrightarrow Cu(NO_3)_2 + NO + H_2O$$

This equation illustrates a complication that is frequently encountered: the nitrogen is in two different oxidation states on the right-hand side of the equation. This is not a serious difficulty, however, because in copper(II) nitrate, as in any nitrate, the nitrogen is in the same oxidation state as in nitric acid. We concern ourselves only with the nitrogen that is in a *different* oxidation state from that of the nitrogen on the left-hand side of the equation:

$$\underset{+5}{HNO_3} + \underset{0}{Cu} \longrightarrow \underset{+2}{Cu(NO_3)_2} + \underset{+2}{NO} + H_2O$$

increase = 2 (Cu, 0 → +2)

decrease in oxidation number = 3 (N, +5 → +2)

As in the previous example, we know to insert 3 before Cu and before $Cu(NO_3)_2$ and to insert 2 before NO. We refrain from inserting 2 before HNO_3, however, because we recognize that the coefficient needed for HNO_3 must be large enough to furnish the six NO_3 groups in $3Cu(NO_3)_2$ in addition to the two nitrogen atoms in 2NO. Thus, the coefficient needed for HNO_3 is 8:

$$3Cu + 8HNO_3 \longrightarrow 3Cu(NO_3)_2 + 2NO + H_2O \qquad \text{(not balanced)}$$

The balancing is completed by inspection: since water is the only product containing hydrogen and there are eight hydrogen atoms on the left, the correct coefficient for H_2O is 4:

$$3Cu + 8HNO_3 \longrightarrow 3Cu(NO_3)_2 + 2NO + 4H_2O \qquad \text{(balanced)}$$

We can check the balancing by noting that there are 24 oxygen atoms on each side.

12-7 Ionic Equations for Oxidation–Reduction Reactions

If a strip of metallic zinc is immersed in a solution containing a copper(II) salt, such as $CuSO_4$ or $Cu(NO_3)_2$, the reaction that occurs can be represented by the equation

$$Zn + Cu^{++} \longrightarrow Zn^{++} + Cu$$

In this reaction zinc is oxidized from the zero state to the +2 state and the copper is reduced from the +2 state to the zero state. The equation we have written is the *net* ionic equation for the reaction. The spectator ions, NO_3^- or SO_4^{--} or whatever, are not included in the equation.

It is common practice to use net ionic equations to represent oxidation–reduction reactions involving ionic substances in solution. Consider, for example, the oxidation of iron(II) sulfate with potassium permanganate in a solution acidified with sulfuric acid. Before the reaction begins, the solutions contain Fe^{++}, SO_4^{--}, K^+, MnO_4^-, H^+, and HSO_4^- ions. In the oxidation–reduction reaction, Fe^{++} ions are converted to Fe^{3+} ions and MnO_4^- ions are reduced to give Mn^{++} ions; also, H^+ ions join with the oxygen from the MnO_4^- ions to form H_2O. The K^+, SO_4^{--}, and HSO_4^- ions do not take part in the reaction and can be omitted from the equation. Thus, the unbalanced net ionic equation has Fe^{++}, MnO_4^-, and H^+ on the left, with Mn^{++}, Fe^{3+}, and H_2O on the right. The balancing of ionic equations for oxidation–reduction reactions can be done by the method described in Section 12-6.

EXAMPLE

Balance the equation

$$Fe^{++} + MnO_4^- + H^+ \longrightarrow Mn^{++} + Fe^{3+} + H_2O$$

After the correct oxidation number has been assigned to each element on each side of the equation, it is evident that iron has been oxidized from the +2 state to the +3 state and that manganese has been reduced from the +7 state to the +2 state:

$$Fe^{++} + MnO_4^- + H^+ \longrightarrow Mn^{++} + Fe^{3+} + H_2O$$

+2 (Fe), +7 (Mn), +2 (Mn), +3 (Fe)

decrease = 5

increase in oxidation state = 1

To have the total increase the same as the decrease, five iron atoms must be oxidized for every manganese atom reduced:

$$5Fe^{++} + MnO_4^- + H^+ \longrightarrow Mn^{++} + 5Fe^{3+} + H_2O$$

(not balanced)

The four oxygen atoms from the MnO_4^- ion result in the formation of four H_2O molecules, and there must be eight H^+ ions on the left to balance the eight hydrogen atoms in the four H_2O molecules on the right:

$$5Fe^{++} + MnO_4^- + 8H^+ \longrightarrow Mn^{++} + 5Fe^{3+} + 4H_2O$$

(balanced)

The balancing of the equation can be checked by noting that the sum of the charges of the ions on the left (17+) equals the sum of the charges of the ions on the right (17+).

In any correctly balanced ionic equation, the algebraic sum of the charges of the ions on the left equals the sum of the charges of the ions on the right.

EXAMPLE 1

Balance the oxidation–reduction equation

$$Al + Ag^+ \longrightarrow Ag + Al^{3+}$$

This equation is already balanced atomically, that is, there is one atom of each kind on each side of the equation. However, in a correctly written equation the charges must also balance. This is accomplished when the changes in oxidation states are considered in the same manner as in the previous examples:

$$Al + Ag^+ \longrightarrow Ag + Al^{3+}$$

0 (Al), +1 (Ag), 0 (Ag), +3 (Al)

decrease = 1

increase in oxidation state = 3

Thus, it takes three silver ions to experience the same total change in oxidation number as one aluminum atom:

$$Al + 3Ag^{+} \longrightarrow 3Ag + Al^{3+}$$

In the balanced equation the sum of the charges on the left (3+) is the same as on the right (3+).

EXAMPLE 2

Balance the equation for the oxidation of chromium(III) hydroxide (which is insoluble in water) by sodium iodate in a solution that has been made basic with sodium hydroxide, given that the following skeleton equation correctly represents the reacting species and the products:

$$Cr(OH)_3 + IO_3^{-} + OH^{-} \longrightarrow I^{-} + CrO_4^{--} + H_2O$$

The first step to be taken is the assignment of the correct oxidation state to each element on each side of the equation. When this is done it becomes evident that chromium is being oxidized from the +3 state to the +6 state while iodine is being reduced from the +5 state to the −1 state:

$$\underset{+3}{Cr}(OH)_3 + \underset{+5}{I}O_3^{-} + OH^{-} \longrightarrow \underset{-1}{I^{-}} + \underset{+6}{Cr}O_4^{--} + H_2O$$

decrease = 6

increase in oxidation state = 3

It can now be deduced that two chromium atoms must be oxidized for every iodine atom that is reduced:

$$2Cr(OH)_3 + IO_3^{-} + OH^{-} \longrightarrow I^{-} + 2CrO_4^{--} + H_2O$$

(not balanced)

The coefficients for both ionic species on the right have now been established, and we note that the sum of their charges is 5−, which must also be the sum of the charges of the ions on the left when the equation is balanced. Therefore, the coefficient for OH^{-} must be 4:

$$2Cr(OH)_3 + IO_3^{-} + 4OH^{-} \longrightarrow I^{-} + 2CrO_4^{--} + H_2O$$

(not balanced)

Since there are now ten hydrogen atoms on the left, the coefficient required for H_2O on the right is 5:

$$2Cr(OH)_3 + IO_3^{-} + 4OH^{-} \longrightarrow I^{-} + 2CrO_4^{--} + 5H_2O$$

(balanced)

We can check the balancing by noting that the number of oxygen atoms on the left ($6 + 3 + 4 = 13$) is the same as on the right ($8 + 5 = 13$).

12-8 Complete Equations from Net Ionic Equations

The net ionic equations we have been writing for oxidation–reduction reactions have the same advantages and disadvantages that characterize net ionic equations for other types of reactions (Section 10-11). They focus attention on the particles that are taking part in the reaction and they omit all spectator ions. It is sometimes preferred to have the spectator ions included so as to have a complete ionic equation, which serves to emphasize that ions of one kind are never found in solution alone. For the purpose of calculating the weights of the reacting substances it may be desirable to write the complete equation in molecular form.

EXAMPLE 1

The net ionic equation for the reaction that occurs when metallic zinc is immersed in a solution of copper(II) salt is as follows:

$$Zn + Cu^{++} \longrightarrow Zn^{++} + Cu$$

Convert this equation into a complete ionic equation, and also into a molecular equation, to show the reaction of zinc with a solution of copper(II) sulfate.

For each Cu^{++} ion in the original copper sulfate solution there is present a SO_4^{--} ion. As each Cu^{++} ion in solution is replaced by a Zn^{++} ion, there is one spectator SO_4^{--} ion for each Zn^{++} ion.

REACTING PARTICLES:	$Zn + Cu^{++}$	$\longrightarrow$	$Cu + Zn^{++}$
SPECTATOR ION:	SO_4^{--}		SO_4^{--}

Therefore, the complete ionic equation is written as follows:

$$Zn + (Cu^{++} + SO_4^{--}) \longrightarrow Cu + (Zn^{++} + SO_4^{--})$$

The parentheses may be omitted if desired. In molecular form the equation reads

$$Zn + CuSO_4 \longrightarrow Cu + ZnSO_4$$

EXAMPLE 2

The following net ionic equation represents the oxidation of aluminum by dichromate ions in the presence of a strong acid:

$$2Al + Cr_2O_7^{--} + 14H^+ \longrightarrow 2Cr^{3+} + 2Al^{3+} + 7H_2O$$

Assuming that the dichromate ions are derived from potassium dichromate and that the strong acid is sulfuric acid, write the complete ionic equation and the molecular equation for the reaction.

The formula for potassium dichromate is $K_2Cr_2O_7$, which means that for every $Cr_2O_7^{--}$ ion in the original solution there are two K^+ ions. The K^+ ions are spectator ions and will be present in the solution after the reaction has occurred. Since the formula for sulfuric acid is H_2SO_4, there is one SO_4^{--} ion in the original solution for every two H^+ ions; thus, for the 14 H^+ ions represented in the equation there are seven SO_4^{--} ions that are spectator ions and will be present in the final solution. Beneath the formulas for the reacting particles in the net ionic equation we will write the formulas for the corresponding spectator ions:

$$\begin{array}{llllll} 2Al + Cr_2O_7^{--} & + 14H^+ & \longrightarrow 2Cr^{3+} & + 2Al^{3+} & + 7H_2O \\ \quad\; 2K^+ & \quad 7SO_4^{--} & \quad\; 3SO_4^{--} & \quad 3SO_4^{--} & \end{array}$$

Note that one SO_4^{--} ion and the two K^+ ions do not appear on the right, but they must be included in the complete ionic equation:

$$2Al + 2K^+ + Cr_2O_7^{--} + 14H^+ + 7SO_4^{--} \longrightarrow$$
$$2Cr^{3+} + 2Al^{3+} + 2K^+ + 7SO_4^{--} + 7H_2O$$

If desired, the positive and negative ions may be grouped into neutral sets:

$$2Al + (2K^+ + Cr_2O_7^{--}) + (14H^+ + 7SO_4^{--}) \longrightarrow$$
$$(2Cr^{3+} + 3SO_4^{--}) + (2Al^{3+} + 3SO_4^{--})$$
$$+ (2K^+ + SO_4^{--}) + 7H_2O$$

With the general ionic equation written in this manner, it is easily converted into a molecular equation:

$$2Al + K_2Cr_2O_7 + 7H_2SO_4 \longrightarrow$$
$$Cr_2(SO_4)_3 + Al_2(SO_4)_3 + K_2SO_4 + 7H_2O$$

12-9 Construction of Ion–Electron Equations

The balanced equation for an oxidation–reduction reaction involving ions in solution may be obtained by adding the ion–electron equations for the two half-reactions. It is useful to learn how to construct ion–electron equations for the oxidation of reducing agents and for the reduction of oxidizing agents so that they may be added to obtain the equations for oxidation–reduction reactions.

A reducing agent must contain an element in an oxidation state that is below the highest state possible for that element. Metals in the elemental state fulfill this requirement. For example, when zinc metal acts as a reducing agent, the zinc atoms undergo oxidation in becoming bipositive ions:

$$Zn \longrightarrow Zn^{++} + 2e^-$$

The reducing action of the ions of a nonmetal often involves their oxidation to the elemental state, as when sulfide ions are oxidized to sulfur atoms:

$$S^{--} \longrightarrow S + 2e^-$$

All ion–electron equations for the oxidation of reducing agents have electrons on the right, since the oxidation involves loss of electrons.

EXAMPLE 1

Write the ion–electron equation for the oxidation of ferrous ion to ferric ion.

To get started, we must know the formulas for the ions. The advantage of naming them iron(II) ion (rather than ferrous ion) and iron(III) ion (instead of ferric ion) is that the formulas are clearly indicated by the names to be Fe^{++} and Fe^{3+}, respectively. The equation, $Fe^{++} \longrightarrow Fe^{3+}$, is already balanced atomically; to balance it electrically, we must add one electron to the right-hand side:

$$Fe^{++} \longrightarrow Fe^{3+} + e^-$$

Thus, we are forced to recognize that the oxidation involves the *loss* of an electron.

EXAMPLE 2

Construct the balanced ion–electron equation for the oxidation of nitrite ion to nitrate ion in acidic solution.

We can start by writing the formula for nitrite ion on the left and the formula for nitrate ion on the right:

$$NO_2^- \longrightarrow NO_3^- \qquad \text{(incomplete)}$$

This incomplete equation shows more oxygen atoms on the right than on the left. Since the solution in which the oxidation is taking place is an acidic aqueous solution, H^+ and H_2O can be included in the equation wherever they are needed. Recognizing that we need one more oxygen atom on the left, we add one H_2O molecule to that side of the equation. Then, to bring hydrogen into balance, we add two H^+ ions to the right-hand side of the equation:

$$NO_2^- + H_2O \longrightarrow NO_3^- + 2H^+ \qquad \text{(incomplete)}$$

The equation is still incomplete because it is not balanced electrically. Since the net charge is 1+ on the right and 1− on the left, we must bring the equation into electrical balance by including two electrons on the right:

$$NO_2^- + H_2O \longrightarrow NO_3^- + 2H^+ + 2e^-$$

It should not escape notice that the presence of two electrons on the right corresponds to an increase of two units in the oxidation number of nitrogen (+3 to +5).

EXAMPLE 3

Construct the balanced ion–electron equation for the oxidation of sulfite ion to sulfate ion in alkaline aqueous solution.

Again we start by writing an incomplete equation that shows more oxygen atoms on the right than on the left:

$$SO_3^{--} \longrightarrow SO_4^{--} \qquad \text{(incomplete)}$$

Since the aqueous solution in which the reaction is taking place is alkaline, OH^- and H_2O can be included in the equation wherever they are needed. Knowing that sulfur is changing from the +4 state to the +6 state, an increase of two units, we will place two electrons on the right; to bring the equation into electrical balance, we must include two OH^- ions on the left:

$$SO_3^{--} + 2OH^- \longrightarrow SO_4^{--} + 2e^- \qquad \text{(incomplete)}$$

Finally, to bring hydrogen into balance, we add one H_2O molecule on the right:

$$SO_3^{--} + 2OH^- \longrightarrow SO_4^{--} + H_2O + 2e^-$$

We can check the balancing of the equation by noting that the total number of oxygen atoms on the left ($3 + 2 = 5$) is the same as on the right ($4 + 1 = 5$).

It must be remembered that a half-reaction involving oxidation of a reducing agent is always accompanied by another half-reaction in which some oxidizing agent undergoes reduction. An oxidizing agent must contain an element in an oxidation state that is higher than the lowest state possible for that element. Nonmetals in the elemental state fulfill this requirement, as when chlorine is reduced to chloride ion:

$$Cl_2 + 2e^- \longrightarrow 2Cl^-$$

The oxidizing action of the ions of a metal frequently involves their reduction to the elemental state, as when copper(II) ions are reduced to copper atoms:

$$Cu^{++} + 2e^- \longrightarrow Cu$$

All ion–electron equations for the reduction of oxidizing agents have electrons on the left, since the reduction involves a gain of electrons.

EXAMPLE 1

Write the ion–electron equation for the reduction of nitric acid to produce NO.

Knowing that HNO_3 is an acid, we realize that H^+ and NO_3^- ions are present in its aqueous solution. We can begin the construction of our equation by writing NO_3^- on the left and NO on the right:

$$NO_3^- \longrightarrow NO \qquad \text{(incomplete)}$$

This incomplete equation shows more oxygen atoms on the left than on the right. Since the solution is acidic, H^+ and H_2O can be included in the equation wherever they are needed. Recognizing that we need two more oxygen atoms on the right, we add two H_2O molecules on that side of the equation. To bring hydrogen into balance, we then add four H^+ ions to the left side of the equation:

$$NO_3^- + 4H^+ \longrightarrow NO + 2H_2O \qquad \text{(incomplete)}$$

Since the net charge is zero on the right (there being only neutral molecules on that side of the equation) and 3+ on the left, we must bring the equation into electrical balance by including three electrons on the left:

$$NO_3^- + 4H^+ + 3e^- \longrightarrow NO + 2H_2O$$

The presence of three electrons on the left in the balanced equation corresponds to a decrease of three units in the oxidation state of nitrogen (+5 to +2).

EXAMPLE 2

Write the ion–electron equation for the reduction of permanganate ion in basic solution to produce insoluble manganese dioxide.

We start by writing the formula for permanganate ion on the left and the formula for manganese dioxide on the right:

$$MnO_4^- \longrightarrow MnO_2 \qquad \text{(incomplete)}$$

Since the solution is basic, OH^- and H_2O can be included in the equation where they are needed. With manganese changing from the +7 state to the +4 state, we know to place three electrons on the left; to bring the equation into electrical balance, we must include four OH^- ions on the right:

$$MnO_4^- + 3e^- \longrightarrow MnO_2 + 4OH^- \qquad \text{(incomplete)}$$

To bring hydrogen into balance, we must include two H_2O molecules on the left:

$$MnO_4^- + 2H_2O + 3e^- \longrightarrow MnO_2 + 4OH^-$$

We can check the balancing of the equation by noting that the number of oxygen atoms on the left is the same as on the right.

12-10 Addition of Ion–Electron Equations

The balanced equation for an oxidation–reduction reaction involving ions in solution can be obtained by adding the balanced ion–electron equation for oxidation and the balanced ion–electron equation for reduction. In the equation for the total reaction, no electrons should appear; that is, the number of electrons lost must equal the number of electrons gained. This sometimes requires an adjustment, as in the case of aluminum metal displacing silver from a solution of a silver salt:

$$Al \longrightarrow Al^{3+} + 3e^-$$

$$Ag^+ + e^- \longrightarrow Ag$$

Before adding these two equations, we must adjust the second one by multiplying all items by 3 so that the number of electrons involved in the reduction of silver ions equals the number lost by the aluminum atom. Then the addition can be carried out with all electrons accounted for:

$$Al + 3Ag^+ \longrightarrow Al^{3+} + 3Ag$$

EXAMPLE 1

Balance the oxidation–reduction equation

$$Al + Cu^{++} \longrightarrow Al^{3+} + Cu$$

The equation is already balanced atomically, for there is one atom of each kind on each side of the equation. However, in the balanced equation the charges must also balance.

In balancing the equation, you may follow this procedure: (1) write the ion–electron equations for the conversion of aluminum atoms to Al^{3+} ions and for the conversion of Cu^{++} ions to copper atoms; (2) note the least common multiple of the numbers of electrons in the two equations; (3) multiply all items in each equation by the number required to make the electrons lost equal to the electrons gained; and (4) add the equations:

$$\begin{aligned} Al &\longrightarrow Al^{3+} + 3e^- \\ Cu^{++} + 2e^- &\longrightarrow Cu \end{aligned}$$

The least common multiple of 3 and 2 is 6. Therefore, multiply the first equation by 2 and the second by 3:

$$\begin{aligned} 2Al &\longrightarrow 2Al^{3+} + 6e^- \\ 3Cu^{++} + 6e^- &\longrightarrow 3Cu \\ \hline 2Al + 3Cu^{++} &\longrightarrow 2Al^{3+} + 3Cu \end{aligned}$$

EXAMPLE 2

Write the balanced equation for the oxidation of iron(II) ions by permanganate ions in acidic solution.

Earlier we wrote the balanced ion–electron equations for the reduction of permanganate ions in acidic medium and for the oxidation of iron(II) ions. All that remains is to add the two equations after multiplying the latter by the number required to make the electrons lost equal to the electrons gained in the former:

$$\begin{aligned} MnO_4^- + 8H^+ + 5e^- &\longrightarrow Mn^{++} + 4H_2O \\ 5 \times (Fe^{++} &\longrightarrow Fe^{3+} + e^-) \\ \hline 5Fe^{++} + MnO_4^- + 8H^+ &\longrightarrow 5Fe^{3+} + Mn^{++} + 4H_2O \end{aligned}$$

EXAMPLE 3

Write the balanced equation for the reaction of copper with dilute nitric acid.

Here it must be known that the oxidation of the copper atoms produces bipositive copper ions and that the reduction of NO_3^- ions produces NO gas. (Under other circumstances, reduction of NO_3^- ions may produce NO_2 gas or elementary N_2 or NH_4^+ ions.) The ion–electron equation for the oxidation of copper is easily constructed:

$$Cu \longrightarrow Cu^{++} + 2e^-$$

but the equation for the reduction of NO_3^- ions must be developed as described in Section 12-9:

$$NO_3^- + 4H^+ + 3e^- \longrightarrow NO\uparrow + 2H_2O$$

We can now obtain the balanced equation for the reaction by adding the balanced equations for the half-reactions (after multiplying each ion–electron equation by the appropriate number):

$$\begin{array}{r} 3 \times (Cu \longrightarrow Cu^{++} + 2e^-) \\ 2 \times (NO_3^- + 4H^+ + 3e^- \longrightarrow NO\uparrow + 2H_2O) \\ \hline 3Cu + 2NO_3^- + 8H^+ \longrightarrow 3Cu^{++} + 2NO\uparrow + 4H_2O \end{array}$$

12-11 Electric Current from Oxidation–Reduction Reactions

Voltaic cells were first described in 1800 by Count Alessandro Volta, the Italian physicist for whom they are named. They are sometimes called galvanic cells for Luigi Galvani, the Italian physiologist whose theories of "animal electricity" inspired Volta's researches. Voltaic cells are commonly used to produce electric current for the operation of cordless radios, electric toothbrushes, flashlights, hearing aids, and many other small appliances requiring low-voltage direct current.

In all voltaic cells, electric current is produced by oxidation–reduction reactions that occur spontaneously within them. When a voltaic cell is in operation, electrons travel through an external circuit instead of transferring directly from the reducing agent to the oxidizing agent. If a zinc strip is immersed in a solution containing Cu^{++} ions, the spontaneous reaction that occurs is accompanied by the release of energy in the form of heat. In the simple voltaic cell diagrammed in Figure 12-4, the zinc strip and the Cu^{++} ions are not in direct contact. Electrons must travel from Zn atoms to Cu^{++} ions by way of the connecting circuit; this flow of electrons constitutes an electric current.

The cell consists of two half-cells. In the zinc half-cell the half-reaction involves the oxidation of zinc atoms to zinc ions:

$$Zn \longrightarrow Zn^{++} + 2e^-$$

In the copper half-cell the half-reaction involves the reduction of Cu(II) ions to Cu atoms:

$$Cu^{++} + 2e^- \longrightarrow Cu$$

Thus, Zn^{++} ions enter the solution in the zinc half-cell, and Cu^{++} ions leave the solution in the copper half-cell. The zinc electrode loses weight and the copper electrode gains weight as the cell reaction continues. Electrons left behind on the zinc electrode by the departing Zn^{++} ions flow through the external circuit toward the copper electrode from which electrons are removed by Cu^{++} ions.

FIGURE 12-4

Voltaic cell made from zinc and copper electrodes immersed in solutions containing Zn^{++} and Cu^{++} ions

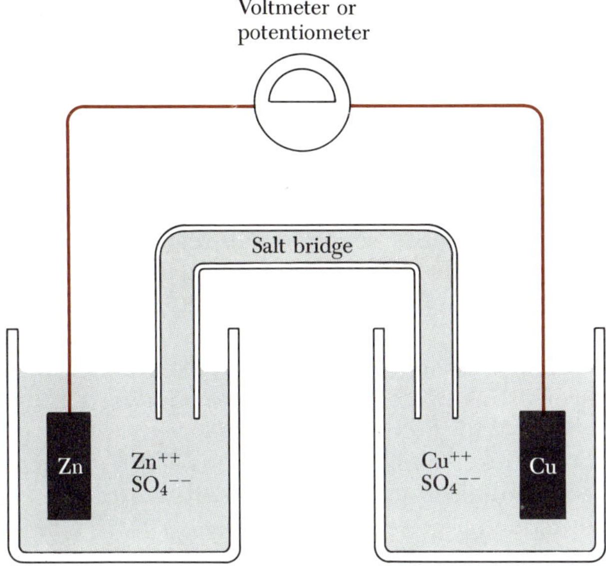

The external circuit may include an ammeter to measure the rate at which the current is flowing, or a voltmeter or potentiometer (as shown in Figure 12-4) to measure the electrical potential developed by the cell, or some other device that would be operated by the electric current. The "salt bridge" connecting the two half-cells contains a solution of an electrolyte, such as KCl. From the solution contained in the salt bridge, Cl^- ions migrate into the zinc half-cell to balance the charge of the Zn^{++} ions produced there; K^+ ions migrate into the copper half-cell to take the place of the Cu^{++} ions that are removed by reduction.

In each half-cell of a voltaic cell, an element is present in two oxidation states. As in electrolytic cells, the two electrodes of a voltaic cell are distinguished by the names *anode* and *cathode*, with the anode being the electrode at which oxidation occurs and the cathode the one at which reduction occurs.

12-12 Reduction Potentials

Every voltaic cell is characterized by the electrical potential (usually expressed in volts) it develops. The voltage (or potential) is of a definite magnitude, depending upon the components of the cell. The voltage of a cell that employs two metals and their ions, such as the zinc–copper cell, depends primarily on the relation of the metals to each other in the activity series, but it also depends on the temperature and on the concentrations of the ions in the two half-cells. The comparison of voltages of different components is made with solutions containing the ions at the standard concentration, 1*M*, at 25°C.

The potential developed by the zinc–copper cell, with Zn^{++} ions and Cu^{++} ions at 1*M* concentration, is 1.10 V at 25°C. This voltage is a measure of the tendency of the cell reaction to occur, i.e., of the "drive" behind the reaction. Since the cell reaction is a composite of an oxidation half-reaction and a reduction half-reaction, we might hope to measure the voltage due to each separate half-reaction. However, a single half-reaction never occurs as an isolated event; therefore, it is impossible to measure the voltage of an individual electrode except on a relative basis.

A nickel–copper cell may be constructed like a zinc–copper cell, with one half-cell consisting of a nickel electrode immersed in a solution containing Ni^{++} ions. With both Ni^{++} and Cu^{++} ions at 1*M* concentration, the potential of a nickel–copper cell at 25°C is 0.59 V. Since this is 0.51 V less than the potential of a zinc–copper cell, we can interpret this to mean that the reducing power of nickel atoms is 0.51 V less than that of zinc atoms or that the oxidizing power of Ni^{++} ions is 0.51 V greater than that of Zn^{++} ions. In other words, the tendency of nickel atoms to become oxidized to Ni^{++} ions is 0.51 V less than the tendency of zinc atoms to become oxidized to Zn^{++} ions, and the tendency of Ni^{++} ions to become reduced to nickel atoms is 0.51 V greater than the tendency of Zn^{++} ions to become reduced to zinc atoms. Therefore, we would predict, correctly, that it would be possible to construct a zinc–nickel cell in which the zinc would act as anode and the nickel as cathode and that the standard potential of this cell at 25°C would be 0.51 V.

By measuring the potentials of properly chosen voltaic cells, we can therefore arrange the metals in order of their activity. For such an activity or electromotive series to be useful, there should be associated with each metal in the series a number such that the difference between the numbers for any two metals would be the potential developed by a cell constructed with these two metals as components. Since one concern of chemists is to classify metals on the basis of their ability or inability to replace hydrogen from acids, a voltage of 0 at 25°C has been assigned arbitrarily to the hydrogen–hydrogen ion half-reaction:

$$2H^+ + 2e^- \rightleftharpoons H_2$$

This half-reaction may proceed in either direction, depending on the nature of the other half-reaction taking place in the cell, as indicated by the double arrows. A reduction half-reaction that has a *greater* tendency to occur than the one written above is assigned a *positive* potential; a reduction half-reaction that has a smaller tendency to occur than this one is assigned a negative potential. Examples of the entries in a table of standard *reduction potentials* at 25°C are the following:

$$Zn^{++} + 2e^- \rightleftharpoons Zn \qquad E° = -0.76\ V$$
$$Ni^{++} + 2e^- \rightleftharpoons Ni \qquad E° = -0.25\ V$$
$$2H^+ + 2e^- \rightleftharpoons H_2 \qquad E° = 0$$
$$Cu^{++} + 2e^- \rightleftharpoons Cu \qquad E° = +0.34\ V$$

These and other reduction potentials are listed in Table 12-1. The sym-

TABLE 12-1

Standard reduction potentials at 25°C

Half-reaction	$E°$ (V)
$K^+ + e^- \rightleftharpoons K$	−2.93
$Ca^{++} + 2e^- \rightleftharpoons Ca$	−2.87
$Na^+ + e^- \rightleftharpoons Na$	−2.71
$Mg^{++} + 2e^- \rightleftharpoons Mg$	−2.37
$Al^{3+} + 3e^- \rightleftharpoons Al$	−1.66
$2H_2O + 2e^- \rightleftharpoons H_2 + 2OH^-$	−0.83
$Zn^{++} + 2e^- \rightleftharpoons Zn$	−0.76
$Cr^{3+} + 3e^- \rightleftharpoons Cr$	−0.74
$Fe^{++} + 2e^- \rightleftharpoons Fe$	−0.44
$Cd^{++} + 2e^- \rightleftharpoons Cd$	−0.40
$PbSO_4 + 2e^- \rightleftharpoons Pb + SO_4^{--}$	−0.36
$Ni^{++} + 2e^- \rightleftharpoons Ni$	−0.25
$Sn^{++} + 2e^- \rightleftharpoons Sn$	−0.14
$Pb^{++} + 2e^- \rightleftharpoons Pb$	−0.13
$2H^+ + 2e^- \rightleftharpoons H_2$	0.00
$Sn^{4+} + 2e^- \rightleftharpoons Sn^{++}$	0.15
$Cu^{++} + 2e^- \rightleftharpoons Cu$	0.34
$I_2 + 2e^- \rightleftharpoons 2I^-$	0.54
$Fe^{3+} + e^- \rightleftharpoons Fe^{++}$	0.77
$Ag^+ + e^- \rightleftharpoons Ag$	0.80
$NO_3^- + 4H^+ + 3e^- \rightleftharpoons NO + 2H_2O$	0.96
$AuCl_4^- + 3e^- \rightleftharpoons Au + 4Cl^-$	1.00
$Br_2 + 2e^- \rightleftharpoons 2Br^-$	1.07
$O_2 + 4H^+ + 4e^- \rightleftharpoons 2H_2O$	1.23
$Cl_2 + 2e^- \rightleftharpoons 2Cl^-$	1.36
$Au^{3+} + 3e^- \rightleftharpoons Au$	1.50
$MnO_4^- + 8H^+ + 5e^- \rightleftharpoons Mn^{++} + 4H_2O$	1.51
$F_2 + 2e^- \rightleftharpoons 2F^-$	2.87

bol E° refers to the voltage when all ions are at the standard concentration, 1*M*.

The voltage of a voltaic cell at 25°C in which all ions are at the standard concentration, 1*M*, can be calculated from the relationship

$$E^\circ_{cell} = E^\circ_{anode} + E^\circ_{cathode}$$

For example, consider a zinc–hydrogen cell; here the zinc is the anode and the hydrogen electrode is the cathode. The value for $E^\circ_{cathode}$ is 0; the value for E°_{anode} is +0.76 V, that is, the standard reduction potential listed for Zn^{++}–Zn in Table 12-1 with the sign reversed because the half-reaction occurring at this electrode is oxidation rather than reduction.

ANODE:	$Zn \rightleftharpoons Zn^{++} + 2e^-$	$E^\circ = +0.76$ V
CATHODE:	$2H^+ + 2e^- \rightleftharpoons H_2$	$E^\circ = 0$

Therefore,

$$E^\circ_{cell} = +0.76 \text{ V} + 0 = 0.76 \text{ V}$$

EXAMPLE 1

What is the voltage of a hydrogen–copper cell?

This cell is referred to as a hydrogen–copper cell rather than a copper–hydrogen cell because here the hydrogen electrode is the anode. The anode is named first in the name for a voltaic cell. Since the copper electrode is the cathode, this is the electrode at which reduction occurs; therefore, the E° value for this electrode is taken directly from Table 12-1 without changing the sign:

ANODE:	$H_2 \rightleftharpoons 2H^+ + 2e^-$	$E^\circ = 0$
CATHODE:	$Cu^{++} + 2e^- \rightleftharpoons Cu$	$E^\circ = +0.34$ V

Therefore,

$$E^\circ_{cell} = 0 + 0.34 \text{ V} = 0.34 \text{ V}$$

EXAMPLE 2

What is the voltage of a nickel–hydrogen or hydrogen–nickel cell, and which is its proper designation?

From the relative positions of hydrogen and nickel in Table 12-1 it is apparent that Ni^{++} ions have less tendency to be reduced than H^+ ions. Therefore, H^+ ions will be reduced and Ni atoms will be oxidized. With oxidation occurring at the nickel electrode, the $E°$ value found in the table for this electrode must have the sign reversed:

ANODE:	$Ni \rightleftharpoons Ni^{++} + 2e^-$	$E° = +0.25\ V$
CATHODE:	$2H^+ + 2e^- \rightleftharpoons H_2$	$E° = 0$

Therefore,

$$E°_{cell} = +0.25\ V + 0 = 0.25\ V$$

Since nickel is the anode, the cell is properly designated a nickel–hydrogen cell.

EXAMPLE 3

What is the standard voltage of a zinc–nickel cell?

The zinc is the anode and nickel is the cathode:

ANODE:	$Zn \rightleftharpoons Zn^{++} + 2e^-$	$E° = +0.76\ V$
CATHODE:	$Ni^{++} + 2e^- \rightleftharpoons Ni$	$E° = -0.25\ V$

Therefore,

$$E°_{cell} = +0.76\ V - 0.25\ V = 0.51\ V$$

EXAMPLE 4

Calculate $E°$ for a zinc–copper cell.

Zinc is the anode and copper is the cathode:

ANODE:	$Zn \rightleftharpoons Zn^{++} + 2e^-$	$E° = +0.76\ V$
CATHODE:	$Cu^{++} + 2e^- \rightleftharpoons Cu$	$E° = +0.34\ V$

Therefore,

$$E°_{cell} = +0.76\ V + 0.34\ V = 1.10\ V$$

EXAMPLE 5

What is $E°$ for a nickel–copper cell?

Assuming that nickel is the anode and copper is the cathode,

ANODE:	$Ni \rightleftharpoons Ni^{++} + 2e^-$	$E° = +0.25\ V$
CATHODE:	$Cu^{++} + 2e^- \rightleftharpoons Cu$	$E° = +0.34\ V$

Therefore,

$$E°_{cell} = +0.25\ V + 0.34\ V = 0.59\ V$$

If we had assumed nickel to be the cathode and copper the

anode, the answer would have been $E^\circ_{cell} = -0.59$ V. No voltaic cell can have a negative voltage! Therefore, we would have known that our assumption was wrong.

12-13 Predicting Oxidation–Reduction Reactions

A table of reduction potentials, such as Table 12-1, may be consulted to determine whether certain proposed reactions are possible. Thus, when we find in the table the two entries

$$Cd^{++} + 2e^- \rightleftharpoons Cd \qquad E^\circ = -0.40 \text{ V}$$
$$Cu^{++} + 2e^- \rightleftharpoons Cu \qquad E^\circ = +0.34 \text{ V}$$

we can predict that the reaction

$$Cd + Cu^{++} \longrightarrow Cd^{++} + Cu$$

will occur spontaneously and that the reaction

$$Cu + Cd^{++} \longrightarrow Cu^{++} + Cd$$

will not occur spontaneously. Immersion of a strip of metallic copper into a solution containing Cd^{++} ions results in no observable reaction, but immersion of a strip of metallic cadmium in a solution containing blue Cu^{++} ions results in the formation of metallic copper accompanied by the decolorization of the solution.

The ability of chlorine to displace bromine from bromides is implied by the relative positions of these two halogens in the table of reduction potentials:

$$Br_2 + 2e^- \rightleftharpoons 2Br^- \qquad E^\circ = +1.07 \text{ V}$$
$$Cl_2 + 2e^- \rightleftharpoons 2Cl^- \qquad E^\circ = +1.36 \text{ V}$$

The E° values indicate that elemental chlorine is more easily reduced than elemental bromine and that the reaction

$$Cl_2 + 2Br^- \longrightarrow Br_2 + 2Cl^-$$

occurs spontaneously while the reverse reaction,

$$Br_2 + 2Cl^- \longrightarrow Cl_2 + 2Br^-$$

does not occur spontaneously. The table also includes the entry

$$MnO_4^- + 8H^+ + 5e^- \rightleftharpoons Mn^{++} + 4H_2O \qquad E^\circ = +1.51 \text{ V}$$

From this we can predict that permanganate ion will oxidize either bromide or chloride.

In general, any reducing agent in the table will reduce any oxidizing agent appearing below it: Fe^{++} will reduce MnO_4^-, Cu will reduce

Fe^{3+}, and so on. You must remember, however, that $E°$ values refer to $1M$ concentrations of ions in all cases and that most chemical reactions are reversible. The greater the spread between two $E°$ values in the table, the more complete will be the reaction between the corresponding substances.

12-14 Types of Voltaic Cell Reactions

A voltaic cell can be designed to utilize the driving force behind the reaction of hydrogen or a metal with a nonmetal, such as chlorine. For example, electric current can be obtained from the reaction of hydrogen with chlorine to form HCl, i.e., H^+ and Cl^- ions, in aqueous solution:

$$H_2(g) + Cl_2(g) \longrightarrow 2H^+ + 2Cl^-$$

One half-cell contains a hydrogen electrode, and the other contains a chlorine electrode. Gas electrodes consist of a gas bubbling about an "inert" metal strip immersed in a solution containing the appropriate ions. The metal strip (usually platinum) provides electrical contact for the half-cell and at the same time catalyzes the establishment of equilibrium between the gas and its ions. The platinum strip is "platinized," that is, coated with a layer of finely divided platinum, called "platinum black" because of its color. The finely divided platinum provides a large surface area for the adsorption of gas molecules.

A standard hydrogen electrode or half-cell consists, then, of a strip of platinized platinum immersed in an aqueous solution in which the concentration of H^+ ions is $1M$, with H_2 gas at a pressure of 1 atm passing continuously over the platinum strip. The voltage of a cell that consists of a standard hydrogen half-cell and a standard chlorine half-cell is found to be 1.36 V at 25°C. In this cell the hydrogen electrode is the anode and the chlorine electrode is the cathode:

$$H_2(g) \longrightarrow 2H^+ + 2e^-$$

$$Cl_2(g) + 2e^- \longrightarrow 2Cl^-$$

The reduction potential of chlorine is therefore given in tables of reduction potentials as +1.36 V.

A voltaic cell can derive its current from a reaction in which neither electrode is directly involved. An example is diagrammed in Figure 12-5. Here the "inert" electrodes could be either platinum or graphite. No reaction occurs until the two electrodes are connected externally. The reaction that then occurs consists in the oxidation of tin(II) ions and the reduction of iron(III) ions. The equations for the half-reactions are as follows:

FIGURE 12-5

Voltaic cell made from platinum electrodes immersed in solutions containing Sn^{++} and Fe^{3+} ions

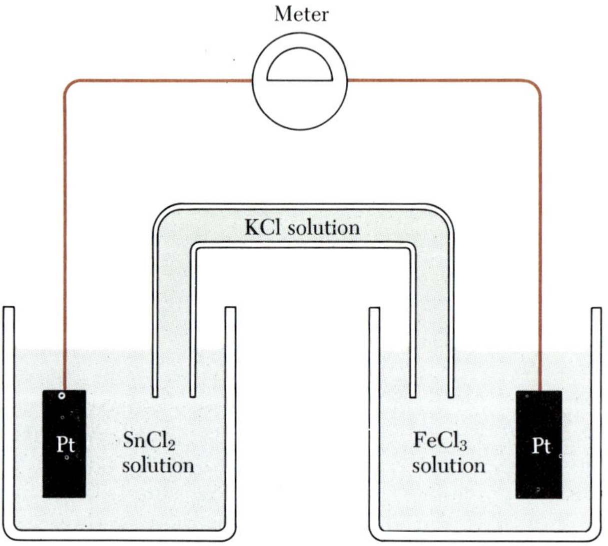

$$\text{ANODE:} \quad Sn^{++} \rightleftharpoons Sn^{4+} + 2e^-$$
$$\text{CATHODE:} \quad Fe^{3+} + e^- \rightleftharpoons Fe^{++}$$

Tin(IV) ions and iron(II) ions are produced by the cell reaction. The voltage developed by the cell is affected somewhat by the concentration of each of the four kinds of ions. If all are present at 1*M* concentration, the cell voltage is 0.62 V. This can be calculated from the following entries in Table 12-1:

$$Sn^{4+} + 2e^- \rightleftharpoons Sn^{++} \qquad E^\circ = +0.15\ \text{V}$$
$$Fe^{3+} + e^- \rightleftharpoons Fe^{++} \qquad E^\circ = +0.77\ \text{V}$$

The standard cell voltage is always the sum of the standard reduction potential for the half-cell in which reduction occurs (cathode) and the standard oxidation potential for the half-cell in which oxidation occurs (anode). The former value is taken directly from Table 12-1, but the latter value is obtained by reversing the sign of the value found in the table:

$$E^\circ_{\text{cell}} = +0.77\ \text{V} - 0.15\ \text{V} = 0.62\ \text{V}$$

The ion–electron equation representing what actually happens in the tin(II)–tin(IV) half-cell must show the oxidation of tin(II) ions when read from left to right:

$$Sn^{++} \rightleftharpoons Sn^{4+} + 2e^-$$

The other ion–electron equation, showing reduction of iron(III) ions, must be doubled before the two equations are added to give the equation for the cell reaction (but this has no effect on the half-cell potential):

CATHODE:	$2Fe^{3+} + 2e^- \rightleftharpoons 2Fe^{++}$	$E° = +0.77$ V
ANODE:	$Sn^{++} \rightleftharpoons Sn^{4+} + 2e^-$	$E° = -0.15$ V
CELL:	$2Fe^{3+} + Sn^{++} \rightleftharpoons 2Fe^{++} + Sn^{4+}$	$E° = +0.62$ V

12-15 Practical Voltaic Cells

A practical voltaic cell must be compact and not too fragile. In principle, however, any reaction involving gain and loss of electrons can be utilized for the production of electric current.

A cell of historical interest is the *gravity cell* (Figure 12-6), which formerly supplied current for telegraph lines. The cathode was a copper grid placed in the bottom of a glass jar half-filled with saturated copper sulfate solution. Over the dense copper sulfate solution was a layer of a dilute solution of zinc sulfate. The anode was a zinc grid or "crowfoot" that was supported at the top of the cell where it would dip into the zinc sulfate solution.

FIGURE 12-6
The gravity cell (Daniell cell)

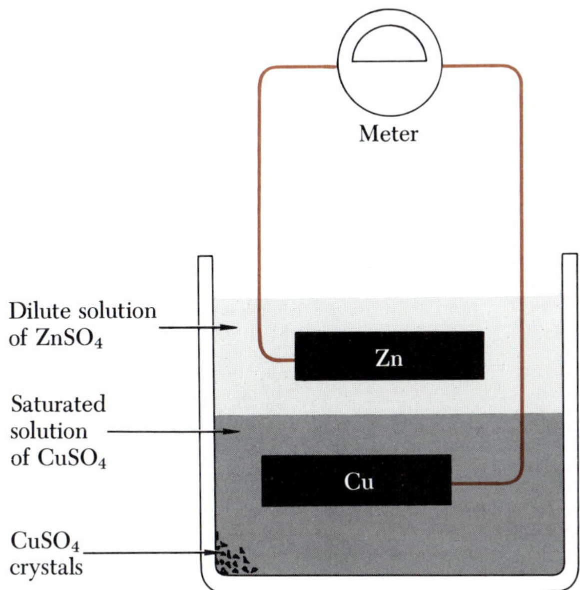

The particular mechanical design of the gravity cell is only one of many that have been employed to utilize the reaction

$$Cu^{++} + Zn \longrightarrow Zn^{++} + Cu$$

All such Cu–Zn cells are called *Daniell cells* because the English physicist J. F. Daniell invented the first one in 1841.

One type of cell that has been developed for specialized purposes is similar to the Daniell cell in principle but employs cadmium and nickel instead of zinc and copper.

For producing small currents at intervals, the so-called dry cell is one of the most important of all voltaic cells. (Small versions of it are used as flashlight batteries.) The potential of a new dry cell is about 1.5 V. This cell (Figure 12-7) is contained within a cylindrical zinc can that serves as the anode. The cathode, a carbon rod, is surrounded by an electrolyte paste (a moistened mixture of ammonium chloride, zinc chloride, and manganese dioxide with a porous inert material such as paper pulp or cotton waste). At the anode, the Zn atoms are oxidized:

$$Zn \longrightarrow Zn^{++} + 2e^-$$

The cathode reaction is more complex; the following equation probably represents the net reaction:

$$2MnO_2 + 2NH_4^+ + 2e^- \longrightarrow 2NH_3 + H_2O + Mn_2O_3$$

The NH_3 that is formed combines with Zn^{++} ions from the anode reaction to form complex ions:

$$Zn^{++} + 4NH_3 \longrightarrow Zn(NH_3)_4^{++}$$

FIGURE 12-7

The dry cell

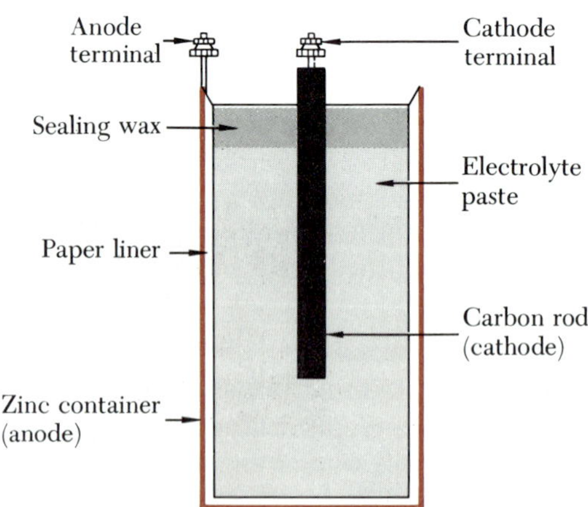

A paper liner separates the zinc can from the paste; therefore, the electrons required for the cathode reaction must pass through the external circuit and enter the cell through the carbon rod. The cell is sealed with sealing wax to prevent loss of water from the paste by evaporation. Water must be present in the paste to permit movement of ions. The only reason for calling the device a "dry cell" is that the sealed zinc can prevents the moist electrolyte from leaking out.

The mercury cell that is used in hearing aids is another type of dry cell.

The lead storage cell, used in automobile batteries, is a voltaic cell that operates reversibly. After it has served as a source of electric current, it can be "charged," i.e., restored to its original condition, by having electrical energy supplied to it from an outside source. The charging is brought about by passing current through the cell in the direction opposite to the current produced by the spontaneous cell reaction. The unique feature of a storage cell is that the reactions taking place at the electrodes when current is passed through it are the reverse of those that occur when the cell acts as a source of current.

The electrodes in a lead storage cell are gratings made of a lead alloy. The openings in one electrode are packed with lead dioxide, and those in the other are filled with spongy lead. The electrolyte is dilute sulfuric acid. When the battery is delivering current, the spongy lead electrode acts as the anode; that is, the atoms of the spongy lead are oxidized to Pb^{++} ion, and they combine with SO_4^{--} ions from the sulfuric acid to coat the electrode with lead sulfate. The lead dioxide electrode acts as cathode, because here the lead dioxide is reduced to Pb^{++} ions, the H^+ ions from the acid combining with the oxygen to form water, and here also the Pb^{++} ions combine with SO_4^{--} ions to coat the electrode with lead sulfate.

ANODE: $$Pb + SO_4^{--} \longrightarrow PbSO_4(s) + 2e^-$$

CATHODE: $$PbO_2 + 4H^+ + SO_4^{--} + 2e^- \longrightarrow PbSO_4(s) + 2H_2O$$

TOTAL CELL REACTION: $$Pb + PbO_2 + 2H_2SO_4 \underset{\text{charging}}{\overset{\text{discharging}}{\rightleftharpoons}} 2PbSO_4 + 2H_2O$$

As indicated by the equation, the concentration of sulfuric acid decreases during the discharge of the cell. This results in a decrease of the density of the electrolyte. Charging the cell regenerates the acid and restores the electrolyte to its higher density. Thus, the condition of charge of a lead storage cell can be determined by the use of a hydrometer to measure the density of the electrolyte.

The potential of a fully charged lead storage cell is a little over 2 V. A 6-V automobile battery contains three cells; a 12-V battery contains six cells.

12-16 Fuel Cells

Although voltaic cells are used for many practical purposes, most electrical energy is obtained from generators operated by steam turbines. The steam is obtained by heating water in a boiler with thermal energy released by burning coal, oil, or gas. The series of changes, chemical energy to thermal energy to mechanical energy to electrical energy, is relatively inefficient. A steam power plant using coal as fuel converts little more than a third of the chemical energy of the coal into electrical energy.

For several decades there has been discussion of the possibility of converting the chemical energy of fuels *directly* into electrical energy in a so-called "fuel cell." Since voltaic cells operate on the basis of spontaneous oxidation–reduction reactions, there is no theoretical reason why a cell should not employ the type of spontaneous oxidation–reduction reaction that occurs in the combustion of a fuel. However, the practical problems encountered in the actual design of fuel cells and in finding satisfactory catalysts for the electrode half-reactions have proved to be quite complex.

One fuel cell with promising possibilities, already employed in spacecraft, utilizes the reaction between hydrogen and oxygen gases under a pressure of 50 atm at 250°C. The electrolyte is aqueous NaOH or KOH. With graphite electrodes, the following reactions occur *in the presence of certain catalysts:*

$$\begin{array}{ll} \text{ANODE:} & 2H_2(g) + 4OH^- \rightleftharpoons 4H_2O + 4e^- \\ \text{CATHODE:} & O_2(g) + 2H_2O + 4e^- \rightleftharpoons 4OH^- \\ \hline & 2H_2(g) + O_2(g) \rightleftharpoons 2H_2O \end{array}$$

The voltage developed by such a cell is about 1 V.

12-17 Reactions in Electrolytic Cells

In the electrolysis of a molten salt, there is only one imaginable half-reaction at either electrode, provided the electrodes are made of materials that do not react when the cell is in operation. For example, the passage of direct current through molten sodium chloride in a cell provided with graphite electrodes can only result in the reduction of Na^+ ions at the cathode and the oxidation of Cl^- ions at the anode (Section 12-2).

If an electrolytic cell contains an aqueous solution of a salt, alternative half-reactions come to mind. Passage of direct current through a cell containing aqueous $CuBr_2$ solution, for example, might conceivably produce either of the following reduction half-reactions at the cathode:

$$Cu^{++} + 2e^- \rightleftharpoons Cu \qquad E° = +0.34 \text{ V}$$
$$2H_2O + 2e^- \rightleftharpoons H_2 + 2OH^- \qquad E° = -0.83 \text{ V}$$

Experience shows that copper, not hydrogen, is the cathode product. We could predict this result by comparing the reduction potentials of the two half-reactions. The half-reaction for the reduction of Cu^{++} ions has a higher $E°$ value than the half-reaction for the reduction of water. Even though the reduction potential for the one varies with the Cu^{++} ion concentration and that for the other varies with the pH, our prediction that the electrolysis should produce copper rather than hydrogen is fulfilled.

Electrolysis of the $CuBr_2$ solution could conceivably produce either of the following oxidation half-reactions at the anode:

$$2Br^- \rightleftharpoons Br_2 + 2e^-$$

$$2H_2O \rightleftharpoons O_2 + 4H^+ + 4e^-$$

Experience shows that bromine, not oxygen, is formed at the anode. This, too, we could predict on the basis of the $E°$ values found in the reduction potential table for reduction half-reactions that are the reverse of the above oxidation half-reactions:

$$Br_2 + 2e^- \rightleftharpoons 2Br^- \qquad E° = +1.07 \text{ V}$$
$$O_2 + 4H^+ + 4e^- \rightleftharpoons 2H_2O \qquad E° = +1.23 \text{ V}$$

Since the tendency of elemental oxygen to undergo reduction spontaneously is greater than that of elemental bromine, we would predict that the oxidation of bromide ions would be more easily brought about than the oxidation of H_2O molecules.

Since H_2O molecules are more readily reduced than such ions as Na^+ or Ba^{++}, electrolysis of solutions of salts of metals of Group IA or Group IIA always produces H_2 gas. The product obtained at the anode depends on the identity of the anion. Chlorides yield chlorine; bromides yield bromine; and iodides yield iodine. But electrolysis of aqueous solutions of sulfates, nitrates, etc. yields oxygen at the anode because H_2O molecules are more easily oxidized than any of the oxyanions.

Industrially, electrolysis is used for the manufacture of aluminum (from bauxite dissolved in molten cryolite); magnesium (from molten $MgCl_2$); sodium (from molten NaCl); fluorine (from molten KHF_2); sodium hydroxide, hydrogen, and chlorine (from aqueous NaCl); and numerous other elements and compounds.

Electrolytic refining is a process used for the production of copper and other metals in their highest state of purity. In the electrolytic refining of copper, the anode is a bar of impure copper and the cathode is a thin sheet of pure copper. These electrodes are immersed in an aqueous solution of $CuSO_4$. The applied voltage is adjusted to about 0.35 so that the copper will go into solution from the anode and plate out upon the cathode.

$$\text{ANODE:} \quad Cu \longrightarrow Cu^{++} + 2e^-$$
$$\text{CATHODE:} \quad Cu^{++} + 2e^- \longrightarrow Cu$$

Active metals present as impurities are also oxidized at the anode, but their positive ions remain in solution and are not reduced at the cathode. Metals less active than copper are not oxidized at the anode but drop to the bottom of the cell as an undissolved sludge. Precious metals, such as silver, gold, and platinum, are recovered from this "anode sludge."

Electroplating is a process by which articles made of metal or an alloy are covered with a thin layer of some other metal. The purpose may be either to prevent corrosion or to make the article more attractive. In the electroplating process the article to be plated is the cathode, a piece of the plating metal is the anode, and they are placed in a solution of a salt of the plating metal. The practical art of electroplating requires a clean surface and careful control of such factors as temperature, voltage, and concentration of the plating bath.

NEW TERMS

Anion: a negative ion.

Anode (in any electrochemical cell, whether voltaic or electrolytic): the electrode at which oxidation occurs.

Cathode (in any electrochemical cell): the electrode at which reduction occurs.

Cation: a positive ion.

Disproportionation: a type of reaction in which an element starts out in an intermediate oxidation state and ends up in a mixture of a higher and a lower state.

Electrochemical cell: a general term that includes both electrolytic cells and voltaic cells.

Electrochemistry: the science that deals with the relation between electric current and chemical change.

Electrolytic cell: an electrochemical cell in which an oxidation–reduction reaction is brought about by an electric current from an outside source.

Faraday: the quantity of electricity equal to the total charge of a mole of monopositive or mononegative ions. It therefore consists of a mole of electrons. Its numerical value, in electrical units, is 96,490 coulombs.

Faraday's Law of Electrolysis: 1 faraday discharges 1 mole of singly charged ions, $\frac{1}{2}$ mole of doubly charged ions, etc. at each electrode in electrolytic cells.

Half-cell: that part of an electrochemical cell in which either the oxidation half-reaction or the reduction half-reaction takes place.

Half-reaction: that part of an oxidation–reduction reaction involving either the oxidation of the reducing agent (with loss of electrons) or the reduction of the oxidizing agent (with gain of electrons).

Ion–electron equation: an equation of the type used to represent the half-reactions occurring at the individual electrodes in an electrochemical cell, in which electrons are gained (at the cathode) or lost (at the anode); for example, $Ag^+ + e^- \longrightarrow Ag$.

Oxidation–reduction reaction ("redox" reaction): a reaction in which one element is oxidized and another element is reduced. (In some instances the atoms that are oxidized and those that are reduced are atoms of the same element.)

Reduction potential (of a particular reduction half-reaction): a measure of the tendency of a specified reduction half-reaction to occur, expressed in terms of the voltage of a cell in which the other half-cell is a hydrogen electrode.

Voltaic action: the production of electric current by means of a chemical reaction.

Voltaic cell (galvanic cell): a device that produces electric current by means of an oxidation–reduction reaction.

EXERCISES

12-1 Distinguish between the meanings of the following pairs of terms: cathode and anode; cation and anion; oxidation and reduction; oxidizing agent and reducing agent; coulomb and ampere; voltaic cell and electrolytic cell.

12-2 How many electrons pass through a cross section of a wire each second while it is carrying a current of 1.000×10^{-14} A?

12-3 What is the total charge, in coulombs, of the following:
(a) 3 moles of electrons
(b) 2 moles of Cl^- ions
(c) 4 moles of Ca^{++} ions
(d) all the ions in 5 moles of $Al_2(SO_4)_3$

12-4 How long would it take a 0.450-A current to plate out 9.64 g of cadmium from $CdSO_4$ solution?

12-5 How long must a current of 0.0140 A pass through an aqueous NaCl solution in order to liberate 0.0160 mole of Cl_2 at the anode?

12-6 A current of 1.30 A is passed for 24.00 hr through a cell containing a solution of zinc sulfate, $ZnSO_4$. What weight of zinc is deposited?

12-7 If a current deposits 2.17 g of copper on the cathode when passed through a copper(II) sulfate solution for 7.75 hr, what is the amperage? (Assume a steady current.)

12-8 A steady current passed through a solution of copper(II) chloride deposits 0.573 g of copper in 23.8 min. (a) What volume of chlorine gas is evolved? (Assume none of the chlorine dissolves.) (b) What weight of copper would be deposited in 1.00 hr?

12-9 Which of the following are oxidation–reduction reactions:
(a) $2Cu_2S(s) + 3O_2(g) \longrightarrow 2Cu_2O(s) + 2SO_2(g)$
(b) $Fe(s) + 2H^+ \longrightarrow Fe^{++} + H_2(g)$
(c) $CaCO_3(s) \longrightarrow CaO(s) + CO_2(g)$
(d) $IO_3^- + 5I^- + 6H^+ \longrightarrow 3I_2 + 3H_2O$
(e) $Ba^{++} + SO_4^{--} \longrightarrow BaSO_4(s)$

12-10 In aqueous solutions in contact with air, iron(II) salts are unstable. Explain. Such solutions become stable when iron nails are placed in them. Explain.

12-11 Diagram a voltaic cell in which the reaction is the displacement of silver from $AgNO_3$ by metallic copper. Write the equation for the half-reaction that takes place in each half-cell, identifying each as oxidation or reduction; then write the equation for the total cell reaction.

12-12 What is the approximate voltage of the cell in the preceding question if the solutions used are 1*M*? Indicate the direction of migration of the ions during the operation of the cell, and show how the electrons flow through the external circuit. Which electrode is the anode and which is the cathode?

12-13 Explain how a hydrometer can be used to determine the state of charge of an automobile battery.

12-14 What would be obtained at each electrode during the electrolysis of 1*M* aqueous solutions of the following compounds, using "inert" electrodes:
(a) sodium fluoride
(b) silver nitrate
(c) copper chloride
(d) sodium hydroxide
(e) sulfuric acid

12-15 Predict whether the following reactions are possible:
(a) displacement of iodine from an iodide by bromine
(b) oxidation of bromide ion by chlorine
(c) reduction of iodine by fluoride ion
(d) oxidation of ferrous ion by permanganate ion
(e) oxidation of iron atoms by silver ions

12-16 Balance the following equations:
(a) $P + HNO_3 + H_2O \longrightarrow H_3PO_4 + NO\uparrow$
(b) $Cu + H^+ + NO_3^- \longrightarrow Cu^{++} + NO_2\uparrow + H_2O$

(c) $H_2SO_4 + HBr \longrightarrow SO_2\uparrow + Br_2 + H_2O$
(d) $H_2SO_4 + HI \longrightarrow H_2S\uparrow + I_2 + H_2O$
(e) $CuS + HNO_3 \longrightarrow Cu(NO_3)_2 + S + NO\uparrow + H_2O$
(f) $MnO_4^- + H_2S + H^+ \longrightarrow Mn^{++} + S + H_2O$
(g) $MnO_4^- + H_2O_2 + H^+ \longrightarrow Mn^{++} + O_2\uparrow + H_2O$

12-17 Complete and balance by constructing balanced half-equations and adding them:
$Cu + Ag^+ \longrightarrow Cu^{++} + Ag$
$BrO_3^- + I^- \longrightarrow Br^- + I_2$ (acidic)
$Cr_2O_7^{--} + I_2 \longrightarrow Cr^{3+} + IO_3^-$ (acidic)
$Cu + NO_3^- \longrightarrow Cu^{++} + NO$ (acidic)
$Co(OH)_2 + Br_2 \longrightarrow Co(OH)_3 + Br^-$ (basic)

REFERENCES

Eblin, L. P., "Oxidation Number in Auto-Redox Reactions," *Journal of Chemical Education*, **28**(4), 221 (1951). A discussion of reactions in which one atom of an element oxidizes another atom of the same element.

Glasstone, S., *Sourcebook on Atomic Energy*, 3rd ed., Van Nostrand, Princeton, N.J., 1967. Chapter 2 includes accounts of the contributions of Faraday and Stoney.

Hildebrand, J. H., and Powell, R. E., *Principles of Chemistry*, Macmillan, New York, 1964, Chapter 21, "Avogadro's Number." An outline of various methods for determining the number of molecules in a mole.

Kieffer, W. F., *The Mole Concept in Chemistry*, Reinhold, New York, 1962 (paperbound). In Chapter 7, "Electrochemistry," the author discusses electrolysis in terms of the faraday as a mole of electrons, avoiding use of "equivalents."

Lockwood, K. L., "Redox Revisited," *Journal of Chemical Education*, **38**(6), 326 (1961). Emphasizes the distinction between stoichiometry and mechanism.

Mysels, K., "The Reactions of the Leclanché Dry Cell," *Journal of Chemical Education*, **32**(12), 638 (1955). A discussion of what is actually known about the chemistry of the dry cell.

Shen, S. Y., "What is the Real Value of a Faraday?" *Chemistry*, **39**(2), 8 (1966). An explanation of the discrepancies in reported values for the faraday. The best value now available is concluded to be 96,490 ± 2 coulombs.

Yalman, R. G., "Writing Oxidation-Reduction Equations," *Journal of Chemical Education*, **36**(5), 215 (1959). A review of textbook methods.

13

Energy and Chemical Change

13-1 Thermochemistry and Thermochemical Equations

Thermochemistry is the branch of chemistry that deals with the thermal changes accompanying both chemical and physical processes, including chemical reactions, changes of state, and formation of solutions. One of the first laws of thermochemistry resulted from the work of Lavoisier and Laplace, who observed (1780) that the quantity of heat required for the decomposition of a compound into its elements is equal to the quantity of heat released when the same quantity of the compound is formed from its elements. We now regard this principle as a special case of the Law of Conservation of Energy, which is a general principle that has been accepted since about 1850. The significant implication of the Law of Lavoisier and Laplace is that we can reverse any thermochemical equation, provided we also reverse the sign of the heat of reaction.

A *thermochemical equation* is an equation that indicates not only the reactants and products but also the magnitude of the heat of reaction, ΔH. As already indicated in Section 8-3, the heat of reaction is the difference between the heat content of the products and the heat content of the reactants. In an exothermic reaction the heat content of the products is less than the heat content of the reactants; the difference, ΔH, is negative and indicates the amount of heat that is released to the surroundings in the process of restoring the products to the original temperature of the reactants. In an endothermic reaction the heat content of the products is greater than the heat content of the reactants; in this case ΔH is positive and indicates the quantity of heat that is absorbed from the surroundings in restoring the products to the temperature the starting materials had before the reaction occurred.

Reactants are defined arbitrarily as the substances written on the left in a chemical equation, and the products are those substances written on the right. When we reverse an equation, the products of the one equation become the reactants in the other and the reactants in the one become the products in the other. If we reverse a thermochemical equation, the heat of reaction remains the same in magnitude but its sign in the one equation is the opposite of its sign in the other. For example,

$$2H_2(g) + O_2(g) \longrightarrow 2H_2O(g) \qquad \Delta H = -115.60 \text{ kcal}$$
$$2H_2O(g) \longrightarrow 2H_2(g) + O_2(g) \qquad \Delta H = +115.60 \text{ kcal}$$

The first of the above equations indicates that the formation of 2 moles of gaseous H_2O from the elements is accompanied by the release of 115.60 kcal of thermal energy. The second equation states that the decomposition of 2 moles of gaseous H_2O into its elements is accompanied by the absorption of 115.60 kcal.

In considering the ΔH value for a reaction, the products and the reactants must be referred to the same temperature. Furthermore, we cannot make legitimate comparisons between ΔH values for different reactions unless all values refer to the same temperature. By common agreement, the reference temperature is 25°C. It is also necessary to establish a reference pressure, because gaseous products will have to expend energy in pushing back the atmosphere to make room for themselves unless the reactants already occupied a like volume. By common agreement the reference pressure is 1 atm. Heats of reaction that are applicable to the standard reference state, 25°C and 1 atm, are called standard heats of reaction and are tagged by the superscript °, that is, they are written $\Delta H°$.

A change of state, such as vaporization or freezing, is accompanied by a $\Delta H°$ value that is characteristic of the particular substance. In writing a thermochemical equation we are required, therefore, to specify that each reactant and product is in the solid, liquid, or gaseous state, as the case may be. The importance of this requirement is illustrated by the following equations for the combustion of hydrogen:

$$2H_2(g) + O_2(g) \longrightarrow 2H_2O(g) \qquad \Delta H° = -115.60 \text{ kcal}$$
$$2H_2(g) + O_2(g) \longrightarrow 2H_2O(l) \qquad \Delta H° = -136.64 \text{ kcal}$$

The difference between the two $\Delta H°$ values, approximately 21 kcal, is the amount of heat required to vaporize 2 moles of H_2O.

For some purposes we find it more convenient to work with other terms that are related to the standard heat of reaction but direct our attention to a specific reactant or to the product of a specific reaction. In the formation of gaseous H_2O from H_2 and O_2, for example, the standard heat of reaction, $\Delta H°$, is -115.60 kcal. We may also say that the *heat of formation* of gaseous H_2O is -57.80 kcal per mole or that the *heat of combustion* of H_2 is -57.80 kcal per mole (dividing the value of $\Delta H°$ by the number of moles of H_2O and H_2, respectively, in the equation). We have already touched upon this in Chapter 8; we will now see how very useful these concepts can be.

13-2 Hess's Law

One of the most useful principles of thermochemistry was stated in 1840 by the Swiss-Russian chemist G. H. Hess. This principle, now known as Hess's Law, states that the heat of a reaction is always the same, whether the reaction occurs in a single step or in a series of steps. Hess's Law is now recognized as another special case of the Law of Conservation of Energy. It enables us to add or subtract thermochemical equations in the same manner as ordinary algebraic equations.

EXAMPLE

The heat of combustion of carbon disulfide has been measured to be -257 kcal/mole. Compare this value with the one obtained by calculation from the $\Delta H°$ values of the following reactions:

$$C(s) + 2S(s) \longrightarrow CS_2(l) \qquad \Delta H° = +21 \text{ kcal}$$
$$C(s) + O_2(g) \longrightarrow CO_2(g) \qquad \Delta H° = -94 \text{ kcal}$$
$$S(s) + O_2(g) \longrightarrow SO_2(g) \qquad \Delta H° = -71 \text{ kcal}$$

We can contrive a set of equations that can be added to give the equation for the combustion of carbon disulfide by reversing the first equation (also reversing the sign of $\Delta H°$), retaining the second equation as it is written, and doubling the third equation (also doubling the $\Delta H°$ value):

$$\begin{aligned} CS_2(l) &\longrightarrow C(s) + 2S(s) & \Delta H^\circ &= -21 \text{ kcal} \\ C(s) + O_2(g) &\longrightarrow CO_2(g) & \Delta H^\circ &= -94 \text{ kcal} \\ 2S(s) + 2O_2(g) &\longrightarrow 2SO_2(g) & \Delta H^\circ &= -142 \text{ kcal} \\ \hline CS_2(l) + 3O_2(g) &\longrightarrow CO_2(g) + 2SO_2(g) \end{aligned}$$

$$\Delta H^\circ = (-21 - 94 - 142) \text{ kcal} = -257 \text{ kcal}$$

The heat of the reaction thus calculated is in agreement with the experimental value.

13-3 Predicting Heats of Reaction

The laws of thermochemistry enable us to predict heats of reaction, as illustrated in Section 13-2 for the heat of combustion of carbon disulfide. The usefulness of these laws becomes most apparent when we use them to calculate heats of reaction that would be difficult or even impossible to determine experimentally.

EXAMPLE 1

Calculate the heat of formation of methane.

This means that we are to calculate the heat of the reaction

$$C(s) + 2H_2(g) \longrightarrow CH_4(g)$$

This reaction cannot be carried out in the laboratory. However, the heats of the following reactions have been determined:

$$\begin{aligned} CH_4(g) + 2O_2(g) &\longrightarrow CO_2(g) + 2H_2O(l) & \Delta H^\circ &= -212.80 \text{ kcal} \\ C(s) + O_2(g) &\longrightarrow CO_2(g) & \Delta H^\circ &= -94.05 \text{ kcal} \\ 2H_2(g) + O_2(g) &\longrightarrow 2H_2O(l) & \Delta H^\circ &= -136.64 \text{ kcal} \end{aligned}$$

From this information we can obtain the heat of formation of methane by calculation. We need to reverse the first equation (changing the sign of ΔH°) and then add the three equations to obtain the desired one:

$$\begin{aligned} CO_2(g) + 2H_2O(l) &\longrightarrow CH_4(g) + 2O_2(g) & \Delta H^\circ &= +212.80 \text{ kcal} \\ C(s) + O_2(g) &\longrightarrow CO_2(g) & \Delta H^\circ &= -94.05 \text{ kcal} \\ 2H_2(g) + O_2(g) &\longrightarrow 2H_2O(l) & \Delta H^\circ &= -136.64 \text{ kcal} \\ \hline C(s) + 2H_2(g) &\longrightarrow CH_4(g) \end{aligned}$$

$$\Delta H^\circ = (212.80 - 94.05 - 136.64) \text{ kcal} = -17.89 \text{ kcal}$$

Thus, the heat of formation of gaseous methane is −17.89 kcal/mole.

EXAMPLE 2

Calculate the heat of formation of carbon monoxide.

This means we are to calculate the heat of the reaction

$$2C(s) + O_2(g) \longrightarrow 2CO(g)$$

The heat of the reaction would be very difficult to determine experimentally because carbon monoxide burns more easily than carbon. If we attempt to burn a sample of carbon in the theoretical amount of oxygen required to form carbon monoxide by the above equation, we actually get a mixture of carbon monoxide, carbon dioxide, and unburned carbon. However, the heats of combustion of carbon (to form carbon dioxide) and of carbon monoxide have both been determined:

$$C(s) + O_2(g) \longrightarrow CO_2(g) \qquad \Delta H^\circ = -94.05 \text{ kcal}$$
$$2CO(g) + O_2(g) \longrightarrow 2CO_2(g) \qquad \Delta H^\circ = -135.26 \text{ kcal}$$

If we double the first equation and reverse the second one, we have two equations that can be added to give the desired one:

$$2C(s) + 2O_2(g) \longrightarrow 2CO_2(g) \qquad \Delta H^\circ = -188.10 \text{ kcal}$$
$$2CO_2(g) \longrightarrow 2CO(g) + O_2(g) \qquad \Delta H^\circ = +135.26 \text{ kcal}$$
$$\overline{2C(s) + O_2(g) \longrightarrow 2CO(g)}$$

$$\Delta H^\circ = (-188.10 + 135.26) \text{ kcal} = -52.84 \text{ kcal}$$

Thus, the heat of formation of carbon monoxide is −26.42 kcal/mole.

EXAMPLE 3

How does the heat content (enthalpy) of diamond compare with that of graphite?

The actual value of H cannot be determined for any substance, but the difference between the H values of diamond and graphite is simply the heat (ΔH) of the reaction

$$C(\text{diamond}) \longrightarrow C(\text{graphite})$$

It is impossible to convert diamond to graphite or graphite to diamond under conditions that enable us to measure the heat of reaction, which is small. However, the heats of combustion have been determined for both diamond and graphite and have been found to be slightly different:

$$C(diamond) + O_2(g) \longrightarrow CO_2(g) \qquad \Delta H° = -94.50 \text{ kcal}$$
$$C(graphite) + O_2(g) \longrightarrow CO_2(g) \qquad \Delta H° = -94.05 \text{ kcal}$$

If we reverse the second equation and add it to the first one, we obtain

$$C(diamond) \longrightarrow C(graphite) \qquad \Delta H° = (-94.50 + 94.05) \text{ kcal} = -0.45 \text{ kcal}$$

Thus, the heat content, or enthalpy, of diamond exceeds that of graphite by 0.45 kcal/mole.

13-4 Thermochemical Calculations with Heats of Formation

The heat of any reaction may readily be calculated by subtracting the sum of the heats of formation of the reactants from the sum of the heats of formation of the products. (This is a corollary to the Law of Conservation of Energy.) All elementary substances taking part in the reaction, either as reactants or products, are omitted because the "heat of formation" is zero, by definition, for an elementary substance. The heat of formation per mole of each substance, of course, must be multiplied by the number of moles involved in the reaction. Table 13-1 lists the standard heats of formation for a number of common compounds.

EXAMPLE

Calculate the heat of the reaction

$$CS_2(l) + 3O_2(g) \longrightarrow CO_2(g) + 2SO_2(g)$$

From Table 13-1 we obtain the following heats of formation:

$$CS_2(l), \quad +21.0 \text{ kcal/mole}$$
$$CO_2(g), \quad -94.0 \text{ kcal/mole}$$
$$SO_2(g), \quad -71.0 \text{ kcal/mole}$$

For the products of the reaction the sum is

$$-94.0 \text{ kcal} + 2(-71.0) \text{ kcal} = -94.0 \text{ kcal} - 142.0 \text{ kcal} = -236.0 \text{ kcal}$$

For the reactants the sum is +21.0 kcal. Subtracting the sum for the reactants from the sum for the products, we find for the heat of reaction that

$$\Delta H° = -236.0 \text{ kcal} - 21.0 \text{ kcal} = -257.0 \text{ kcal}$$

TABLE 13-1

Standard heats of formation

Compound	ΔH° (kcal/mole at 25°C and 1 atm)
$H_2O(l)$	−68.32
$H_2O(g)$	−57.80
$H_2S(g)$	−4.82
$NH_3(g)$	−11.04
$CH_4(g)$	−17.89
$CO(g)$	−26.42
$CO_2(g)$	−94.05
$CS_2(l)$	+21.0
$CCl_4(l)$	−33.34
$C_2H_6(g)$	−20.24
$C_2H_4(g)$	+12.50
$C_2H_2(g)$	+54.19
$C_6H_6(l)$	+11.72
$HF(g)$	−64.20
$HCl(g)$	−22.06
$HBr(g)$	−8.66
$HI(g)$	+6.20
$SO_2(g)$	−70.96

A converse application is to calculate the heat of formation of a compound from the heats of formation of the other substances taking part in a reaction. This requires, of course, that the heat of the reaction be determined experimentally.

EXAMPLE

The heat of combustion of acetylene has been found to be −310.6 kcal/mole. Calculate its heat of formation.

Writing the equation for the combustion of acetylene,

$$2C_2H_2(g) + 5O_2(g) \longrightarrow 4CO_2(g) + 2H_2O(l) \qquad \Delta H^\circ = -621.2 \text{ kcal}$$

we see that we need to obtain the heats of formation of carbon dioxide and water from Table 13-1:

$$CO_2(g), \quad -94.05 \text{ kcal/mole}$$
$$H_2O(l), \quad -68.32 \text{ kcal/mole}$$

For the products of our reaction the sum of the heats of formation is

$$4(-94.05)\text{ kcal} + 2(-68.32)\text{ kcal} = -376.20\text{ kcal} - 136.64\text{ kcal}$$
$$= -512.84\text{ kcal}$$

For the reactants the sum of the heats of formation is $2x$, where x is the heat of formation per mole of acetylene. Subtracting the sum for the reactants from the sum for the products and setting the difference equal to the heat of the reaction, we have

$$\Delta H^\circ = -512.84\text{ kcal} - 2x = -621.2\text{ kcal}$$

Therefore,

$$2x = (621.2 - 512.84)\text{ kcal}$$
$$= +108.4\text{ kcal}$$
$$x = +54.2\text{ kcal}$$

13-5 Bond Energies

The heat of formation of a compound is closely related to the attraction between the atoms within the molecules of the compound. Heat of formation is not, however, a direct measure of bond strength because of the complication that both bond forming and bond breaking are involved in a chemical reaction. The heat of formation of water is negative because *more* energy is released in the formation of H—O bonds than is required to break the necessary number of H—H and O—O bonds (Section 11-3). On the other hand, the reaction of elementary hydrogen with elementary iodine is endothermic, that is, the heat of formation of hyrogen iodide is positive, because the energy released in the formation of H—I bonds is *less* than the energy required to break the H—H bonds in the hydrogen molecules and the I—I bonds in the iodine molecules. Thus, heat of formation is not a simple, direct measure of bond strength because it involves *differences* between bond-breaking and bond-forming energies.

It is sometimes desired to obtain an estimate for the heat of a reaction in which the heat of formation of one of the products is not known. In this event it is necessary to make an *estimate* of the unknown heat of formation. Such an estimate can be arrived at by use of bond energies.

The definition of bond energy is easiest for diatomic molecules. The bond energy for H_2, for example, is the energy required to break the bond between the H atoms in a mole of H_2 molecules. This energy can be measured by spectroscopic methods. It is alluded to in Section 8-4, where the statement is made that energy in the amount of 104 kcal is absorbed in the dissociation of 1 mole of H_2 molecules.

Bond energies for fluorine, chlorine, bromine, iodine, and the corresponding hydrogen halides are listed in Table 13-2. From this information we can predict the heats of formation of the hydrogen halides from their elements. For example, the reaction

$$H_2(g) + Cl_2(g) \longrightarrow 2HCl(g)$$

requires breaking an H—H bond and a Cl—Cl bond, but two H—Cl bonds are formed. Therefore, the heat evolved in the formation of 2 moles of HCl from its elements is (2×102) kcal $-$ $(104 + 57)$ kcal $= 43$ kcal. In other words, we can predict from the bond energies that the heat of formation of HCl from its elements is approximately -22 kcal, which agrees with the experimental value.

Some of the bond energies listed in Table 13-2 are only average values. For example, the heat absorbed in the process

$$CH_4(g) \longrightarrow C(g) + 4H(g)$$

is 397 kcal per mole of CH_4. Since there are four C—H bonds in each CH_4 molecule, the average C—H bond energy is one-fourth of this value or 99 kcal per mole of C—H bonds. This value is the one to be used in estimating the heats of reactions in which all four C—H bonds are broken. However, spectroscopic determination of the bond-dissociation energies of the individual bonds shows that the values for the different steps in the atomization of a CH_4 molecule are not identical.

$$CH_4(g) \longrightarrow CH_3(g) + H(g) \qquad \Delta H = 103 \text{ kcal}$$
$$CH_3(g) \longrightarrow CH_2(g) + H(g) \qquad \Delta H = 87 \text{ kcal}$$
$$CH_2(g) \longrightarrow CH(g) + H(g) \qquad \Delta H = 125 \text{ kcal}$$
$$CH(g) \longrightarrow C(g) + H(g) \qquad \Delta H = 81 \text{ kcal}$$

The sum of the four bond-dissociation energies, 396 kcal, is in good

TABLE 13-2

Average bond energies

Bond	*Energy* (kcal/mole)	*Bond*	*Energy* (kcal/mole)
H—H	104	O—H	111
F—F	36	S—H	88
Cl—Cl	57	N—H	93
Br—Br	46	C—H	99
I—I	36	C—C	83
H—F	135	C═C	148
H—Cl	102	C≡C	198
H—Br	86	O—O	33
H—I	71	N—N	38

agreement with the thermochemically calculated value of 397 kcal. The bond-dissociation energy of a bond is the energy required to break that bond alone. For diatomic molecules the bond energy and the bond-dissociation energy are the same, but for polyatomic molecules they are different. The bond energy should be called the *average* bond energy when reference is made to polyatomic molecules. The average bond energy is the sum of the successive bond-dissociation energies divided by the number of bonds.

The Pauling method for establishing a scale of electronegativities (Section 5-7) involves the use of bond energies. The relation between the bond energy concept and the electronegativity concept can be seen by comparing the dissociation energies of H_2, Cl_2, and HCl. The H—H bond energy is 104 kcal per mole. In the H_2 molecule the sharing of the electron pair by the two H atoms is equal, so we can reasonably assume that each H atom contributes half the bond energy, or 52 kcal per mole of H—H bonds. Similarly, from the dissociation energy of Cl_2, 57 kcal per mole, we can assume that each Cl atom contributes 28.5 kcal per mole of Cl—Cl bonds. If the bond in HCl were nonpolar, the bonding electron pair would be shared equally by the H and Cl atoms. Each H atom would contribute 52 kcal per mole of H—Cl bonds, and each Cl atom would contribute 28.5 kcal per mole of H—Cl bonds, making the total bond energy 52 kcal + 28.5 kcal = 80.5 kcal. Actually, the H—Cl bond energy is 102 kcal per mole. The electrons are *not* equally shared—the bond is polar and is stronger than a nonpolar bond, where the sharing is equal. The additional binding energy is assumed to depend on the difference between the electron-attracting abilities of the bonded atoms. In other words, there is an *ionic* contribution to the bond energy that is in addition to the covalent contribution. The heat of formation of a hydrogen halide, HX (where X stands for any halogen), is a measure of how much stronger the H—X bond is (because of its partial ionic character) than the average of H—H and X—X.

When we make similar calculations for the other hydrogen halides, we find that the differences between the calculated values and the observed values decrease regularly as follows:

	Obs.	*Calc.*	
HF	135	− 70 =	65
HCl	102	− 80 =	22
HBr	86	− 75 =	11
HI	71	− 70 =	1

The fact that the discrepancy is greatest for HF and least for HI implies that the sharing of electrons is most unequal in the HF molecule and most nearly equal in the HI molecule. This thought is often expressed by saying that in HF the partial ionic character of the bond is greatest whereas in HI it is least. Pauling's electronegativity values were selected by a complex empirical procedure to account for the actual bond energies.

13-6 The Direction of Chemical Change

One of the most important questions with which a chemist is faced is whether or not a particular reaction can take place. Such a question, of course, can often be answered by experiment, but the chemist would prefer not to waste time and effort on impossible reactions. Furthermore, a negative experimental result may be misleading because it does not prove that a reaction cannot occur. It may only mean that the reaction has taken place too slowly to be detected—in which case a good catalyst might lead to success. The ideal is to establish criteria for determining whether a reaction *can* occur, regardless of rate.

It is really an oversimplification to ask whether a reaction *can* occur. The proper question to ask is not "Is this reaction possible?" but rather "How far will this reaction proceed before equilibrium is reached?" The value of the equilibrium constant tells us the proportions in which the different reactants and products must be present to constitute an equilibrium mixture. If the reactants and products are brought together initially in equilibrium proportions, no net reaction will occur spontaneously. If reactants and products are brought together in proportions other than equilibrium proportions, it will be possible for a reaction to proceed far enough in the required direction to attain the point of equilibrium—but the reaction will not proceed past this point unless it is "forced" to do so.

Spontaneous changes are changes that can occur within systems when the systems are left to themselves. We are familiar with the fact that water runs downhill, that heat flows from hotter bodies to colder bodies, that gases expand into regions of lower pressure, and that chemical reactions proceed toward equilibrium. We use the term *nonspontaneous change* in referring to the reverse of a spontaneous change. Nonspontaneous changes do not occur when systems are left to themselves. Water does not run uphill, heat does not flow from a cold body into a hotter one, gases do not compress themselves, and chemical reactions that have reached equilibrium do not proceed further.

A nonspontaneous process is not properly referred to as "impossible" because we can induce a nonspontaneous change to take place within a system by supplying energy from outside the system. In such a case we say we are doing work upon the system. We can use outside energy to pump water uphill, to transfer heat from a colder region to a hotter region (as in a refrigerator), to compress a gas (as when we inflate a tire), or to make chemical reactions "run backward" (as in electrolytic cells).

A chemical reaction that has a large equilibrium constant is one that proceeds a large fraction of the way toward completion before attaining equilibrium. A reaction with a small equilibrium constant attains equilibrium when only a small fraction of the starting materials has been transformed into reaction products. The question "Is this reaction possible?" can be taken to mean "Does this reaction have an equilibrium constant that is large enough to make the reaction feasible?"

13-7 Spontaneous Change and the Concept of Entropy

Intuitively, we tend to feel that the ability of substances to react with each other should be related in some way to the energy change that accompanies the reaction. The thermochemical studies of Julius Thomsen (1854) in Denmark and Marcellin Berthelot (1867) in France led them to assume that the heat of reaction, ΔH, is a measure of "chemical affinity," and that only exothermic reactions could occur spontaneously. However, the "Thomsen-Berthelot Principle" was wrong. It has been definitely established that many endothermic reactions are spontaneous, provided the temperature is sufficiently great. The relation between chemical change and energy change is more subtle than was at first realized.

Our experience tells us that some natural processes occur spontaneously with no energy change involved. For example, suppose we have a bulb filled with helium connected, through a stopcock, to another bulb filled with argon (Figure 13-1). When the stopcock is opened, each gas diffuses into the other bulb until eventually a uniform mixture is present in both bulbs. This spontaneous process occurs in spite of the fact that ΔH is zero. It is conceivable that the chance motion of the molecules might, at some later time, result in all the helium molecules being in one bulb again and all the argon molecules in the other. Our experience leads us not to expect this to happen soon, though, because we realize that the probability of the occurrence is vanishingly small. With all the helium molecules in one bulb and all the argon molecules in the other, a kind of *order* prevails that no longer exists after each bulb contains a mixture of the same composition as the mixture in the other bulb. We cannot say that the mixing process occurs because of the attainment of a lower energy state because no heat is released or absorbed during the mixing. Rather, it occurs because the system attains a state of increased *disorder*, which is a state characterized by inherently greater *probability*. We may generalize and say that left to itself, a system tends to seek a state of maximum disorder.

FIGURE 13-1

A spontaneous process resulting in increased disorder

(a) Helium and argon in separate bulbs, stopcock closed. (b) Uniform mixture in each bulb, stopcock open.

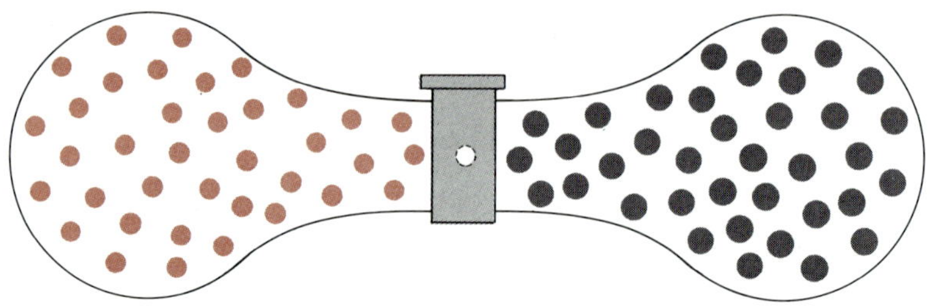

(a) Helium and argon in separate bulbs, stopcock closed

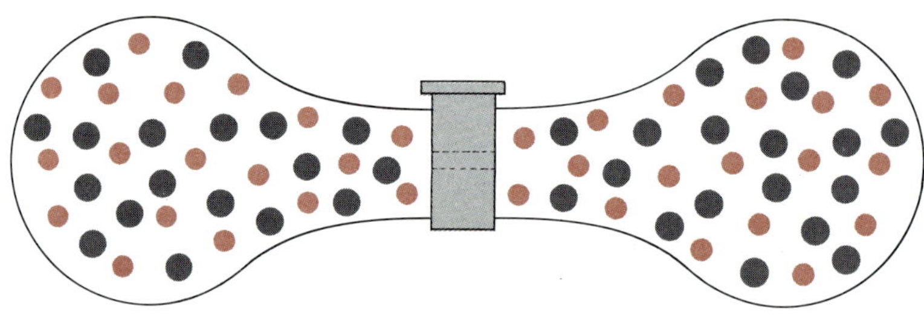

(b) Uniform mixture in each bulb, stockcock open

The term *entropy* is used to refer to the degree of randomness or disorder in a system and thus to the probability of the existence of the system in that particular state or condition. When a system undergoes a change that includes an increase in disorder, we say that the entropy of the system has increased. Using S to represent entropy and ΔS to represent change in entropy, we can say that ΔS is positive when there is an increase in disorder.

The tendency toward increase in entropy, i.e., the tendency toward maximum disorder, can be observed when two similar liquids are brought together. For example, heptane molecules, C_7H_{16}, and octane molecules, C_8H_{18}, are so much alike that the attraction between heptane and octane molecules is approximately the same as the attraction of heptane molecules for other heptane molecules or of octane molecules for other octane molecules. When the two liquids are brought together there is no operative factor to prevent them from mixing to form a homogeneous mixture, i.e., a solution. The mixing results from the thermal activity of the molecules. By mixing, they achieve maximum disorder—the state of highest probability. Once again, no significant energy change is involved.

The "driving force" behind many chemical reactions is the tendency toward increased entropy. The change in entropy for a chemical reaction is given by the equation

$$\Delta S_{\text{reaction}} = S_{\text{products}} - S_{\text{reactants}}$$

We can generalize and say that there is a *tendency* for reactions to occur that are accompanied by an increase in entropy within the system, just as there is a tendency for reactions to occur that are exothermic.

A reaction for which ΔS is positive and ΔH is negative is a spontaneous reaction, regardless of temperature. If a reaction is characterized by a negative ΔS and a positive ΔH, it is nonspontaneous at any temperature. These sweeping statements need to be modified only to the extent of allowing for the complication that ΔH and ΔS may themselves have somewhat different values at different temperatures. However, at whatever temperature we are operating, in order for a reaction to be spontaneous, either ΔH must be negative or ΔS must be positive.

13-8 Gibbs Free Energy

The general study of the relations between energy and chemical change is called *chemical thermodynamics*. Thermochemistry is included, but chemical kinetics (see Chapter 11) is not. The science of chemical thermodynamics was founded by the notable American mathematician J. Willard Gibbs. He was the first to hold the modern view that a spontaneous reaction is one that is capable of doing useful work.

Most chemical reactions are carried out in open containers. If the products occupy a greater volume than the reactants, work is required to push back the atmosphere. Since this work of expansion cannot be utilized for any other purpose, it is excluded from consideration when we use the term *useful work* in connection with a reaction taking place at constant temperature and constant pressure. Useful work would be, for example, the electrical work done by a voltaic cell. How nearly the useful work actually performed approaches the maximum of which the reaction is capable depends on the operating efficiency of the device by means of which the work is done. Our concern here is with the amount of useful work of which the reaction is capable. If the reaction is permitted to occur without doing any useful work, the system nevertheless has lost its capacity for doing useful work in the same amount as if the work had been done. By analogy, water running downhill loses its capacity for doing useful work by an amount that depends on how far it falls, and this in no way is related to whether or not it impinges against the blades of a turbine as it falls.

Every substance is assumed to have a certain amount of internal energy of which only a portion is extractable as useful work. The portion of the heat content, H, that is potentially extractable as useful work in a constant-pressure process is referred to as the *Gibbs free energy, G.* The maximum amount of useful work that can be obtained from a process taking place at constant temperature and constant pressure is the difference between the Gibbs free energy of the products and the Gibbs free energy of the reactants:

$$\Delta G_{\text{reaction}} = G_{\text{products}} - G_{\text{reactants}}$$

The quantity ΔG is called the Gibbs free energy of the reaction because it refers to energy that is available to do useful work. The free energy of a reaction in which 1 mole of a compound is formed from its elements is called the *free energy of formation* of the compound.

For reactions taking place at constant temperature and constant pressure, the quantity ΔG is the ultimate criterion for judging whether they will occur spontaneously. For a process in which useful work is done or could be done, ΔG is negative and equal in amount to the *maximum* useful work that could be done. Therefore, the criteria for spontaneity are as follows:

1. A reaction is spontaneous if ΔG is negative.
2. A reaction is nonspontaneous, that is, the reverse reaction is spontaneous, if ΔG is positive.
3. A system is in a state of equilibrium if $\Delta G = 0$.

It must be emphasized that this criterion of spontaneity says nothing about how long it may take for a spontaneous process to occur.

Like ΔH, the ΔG value for a reaction is independent of the number of steps taken to convert the original reactants to the final products. The ΔG value for a reaction can be calculated by adding the ΔG values for the individual steps, but it is important that each ΔG should pertain to the same temperature and pressure. The ΔG value for a reaction at 25°C (298°K) and 1 atm is labeled ΔG°_{298} and is called the *standard free energy* of the reaction.

EXAMPLE

Calculate the standard free energy of formation for sulfur dioxide from the following data:

$$SO_2(g) + Cl_2(g) \longrightarrow SO_2Cl_2(g) \qquad \Delta G^\circ_{298} = -2.27 \text{ kcal}$$

$$\text{S(rhombic)} + O_2(g) + Cl_2(g) \longrightarrow SO_2Cl_2(g) \qquad \Delta G^\circ_{298} = -74.06 \text{ kcal}$$

Reversing the first equation and adding it to the second one, we obtain the desired result:

$$S(\text{rhombic}) + O_2(g) \longrightarrow SO_2(g)$$

$$\Delta G^\circ_{298} = -74.06 \text{ kcal} + 2.27 \text{ kcal} = -71.79 \text{ kcal}$$

13-9 Free Energies of Formation

Free energies of reactions are obtainable by spectroscopic, thermochemical, and electrochemical methods. Tables of free energies of formation have been constructed that are comparable to tables of heats of formation. From these tables the free energy of a proposed reaction can be calculated by subtracting the sum of the free energies of formation of the reactants from the sum of the free energies of formation of the products. (This is another corollary of the Law of Conservation of Energy.)

EXAMPLE

Calculate the free energy of the reaction

$$CH_4(g) + 4Cl_2(g) \longrightarrow CCl_4(l) + 4HCl(g)$$

From Table 13-3 we obtain the following free energies of formation:

$$\begin{aligned} CH_4(g), &\quad -12.1 \text{ kcal/mole} \\ CCl_4(l), &\quad -16.4 \text{ kcal/mole} \\ HCl(g), &\quad -22.8 \text{ kcal/mole} \end{aligned}$$

For the products the sum is

$$\begin{aligned} -16.4 \text{ kcal} + 4(-22.8 \text{ kcal}) &= -16.4 \text{ kcal} - 91.2 \text{ kcal} \\ &= -107.6 \text{ kcal} \end{aligned}$$

Subtracting the free energy of formation of methane from this sum, we find for the reaction that

$$\begin{aligned} \Delta G^\circ = -107.6 \text{ kcal} - (-12.1 \text{ kcal}) &= -107.6 \text{ kcal} + 12.1 \text{ kcal} \\ &= -95.5 \text{ kcal} \end{aligned}$$

Since the calculated ΔG° value is a large negative quantity, the proposed reaction is feasible. (Chlorine *does* react with methane.)

Most of the compounds in Table 13-3 are shown to have negative free energies of formation. This means that most of them are stable with respect to decomposition into their elements at 25°C and 1 atm. For example, the value for the free energy of formation of ammonia (−4.0

TABLE 13-3

Standard free energies of formation

Compound	ΔG° (kcal/mole at 25°C and 1 atm)
$H_2O(l)$	−56.7
$H_2O(g)$	−54.6
$H_2S(g)$	−7.9
$NH_3(g)$	−4.0
$CH_4(g)$	−12.1
$CO(g)$	−32.8
$CO_2(g)$	−94.3
$CS_2(l)$	+15.2
$CCl_4(l)$	−16.4
$C_2H_6(g)$	−7.9
$C_2H_4(g)$	+16.3
$C_2H_2(g)$	+50.0
$C_6H_6(l)$	+29.8
$HF(g)$	−64.7
$HCl(g)$	−22.8
$HBr(g)$	−12.7
$HI(g)$	+0.3
$SO_2(g)$	−71.8

kcal per mole) means that the decomposition of ammonia into its elements is *not* a spontaneous reaction at room temperature and atmospheric pressure. Quite the contrary: the formation of ammonia from its elements is a spontaneous reaction, and it ought to be possible to produce ammonia by reacting hydrogen with nitrogen. In practice, the reaction is so slow at 25°C and 1 atm that no detectable quantities of ammonia are formed even after a mixture of hydrogen and nitrogen gases has stood for many years. Higher temperatures, higher pressures, and a good catalyst are necessary for the success of the synthesis (Section 11-11). However, Fritz Haber never would have attempted to make the synthesis a success if the value calculated for ΔG under the proposed conditions had been a large positive number.

13-10 The Relation between ΔG and ΔH

One of the most important relationships in chemical thermodynamics is given by the equation

$$\Delta G = \Delta H - T\Delta S$$

where T is the absolute temperature. The sign of the free energy, ΔG, tells whether or not a constant-temperature, constant-pressure reaction is spontaneous at a particular temperature and pressure. For most chemical reactions at ordinary temperatures, the sign of ΔG is the same as the sign of ΔH because $T\Delta S$ is smaller than ΔH. It is usually true, therefore, that exothermic reactions are spontaneous and endothermic reactions are nonspontaneous at ordinary temperatures, as Thomsen and Berthelot stated. At *higher* temperatures, however, the larger value of T may make $T\Delta S$ sufficiently large that the sign of ΔS will determine the sign of ΔG. This explains why the Thomsen-Berthelot Principle was wrong. The *lower* the temperature the more likely it becomes that the sign of ΔH will determine the sign of ΔG.

With ΔS expressed in calories per degree, the product $T\Delta S$ is expressed in calories. With ΔH also expressed in calories, the calculated ΔG is given in calories.

The equation embodies the observation that the change in entropy within a system cannot, by itself, determine whether a reaction is spontaneous. On the other hand, neither is ΔH the sole factor in determining whether a reaction is spontaneous. Consider, for example, the following reaction:

$$C(s) + H_2O(g) \longrightarrow CO(g) + H_2(g) \qquad \Delta H^\circ = +31 \text{ kcal}$$

This reaction is nonspontaneous at 25°C, as might be anticipated from the positive value of ΔH° and as can be confirmed by calculating ΔG° for the reaction, which is approximately +22 kcal. There is a high degree of order in the crystalline structure of carbon, and it is apparent that the reaction involves an increase in the disorder of the system since both products of the reaction are gases. Thus, ΔS is positive for the reaction. Consequently, at sufficiently high temperatures the magnitude of $T\Delta S$ becomes greater than that of ΔH, with the result that ΔG becomes negative and the reaction becomes spontaneous. Hydrogen gas is, indeed, obtained commercially by passing steam over coke at 1000°C (Section 8-5).

In general, ΔS will be positive if large particles are decomposing into smaller ones. At very high temperatures all complex molecules cease to exist. Above 100,000°K even the atoms "decompose" into ions and electrons. Above 1,000,000°K the atoms are stripped down to their nuclei. The nuclei disintegrate at still higher temperatures.

13-11 Free Energy and Cell Potentials

The electrochemical method for obtaining the standard free energy of a reaction is based on a significant equation that correlates thermodynamics with electrochemistry. This equation states that

$$\Delta G^\circ = -nFE^\circ$$

where E° is the standard potential for the cell in which the reaction takes place, n is the number of moles of electrons involved in the cell reaction, and F is the value of the faraday (96,490 coulombs). With E° in volts and F in coulombs, the calculated value for ΔG° will be in volt-coulombs, i.e., joules. To convert to kilocalories, we divide by 4184, the number of joules in 1 kcal.

EXAMPLE

Calculate the standard free energy of the reaction

$$Cu(s) + 2Ag^+ \rightleftharpoons Cu^{++} + 2Ag(s)$$

From Table 12-1 we find that

$$\begin{aligned} E^\circ_{cell} &= 0.80\ V - 0.34\ V \\ &= 0.46\ V \end{aligned}$$

For this reaction $n = 2$ moles.

$$\begin{aligned} \Delta G^\circ &= -\frac{2\ \text{moles} \times 96{,}490\ \text{coulombs/mole} \times 0.46\ V}{4184\ \text{V-coulombs/kcal}} \\ &= -21.2\ \text{kcal at } 25^\circ C \end{aligned}$$

The large negative value for ΔG° means that the reaction is spontaneous and will proceed well toward completion before reaching equilibrium.

NEW TERMS

Bond-dissociation energy: the energy required to break a particular bond. It is the same as (average) bond energy if a molecule is diatomic.

Bond energy: the energy required to break the bond between two atoms. Average bond energy is the total energy required to break a set of similar bonds divided by the number of bonds. Bond energy always means average bond energy.

Chemical thermodynamics: the general study of the relations between energy and chemical change. (Thermochemistry is included, but chemical kinetics is not. Thermodynamics is concerned with what is *possible*, kinetics is concerned with *how*, and *how fast*, a process actually takes place.)

Entropy: the thermodynamic measure of the disorder within a system.

Gibbs free energy: internal energy that is extractable as useful work.

Hess's Law: the quantity of heat evolved or absorbed in a reaction is the same when it occurs in a series of steps as when it occcurs in a single step.

Law of Lavoisier and Laplace: the quantity of heat required for the decomposition of a compound into its constituent elements is the same as the quantity of heat released when the same quantity of the compound is formed from its elements.

Spontaneous change: a change that can occur within a system, regardless of rate, without the use of an external driving force.

Thermochemical equation: a chemical equation that indicates the magnitude of the heat of reaction.

Thermochemistry: the branch of chemistry that deals with the changes in heat energy that accompany chemical reactions and such physical transformations as melting and evaporation.

Thomsen-Berthelot Principle: a fallacious idea, no longer accepted, according to which only exothermic processes are spontaneous.

EXERCISES

13-1 Using the data in Table 13-1, calculate the heats of the following reactions:
(a) $CH_4(g) + 4Cl_2(g) \longrightarrow CCl_4(l) + 4HCl(g)$
(b) $2HI(g) + Cl_2(g) \longrightarrow 2HCl(g) + I_2(s)$
(c) $3C_2H_2(g) \longrightarrow C_6H_6(l)$
(d) $C_2H_2(g) + H_2(g) \longrightarrow C_2H_4(g)$
(e) $C_2H_2(g) + 2H_2(g) \longrightarrow C_2H_6(g)$
(f) $2C_2H_2(g) + 5O_2(g) \longrightarrow 4CO_2(g) + 2H_2O(g)$

13-2 Given $2SO_2(g) + O_2(g) \longrightarrow 2SO_3(g)$, $\Delta H = -43$ kcal, calculate the heat of formation of sulfur trioxide from its elements. (Consult Table 13-1.)

13-3 Using Table 13-1, calculate the molar heats of combustion of C_2H_6, C_2H_4, and C_2H_2, and then calculate the amounts of heat released upon burning 1 g of each.

13-4 Tungsten is obtained by the reaction $3H_2(g) + WO_3(s) \longrightarrow W(s) + 3H_2O(g)$. Calculate ΔH for the reaction, given that the heat of formation of $WO_3(s)$ is -200.84 kcal/mole. (See Table 13-1.)

13-5 Given the following thermochemical equations,

$$H_2(g) + I_2(s) \longrightarrow 2HI(g) \qquad \Delta H^\circ_{298} = +12.4 \text{ kcal}$$
$$H_2(g) + I_2(g) \longrightarrow 2HI(g) \qquad \Delta H^\circ_{298} = -2.7 \text{ kcal}$$

calculate the heat of sublimation of iodine.

13-6 Given the following thermochemical equations,

$$CaC_2(s) + 2H_2O(l) \longrightarrow Ca(OH)_2(s) + C_2H_2(g) \qquad \Delta H = -29.97 \text{ kcal}$$
$$CaO(s) + H_2O(l) \longrightarrow Ca(OH)_2(s) \qquad \Delta H = -15.58 \text{ kcal}$$

and given also that the heats of formation of $CaO(s)$ and $C_2H_2(g)$ are

-151.90 kcal/mole and $+54.19$ kcal/mole, respectively, calculate the heat of formation of calcium carbide, $CaC_2(s)$.

13-7 Using information given in Table 13-2, compute what the bond energy would be for ICl (iodine monochloride) if there were equal sharing of the electrons. Would you expect the experimental value for the bond energy to be larger or smaller than the value you have calculated?

13-8 Using Table 13-3, calculate the standard free energies of the following reactions:
(a) $2CO(g) + O_2(g) \longrightarrow 2CO_2(g)$
(b) $2HBr(g) + Cl_2(g) \longrightarrow 2HCl(g) + Br_2(g)$
(c) $2C_6H_6(l) + 15O_2(g) \longrightarrow 6CO_2(g) + 3H_2O(l)$

13-9 From information given in Table 13-3, determine whether the following reactions are spontaneous:
(a) $2NH_3(g) \longrightarrow N_2(g) + 3H_2(g)$
(b) $C_2H_2(g) \longrightarrow 2C(s) + H_2(g)$
(c) $C_2H_6(g) \longrightarrow 2C(s) + 3H_2(g)$
(d) $Cl_2(g) + 2HI(g) \longrightarrow 2HCl(g) + I_2(s)$

13-10 At its boiling point a liquid is in equilibrium with its vapor when the vapor is at a pressure of 1 atm. Therefore, $\Delta G = 0$ and $\Delta H = T\Delta S$. Trouton's Rule (Section 7-6) states, in effect, that the entropy of vaporization is about 22 cal/mole-deg. In terms of the entropy concept, explain the failure of water to conform to Trouton's Rule (see Section 8-7).

13-11 Ice melts in a warm room even though ΔH is positive. How do you account for the fact that the process is spontaneous? Why is it not spontaneous below 0°C?

13-12 From the information given in Table 12-1, calculate the standard free energy of the reaction

$$2Na + 2H_2O \longrightarrow 2Na^+ + 2OH^- + H_2\uparrow$$

and determine whether the reaction is spontaneous.

13-13 The standard free energy of the reaction

$$Fe^{++} + Cd \longrightarrow Fe + Cd^{++}$$

is 1.8 kcal. From this information, determine which electrode would be the cathode and which would be the anode in an iron–cadmium cell.

13-14 For the reaction $2NO_2(g) \rightleftharpoons N_2O_4(g)$, the equilibrium constant is 172 at 25°C and 79.4 at 35°C. Is the dimerization of NO_2 exothermic or endothermic? Explain. (Suggestion: Review Chapter 11.)

REFERENCES

Allen, J. A., *Energy Changes in Chemistry*, Allyn & Bacon, Boston, 1966 (paperbound). An elementary presentation of basic ideas of thermodynamics and kinetics.

Bent, H. A., *The Second Law*, Oxford University Press, New York, 1965 (paperbound). An extended, enjoyable consideration of entropy.

Campbell, J. A., *Why Do Chemical Reactions Occur?* Prentice-Hall, Englewood Cliffs, N.J., 1965 (paperbound). Stimulating discussion of thermodynamic topics.

Luder, W. F., *A Different Approach to Thermodynamics*, Reinhold, New York, 1967 (paperbound). Emphasizes one theme: spontaneous chemical reactions.

Mahan, B. H., *Elementary Chemical Thermodynamics*, Benjamin, New York, 1963 (paperbound). Thermodynamic principles and some applications.

Nash, L. K., *Chemical Thermodynamics*, Addison-Wesley, Reading, Mass., 1962 (paperbound). The first and second laws of thermodynamics and their consequences.

Pimentel, G. C., and Spratley, R., *Understanding Chemical Thermodynamics*, Holden-Day, San Francisco, 1969 (paperbound). A monograph intended as a supplement to texts in introductory chemistry courses.

Strong, L. E., and Stratton, W. J., *Chemical Energy*, Reinhold, New York, 1965 (paperbound). Presents thermodynamics without calculus.

Wilson, M., and the Editors of Life, *Energy*, Time, Inc., New York, 1963. A description, accompanied by drawings and photographs, of energy in its many forms.

14

Organic Compounds

14-1 Organic Chemistry

The number of compounds containing carbon is so great that the study of their preparation and properties constitutes a special division of chemistry, which for historical reasons is called *organic chemistry*.

Man's basic needs for food, fuel, and clothing are all related to organic chemistry. The three important classes of organic compounds in foods are fats, carbohydrates, and proteins, and their digestion involves organic chemical reactions. These foodstuffs are of biological origin. The important fuels—coal, wood, natural gas, and petroleum—are also of biological origin. Cotton, wool, and silk are natural organic substances used for clothing, and they too are of biological origin.

Originally, no distinction was made between the chemistry of substances of mineral origin and the chemistry of substances of biological origin. As more and more compounds were isolated from natural sources, the differences between these two groups of compounds became so apparent that chemists began to differentiate between them by the terms *organic* for compounds of biological origin and *inorganic* for compounds of mineral origin. All compounds fell into one or the other of these two categories. The term *organic chemistry* was first used in 1808 by the Swedish chemist J. J. Berzelius.

The essential difference between organic and inorganic compounds —that organic compounds contain carbon whereas inorganic compounds do not—was first recognized by the German chemist Leopold Gmelin in 1848. When laboratory syntheses began to produce carbon compounds that had not been found in nature, they were nevertheless classed as organic compounds and thus the term "organic" was extended beyond its original meaning. Today the number of carbon compounds obtained by chemical synthesis in laboratories and chemical factories far exceeds the number that have been isolated from natural products. The term "organic chemistry" will probably continue to be used, but the field of study might more suitably be called simply "the chemistry of compounds containing carbon."

We now know that any organic compound can be derived, in one or more real or imaginary steps, from the corresponding hydrocarbon, a compound containing only the two elements carbon and hydrogen. This fact is emphasized in an alternative definition that describes organic chemistry as "the chemistry of the hydrocarbons and their derivatives."

The study of the chemical reactions occurring in living organisms is now called *biochemistry* and is the subject of Chapter 15.

14-2 Methane, the Simplest Hydrocarbon

There are thousands of compounds consisting of carbon and hydrogen alone. These compounds are called *hydrocarbons*. Only one of these thousands contains a single carbon atom per molecule. Its molecular formula is the same as its empirical formula, CH_4, and it is called methane.

Methane constitutes the major portion of the natural gas obtained from gas wells and used as a fuel. Quantities of it occur in solution in the complex hydrocarbon mixture known as petroleum, which is obtained from oil wells. Methane is formed in nature by the action of anaerobic bacteria on cellulose and other organic matter. In marshes and swamps the gas bubbles that are often seen rising to the surface of the water contain methane, carbon dioxide, and nitrogen. An old and popular name for methane is "marsh gas." It is also present in coal mines, where it is one of the causes of explosions and is known as "fire damp."

Methane, a nonpolar substance, does not dissolve in water because its molecules are unable to form hydrogen bonds with water molecules. Condensation of the gas to liquid requires cooling to −160°C at standard pressure. The gas is colorless and odorless.

The four bonds in the methane molecule are identical. This can be inferred from certain types of chemical evidence. By reaction with bromine, methane can be converted to bromomethane (methyl bromide):

$$CH_4 + Br_2 \longrightarrow CH_3Br + HBr$$

In this type of reaction, known as *substitution*, a hydrogen atom is removed from a hydrocarbon and another atom or group is inserted in its place. In the above reaction, a bromine atom has been substituted for a hydrogen atom. If bromomethane is treated with zinc in the presence of acid, methane is regenerated. No matter how many times the cycle of substitution and regeneration is repeated, the same bromomethane is obtained; that is, the substitution product each time has properties that are identical. Unless a particular one of the four hydrogens is substituted preferentially each time (a possibility that we shall keep in mind), we must conclude that it makes no difference which of the four hydrogen atoms is replaced and that all four hydrogen atoms are equivalent with all bond angles equal.

There are three possible arrangements of the atoms that would make all four hydrogen atoms equivalent and also make all bond angles equal. In the first arrangement, Figure 14-1a, the molecule is planar, i.e., flat, with a hydrogen atom at each corner of a square and the carbon atom at the center. In the second arrangement, Figure 14-1b, the molecule has the shape of a pyramid with a square base; the carbon atom is at the apex, and a hydrogen atom is at each corner of the base. In the third

FIGURE 14-1

Three spatial arrangements that would make all four hydrogen atoms in the methane molecule equivalent

(a) Planar. (b) Pyramidal. (c) Tetrahedral.

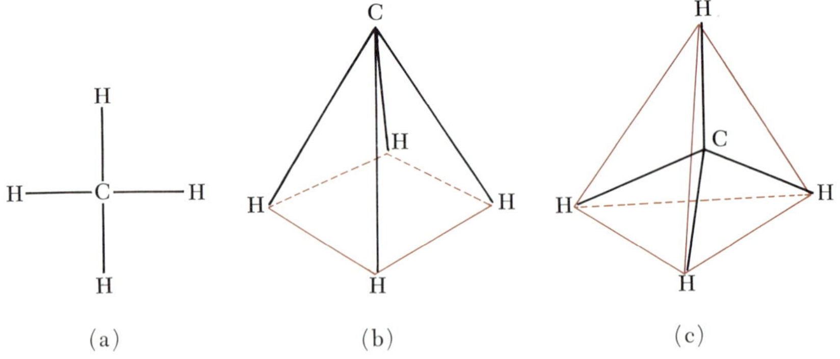

FIGURE 14-2

The two kinds of CH_2Br_2 molecules that would exist if the molecules were flat

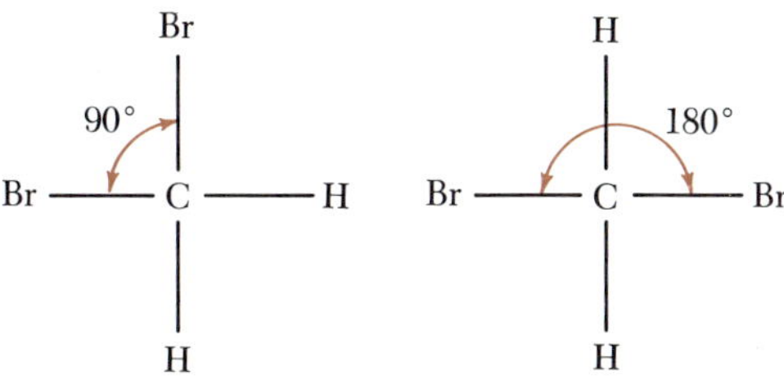

arrangement, Figure 14-1c, the molecule is in the shape of a regular tetrahedron, with the carbon atom at the center and a hydrogen atom at each of the four corners.

The shape of the methane molecule, i.e., planar, pyramidal, or tetrahedral, can be inferred from additional chemical evidence. Further reaction of bromine with bromomethane forms dibromomethane (methylene bromide):

$$CH_3Br + Br_2 \longrightarrow CH_2Br_2 + HBr$$

There is only one compound with the molecular formula CH_3Br, and there is only one CH_2Br_2. If the CH_4 molecule (and the CH_3Br molecule) were flat, then presumably a CH_2Br_2 molecule would also be flat. But if this were true, then two kinds of CH_2Br_2 molecule could exist, one with a 90° angle between C—Br bonds and the other with a 180° angle between C—Br bonds (Figure 14-2). If two kinds of CH_2Br_2 molecule existed, it would presumably be possible to separate them by some means. However, all of the experimental evidence indicates that there is only one CH_2Br_2, and we take this to mean that the CH_4 molecule is not flat. We also infer that it is not pyramidal, because if it were there could be two kinds of CH_2Br_2 molecule, one with the two bromine atoms

FIGURE 14-3

The two kinds of CH_2Br_2 molecules that would exist if the molecules were pyramidal

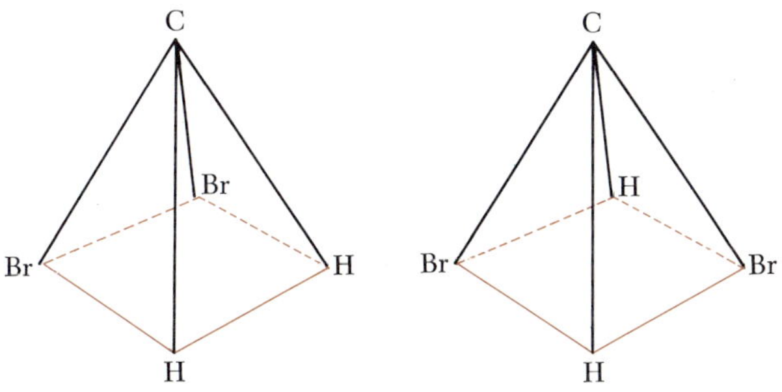

at neighboring corners of the base and the other with the bromine atoms at diagonally opposite corners of the base (Figure 14-3). However, the third possibility, a tetrahedral CH_4 molecule, would permit only one CH_2Br_2 to exist. The tetrahedral arrangement of CH_2Br_2 is illustrated in Figure 14-4; we can see that the spatial relationship of the second bromine to the first is the same in each case. This evidence that the CH_4 molecule is tetrahedral reinforces the earlier conclusion that the four C—H bonds in methane are identical and removes our suspicion that a particular hydrogen atom in methane might be substituted preferentially.

Further reaction of bromine with dibromomethane forms tribromomethane (bromoform):

$$CH_2Br_2 + Br_2 \longrightarrow CHBr_3 + HBr$$

The fact that only one compound is known with the formula $CHBr_3$ does not, in itself, constitute evidence for or against the tetrahedral shape of the CH_4 molecule. However, the nonpolar character of tetrabromomethane (carbon tetrabromide), which is formed when tribromomethane undergoes further reaction with bromine, indicates a symmetrical molecule:

$$CHBr_3 + Br_2 \longrightarrow CBr_4 + HBr$$

Since bromine is more electronegative than carbon, each C—Br bond is polar. If the CBr_4 molecule were pyramidal, it would also be polar. However, since CBr_4 is nonpolar, the four C—Br bonds must be symmetrically arranged in space so that their polarities cancel. Methane itself is nonpolar, and from this fact alone we could have concluded that the molecule is either planar or tetrahedral.

FIGURE 14-4

The three identical arrangements of atoms in the tetrahedral CH_2Br_2 molecule

The arrows indicate the spatial relationship between the two bromines and carbon, which is the same in each case.

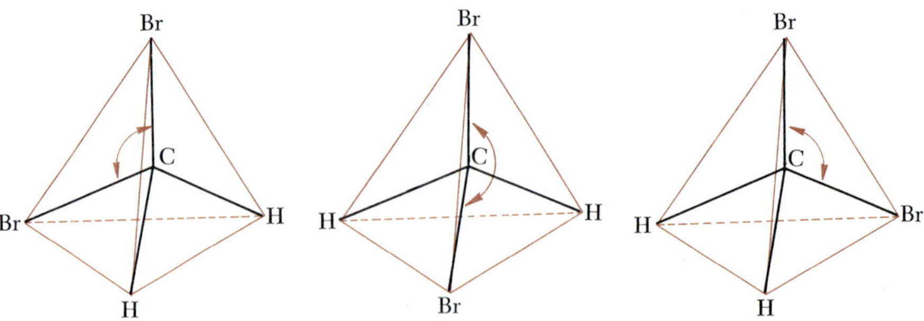

In 1874 the Dutch chemist Jacobus van't Hoff and the French chemist Jules Le Bel concluded, independently of each other and on purely logical grounds, that the methane molecule is tetrahedral. Their deduction has been verified by modern physical methods of investigating molecular structure, such as X-ray diffraction, electron diffraction, and molecular spectroscopy. The theory of hybrid bond orbitals (Section 5-9) provides an explanation for the tetrahedral structure of the methane molecule.

When we speak of "the tetrahedral carbon atom," we mean merely that when a carbon atom is bonded to four other atoms, the spatial distribution of the four bonds is symmetrical in three dimensions. In the CH_4 molecule the angles between the bonds are 109.5°. There are six such bond angles, and they are all the same. Each C—H bond length (distance between nuclei) is 1.09 Å.

On a flat surface, such as a sheet of paper or a blackboard, the bonds connecting a carbon atom to four other atoms are usually written

```
  |
—C—
  |
```

as when we write the formula for methane,

```
    H
    |
H—C—H
    |
    H
```

You must remember that the atoms are *not* all in the same plane.

14-3 Ethane and Propane

Methane is the first member of a series of hydrocarbons with the general formula C_nH_{2n+2}, where n is the number of carbon atoms in the molecule. The C_nH_{2n+2} series is officially called the *alkane series*. It is also known as the *methane series* or the *paraffin series* of hydrocarbons. The name *paraffin*, which means "having little affinity," was originally applied to the white, waxy substance, a mixture of C_nH_{2n+2} hydrocarbons of high molecular weights, that is now commonly known as paraffin wax.

The second member of the series is another gas, ethane, which boils at −89°C. The chemical analysis of ethane leads to CH_3 as its empirical formula. Determination of its molecular weight shows that its molecular formula is C_2H_6. Its *extended structural formula* is

$$
\begin{array}{ccccccc}
 & & H & & H & & \\
 & & | & & | & & \\
H & — & C & — & C & — & H \\
 & & | & & | & & \\
 & & H & & H & &
\end{array}
$$

The four bonds that unite each carbon to three hydrogen atoms and to the other carbon are tetrahedrally arranged. The molecule may be said to contain two overlapping tetrahedra. Each carbon atom is at the center of one tetrahedron and at one corner of the other tetrahedron.

Carbon atoms are unique in their ability to become bonded to one another, as in the ethane molecule. This is the primary reason for the existence of such a large number of organic compounds.

For convenience in writing, ethane is often represented by the *condensed structural formula* CH_3—CH_3. Here the line refers to the bond between the two carbons, just as if the formula were written H_3C—CH_3, as is sometimes done. Condensed structural formulas are frequently written without the use of lines to represent bonds; thus, ethane can be written simply as CH_3CH_3.

In ethane the bond angles (109.5°) and C—H bond lengths (1.09 Å) are the same as in methane. The C—C bond length is 1.54 Å, the same as in diamond. Virtually the same values hold for all alkanes. Therefore, it is usually not necessary to call attention to the actual shape of a particular alkane molecule.

There is a possible variation in the shape of an ethane molecule that we did not encounter with methane. The molecule could have an arrangement like that indicated in Figure 14-5a, or one like that in Figure 14-5b, or any intermediate arrangement. Formulas of the type given in Figure 14-5 are called *projection formulas* and represent the molecule as viewed from one end, with one carbon behind the other. The nearer carbon atom and the three atoms attached to it are repre-

FIGURE 14-5

Two conformations of ethane

(a) Eclipsed conformation. (b) Staggered conformation.

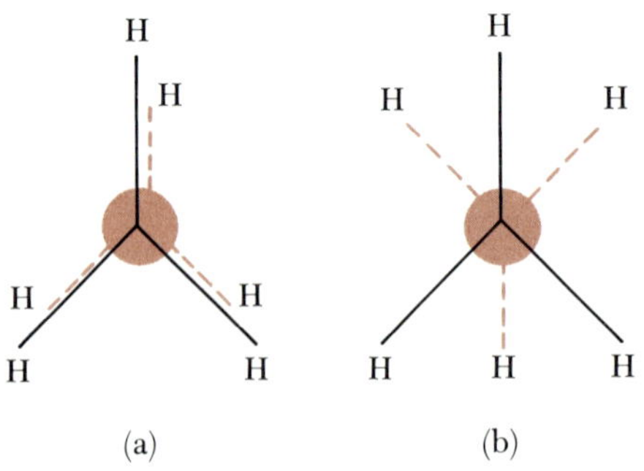

sented by the inverted Y; the more distant carbon and the atoms attached to it are indicated by the circle and the broken lines.

Any of the possible arrangements of the ethane molecule (two of which are shown in Figure 14-5) could be changed to another simply by rotating one end of the molecule about the C—C bond. Since there is only one compound with the formula $CH_2X—CH_2X$ (where X is Cl, Br, etc.), we must conclude that the molecule has considerable freedom to rotate about this bond.

These different arrangements that can be converted into one another by rotating a part of a molecule without breaking any bonds are called *conformations*. In Figure 14-5 the arrangement in (a) is called the *eclipsed* conformation of ethane; the arrangement in (b) is called the *staggered* conformation.

There is some evidence that when the molecule is in the eclipsed conformation the electron pairs by which hydrogen atoms are bonded to carbon atoms repel one another more strongly than in any other conformation. In the staggered conformation the repulsion is least because the distances between electron pairs are greatest. Consequently, the eclipsed conformation of ethane tends to convert spontaneously into the staggered conformation. However, a substantial fraction of the collisions of ethane molecules with one another are sufficiently energetic, even at room temperature, to cause rotation about the C—C bond, with a molecule passing through the eclipsed conformation as it goes from one staggered conformation to the next.

The third member of the alkane series is propane, C_3H_8, which boils at −42°C at 1 atm but is easily kept in the liquid state under pressure. It is marketed in tanks as LP (liquid propane) gas. Although its condensed structural formula is written as $CH_3—CH_2—CH_3$, the three carbon atoms do not actually lie on a straight line. The tetrahedral distribution of all the atoms bonded to a given carbon demands that the carbon atoms have a zigzag arrangement in all conformations. Nevertheless, propane is referred to as a *straight-chain hydrocarbon* because its carbon atoms are joined in a continuous chain. All members of the alkane series are *aliphatic hydrocarbons*, that is, they have an open-chain structure, with no ring formations. They are also classed as *saturated hydrocarbons* because their hydrogen content is at a maximum. Closed-chain hydrocarbons (those containing rings of carbon atoms) are discussed in Section 14-13.

14-4 Isomerism and the Properties of the Alkanes

There are two ways of writing the structural formula for a four-carbon alkane, C_4H_{10}:

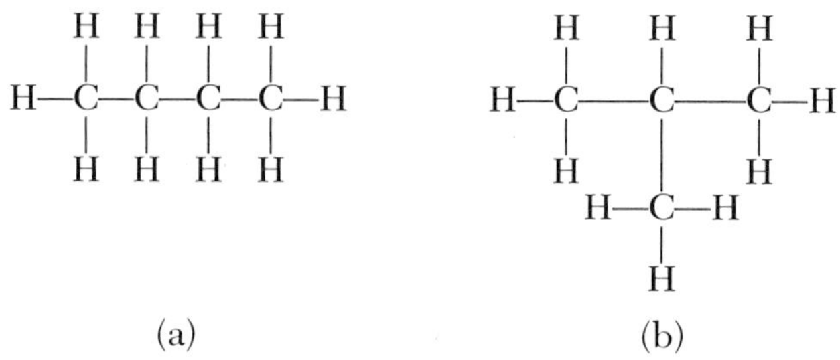

No amount of rotating or twisting would enable a molecule having one of these structures to become transformed into a molecule having the other structure without the breaking of bonds. In (a), no carbon atom is bonded to more than two other carbon atoms, but one of the carbon atoms in (b) is bonded to all of the other three carbon atoms. This is not a mere difference in conformation; it is a structural difference known as *isomerism*. Two or more different compounds having the same number and kinds of atoms per molecule are called *isomers*. The two compounds represented above have the same percentage composition, molecular weight, and molecular formula (C_4H_{10}), but they have different physical and chemical properties. The substance with the *branched chain*, (b), boils at −12°C and is known as isobutane; the compound with the so-called straight (i.e., unbranched) chain, (a), boils at 0°C and is known as *n*-butane ("normal butane"). The two isomeric butanes have different melting points, densities, solubilities, and reactivities toward various chemical reagents. Thus, isobutane and *n*-butane are different compounds, even though they have the same molecular formula.

Comparison of the molecular formulas of methane, ethane, propane, and the butanes reveals that each succeeding member of the alkane series has one more carbon atom and two more hydrogen atoms. A *homologous series* is defined as a series of compounds in which each member differs from the preceding member by the increment CH_2. The systematic study of the thousands of carbon compounds is simplified greatly by grouping them into homologous series.

There is a *general formula* for all members of a particular homologous series, for example, C_nH_{2n+2} for the alkanes. The members of a homologous series share certain general properties as a result of their similar molecular structures. Thus, all alkanes are held together entirely by single covalent bonds that are either nonpolar, because they join two atoms of the same kind, C—C, or are only slightly polar because they join two atoms that differ only slightly in electronegativity, C—H. Furthermore, the four bonds of each carbon are directed tetrahedrally, i.e., symmetrically, so that alkane molecules are nonpolar.

The boiling point increases regularly with increasing molecular weight if we compare the *n*-alkanes. However, branched-chain isomers have lower boiling points than the straight-chain isomers.

In accordance with the rule "like dissolves like," the alkanes are soluble in nonpolar solvents, such as benzene and carbon tetrachloride, and insoluble in polar solvents, such as water. The liquid alkanes serve as solvents for compounds of low polarity, but they do not dissolve highly polar compounds.

The chemical reactions of the alkanes are comparatively few. They include substitution and combustion. Combustion of hydrocarbon fuels gives water and carbon dioxide or carbon monoxide, depending on the adequacy of the oxygen supply:

$$CH_4 + 2O_2 \longrightarrow CO_2 + 2H_2O$$

$$2CH_4 + 3O_2 \longrightarrow 2CO + 4H_2O$$

$$2C_8H_{18} + 25O_2 \longrightarrow 16CO_2 + 18H_2O$$

$$2C_8H_{18} + 17O_2 \longrightarrow 16CO + 18H_2O$$

14-5 Names for the Alkanes

All members of the alkane series have names ending with *-ane*. Except for the first four members, the names are derived from prefixes indicating the number of carbon atoms. Thus, C_5H_{12} is pentane, C_6H_{14} is hexane, C_7H_{16} is heptane, C_8H_{18} is octane, C_9H_{20} is nonane, etc.

As the number of atoms increases, the number of possible arrangements of those atoms increases rapidly. There are three isomeric pentanes, five hexanes, and so on.

We have seen that the two butanes can be distinguished by assigning the names *n*-butane to the isomer with the straight chain and isobutane to the one with the branched chain. Since there are three isomeric pentanes, there is need for a third name. The name *n*-pentane has been assigned to the isomer with the unbranched chain, isopentane to the one with the least branching, and neopentane to the one with the most branching.

$$CH_3—CH_2—CH_2—CH_2—CH_3$$

n-pentane
bp 36°C

$$CH_3—CH_2—\underset{\underset{CH_3}{|}}{CH}—CH_3$$

isopentane
bp 28°C

$$CH_3—\overset{\overset{CH_3}{|}}{\underset{\underset{CH_3}{|}}{C}}—CH_3$$

neopentane
bp 10°C

With five isomeric hexanes, the problem of assigning names to individual hydrocarbons becomes more complicated. It is obvious that a systematic method of nomenclature is necessary if each compound is to be given a suitable name. Several systems have been used for naming hydrocarbons. All of them require that we have names for the various groups of atoms (CH_3, C_2H_5, etc.) that are present in the hydrocarbon molecules. Each group is itself a hydrocarbon less one hydrogen atom. These C_nH_{2n+1} groups are called *alkyl* groups. Each alkyl group is named by dropping *-ane* from the name of the hydrocarbon and adding *-yl*. Thus, CH_3 is the *methyl* group and C_2H_5 is *ethyl*.

The most widely used system for naming hydrocarbons and their derivatives was adopted in 1892 at the International Congress held at Geneva, Switzerland, and is often referred to as the *Geneva system*. It has been extended from time to time by the International Union of Pure and Applied Chemistry (IUPAC), whose responsibility it is to determine international rules of nomenclature for all branches of chemistry. The IUPAC rules for naming alkanes may be summarized as follows:

1. Each alkane has a name ending in *-ane*.
2. The normal or "straight-chain" (i.e., continuous-chain) alkanes are referred to by their common names: methane, ethane, propane, butane, pentane, hexane, heptane, octane, etc.
3. Branched-chain hydrocarbons are named as derivatives of normal hydrocarbons, with the longest "normal chain" in the molecule regarded as the parent hydrocarbon. The names of the branches are attached as prefixes to the parent name. For example:

$$\begin{array}{ccccc} CH_3 & — & CH & — & CH_3 \\ & & | & & \\ & & CH_3 & & \end{array}$$

methylpropane

$$\begin{array}{ccccc} & & CH_3 & & \\ & & | & & \\ CH_3 & — & C & — & CH_3 \\ & & | & & \\ & & CH_3 & & \end{array}$$

dimethylpropane

$$\begin{array}{ccccccc} & & CH_3 & & CH_3 & & \\ & & | & & | & & \\ CH_3 & — & CH & — & CH & — & CH \\ & & | & & | & & \\ & & CH_3 & & CH_3 & & \end{array}$$

tetramethylbutane

$$\begin{array}{ccccccccc} CH_3 & — & CH_2 & — & CH & — & CH_2 & — & CH_3 \\ & & & & | & & & & \\ & & & & CH_2 & & & & \\ & & & & | & & & & \\ & & & & CH_3 & & & & \end{array}$$

ethylpentane

4. When it is necessary to indicate the positions of the branches, this is done by numbering the carbon atoms of the parent alkane and using the number that indicates the position of a branch as part of the name. The numbering of the parent alkane is done from the end that gives the branches the smaller numbers. For example:

$$\underset{\displaystyle CH_3}{\overset{}{CH_3\underset{|}{C}HCH_2CH_2CH_3}}$$

2-methylpentane
(*not* 4-methylpentane)

$$\underset{\displaystyle CH_3}{CH_3CH_2\underset{|}{C}HCH_2CH_3}$$

3-methylpentane

$$\begin{array}{l} CH_3CHCH_2CHCH_2CH_3 \\ \quad\;\; | \qquad\;\; | \\ \quad CH_3 \quad CH_2 \\ \qquad\qquad\;\; | \\ \qquad\qquad CH_3 \end{array}$$

2-methyl-4-ethylhexane

5. Consecutive numbers in a name are separated by commas. If two branches are on the same carbon atom, the number is repeated. For example:

$$\begin{array}{l} CH_3—CH—CH—CH_3 \\ \qquad\;\; | \qquad\; | \\ \qquad CH_3 \;\; CH_3 \end{array}$$

2,3-dimethylbutane

$$\begin{array}{l} \qquad\quad CH_3 \\ \qquad\qquad | \\ CH_3—C—CH_2—CH_3 \\ \qquad\qquad | \\ \qquad\quad CH_3 \end{array}$$

2,2-dimethylbutane

All alkanes having a total of, say, five carbon atoms are pentanes, regardless of the fact that they are named as derivatives of simpler alkanes. Thus, dimethylpropane is a pentane because it has five carbon atoms (one in each of two methyl groups and three in the "normal chain" indicated by the parent name). Similarly, 2-methyl-4-ethylhexane is a nonane because it has a total of nine carbon atoms.

14-6 Unsaturated Aliphatic Hydrocarbons

The *alkenes*, with the general formula C_nH_{2n}, are the hydrocarbons with a double bond. Alkenes, also known as *olefins*, can be formed by removal of hydrogen from alkanes. Indeed, dehydrogenation is a widely used method for the industrial preparation of alkenes, two hydrogen atoms from adjacent carbon atoms of an alkane molecule being removed.

The simplest member of this homologous series is ethene, C_2H_4, more commonly known as ethylene. Ethene is manufactured from ethane by dehydrogenation, with chromium(III) oxide serving as a catalyst, at 500°C:

$$\begin{array}{c} H\;\;H \\ |\;\;\;| \\ H—C—C—H \\ |\;\;\;| \\ \boxed{H\;\;H} \end{array} \xrightarrow{-2[H]} \begin{array}{c} H \qquad\quad H \\ \diagdown \quad \diagup \\ C{=}C \\ \diagup \quad \diagdown \\ H \qquad\quad H \end{array}$$

The removal of the two hydrogen atoms is represented by the broken lines. In the ethene molecule the two carbon atoms are bonded by two shared pairs of electrons, i.e., a double bond. The electron-dot formula, the extended structural formula, and the condensed structural formula are as follows:

$$\begin{matrix} H & & H \\ & C::C & \\ H & & H \end{matrix} \qquad \begin{matrix} H & & H \\ & \diagdown \quad \diagup & \\ & C{=}C & \\ & \diagup \quad \diagdown & \\ H & & H \end{matrix} \qquad H_2C{=}CH_2$$

The ethene molecule has all six atoms in the same plane, and all bond angles are 120°. The carbon-to-carbon distance is 1.34 Å, as compared to 1.54 Å in ethane.

Since alkenes contain less hydrogen than alkanes, they are referred to as *unsaturated* aliphatic hydrocarbons; that is, they have the same open-chain structure as the alkanes but contain less than the maximum quantity of hydrogen. Their unsaturation is responsible for their ability to undergo the type of reaction known as *addition*. In the addition of hydrogen to ethene, one hydrogen atom is added to each doubly bonded carbon, converting the double bond to a single bond. The product of the reaction is ethane:

$$\begin{matrix} H & & H \\ & C{=}C & \\ H & & H \end{matrix} + H_2 \longrightarrow \begin{matrix} & H & H & \\ & | & | & \\ H- & C- & C & -H \\ & | & | & \\ & H & H & \end{matrix}$$

Hydrogenation of ethene is, of course, the reverse of the dehydrogenation of ethane. Addition reactions are characteristic of compounds containing the carbon–carbon double bond. Many compounds can be readily obtained by addition of properly chosen reagents to an alkene. For example:

$$\begin{matrix} H & & H \\ & C{=}C & \\ H & & H \end{matrix} \begin{cases} \xrightarrow{+\,Cl_2} \begin{matrix} & H & H & \\ & | & | & \\ H- & C- & C & -H \\ & | & | & \\ & Cl & Cl & \end{matrix} \\ \xrightarrow[+\,HCl]{} \begin{matrix} & H & H & \\ & | & | & \\ H- & C- & C & -H \\ & | & | & \\ & H & Cl & \end{matrix} \end{cases}$$

When dehydrogenation of a higher alkane occurs, two hydrogen atoms can be removed from adjacent carbon atoms in more than one way. Dehydrogenation of propane gives the same product, propene (commonly called propylene), regardless of which two carbon atoms lose hydrogen atoms:

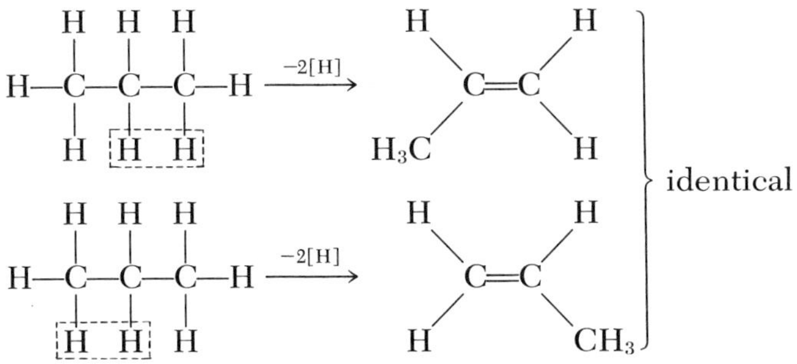

```
    H  H  H                   H         H  ⎫
    |  |  |                    \       /   ⎪
H—C—C—C—H  —2[H]→            C═C       ⎪
    |  |  |                    /       \   ⎪
    H [H  H]                H3C         H  ⎪
                                           ⎬ identical
    H  H  H                   H         H  ⎪
    |  |  |                    \       /   ⎪
H—C—C—C—H  —2[H]→            C═C       ⎪
    |  |  |                    /       \   ⎪
   [H  H] H                   H         CH3⎭
```

Thus, there is only one propene, just as there is only one propane.

When we consider the structural formula for a four-carbon alkene, C_4H_8, we find that there are several ways of writing it. First of all, the molecule may have the straight-chain structure of *n*-butane or the branched-chain structure of isobutane. Next we note that in an unbranched chain the double bond can be between the first and second carbons or between the second and third carbons. So far we have three structures:

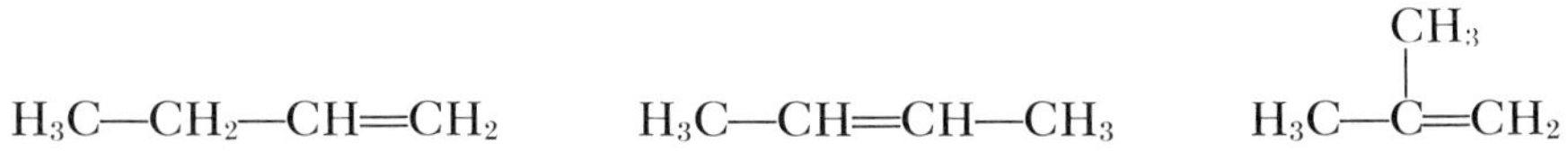

```
                                                       CH3
                                                       |
H3C—CH2—CH═CH2        H3C—CH═CH—CH3        H3C—C═CH2
   1-butene                2-butene             methylpropene
```

Actually, there are four alkenes having the molecular formula C_4H_8, because there are two 2-butenes.

To understand why there are two 2-butenes, we must recall that all six atoms of the ethene molecule lie in the same plane. The one end of the molecule is not free to rotate about the carbon–carbon bond as is the case with ethane. *A double bond does not permit freedom of rotation.* Therefore, a portion of any alkene molecule is flat; that is, the two doubly bonded carbon atoms and the four atoms bonded to them all lie in the same plane. This is the case in 2-butene, but there are two different ways in which the atoms attached to the doubly bonded carbon atoms can be arranged in space. In one arrangement, Figure 14-6a, the CH_3 groups lie on the same side of the molecule; in the other arrangement, Figure 14-6b, the CH_3 groups lie on opposite sides. The one compound is designated by the prefix *cis*- (Latin: on this side) and the other by *trans*- (Latin: on the other side).

The type of isomerism that distinguishes methylpropene from the other butenes is *skeletal* isomerism, and the type that distinguishes 1-butene from the 2-butenes is *positional* isomerism. The type distinguishing *cis*-2-butene from *trans*-2-butene is *geometric* isomerism. Geometric isomers have the same structural constitution, that is, they

FIGURE 14-6

The spatial configurations of the 2-butenes

(a) *cis*-2-butene. (b) *trans*-2-butene.

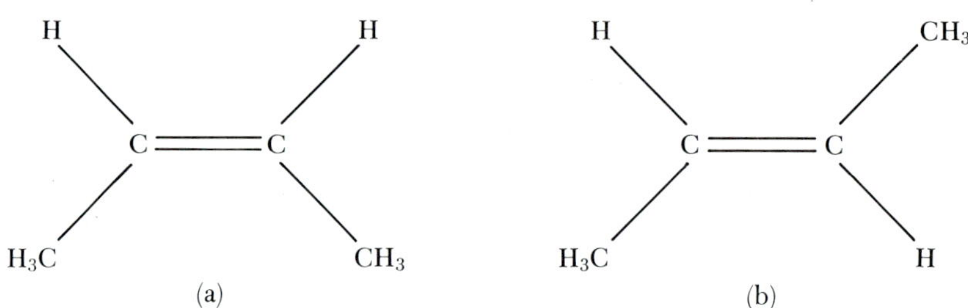

are alike with respect to which atoms are joined to which, but they differ from one another in the way the atoms are oriented in space. The increased possibilities of isomerism among the alkenes result in the existence of more pentenes than pentanes, more hexenes than hexanes, etc.

The IUPAC rules for naming alkenes are as follows:

1. The ending *-ane* in the name of the corresponding alkane is replaced by *-ene*: ethene, propene, butene, pentene, etc.
2. The parent compound is taken to be the longest normal chain *containing the double bond.*
3. The carbon atoms of the parent compound are numbered from the end *nearest* the double bond.
4. The position of the double bond is indicated by the smaller of the two numbers that refer to the carbon atoms to which it is attached. Examples:

$CH_3CH_2CH{=}CH_2$
1-butene
(*not* 2-butene)

$CH_3CH_2CH{=}CHCH_3$
2-pentene
(*not* 3-pentene)

5. The names and positions of side chains are indicated in the same manner as in naming alkanes. Examples:

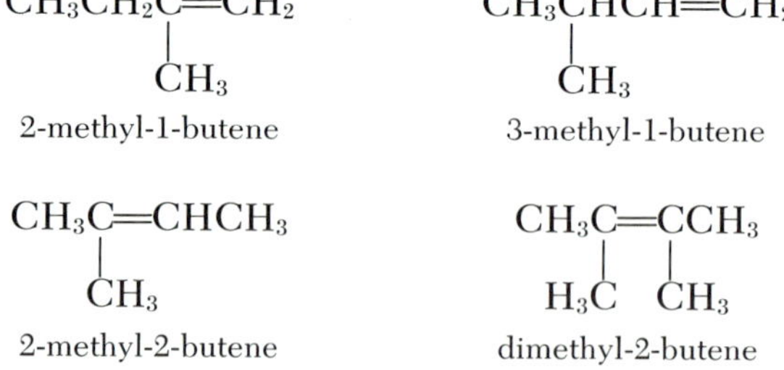

In the petroleum industry, alkenes are obtained by the "cracking" of alkanes, i.e., decomposition brought about by heat. The products of cracking are alkenes, hydrogen, and alkanes smaller than the starting alkanes. For example, *n*-butane yields the mixtures shown below:

$$CH_3—CH_2—CH_2—CH_3 \xrightarrow{400–600°C} \begin{cases} H_2 + C_4H_8 \\ CH_4 + C_3H_6 \\ C_2H_6 + C_2H_4 \end{cases}$$

The alkenes of low molecular weight are important raw materials for the chemical industry.

Alkynes are aliphatic hydrocarbons with the general formula C_nH_{2n-2} and have a carbon–carbon triple bond as their distinguishing structural feature. The simplest alkyne is ethyne, C_2H_2, well known by the older name acetylene. All four atoms of the acetylene molecule lie along the same straight line; that is all bond angles are 180°: H—C≡C—H. The length of the carbon–carbon triple bond is 1.20 Å and that of the C—H bonds is 1.06 Å. Like the alkenes, and quite unlike the alkanes, the alkynes are highly reactive because of their unsaturation.

Acetylene, the alkyne of greatest industrial importance, is used as a fuel for the oxyacetylene torch. Large quantities of acetylene are used each year as the starting material for the synthesis of many important compounds, including acetic acid. Acetylene can be prepared by the action of water on calcium carbide, CaC_2:

$$CaC_2 + 2HOH \longrightarrow Ca(OH)_2 + HC{\equiv}CH$$

There are many hydrocarbons that contain more than one point of unsaturation. The *alkadienes* (or *diolefins*) are hydrocarbons that contain two carbon–carbon double bonds. They are named in the same way as alkenes, except that the suffix *-diene* (pronounced "die een") is used to indicate two double bonds, and each name has two numbers that indicate the positions of the two double bonds (except for propadiene, in which there can be no variation in the positions of the bonds). One of the most important alkadienes, used in the manufacture of synthetic rubber, has the formula

$$CH_2{=}CH—CH{=}CH_2$$

and is called 1,3-butadiene (the numbers indicate that the double bonds follow the first and third carbon atoms). When both double bonds involve the same carbon atom, the compounds are usually unstable.

The alkadienes are isomeric with the alkynes; that is, they have the same general formula, C_nH_{2n-2}. Unsaturated hydrocarbons, of course, may contain three or more double bonds or both double and triple bonds. For example,

$$CH_2{=}CH—CH{=}CH—CH{=}CH_2$$

1,3,5-hexatriene

$$HC{\equiv}C—CH{=}CH_2$$

1-buten-3-yne *or* vinyl acetylene

Alkenes, alkynes, and alkadienes, although all unsaturated species, are classified with the saturated alkanes as aliphatic hydrocarbons because of their open-chain structure.

14-7 Polymers

Polymers are compounds of high molecular weight that result from the chemical combination of a large number of identical simple molecules. Polyethylene is a well-known polymer that is used in the manufacture of such everyday items as plastic bottles and vegetable bags. It is flexible, tough, resistant to water, and an excellent electrical insulator. It is synthesized by the polymerization of ethylene:

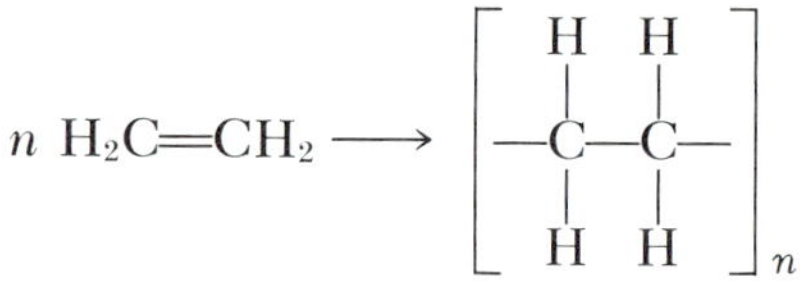

$$n\ H_2C{=}CH_2 \longrightarrow \left[\begin{array}{c} \text{H}\quad\text{H} \\ |\quad\ | \\ -\text{C}-\text{C}- \\ |\quad\ | \\ \text{H}\quad\text{H} \end{array}\right]_n$$

n molecules of ethylene — one molecule of polyethylene

The molecular weight of the polyethylene depends on *n*, the number of ethylene units that have combined. This number is difficult to control with precision, and *n* might even have a different numerical value for each polyethylene molecule in a sample of the polymer. The average value of the molecular weight of commercial polyethylene is of the order of a few million; thus, polyethylene is eminently entitled to be classified as a *high polymer*. The lack of crystallinity typical of most high polymers is due in part to the varying sizes and shapes of the molecules.

Polypropylene, a polymer whose use is increasing, is made by polymerization of propylene:

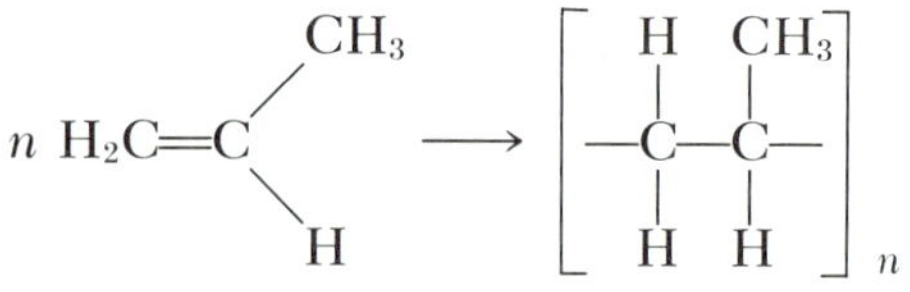

$$n\ H_2C{=}C\begin{array}{l} \diagup CH_3 \\ \\ \diagdown H \end{array} \longrightarrow \left[\begin{array}{c} \text{H}\quad\text{CH}_3 \\ |\quad\ | \\ -\text{C}-\text{C}- \\ |\quad\ | \\ \text{H}\quad\text{H} \end{array}\right]_n$$

n molecules of propylene — one molecule of polypropylene

Various other compounds containing the ethylenic structure are used in the manufacture of high polymers. Examples are vinyl chloride, $H_2C{=}CHCl$, used in the synthesis of poly(vinyl chloride) for garden hoses, plastic tile, etc., and tetrafluoroethylene, $F_2C{=}CF_2$, used to make Teflon, a polymer possessing remarkable resistance to solvents and acids.

Not all polymers are synthetic. Among the many polymeric substances that occur in nature are rubber, cellulose, and the proteins. Rubber consists of long, coiled molecules composed of large numbers of isoprene units,

$$\begin{array}{c} \quad CH_3 \\ \quad | \\ -CH_2-C{=}CH-CH_2- \end{array}$$

linked end to end. The first synthetic rubber was not an exact duplicate of natural rubber, but in 1958 the development of synthetic polyisoprene rubber was announced by several American manufacturers.

Synthetic polymers are frequently classified according to their uses. If they are stretchable, like rubber, they are called *elastomers*. Some can be extruded through spinnerets to produce synthetic *fibers*. The term *plastics* is applied to the synthetic polymers from which a variety of commercial products are formed, including plastic wrappings and molded objects of various shapes and sizes.

The plastics industry uses the term *resin* for a synthetic polymeric substance as it is initially obtained from the reaction mixture, and it uses the term *plastic* for the material developed from the resin by addition of other substances.

14-8 Alcohols

Organic compounds are divided into classes having similar properties and similar structures. The *alcohols* comprise one of the most important of these classes. Several of the alcohols of low molecular weight are liquids that can dissolve molecular compounds readily and ionic compounds somewhat less readily. Any alcohol is a potential starting material for the preparation of other organic compounds.

The simplest of the alcohols is methyl alcohol, also known as "wood alcohol" because it can be obtained by heating dried hardwood (oak, hickory, etc.) in the absence of air. By chemical analysis and molecular weight determinations, its molecular formula has been established to be CH_4O. As its formula suggests, methyl alcohol can be obtained by controlled oxidation of methane from natural gas, but a far more important synthesis, used since 1920, consists in the catalytic hydrogenation of carbon monoxide:

$$CO + 2H_2 \longrightarrow CH_3OH$$

In the above equation we have represented methyl alcohol by its condensed structural formula. The extended structural formula and the electronic formula are as follows:

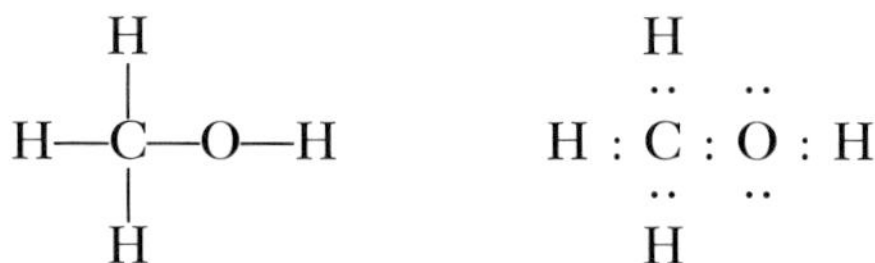

The important feature of the structure of this or any other alcohol molecule is that one of the hydrogen atoms is bonded to an oxygen instead of to a carbon. We may think of a methyl alcohol molecule as consisting of an OH (hydroxyl) group and a CH_3 (methyl) group.

Since the OH group is the one structural feature common to all alcohols, it is called the *functional group* characteristic of alcohols. An alcohol that contains only one OH group per molecule is known as a *monohydric* alcohol. Any alcohol derived from an alkane is called a *saturated* alcohol. The general formula for the saturated monohydric alcohols is $C_nH_{2n+1}OH$, which may be written more briefly as ROH, where R is an *alkyl* group, C_nH_{2n+1}.

The IUPAC method for naming alcohols consists in the use of one-word names based on the names of the corresponding hydrocarbons and having the suffix *-ol*. By this method the name for CH_3OH is *methanol.*

All alcohols contain the OH group, but they are nonelectrolytes; their aqueous solutions are neutral and do not conduct electric current. Alcohols react with sodium to give hydrogen and sodium alkoxide, just as water reacts with it to give hydrogen and sodium hydroxide:

$$2HOH + 2Na \longrightarrow \underset{\text{sodium hydroxide}}{2Na^+OH^-} + H_2$$

$$2CH_3OH + 2Na \longrightarrow \underset{\text{sodium methoxide}}{2Na^+OCH_3^-} + H_2$$

$$2ROH + 2Na \longrightarrow \underset{\text{sodium alkoxide}}{2Na^+OR^-} + H_2$$

A good method for disposing of sodium metal wastes is to dissolve them in alcohol, with which the sodium reacts more slowly than with water.

Methanol is a highly toxic liquid; even the inhalation of its vapor is hazardous. As in the case of water, the fact that methanol exists as a liquid despite its low molecular weight is explained in terms of hydrogen bonding, which can occur between the H of an OH group and the O of another OH group:

```
        R           R           R
        |           |           |
-----H—O-----H—O-----H—O-----
```

Here we have used R to represent the methyl group. The H atoms of the methyl group, like those in methane itself, are unable to form hydrogen bonds. Thus, there is no way that a methanol molecule could be hydrogen-bonded to *three* other molecules at the same time, as is the case with a water molecule.

Although methane is insoluble in water, methanol is miscible with water in all proportions. This, again, is due to hydrogen bonding. A hydrogen bond can form between a molecule of methanol and a molecule of water just as one can form between two molecules of the alcohol or between two molecules of water.

The second member of the saturated monohydric series of alcohols is ethanol (ethyl alcohol), also known as "grain alcohol" since it can be obtained by the fermentation of grain (for example, rye or corn). It is the alcohol contained in alcoholic beverages, where it is the product of the fermentation of the low molecular weight sugar called glucose. The fermentation process is catalyzed by the biological catalyst or enzyme known as zymase; carbon dioxide is also a product of the reaction:

$$C_6H_{12}O_6 \xrightarrow{\text{zymase}} 2CO_2 + 2C_2H_5OH$$

Ethanol bears the same formal relationship to ethane that methanol bears to methane; that is, one of the H atoms of ethane has been replaced by an OH group:

$$\begin{array}{ccccccccc} & & H & & H & & & & \\ & & | & & | & & & & \\ H & — & C & — & C & — & O & — & H \\ & & | & & | & & & & \\ & & H & & H & & & & \end{array}$$

The structure is clearly implied by the condensed structural formula CH_3CH_2OH, or even by C_2H_5OH. We may think of an ethanol molecule as consisting of an ethyl group and a hydroxyl group.

Because of hydrogen bonding, ethanol has a higher boiling point than any hydrocarbon with a comparable molecular weight and is miscible with water in all proportions. As we consider successive members of this homologous series, the alkyl group C_nH_{2n+1} becomes larger in increments of CH_2 with the result that the higher alcohols are less soluble in water because they are more like hydrocarbons.

The third member of the series should be an alcohol with the formula C_3H_7OH. Actually, there are two such alcohols, one with the OH group attached to an end carbon and the other with the OH group attached to the middle carbon of the chain:

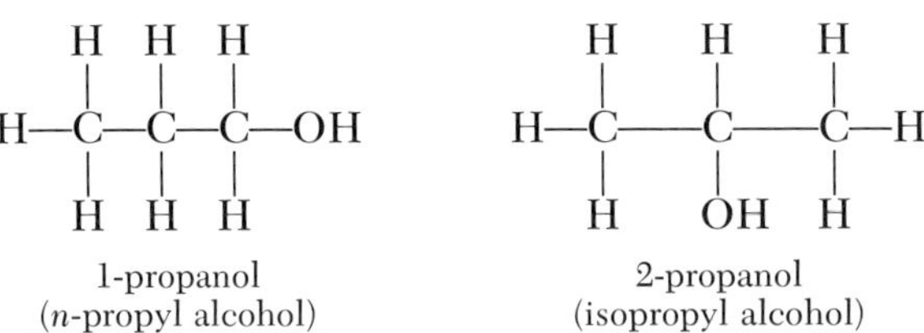

1-propanol
(*n*-propyl alcohol)
boiling point, 97°C

2-propanol
(isopropyl alcohol)
boiling point, 81°C

These two alcohols provide an illustration of positional isomerism. Normal propyl alcohol, as it is commonly called, can be represented by the condensed structural formula $CH_3CH_2CH_2OH$, and isopropyl alcohol can be represented as $CH_3CHOHCH_3$, but to represent either of them as C_3H_7OH is ambiguous.

Many of the alcohols are useful compounds, but methanol and ethanol are produced in the largest quantities. Methanol is used for the manufacture of formaldehyde and other chemicals, as a radiator antifreeze, as a solvent and as a denaturant for ethanol. Denaturing is the addition of substances that make ethanol unfit to drink. Ethanol for industrial use must be denatured if payment of the alcoholic beverage tax is to be avoided.

Ethanol is used in the manufacture of a great variety of dyes, medicines, perfumes, and insecticides. Drugs that are insoluble in water are often used in ethanol solutions called *tinctures*. Much of the industrial ethanol is now obtained synthetically by the catalyzed combination of water with ethylene resulting from the cracking of petroleum:

$$CH_2{=}CH_2 + HOH \longrightarrow CH_3{-}CH_2OH$$

Isopropyl alcohol is well known as a component of "rubbing alcohol." Higher alcohols are used where a definite number of carbon atoms per molecule is required for the purpose at hand; for example, the manufacture of butyl acetate requires butyl alcohol.

14-9 Ethers

Ethers have the general formula ROR′. The simplest ether is CH_3OCH_3, methyl ether. The next ether in order of molecular weight is $CH_3OC_2H_5$, methyl ethyl ether; the next is $C_2H_5OC_2H_5$, ethyl ether; etc. If the two alkyl groups are different, the ether is named by identifying the two alkyl groups; if the alkyl groups are identical, it is only necessary to name the group once. The name *diethyl ether*, for example, is correct, but the name *ethyl ether* is adequate.

Since one molecule of an ether cannot form hydrogen bonds with another molecule, the ethers are much more volatile than alcohols of corresponding molecular weight. For example, the boiling point of methyl ether is -24°C, which is 102° lower than the boiling point of ethanol, 78°C. These two compounds have the same molecular weight and the same molecular formula, C_2H_6O; that is, methyl ether and ethanol are isomers. Since they contain different functional groups, they are called *functional isomers*. The group of atoms that is characteristic of all ethers is present in methyl ether,

$$-\overset{|}{\underset{|}{C}}-O-\overset{|}{\underset{|}{C}}-$$

while in methyl alcohol there is the group of atoms always present in any alcohol, the OH group.

Ethyl ether, the best known of the ethers, is the anesthetic usually referred to simply as "ether." It is obtained by heating ethanol with concentrated sulfuric acid under carefully controlled conditions. Sulfuric acid is an excellent dehydrating agent. *Intermolecular dehydration* of an alcohol, i.e., removal of H and OH from different alcohol molecules, is required for the production of an ether:

$$2C_2H_5OH \longrightarrow (C_2H_5)_2O + H_2O$$

Intramolecular dehydration of an alcohol, i.e., removal of H and OH from adjacent carbon atoms of the same alcohol molecule, produces an alkene:

$$CH_3CH_2OH \longrightarrow CH_2{=}CH_2 + H_2O$$

Production of ethyl ether from ethanol requires that the dehydration reaction be carried out under conditions that minimize or preclude the formation of ethylene.

14-10 Carboxylic Acids

The first three members of the homologous series of compounds known as the aliphatic carboxylic acids are the following:

$HCOOH$	CH_3COOH	CH_3CH_2COOH
formic acid	acetic acid	propionic acid

The functional group characteristic of these acids is the COOH group, which is called the *carboxyl* group. It has the structural formula

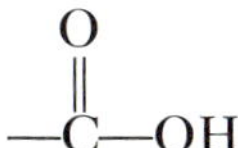

We may write RCOOH as the general formula for aliphatic carboxylic acids if we allow R to be a hydrogen atom instead of an alkyl group in the first member of the series, formic acid.

The solubility of the carboxylic acids in water is due to hydrogen bonding between acid and water. Long-chain acids are not very soluble, because the acids become more like hydrocarbons as the carbon chain lengthens. The acidity of aqueous solutions of compounds of this series is due to partial ionization. For example:

$$\underset{\text{}}{H_2O} + \underset{\text{acetic acid}}{CH_3COOH} \rightleftharpoons \underset{\text{acetate ion}}{CH_3COO^-} + \underset{\text{hydronium ion}}{H_3O^+}$$

The hydrogen atom that is attached to the oxygen atom in the carboxyl group is responsible for the acidity.

The aliphatic monocarboxylic acids are known as *fatty acids* because some of them, both saturated and unsaturated, are obtained from animal fats and from animal and vegetable oils. Since the carboxylic acids were first isolated from natural sources, they were given common names indicating what these sources were. For example:

Formic acid (1 carbon atom)	Latin: *formica*, ant
Acetic acid (2 carbons)	Latin: *acetum*, vinegar
Butyric acid (4 carbons)	Latin: *butyrum*, butter
Caproic acid (6 carbons)	Latin: *caper*, goat
Palmitic acid (16 carbons)	Latin: *palma*, palm
Stearic acid (18 carbons)	Greek: *stear*, tallow
Arachidic acid (20 carbons)	Latin: *arachis*, peanut

A mixture of stearic and palmitic acids is used in the manufacture of candles. Magnesium and zinc stearates are used in face powders; these compounds are salts of stearic acid.

Carboxylic acids react with alcohols to form *esters*; therefore, the reaction is called *esterification*. For example:

$$\underset{\text{formic acid}}{H\overset{\overset{\displaystyle O}{\|}}{C}OH} + \underset{\text{methyl alcohol}}{HOCH_3} \longrightarrow \underset{\text{methyl formate}}{H\overset{\overset{\displaystyle O}{\|}}{C}OCH_3} + H_2O$$

$$\underset{\text{acetic acid}}{CH_3\overset{\overset{\displaystyle O}{\|}}{C}OH} + \underset{\text{ethyl alcohol}}{HOC_2H_5} \longrightarrow \underset{\text{ethyl acetate}}{CH_3\overset{\overset{\displaystyle O}{\|}}{C}OC_2H_5} + H_2O$$

The general formula for esters, RCOOR′, indicates that an ester differs from an acid only in having an alkyl group in the place of the carboxylic hydrogen. (The two alkyl groups, R and R′, may be, but usually are not, identical.) It can also be said that a salt differs from an acid only in having a metal or other positive ion in the place of the carboxylic hydrogen. Esters are named as if they were alkyl salts of the carboxylic acids.

The reaction of an acid with an alcohol may superficially resemble neutralization of an acid by a base. However, salts are ionic, esters are covalent; neutralization is almost instantaneous, esterification is slow;

neutralization goes to completion, esterification does not go to completion. It has been established experimentally, by means of studies with isotopes, for example, that in esterification the OH from the acid and the H from the alcohol unite to form water; this is not analogous to what happens in neutralization.

Esters have some of the most pleasant fragrances known. Many of them are responsible for distinctive fruit flavors; two examples are amyl acetate, $CH_3COOC_5H_{11}$, in bananas and octyl acetate, $CH_3COOC_8H_{17}$, in oranges.

One of the most important reactions of esters is *saponification*, a reaction with a strong base in aqueous solution. For example:

$$\underset{\text{ethyl acetate}}{CH_3\overset{\overset{\large O}{\|}}{C}OC_2H_5} + Na^+OH^- \longrightarrow \underset{\text{sodium acetate}}{CH_3\overset{\overset{\large O}{\|}}{C}O^-Na^+} + \underset{\text{ethyl alcohol}}{HOC_2H_5}$$

Saponification of an ester, RCOOR′, with sodium hydroxide produces the sodium salt of the parent acid, $RCOO^-Na^+$, and liberates the parent alcohol, R′OH. The original meaning of the word *saponification* (Latin: *saponis*, soap) was "soap making." *Soaps* are alkali metal salts obtained by the reaction of a strong base such as NaOH, with fats, which are *glycerides*, i.e., esters of the trihydric alcohol called glycerol.

Ordinary soap is a mixture of sodium salts of long-chain monocarboxylic acids. It is a mixture because it is made from fats that are mixtures of different glycerides. Important components of ordinary soap are sodium palmitate, $C_{15}H_{31}COONa$; sodium stearate, $C_{17}H_{35}COONa$; and sodium oleate, $C_{17}H_{33}COONa$. Hard soaps are made from fats that are mostly saturated; soft soaps are made from highly unsaturated fats.

Each individual alkali metal salt of a carboxylic acid is called a soap, provided it contains from 10 to 18 carbon atoms. If the chain is less than 10 carbon atoms long, the salt is unable to emulsify oil in water. If the chain contains more than 18 carbon atoms, the salt is not sufficiently soluble in water to act as an effective detergent. Soaps of low molecular weight are most soluble in water. Potassium soaps are more soluble than sodium soaps (but also more costly to make).

The anions of the salts present in soaps, for example, the stearate ion,

$$CH_3CH_2CH_2CH_2CH_2CH_2CH_2CH_2CH_2CH_2CH_2CH_2CH_2CH_2CH_2CH_2CH_2COO^-$$

have long hydrocarbon "tails" and polar "heads." The polar ends, the carboxylate groups, tend to make the anions water soluble, and the nonpolar chains tend to make them oil soluble. Although "oil and water don't mix" (that is, they cannot dissolve in each other to form a true solution), oil can be emulsified by being shaken with an aqueous solution of soap. The soap anions surround the oil droplets with their hydrocarbon

FIGURE 14-7

Oil droplet in an emulsion

The soap anions surround each droplet, hydrocarbon chains in the oil, polar heads (COO^- groups) in the water. Soap cations are in the water, at varying distances from the charged droplet.

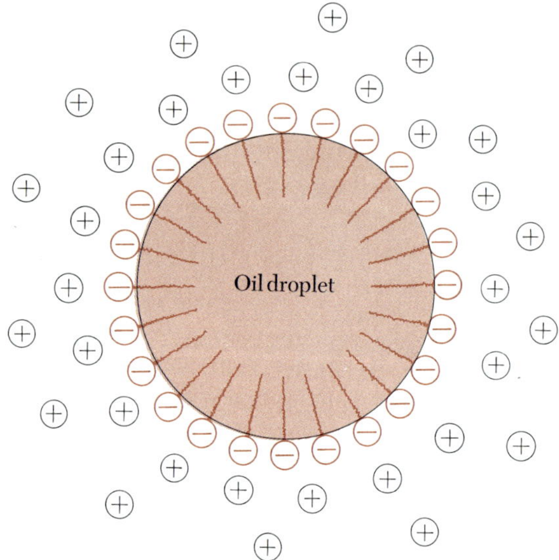

chains oriented inward and their polar heads extending outward into the surrounding water (Figure 14-7). The resulting emulsion is not a true solution because the oil is not molecularly dispersed in the water.

Soap solutions have a surface tension much less than that of water, and this makes the water "wetter," that is, the wetting power of a soap solution is greater than that of water alone. Soap solutions therefore are able to envelop grease and dirt particles and remove them from the surface to which they are attached. Thus, soaps serve as detergents through a combination of wetting action and emulsifying action.

Soaps cannot be used in acidic solution because the water-insoluble fatty acids precipitate out as a scum:

$$C_{17}H_{35}COO^- + H^+ \longrightarrow C_{17}H_{35}COOH \downarrow$$

The soap anions are also precipitated by the calcium, magnesium, and iron(II) ions present in hard water:

$$2C_{17}H_{35}COO^- + Ca^{++} \longrightarrow (C_{17}H_{35}COO)_2Ca \downarrow$$

This is the cause of bathtub "ring" and other undesirable effects of using soap in hard water.

14-11 Synthetic Detergents

Detergent properties are shown not only by soaps but also by other long-chain compounds that have a polar group at one end. An example of a detergent that is not a soap is sodium lauryl sulfate, $NaC_{12}H_{25}SO_4$.

The making of soap is one of the oldest of the chemical arts, but the synthetic detergents ("syndets") have been developed since 1930. Since 1954 the sale of syndets has exceeded that of soaps.

The fat most commonly used for the production of detergents is coconut oil. The first step in the process is the reduction of the fat. The reduction of glycerides, like the reduction of other esters, is brought about either with hydrogen and a catalyst or with sodium metal and alcohol (this now being one of the major uses of sodium metal):

$$\begin{array}{l} RCOOCH_2 \\ \quad\;\; | \\ R'COOCH \\ \quad\;\; | \\ R''COOCH_2 \end{array} \xrightarrow{\text{reduction}} \left\{ \begin{array}{l} RCH_2OH \\ R'CH_2OH \\ R''CH_2OH \end{array} \right\} + \begin{array}{l} CH_2OH \\ \; | \\ CHOH \\ \; | \\ CH_2OH \end{array}$$

a glyceride — mixture of primary alcohols — glycerol

Note that glycerol is a by-product of this first step in the manufacture of syndets. Even so, the soap and syndet industries combined are unable to supply all the glycerol required for its many uses.

Reduction of coconut oil yields a high proportion of lauryl alcohol. It is first converted to lauryl hydrogen sulfate, which is an ester of sulfuric acid:

$$\underset{\text{lauryl alcohol}}{CH_3(CH_2)_{10}CH_2OH} + \underset{\text{sulfuric acid}}{HOSO_2OH} \longrightarrow \underset{\text{lauryl hydrogen sulfate}}{CH_3(CH_2)_{10}CH_2OSO_2OH} + H_2O$$

Treatment of lauryl hydrogen sulfate with sodium hydroxide yields sodium lauryl sulfate:

$$CH_3(CH_2)_{10}CH_2OSO_2OH + Na^+OH^- \longrightarrow CH_3(CH_2)_{10}CH_2OSO_3^-Na^+ + H_2O$$

Sodium lauryl sulfate acts in essentially the same manner as soap when it is used as a detergent. The lauryl sulfate ion, like the stearate ion, has a long hydrocarbon "tail" and a polar "head." The advantage of sodium alkyl sulfates over sodium carboxylates is that they do not form precipitates with the calcium, magnesium, and iron(II) ions present in hard water.

Sodium lauryl sulfate was the first of the synthetic detergents to be placed on the market. A large number of compounds useful as detergents have since been developed. Syndets vary considerably in their chemical structure. Unfortunately, those with highly branched chains are not readily destroyed by microorganisms. The wide use of detergents of this type resulted in their appearance in domestic water supplies and caused serious foaming problems in sewage disposal plants. Since 1965 the manufacture of detergents has been limited to *biodegradable* compounds, i.e., those with unbranched hydrocarbon chains.

14-12 Ketones and Aldehydes

The number of hydrogen atoms attached to the same carbon atom as the OH group determines some of the important chemical properties and uses of alcohols. By this method of classification, alcohols are termed primary, secondary, or tertiary. A *primary* alcohol has two hydrogen atoms attached to the alcohol carbon; that is, it contains the CH_2OH group. Examples are ethanol, CH_3CH_2OH, and 1-propanol (*n*-propyl alcohol), $CH_3CH_2CH_2OH$. A *secondary* alcohol has only one hydrogen atom attached to the alcohol carbon because two carbon atoms are attached to it. An example is 2-propanol (isopropyl alcohol), $CH_3CHOHCH_3$. A *tertiary* alcohol has three carbon atoms attached to the alcohol carbon and therefore has *no* hydrogen atom attached to it. An example is 2-methyl-2-propanol (tertiary butyl alcohol):

$$\begin{array}{c} CH_3 \\ | \\ CH_3—C—OH \\ | \\ CH_3 \end{array}$$

The general formula for tertiary alcohols is R_3COH, where the three alkyl groups may all be different or identical; for secondary alcohols, the general formula is R_2CHOH; for primary alcohols, RCH_2OH.

The differences in the three classes of alcohols show up in the presence of mild oxidizing agents in aqueous solution. Let us consider, for example, the oxidation of 2-propanol by permanganate ions in aqueous solution. The oxidation of the 2-propanol produces an important organic compound known as acetone, and the reduction of the purple permanganate produces a brown precipitate, MnO_2.

$$3CH_3CHOHCH_3 + 2MnO_4^- \longrightarrow 3CH_3COCH_3 + 2MnO_2\downarrow + 2H_2O + 2OH^-$$

Acetone is the simplest member of a class of compounds known as *ketones*, which have the general formula

$$\begin{array}{c} O \\ \| \\ R—C—R' \end{array}$$

where R and R′ are alkyl groups that may or may not be identical. Mild oxidation of a secondary alcohol always produces a ketone.

The oxidation of ethanol with aqueous permanganate produces acetic acid, and the oxidation of 1-propanol produces propionic acid. Oxidation of any primary alcohol produces a carboxylic acid because an oxygen atom takes the place of the two hydrogen atoms attached to the alcohol carbon:

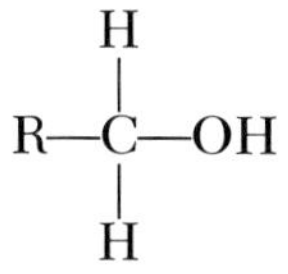

$$\underset{\text{primary alcohol}}{R{-}\overset{\displaystyle H}{\underset{\displaystyle H}{\overset{|}{\underset{|}{C}}}}{-}OH} \qquad\qquad \underset{\text{carboxylic acid}}{R{-}\overset{\displaystyle O}{\overset{\|}{C}}{-}OH}$$

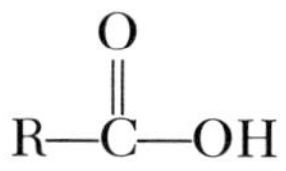

However, if the oxidation of a primary alcohol is carefully regulated, an intermediate product known as an *aldehyde* can be obtained. Thus, careful oxidation of ethanol produces acetaldehyde,

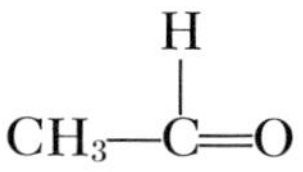

$$CH_3{-}\overset{\displaystyle H}{\overset{|}{C}}{=}O$$

Aldehydes have the general formula

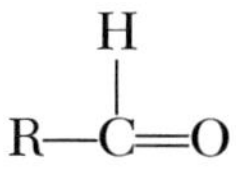

$$R{-}\overset{\displaystyle H}{\overset{|}{C}}{=}O$$

The differences among primary, secondary, and tertiary alcohols with respect to mild oxidation may be summarized as follows:

PRIMARY ALCOHOL:

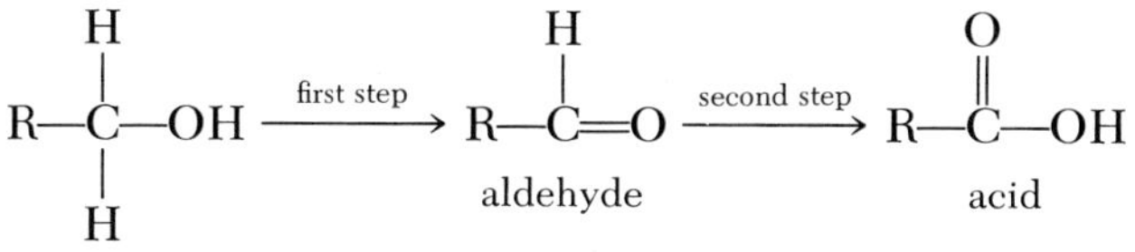

$$R{-}\overset{\displaystyle H}{\underset{\displaystyle H}{\overset{|}{\underset{|}{C}}}}{-}OH \xrightarrow{\text{first step}} \underset{\text{aldehyde}}{R{-}\overset{\displaystyle H}{\overset{|}{C}}{=}O} \xrightarrow{\text{second step}} \underset{\text{acid}}{R{-}\overset{\displaystyle O}{\overset{\|}{C}}{-}OH}$$

SECONDARY ALCOHOL:

$$R{-}\overset{\displaystyle H}{\underset{\displaystyle OH}{\overset{|}{\underset{|}{C}}}}{-}R \longrightarrow \underset{\text{ketone}}{R{-}\overset{\displaystyle O}{\overset{\|}{C}}{-}R}$$

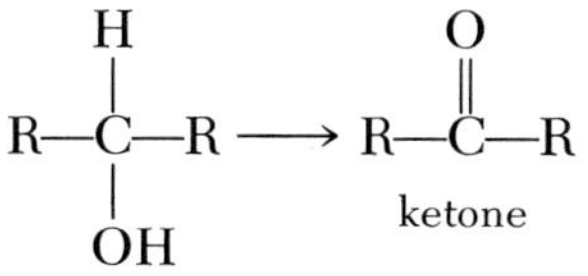

TERTIARY ALCOHOL:

Oxidation is impossible without breaking the carbon chain. Strong oxidizing agents can oxidize a tertiary alcohol, but since this involves breaking the carbon chain, the products have fewer carbon atoms than the original alcohol.

The common name of an aldehyde is determined by the name of the acid into which it is readily converted by oxidation. For example,

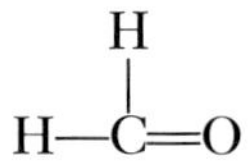

$$\begin{array}{c} \text{H} \\ | \\ \text{H—C=O} \end{array}$$

is formaldehyde because its oxidation yields formic acid; likewise,

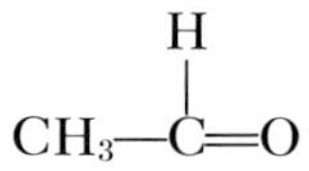

$$\begin{array}{c} \text{H} \\ | \\ \text{CH}_3\text{—C=O} \end{array}$$

is acetaldehyde because it is oxidizable to acetic acid; and so on.

Formaldehyde is a gas, used as a germicide and fungicide, and is commonly handled as an aqueous solution. Acetaldehyde is also a gas, now available synthetically by the catalyzed hydration of acetylene:

$$H—C{\equiv}C—H + H_2O \xrightarrow{HgSO_4} CH_3CHO$$

Acetaldehyde is used chiefly in making acetic acid. It will be recalled that acetylene is made by the reaction of water with calcium carbide (Section 14-7). Calcium carbide is prepared by the reaction between calcium oxide and coke at electric furnace temperatures:

$$CaO + 3C \xrightarrow{2000^\circ} CaC_2 + CO$$

Calcium oxide (quicklime) is manufactured from limestone in lime kilns:

$$CaCO_3 \longrightarrow CaO + CO_2$$

Coke is obtained from coal. Thus, acetic acid is obtained in a few simple steps from three abundant raw materials: coal, limestone, and water.

The common name of a ketone usually consists of the names of its two alkyl groups followed by the word *ketone*. Thus,

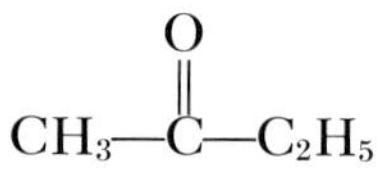

$$\begin{array}{c} \text{O} \\ \| \\ \text{CH}_3\text{—C—C}_2\text{H}_5 \end{array}$$

is methyl ethyl ketone. If the two alkyl groups are identical, the group is identified only once. For example,

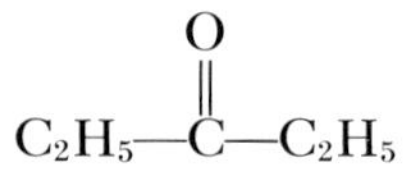

$$\begin{array}{c} \text{O} \\ \| \\ \text{C}_2\text{H}_5\text{—C—C}_2\text{H}_5 \end{array}$$

is ethyl ketone. The simplest of the ketones, CH_3COCH_3, has long been known as acetone.

Acetone is the most important ketone. It is used widely as a solvent and as the starting material for the manufacture of other chemicals. Water and nonpolar solvents are miscible with acetone. Methyl ethyl ketone, known industrially by the initials MEK, is used as a solvent for removing wax from lubricating oil. It is also familiar as fingernail polish remover.

14-13 Rings of Carbon Atoms

Petroleum from certain areas—for example, California—is rich in hydrocarbons known to the petroleum industry as *naphthenes.* In the molecules of these hydrocarbons, the carbon atoms are arranged to form rings instead of open chains. All the C—C bonds are single bonds, as in the alkanes, so that much of the chemistry of these compounds is essentially the same as that of the alkanes. To the chemist the members of this series are known as *cycloparaffins* or *cycloalkanes.* Any hydrocarbon that has one or more rings is a *cyclic hydrocarbon.*

The general formula for the cycloalkanes is C_nH_{2n}. They are named by adding the prefix cyclo- to the name of the corresponding alkane, i.e., the alkane having the same number of carbon atoms. The simplest possible cycloalkane is cyclopropane, C_3H_6, a potent anesthetic gas. In the cyclopropane molecule the three carbon atoms form a triangle. The bond angle is 60°, which differs sufficiently from the tetrahedral angle to make the structure of the molecule unstable. For example, cyclopropane undergoes a ring-opening reaction with bromine, or with hydrogen in the presence of a catalyst:

$$\text{cyclo-}(H_2C{-}CH_2{-}CH_2) + Br_2 \longrightarrow H_2C(Br){-}CH_2{-}CH_2(Br)$$

$$\text{cyclo-}(H_2C{-}CH_2{-}CH_2) + H_2 \xrightarrow[80°C]{\text{with catalyst}} H_3C{-}CH_2{-}CH_3$$

The second member of the series is cyclobutane, C_4H_8, in which the four carbon atoms form a square; that is, the bond angles are 90°. Here the deviation of the bond angle from the tetrahedral angle is not as great as in cyclopropane. Cyclobutane does not undergo the ring-opening reaction with bromine. Addition of hydrogen occurs in the presence of a catalyst, but it requires a higher temperature (200°C) than that required for the hydrogenation of cyclopropane (80°C):

$$\begin{matrix} H_2C{-}CH_2 \\ | \quad\;\; | \\ H_2C{-}CH_2 \end{matrix} + H_2 \xrightarrow[200°C]{\text{with catalyst}} H_3C{-}CH_2{-}CH_2{-}CH_3$$

In cyclopentane a nearly planar arrangement of the carbon atoms permits the bond angles to have the tetrahedral value. All rings larger than this are *puckered* so that each bond angle can be 109.5°. The puckered ring of cyclohexane is shown in Figure 14-8 in the "chair" conformation

FIGURE 14-8

Two conformations of cyclohexane

(a) Chair conformation. (b) Boat conformation.

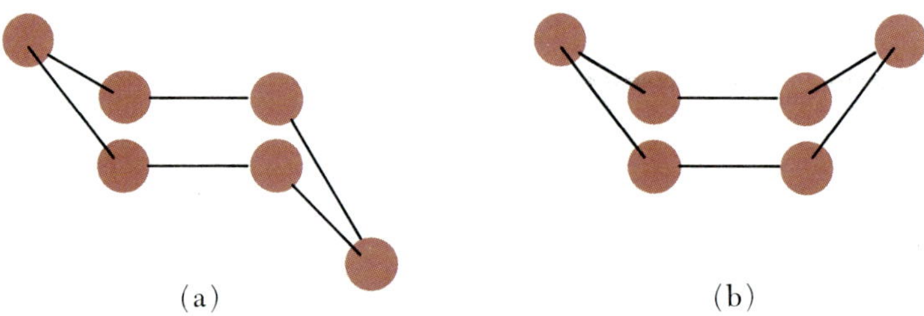

and in the "boat" conformation. If the cyclohexane ring were flat, the bond angles would be 120°. For larger rings a planar arrangement would make the bond angle differ even more from the tetrahedral angle.

Regardless of whether the rings are flat or puckered, cycloalkanes are often represented by simple geometric figures. Thus, the following are accepted "formulas":

The pentagon represents a ring of five carbon atoms, each singly bonded to two other carbons and also to two hydrogen atoms; the hexagon represents a molecule of similar structure, with six carbon atoms in the ring. Likewise, a heptagon represents cycloheptane, C_7H_{14}, and an octagon represents cyclooctane, C_8H_{16}.

Rings containing as many as 30 carbon atoms have been prepared. Except for the first two members, the chemical behavior of the cycloalkanes is very similar to that of the alkanes.

As we would expect, it is possible for one or more double bonds to be present in a ring of carbon atoms, as in cyclopentene:

CH
H_2C CH or
H_2C—CH_2

An example of a cyclic hydrocarbon with two double bonds is 1,3-cyclopentadiene:

CH
H_2C CH or
HC=CH

In general, the properties of cyclic compounds resemble those of aliphatic (open-chain) compounds. Cycloalkenes show the addition reactions typical of alkenes, and cycloalkadienes have the ability to add twice as much H_2 or Br_2 per mole as cycloalkenes. The cycloalkanes, cycloalkenes, and cycloalkadienes are classified together as alicyclic hydrocarbons.

14-14 The Benzene Ring

Benzene, C_6H_6, was discovered in 1825 by Michael Faraday. Its low hydrogen content suggests a high degree of unsaturation, but benzene exhibits considerable inertness toward those chemical reagents that ordinarily add to carbon–carbon double bonds. The properties of benzene have been investigated with unusual thoroughness because of its paradoxical nature and the importance of the many compounds derived from it.

In 1865 August Kekulé, in Germany, proposed a ring structure for benzene that differs from the structures of cyclohexane, cyclohexene, and cyclohexadiene only in having a higher degree of unsaturation:

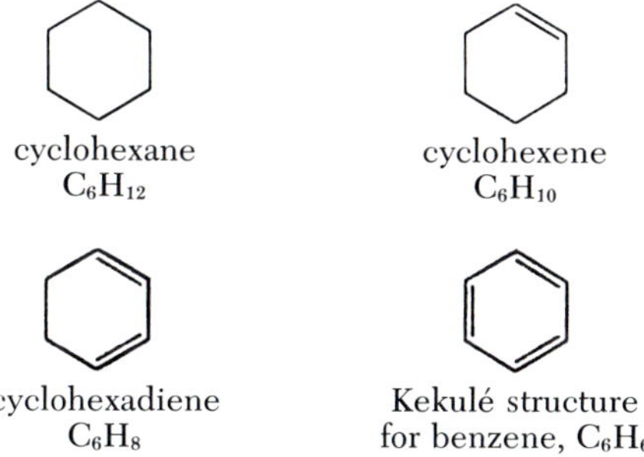

The Kekulé proposal was a great step forward, but it did not account for the stability of the ring and its reluctance to participate in addition reactions.

Modern methods for investigating structure, such as X-ray determination, have shown that the benzene molecule is flat and that the six carbon–carbon bonds are of equal length, 1.39 Å. If there were three single and three double bonds, we would expect their lengths to be 1.54 Å and 1.34 Å, respectively, since these are the bond lengths in alkanes and alkenes. The fact that the bonds in benzene seem to be intermediate between single and double bonds has received its most satisfactory explanation by the application of the theory of resonance.

Two equally reasonable structures that differ only in the location of electrons can be written for the benzene molecule:

The "real" structure is neither of these but is a resonance hybrid (Section 5-12) that is intermediate in character between the two. All six carbon–carbon bonds are equivalent. Each is more than single and less than double, i.e., a hybrid bond that cannot satisfactorily be represented on paper or with models.

One method for attempting to represent the structure of benzene consists in writing the two contributing structures with braces enclosing them and a two-headed arrow separating them:

This symbolism must not be taken to mean that molecules of two structures exist in equilibrium, but that only one kind of molecule exists: a hybrid of the two contributing structures.

A simplified symbolism frequently employed for the benzene ring consists of a hexagon with a circle inside:

benzene　　bromobenzene　　1,2-dibromobenzene

Each corner of the hexagon represents a carbon atom with a hydrogen atom bonded to it; if some other atom or group has replaced the hydrogen atom, it is shown in the formula. The purpose of the circle is to distinguish the benzene ring from the cyclohexane ring.

Benzene is the simplest hydrocarbon containing a benzene ring. When benzene is heated to a high temperature, it loses hydrogen and forms biphenyl:

$$2C_6H_6 \xrightarrow{800°C} C_6H_5—C_6H_5 + H_2$$

The biphenyl molecule consists of two benzene rings connected by a single bond. It is also possible for two benzene rings to share two carbon atoms; this is the situation in naphthalene (which constitutes "mothballs").

biphenyl, $C_{12}H_{10}$ naphthalene, $C_{10}H_8$

There are, of course, compounds containing more than two benzene rings. Any hydrocarbon containing one or more benzene rings is referred to as an *aromatic hydrocarbon.*

Many of the aromatic hydrocarbons contain both aliphatic and aromatic units. The simplest of these is toluene, $C_6H_5—CH_3$. Ethylbenzene and styrene are also hydrocarbons of this type.

toluene ethylbenzene styrene

The group C_6H_5, obtained by dropping a hydrogen atom from the benzene ring, is called the *phenyl* group. It is frequently represented by the Greek letter Φ (phi). Thus, bromobenzene can be written ΦBr and can be called phenyl bromide. The group $C_6H_5CH_2$, obtained by dropping a hydrogen from the methyl group of toluene, is called the *benzyl* group. Thus, ΦCH_2Br is benzyl bromide.

All groups obtained by dropping a hydrogen atom from an aromatic nucleus (benzene ring) are called *aryl* groups. Examples of aryl groups are the phenyl group and the naphthyl group, $C_{10}H_7$. The benzyl group is not an aryl group.

Coal is the major source of aromatic compounds. The coking of bituminous coal yields as a by-product a viscous material of unpleasant odor known as coal tar. Distillation of coal tar yields aromatic compounds that include hydrocarbons such as benzene, toluene, and naphthalene.

Aromatic compounds include benzaldehyde, ΦCHO; benzoic acid, ΦCOOH; phenyl acetate, $CH_3COO\Phi$; methyl benzoate, $\Phi COOCH_3$; phenyl benzoate, ΦCOOΦ; and so on.

Phenols are compounds that have an OH group attached directly to an aromatic ring. They have the general formula ArOH, where Ar is an *aryl* group: a phenyl group, a substituted phenyl group, a naphthyl group, etc. When it is necessary to make the distinction, the hydroxyl group in phenols is called a *phenolic hydroxyl* while that in alcohols is called an *alcoholic hydroxyl.* Benzyl alcohol, ΦCH_2OH, is not a phenol because the OH group is not attached directly to the benzene ring.

The simplest member of the phenol family is *phenol*, ΦOH. This compound is a powerful antiseptic, popularly known as carbolic acid. The hydroxy derivatives of toluene are known as *cresols.*

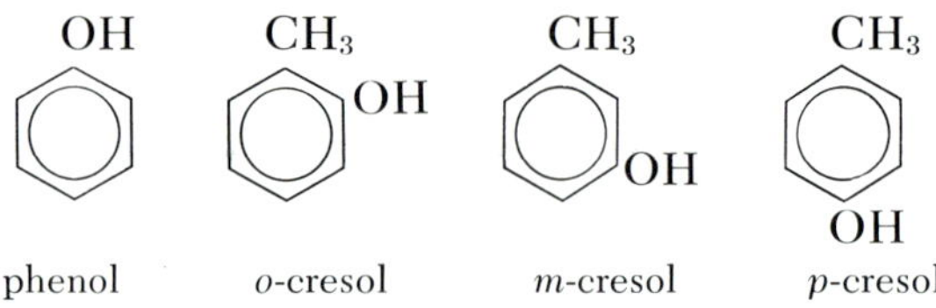

In naming a compound that has more than one group attached to a benzene ring, we must indicate their relative positions. A disubstituted benzene has three possible isomers that are differentiated by the prefixes *ortho-* (*o*-), *meta-* (*m*-), and *para-* (*p*-). The *ortho* isomer is the compound in which two *adjacent* hydrogen atoms of the benzene ring have been replaced. The *meta* isomer is obtained by replacement of two *alternate* hydrogen atoms; and the *para* isomer, by replacement of two *opposite* hydrogen atoms.

Phenol is somewhat soluble in water in consequence of hydrogen bonding between phenol and water molecules. Higher phenols are not water soluble. Although colorless when pure, the phenols are easily oxidized to give colored oxidation products.

Aryl groups have more attraction for electrons than hydrogen has, whereas alkyl groups have slightly less attraction than hydrogen. Consequently, phenols are considerably more acidic than water, although alcohols are less acidic than water. The ionization constant for phenol is 1.1×10^{-10}. The water-insoluble phenols dissolve in dilute solutions of strong bases to form salts of phenols:

$$ArOH(s) + OH^- \rightleftharpoons ArO^- + H_2O$$

Despite their significant differences, alcohols and phenols resemble each other to a limited extent since both contain the OH group. Both alcohols and phenols form ethers and esters.

NEW TERMS

Addition: a type of reaction characteristic of compounds containing double (or triple) bonds, in which two H atoms (or two Br atoms, etc.) are added to the double bond, i.e., one H atom to each doubly bonded carbon, converting the double bond to a single bond.

Alcohol: any compound, ROH, that contains an OH group and a hydrocarbon group; one example is methyl alcohol (methanol), CH_3OH.

Aldehyde: any compound that contains a $\overset{\displaystyle H}{\overset{|}{-C{=}O}}$ group.

Alicyclic hydrocarbons: hydrocarbons that contain rings but resemble aliphatic hydrocarbons in chemical properties. The name is a combination of *aliphatic* and *cyclic*. Alicyclic hydrocarbons include cycloalkanes, cycloalkenes, cycloalkadienes, etc. (but not the aromatic hydrocarbons).

Aliphatic hydrocarbons: hydrocarbons that have an open-chain structure (i.e., no rings). They include the alkanes, alkenes, alkadienes, and alkynes.

Alkadiene: any open-chain hydrocarbon that contains two double bonds. The alkadiene series is also known as the diolefin series. The general formula is C_nH_{2n-2}.

Alkane: any hydrocarbon that is a member of the homologous series with the general formula C_nH_{2n+2}. The alkane series is also known as the methane series and as the paraffin series.

Alkene: a hydrocarbon containing one double bond. The general formula for alkenes is C_nH_{2n}. The alkene series is also known as the ethylene series and as the olefin series.

Alkyl group: any group, C_nH_{2n+1}, in which the number of hydrogen atoms is one more than twice the number of carbon atoms, for example, the methyl group, CH_3. An alkyl group is named according to its corresponding hydrocarbon.

Alkyne: a hydrocarbon containing a triple bond. The general formula for alkynes is C_nH_{2n-2}. The alkyne series is also known as the acetylene series.

Aromatic hydrocarbon: a hydrocarbon containing one or more benzene rings. (The three general classes of hydrocarbons are the aliphatic hydrocarbons, the alicyclic hydrocarbons, and the aromatic hydrocarbons.)

Aryl group: any group obtained by dropping a hydrogen atom from an aromatic nucleus; two examples are the phenyl group, C_6H_5, and the naphthyl group, $C_{10}H_7$.

Biochemistry: the branch of chemistry that deals with chemical reactions occurring in living organisms.

Carboxylic acid: any compound, RCOOH, that contains the carboxyl group, $-\overset{\overset{\displaystyle O}{\|}}{C}-OH$.

Conformations: the different arrangements that can be converted into one another by rotating a part of a molecule without breaking any bonds.

Cyclic hydrocarbon: a hydrocarbon in which carbon atoms are arranged in one or more rings.

Ester: any compound conforming to the general formula $R-\overset{\overset{\displaystyle O}{\|}}{C}-O-R'$.

Esterification: the reaction of a carboxylic acid with an alcohol (so-called because an ester is formed).

Ether: any compound conforming to the general formula R—O—R′.

Fatty acids: aliphatic monocarboxylic acids.

Functional group: a group of atoms occurring in the molecules of all compounds of a given class, for example, the OH group in alcohols.

Functional isomers: two or more different compounds whose molecules contain the same number and kinds of atoms but differ as to which atoms are joined to which with the result that different functional groups are present. Examples: methyl ether and ethanol; acetic acid and methyl formate.

Geometric isomers: two or more different compounds whose molecules contain the same number and kinds of atoms and are alike as to which atoms are joined to which but differ as to the orientation in space of groups attached to a double bond or a ring. Example: *cis*-2-butene and *trans*-2-butene. Geometric isomerism is also known as *cis–trans* isomerism.

Homologous series: a series of compounds in which each member differs from the preceding member in having one more carbon atom and two more hydrogen atoms. Examples of homologous series include the alkanes, the alkenes, the cycloalkanes, and the cycloalkenes.

Hydrocarbons: compounds consisting of hydrogen and carbon alone.

Isomers: two or more different compounds that have the same number and kinds of atoms per molecule.

Ketone: any compound that conforms to the general formula $R{-}\overset{\overset{\displaystyle O}{\|}}{C}{-}R'$, where R and R′ are both alkyl or aryl groups (that is, neither can be a hydrogen atom).

Organic chemistry: the chemistry of the hydrocarbons and their derivatives.

Phenol: any compound, ArOH, where Ar is an aryl group, that contains an OH group attached to a benzene ring.

Polymers: compounds of high molecular weight that result from the chemical combination of a large number of identical simple molecules, for example, polyethylene.

Positional isomers: two or more different compounds whose molecules contain the same number and kinds of atoms but differ as to the position of some distinguishing structural feature such as the functional group. Examples: 1-propanol and 2-propanol; *o*-cresol, *m*-cresol, and *p*-cresol.

Primary alcohol: an alcohol that contains a CH_2OH group, that is, has two hydrogen atoms bonded to the same carbon as the OH group.

Saponification: the reaction of a strong base with an ester (so-called because soap is formed by this type of reaction).

Saturated hydrocarbons: hydrocarbons that contain no double or triple bond. They are said to be saturated because their hydrogen content is at a maximum.

Secondary alcohol: an alcohol that has two carbon atoms (and only one hydrogen atom) bonded to the same carbon as the OH group.

Skeletal isomers: two or more different compounds whose molecules contain the same number and kinds of atoms but differ as to the branching of the carbon chains. Example: isobutane and *n*-butane.

Soaps: salts obtained by the reaction of a strong base with fats; alkali metal salts of long-chain monocarboxylic acids.

Substitution: the removal of a hydrogen atom from a hydrocarbon and the insertion of another atom or group in its place. Elementary hydrogen is not a product of the reaction, as it would be if the reaction were a displacement reaction.

EXAMPLE OF SUBSTITUTION: $CH_4 + Br_2 \longrightarrow CH_3Br + HBr$
EXAMPLE OF DISPLACEMENT: $2HCl + Zn \longrightarrow ZnCl_2 + H_2$

Tertiary alcohol: an alcohol that has three carbon atoms (and therefore no hydrogen) bonded to the same carbon as the OH group.

EXERCISES

14-1 How is it known that the four bonds in methane are identical?

14-2 What evidence is there that the methane molecule is tetrahedral? How do we rule out the possibility of its being flat or pyramidal?

14-3 Suppose two isomeric compounds with the molecular formula CH_2Br_2 were discovered. Would the discovery be considered significant? Why or why not?

14-4 What is the general formula for each of the following classes of hydrocarbons: alkanes, alkenes, alkynes, alkadienes, cycloalkanes, cycloalkenes?

14-5 Compare the following types of formulas: empirical, molecular, extended structural, condensed structural, projection.

14-6 Define these terms: conformations, eclipsed conformation, staggered conformation, isomerism.

14-7 Write the structural formulas for all the isomeric pentanes, pentenes, and hexanes. Assign the correct IUPAC name to each compound.

14-8 Do you believe the existence of cyclobutyne is possible? Why or why not?

14-9 Define the following terms so as to distinguish among them: aliphatic, alicyclic, and aromatic hydrocarbons.

14-10 Of what importance is hydrogen bonding in determining the properties of alcohols? Ethers? Carboxylic acids? Ketones?

14-11 Explain the use of the terms *primary, secondary*, and *tertiary* in the classification of alcohols.

14-12 Illustrate four types of isomerism.

14-13 Write the structural formulas for all the possible compounds with the molecular formula C_3H_8O.

14-14 How does esterification resemble neutralization? How do the two reactions differ?

14-15 Write the structural formulas for three compounds that share the molecular formula $C_2H_4O_2$.

14-16 The boiling point of methyl acetate (57°C) is less than that of either its parent alcohol (CH_3OH, 65°C) or its parent acid (CH_3COOH, 118°C); but the molecular weight of the methyl acetate is, of course, considerably greater than that of either of its parents. Explain the low boiling point of methyl acetate.

14-17 Why are all six carbon–carbon bonds in benzene considered identical?

14-18 Organic chemistry is sometimes defined as the chemistry of carbon compounds and sometimes as the chemistry of hydrocarbons and their derivatives. What, if any, real difference is there between these definitions?

REFERENCES

Benfey, O. T., *From Vital Force to Structural Formulas*, Houghton Mifflin, Boston, 1964 (paperbound). Important developments in the early history of organic chemistry.

Flood, W. E., *Dictionary of Chemical Names*, Philosophical Library, New York, 1963. An inquiry into the origins of chemical names for those who find pleasure in things not strictly utilitarian.

Hart, H., and Schuetz, R. D., *Organic Chemistry*, 3rd ed., Houghton Mifflin, Boston, 1966. A brief text for students to acquire a knowledge of organic chemistry with a minimum of detail.

Herz, W., *The Shape of Carbon Compounds*, Benjamin, New York, 1963 (paperbound). One of the volumes in the "General Chemistry Monograph Series." An introduction to organic chemistry with emphasis on structure.

Mark, H. F., and the Editors of Life, *Giant Molecules*, Time, Inc., New York, 1966. An explanation of natural and synthetic polymers for interested persons who have not had formal training in science. Supplemented by elegant pictorial essays.

Pauling, L., and Hayward, R., *The Architecture of Molecules*, Freeman, San Francisco, 1964. Color drawings that show left-handed and right-handed molecules and many other aspects of molecular structure.

Rossini, F. D., "Hydrocarbons in Petroleum," *Journal of Chemical Education*, **37**(11), 554 (1960). An account of present knowledge of what petroleum is chemically—the relative amounts of different classes of hydrocarbons and of individual compounds within given classes.

15

Biochemical Compounds

15-1 The Scope of Biochemistry

Biochemistry is the branch of chemistry that is concerned with the chemical processes that occur in living plants and animals. These processes include photosynthesis in plants, the digestion and utilization of food by animals, the synthesis within living organisms of the many compounds and structural units of which they are constituted, and the production of the energy required to maintain life. Even the complex phenomena of growth, reproduction, aging, and disease are included within the domain of biochemical research.

Most of the important biochemical compounds are included in the four classes known as carbohydrates, lipids, proteins, and nucleic acids. Although most of this chapter has been divided into groups of sections dealing with compounds in these four categories, we will make a worthwhile digression to speak first of that important group of protein substances known as *enzymes*, which enter into so much of the discussion throughout the entire chapter.

15-2 Enzymes

When chemical reactions that take place within living organisms are duplicated in the laboratory, high temperatures or very strong reagents are usually required for them to occur at reasonable speeds. In general, neither of these conditions is compatible with life. Living cells produce substances whose function is to accelerate these reactions at a temperature compatible with life and under conditions no more extreme than mildly acidic or moderately alkaline. These biocatalysts, once called "ferments," are now known as *enzymes*.

Enzymes are made only by living cells, and all living cells contain them. Many have been extracted from their sources and utilized for catalyzing reactions in test tubes. It has been repeatedly demonstrated that an enzyme will selectively catalyze a particular reaction to the exclusion of other reactions that might take place between the same materials. Their biochemical importance is a consequence of their ability to direct the reactions that occur within an organism and of the specificity of their directive effect.

A similar specificity of catalytic powers is well known among inorganic catalysts. For example, platinum catalyzes the reaction of carbon monoxide with hydrogen to form methane and water, but zinc chromate catalyzes the combination of these two gases to form methanol.

Reactions catalyzed by enzymes, like other chemical reactions, lead to a state of chemical equilibrium, even though the reactions often appear to go to completion because the equilibrium is so far over to one side. Actually, the reactions can go either way, depending on concentrations. If the living cell excretes a product, or stores it in a separate compartment, the product is removed from the equilibrium system so that the enzyme-catalyzed reaction can continue in the one direction—either the direction of synthesis or the direction of decomposition, as the case may be.

Living cells often have means of disposing of the end products of a reaction: they may diffuse away or they may be stored in a special place or they may be immediately converted into something else by another reaction. In such cases enzymes may appear to promote a reaction in only one direction. In the laboratory, many substances have been synthesized by the action of the same enzymes that would have hydrolyzed them under other concentration conditions. The same equilibrium point has been reached from both directions. Although nothing appears to be happening at the equilibrium point, both synthesis and hydrolysis are in reality taking place—at exactly the same rate.

Many enzymes have been obtained in crystalline form. In every case the crystals have been proved to be proteins. It is now considered probable beyond a reasonable doubt that all enzymes are proteins. The general properties of enzymes, such as their behavior toward acid, alkali, and heat, indicate that they have a protein nature.

The rates of enzyme-catalyzed reactions are affected by temperature, by pH, by the concentration of the enzyme, and by the concentration of the substance undergoing chemical change under the influence of the enzyme. The generalization that the speeds of chemical reactions increase with rising temperatures holds true, within limits, for reactions catalyzed by enzymes. However, enzymes are inactivated at high temperatures. There is, therefore, an optimum temperature for each enzyme at which its catalytic activity is greatest.

Enzymes were originally given whatever names pleased their discoverers. Among the old names that have persisted are ptyalin, pepsin, trypsin, and rennin. It was later agreed that enzymes would be named by adding the suffix *-ase* to the name of the *substrate*, the particular compound or class of compounds acted upon by that enzyme. Thus, carbohydrases hydrolyze carbohydrates, lipases hydrolyze fats, etc.

CARBOHYDRATES

15-3 The Composition of Carbohydrates

Carbohydrates are produced in green plants from carbon dioxide and water by photosynthesis and constitute one of the important classes of foods for animals. Compounds of this class are composed of carbon, hydrogen, and water. The term *carbohydrate* originated from superficial consideration of the formulas for many compounds of this class, which seemed to indicate that they were hydrates of carbon, there being twice as many hydrogen atoms as oxygen atoms in the familiar carbohydrates. For example, glucose (known commercially as "dextrose") is $C_6H_{12}O_6$ or $C_6(H_2O)_6$, and sucrose (ordinary table sugar) is $C_{12}H_{22}O_{11}$ or $C_{12}(H_2O)_{11}$. Structurally, the carbohydrates are in no sense hydrates of carbon, that is, they do not contain water as such—and carbohydrates are now known in which the ratio of hydrogen to oxygen is not 2:1.

Chemically, carbohydrates are defined as the class of compounds that includes polyhydroxyaldehydes, polyhydroxyketones, and substances that can be made to react with water to produce polyhydroxyaldehydes or polyhydroxyketones. They include simple sugars, starch, dextrins, glycogen, and cellulose and are classified as monosaccharides, oligosaccharides, and polysaccharides, according to the degree of their complexity.

A *monosaccharide* is a carbohydrate that cannot be made to react with water to give simpler carbohydrates. The simplest monosaccharides have three carbon atoms per molecule ($C_3H_6O_3$), but the most important ones are five-carbon and six-carbon sugars. Examples include ribose, $C_5H_{10}O_5$; deoxyribose, $C_5H_{10}O_4$; glucose, $C_6H_{12}O_6$; and fructose, $C_6H_{12}O_6$. *Sugars* are defined as water-soluble carbohydrates that are more or less sweet; they have names ending with *-ose* (but so do some other compounds that are not sugars, such as cellulose).

Oligosaccharides are carbohydrates that can be made to react with water to give a small number of monosaccharide molecules. Most of the oligosaccharides that occur in nature are disaccharides, trisaccharides, or tetrasaccharides. An example is sucrose, the disaccharide obtained from sugar cane and sugar beets, which hydrolyzes to give the monosaccharides glucose and fructose:

$$\underset{\text{sucrose}}{C_{12}H_{22}O_{11}} + H_2O \longrightarrow \underset{\text{glucose}}{C_6H_{12}O_6} + \underset{\text{fructose}}{C_6H_{12}O_6}$$

Polysaccharides are carbohydrates that yield a large number of monosaccharide molecules when they undergo hydrolysis. Thus, polysaccharides are polymers composed of monosaccharide units. For example, starch and cellulose are polymers of glucose. Glucose is the sole product obtained by the complete hydrolysis of either starch or cellulose:

$$(C_6H_{10}O_5)_n + nH_2O \longrightarrow nC_6H_{12}O_6$$

A molecule of starch may contain from 200 to 3000 glucose units, but several thousand glucose units are present in a cellulose molecule.

15-4 Classification of Monosaccharides

Monosaccharides may be classified according to the number of carbon atoms present in one molecule. A *triose* is a monosaccharide that contains three carbon atoms per molecule; a *tetrose* contains four carbon atoms per molecule; a *pentose*, five; a *hexose*, six; and a *heptose*, seven.

Monosaccharides are also classified according to whether they have an aldehyde or a ketone structure. A polyhydroxyaldehyde is called an *aldose*; a polyhydroxyketone is called a *ketose*.

A single word may be used to indicate both the number of carbon atoms and the type of structure that is present. For example, an aldopentose is a five-carbon monosaccharide that contains an aldehyde group, and a ketohexose is a six-carbon monosaccharide that has a ketone structure.

According to the definition for carbohydrates, there must be present at least two OH groups and a $\rangle C{=}O$ group in the same molecule. Since the OH groups are always attached to different carbon atoms, the simplest compounds that fit the definition are $CH_2OH—CHOH—CHO$, glyceraldehyde, and $CH_2OH—CO—CH_2OH$, dihydroxyacetone.

Almost all important monosaccharides are pentoses or hexoses.

15-5 Monosaccharides of Biochemical Importance

Glucose, $CH_2OH—CHOH—CHOH—CHOH—CHOH—CHO$, is the aldohexose that is produced by photosynthesis in green plants, and it is the sugar into which other carbohydrates are converted in the bodies of animals. The oxidation of glucose is the source of the energy required for the operation of muscles and glands and for the transmission of signals through the nervous system. Glucose is present to the extent of about 1 mg per ml in the blood of normal mammals and is therefore known as *blood sugar*. It also occurs in most sweet fruits; it was first found in ripe grapes, from which it received the name *grape sugar*.

The glucose molecule can exist in three forms that are readily interconvertible (Figure 15-1). In aqueous solution all three forms are present in a state of dynamic equilibrium. The equilibrium mixture is approximately 36 percent alpha(α)-glucose, 64 percent beta(β)-glucose, and 0.02 percent open-chain form.

The formulas employed in Figure 15-1 are *perspective formulas*. In the formula for each cyclic structure the ring is shown as a hexagon slightly tilted out of the plane of the paper, with thick lines representing the bonds nearest the viewer and thin lines representing those that are farthest away. Solid lines indicate positions above the ring, and broken lines indicate positions below the ring. The only difference between

FIGURE 15-1

The three forms of glucose

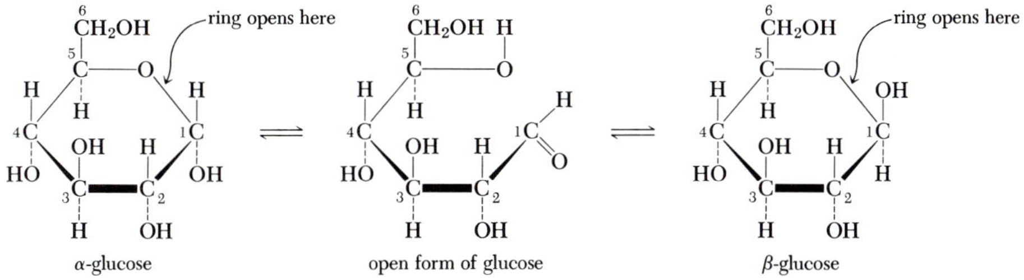

ring opens here — α-glucose ⇌ open form of glucose ⇌ β-glucose — ring opens here

FIGURE 15-2

Condensed perspective formulas for the cyclic forms of glucose and galactose

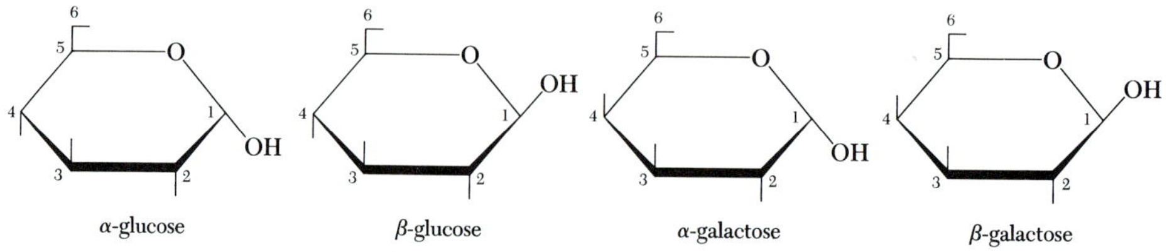

α-glucose β-glucose α-galactose β-galactose

α-glucose and β-glucose is that the OH group attached to carbon-1 projects downward in α-glucose and upward in β-glucose ("upward" meaning the direction in which carbon-6 projects from the ring). Carbohydrate rings are not actually planar. The rings are buckled so as to permit the natural bond angles.

It is convenient to represent the structures of α-glucose and β-glucose with condensed or simplified perspective formulas, as in Figure 15-2. Here, the ring carbon atoms are omitted, as are the hydrogen atoms attached to them. It is understood that a carbon atom is located at each point where lines intersect and that the short lines projecting upward or downward from the ring represent OH groups.

Galactose has the same molecular formula, $C_6H_{12}O_6$, as glucose. It does not occur as such in nature, but the hydrolysis of lactose (milk sugar) produces one molecule of galactose and one molecule of glucose. The only structural difference between galactose and glucose is the orientation of the OH group attached to carbon-4, as may be seen by comparing the formulas in Figure 15-2.

The sweetest of all sugars is *fructose*, which is also known as *levulose* or *fruit sugar*. Fructose, glucose, and sucrose occur together in fruit juices. Honey is a supersaturated solution of these sugars. Fructose is a ketohexose. Like glucose and galactose, it can exist in an open-chain form and in two cyclic forms. However, the rings present in the cyclic forms of fructose are five-membered rings (Figure 15-3).

FIGURE 15-3

Open form and cyclic forms of fructose

α-fructose open form of fructose β-fructose

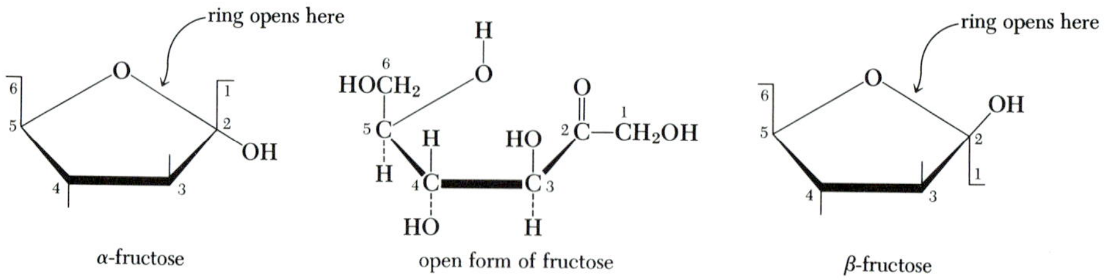

FIGURE 15-4

Ribose and deoxyribose

The five-membered rings that occur in nucleic acids.

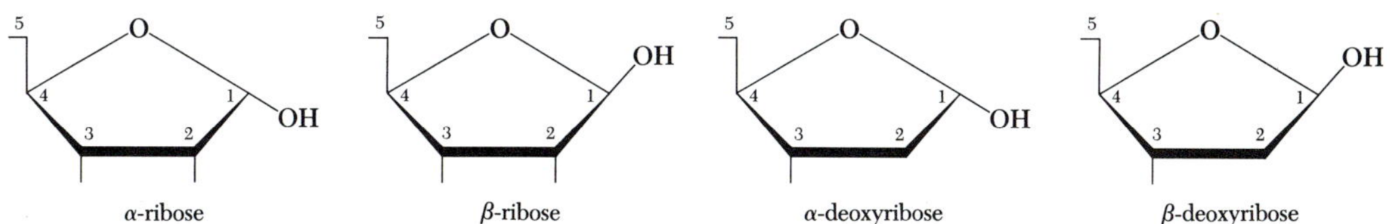

Ribose, $CH_2OH—CHOH—CHOH—CHOH—CHO$, and deoxyribose, $CH_2OH—CHOH—CHOH—CH_2—CHO$, are aldopentoses that are of great importance because they occur in all living cells as parts of nucleic acid molecules. Ribose occurs in combination in ribonucleic acids (the RNA's), which are involved in the biosynthesis of proteins. Deoxyribose is present in combination in deoxyribonucleic acids (the DNA's), which appear to be the carriers of the genetic code in chromosomes (see Section 15-21). Ribose and deoxyribose occur in open-chain form and in cyclic structures containing five-membered and six-membered rings. In the nucleic acids (Section 15-20) they occur in five-membered rings (Figure 15-4).

15-6 Important Disaccharides

The disaccharides have the molecular formula $C_{12}H_{22}O_{11}$. The most important are sucrose, lactose, and maltose.

Sucrose is ordinary table sugar, obtained commercially from sugar cane and sugar beets. It also occurs in sorghum cane and in the sap of the sugar maple. *Lactose* is synthesized from the glucose of the blood in the mammary glands. Commercially, lactose is obtained as a by-product of cheese-making; it is recovered from whey, which is the aqueous solution remaining after the milk proteins have been coagulated. It is used in the manufacture of infant foods. *Maltose*, also known as malt sugar, is formed by the hydrolysis of starch through the action of the enzyme diastase, which is present in malt (sprouted barley). Commerically, it is obtained for use in infant foods and in malted milk by the partial hydrolysis of cornstarch with acid:

$$\underset{\text{starch}}{(C_6H_{10}O_5)_n} + \tfrac{1}{2}nH_2O \xrightarrow[\text{or acid}]{\text{diastase}} \underset{\text{maltose}}{\tfrac{1}{2}nC_{12}H_{22}O_{11}}$$

When the hydrolysis of cellulose is carried to the same stage of degradation, the product is a different disaccharide called *cellobiose*.

Hydrolysis of sucrose occurs in the presence of acids or of the enzyme called sucrase. In the hydrolysis reaction, each sucrose molecule yields one molecule of glucose and one molecule of fructose. Hydrolysis of lactose, by acid or by the enzyme lactase, yields equal amounts of glucose and galactose. When maltose is subjected to hydrolysis, by acid or by the enzyme maltase, each maltose molecule yields two molecules of glucose. Hydrolysis of cellobiose also yields two molecules of glucose, but this requires a different enzyme.

$$C_{12}H_{22}O_{11} + H_2O \xrightarrow[\text{or acid}]{\text{enzyme}} C_6H_{12}O_6 + C_6H_{12}O_6$$

$$\text{Sucrose} + H_2O \longrightarrow \text{glucose} + \text{fructose}$$

$$\text{Lactose} + H_2O \longrightarrow \text{glucose} + \text{galactose}$$

$$\text{Maltose} + H_2O \longrightarrow \text{glucose} + \text{glucose}$$

$$\text{Cellobiose} + H_2O \longrightarrow \text{glucose} + \text{glucose}$$

FIGURE 15-5

Disaccharide structures

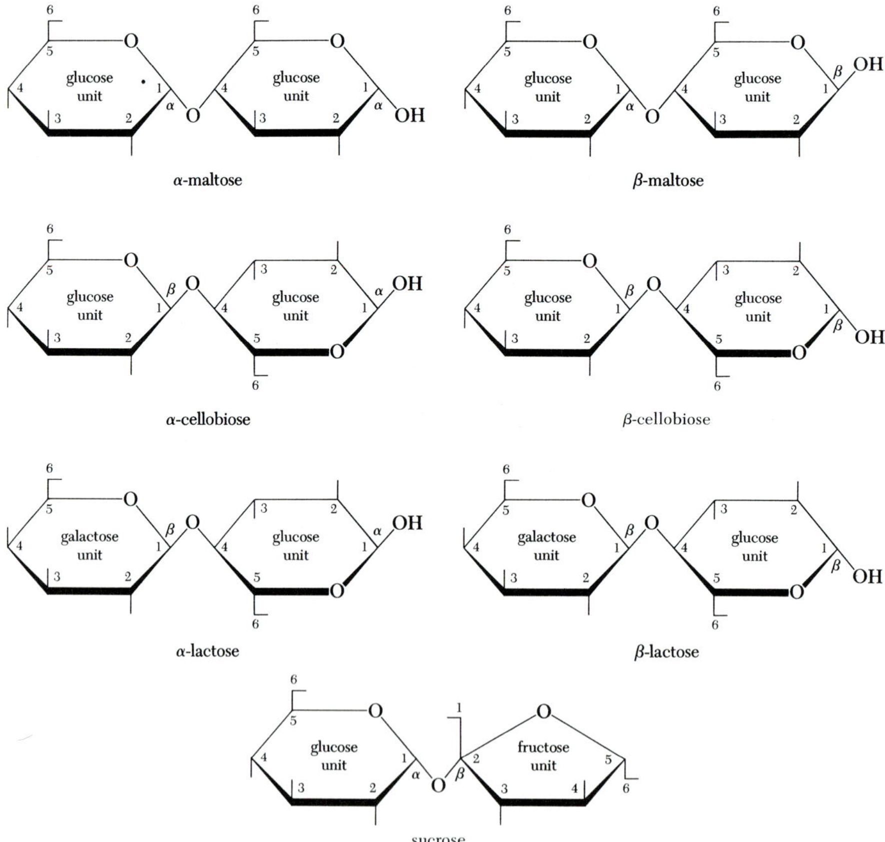

Each disaccharide molecule consists of two cyclic monosaccharide units linked together through an oxygen bridge (Figure 15-5). In maltose and in cellobiose both units are glucose units, with the linking oxygen connected to carbon-1 of one ring and carbon-4 of the other ring. The only structural difference between these two disaccharides is that maltose has α linkages and cellobiose has β linkages. This means that in maltose the linking oxygen is oriented "downward" as opposed to the "upward" orientation of carbon-6 of the left ring, while in cellobiose the linking oxygen is oriented "upward," i.e., in the same direction as carbon-6 of the left ring. In lactose there is a β linkage between carbon-1 of the galactose ring and carbon-4 of the glucose ring. In sucrose the linkage is between carbon-1 of the glucose ring and carbon-2 of the fructose ring; the glucose unit is α-linked and the fructose unit is β-linked.

15-7 Polysaccharides

Much of the glucose produced by photosynthesis in green plants is converted into *starch* and *cellulose*. Starch is stored as reserve food in the roots, tubers, and seeds of plants. Cellulose is the structural material of which the fibrous tissue in the cell walls of plants is composed. Cotton, which is a seed hair, consists of cellulose in nearly pure form. Flax, jute, hemp, and wood also are sources of cellulose fibers that are used as raw materials in the paper and textile industries. The chief use of starch is as food. Commercial sources of starch include corn, wheat, rice, potatoes, cassava, and sago.

Starch and cellulose are *glucosans*, i.e., polymers of glucose. Both consist of long chains of glucose units linked together (Figure 15-6). In starch the glucose units are α-linked, so that maltose is obtained by its incomplete hydrolysis. In cellulose the glucose units are β-linked, so that cellobiose is the product of incomplete hydrolysis. Complete hydrolysis of either starch or cellulose gives glucose as the only product. Starch may be regarded as a polymer of maltose, and cellulose as a polymer of cellobiose. The same idea can be expressed by characterizing starch as a polymer made up of α-glucose units and cellulose as a polymer composed of β-glucose units.

Starch is composed of both branched and unbranched chains of α-glucose units. The unbranched α-glucose polymer is called *amylose*; the branched polymer is called *amylopectin*. In amylopectin there are both 1,4 and 1,6 linkages, which means that some linkages are between carbon-1 of one ring and carbon-4 of the other and other linkages are between carbon-1 of one ring and carbon-6 of the other. Molecular weight determinations indicate that there are 200 or more α-glucose

FIGURE 15-6

The structural features of starch and cellulose

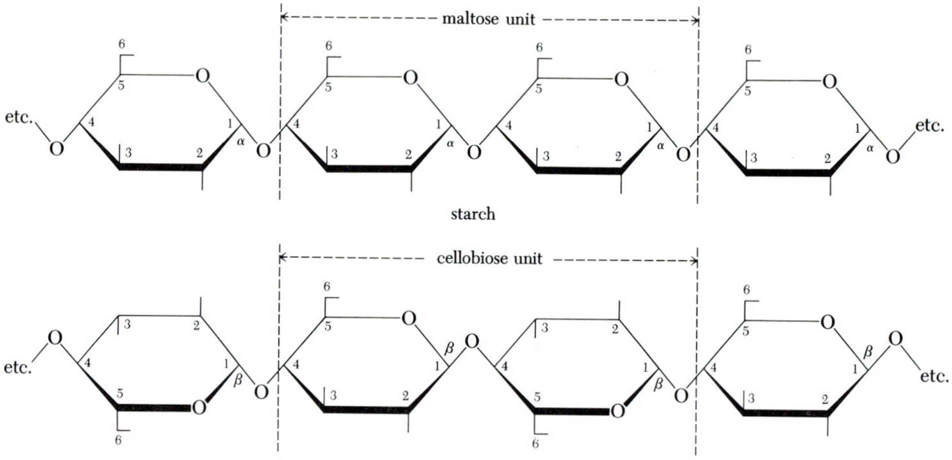

maltose unit

starch

cellobiose unit

cellulose

units in each amylose chain and at least 1000 units per molecule of amylopectin.

Cellulose molecules contain several thousand β-glucose units in unbranched chains. The long chains lie side by side, held together by hydrogen bonds between neighboring OH groups. These bundles of parallel molecules are twisted into ropelike structures. The rigidity of wood results from the fact that the cellulose "ropes" are embedded in lignin, a complex polymeric resin whose structure has not been definitely established. In the manufacture of paper from wood it is necessary to remove the lignin by dissolving it with such reagents as calcium bisulfite, $Ca(HSO_3)_2$.

Glycogen, often called *animal starch*, is the form in which glucose is stored in the animal body. It is readily converted into glucose if the blood sugar level falls below normal. The structure of glycogen is very similar to that of amylopectin, except that it is even more highly branched and has a greater molecular weight.

15-8 Carbohydrates in Food

The distinguishing characteristic of green plants is their ability to synthesize their own components from simple substances such as carbon dioxide and water. The synthesis is called *photosynthesis* because it utilizes energy from sunlight. It occurs in the leaves in the presence of the green pigment called *chlorophyll*.

One of the most important products of photosynthesis is glucose. Its formation can be represented by the equation

$$6CO_2 + 6H_2O \xrightarrow{\text{light}} C_6H_{12}O_6 + 6O_2 \qquad \Delta H = 673 \text{ kcal}$$

The equation does not represent the mechanism of the reaction, which is not yet fully understood but is known to involve many steps.

Carbohydrates in the food of animals are oxidized to carbon dioxide and water—or stored as glycogen or fat for later oxidation—with the release of energy. The body maintains its temperature by the steady production of heat, and it also needs energy for moving muscles, transmitting messages through the nervous system, etc.

In accordance with the Law of Conservation of Energy, the complete combustion of 1 mole of glucose liberates the same amount of energy as was entrapped from the energy of sunlight during its formation:

$$C_6H_{12}O_6 + 6O_2 \longrightarrow 6CO_2 + 6H_2O \qquad \Delta H = -673 \text{ kcal}$$

The body does not subject glucose to combustion, but the glucose undergoes oxidation in a series of many steps; CO_2 and H_2O are the ultimate products, and 673 kcal per mole of glucose is ultimately released.

Glucose taken in as food is absorbed directly from the small intestine into the bloodstream, where its normal concentration is about 1 mg per ml. Starch taken in as food is acted upon in the mouth by ptyalin, a salivary enzyme that hydrolyzes starch to maltose. There is little time for ptyalin to act on starch in the mouth, but its action persists in the stomach until the swallowed food masses are slowly penetrated by the gastric juice. There are no starch-digesting enzymes in the stomach, and ptyalin is inactivated by the hydrochloric acid in the gastric juice. However, starch digestion is resumed in the small intestine where the enzyme amylopsin is provided by the pancreatic juice to complete the hydrolysis of starch to maltose. Among the enzymes present in the intestinal juice are sucrase, lactase, and maltase, which hydrolyze sucrose, lactose, and maltose, respectively. The main source of sucrose is cane sugar or beet sugar in our food; lactose is present in milk; maltose is formed by the partial digestion of starch (discussed above) and is present in certain foods.

Cellulose is not digested by higher animals, because there are no enzymes present in their systems that are capable of hydrolyzing the β-linkages of the cellulose structure. The enzymes that hydrolyze the α-linkages of starch are without effect on the β-linkages of cellulose. However, ruminants—sheep, cattle, goats, deer, and other cud-chewing animals—are able to make use of the cellulose in their diet. They swallow their food without thoroughly chewing it, and it enters the first stomach cavity, the rumen, which contains microorganisms that produce cellulases, enzymes capable of hydrolyzing cellulose. Later, while resting, the animals regurgitate the food mass (the cud) and chew it, bringing salivary enzymes into contact with the starch components of the food. When the chewed food is swallowed, it enters another stomach cavity that contains acid gastric juice.

The alimentary tracts of termites also contain microorganisms that produce cellulases, which enable them to digest wood. This is why termites are so destructive.

The monosaccharides (glucose, galactose, and fructose) formed by the digestion of disaccharides are absorbed through the intestinal walls into the bloodstream and carried to the liver. Here fructose and galactose are converted into glucose, which is converted into glycogen for storage. If this does not suffice to keep the blood sugar level down to normal, glucose may be converted into fat and stored in the fat depots of the body. During fasting the liver breaks down glycogen to supply glucose to the blood; if necessary for maintaining the blood sugar level, glucose will be formed from the stored fats—or even from the amino acids formed by the digestion of proteins (see Section 15-16).

The ultimate oxidation of glucose in the cells of the body involves complicated metabolic reactions, each step regulated by specific enzymes. In several of these steps a so-called high-energy compound is one of the products. Biochemical high-energy compounds are usually complex phosphates that release large amounts of energy when subsequently hydrolyzed. In carbohydrate metabolism, one of the most important substances in this category is the compound adenosine triphosphate (ATP). If we let R represent the organic portion of the ATP molecule, we can write the following equation to represent the hydrolysis of ATP, forming adenosine diphosphate (ADP):

$$\begin{aligned} &RO{-}POOH{-}O{-}POOH{-}O{-}PO(OH)_2 + H_2O \longrightarrow \\ &\qquad RO{-}POOH{-}O{-}PO(OH)_2 + H_3PO_4 \\ &\qquad \Delta G = -8.0 \text{ kcal} \end{aligned}$$

This reaction is induced by enzyme action and is usually coupled with another reaction that could not otherwise take place because of unfavorable energy relationships; for example, the synthesis of a protein.

The release of utilizable energy through the biochemical oxidation of glucose is coupled to the reconversion of ADP to ATP. Thus, ATP is related to the biological "machine" much as electricity is related to an electric motor.

The ADP produced by the hydrolysis of ATP may itself undergo hydrolysis, forming adenylic acid and releasing additional energy:

$$\begin{aligned} &RO{-}POOH{-}O{-}PO(OH)_2 + H_2O \longrightarrow RO{-}PO(OH)_2 + H_3PO_4 \\ &\qquad \Delta G = -6.5 \text{ kcal} \end{aligned}$$

Adenylic acid, also known as adenosine monophosphate (AMP), is given further attention in Section 15-20.

LIPIDS

15-9 Types of Lipids

Lipids comprise the fats and a variety of other substances whose properties are similar in some degree to those of fats. Among these properties are solubility in the "fat solvents" such as acetone, alcohol, ether, and chloroform; insolubility in water; and a greasy feel. Lipids include, among others, the following types of substances:

1. Waxes: long-chain esters of fatty acids with monohydric alcohols.
2. Fats: esters of fatty acids with glycerol.
3. Phospholipids: substituted fats containing phosphorus and nitrogen.
4. Glycolipids: compounds containing both carbohydrate and lipid portions.
5. Lipoproteins: complex substances in which lipids are associated with proteins.

15-10 Waxes

Waxes are esters derived from long-chain monohydric alcohols and long-chain fatty acids. Both the parent acids and the parent alcohols of the natural waxes have 16 or more carbon atoms. Actually, any natural wax is a mixture of different esters; other types of compounds, such as long-chain alcohols and long-chain hydrocarbons, are frequently present also. Beeswax, the material from which bees build honeycomb, is chiefly myricyl palmitate, $C_{15}H_{31}COOC_{30}H_{61}$.

Commercially, the most important natural wax is carnauba wax, which occurs as a coating on the leaves of the Brazilian wax palm. Because of its hardness and its imperviousness to water, it is used in floor polishes, automobile polishes, and carbon paper coatings. Its chief ingredients are esters of C_{24} and C_{28} acids with C_{32} and C_{34} alcohols. Also present are considerable quantities of polymeric esters of ω-hydroxy acids, $HO(CH_2)_nCOOH$, where n may be any number from 17 to 29. (Omega, ω, the last letter of the Greek alphabet, is used to indicate that the OH group is attached to the carbon atom furthest removed from the COOH group.) The unique physical properties of carnauba wax are probably due to the presence of high molecular weight esters of this type.

15-11 Fats and Oils

One of the important classes of foodstuffs is the group of substances known as the fats. These organic compounds occur naturally in the bodies of animals and in the fruits and seeds of plants. A fat that is liquid at ordinary temperatures is called an *oil*.

The fats in our diet are obtained from such sources as cream, butter, lard, tallow, corn oil, and cottonseed oil. In addition to their importance in foods, some substances of this type are used as drying oils in paints and varnishes (for example, linseed oil). Of great economic significance is the utilization of fats and oils in the manufacture of soaps and synthetic detergents.

The vegetable oils that are produced in greatest quantity (more than 1 billion pounds a year) are coconut oil, cottonseed oil, soybean oil, linseed oil, olive oil, palm oil, and peanut oil. Other vegetable oils produced in large quantity (more than 200 million pounds a year) are corn oil, sesame oil, castor oil, and sunflower seed oil.

The fats and oils used as foods are esters of monocarboxylic acids. All are derived from the same trihydric alcohol, glycerol, and are known as *glycerides*. All edible oils are of this type and are not to be confused with the hydrocarbon fuel oils and lubricating oils that are obtained from petroleum.

Differences between different fats are due to the fact that they are derived from different acids. With very few exceptions the acids are straight-chain compounds. Only those containing an even number of carbon atoms occur in significant amounts. Some of the acids are saturated; others contain one or more double bonds.

The most important saturated acids occurring in fats are the following:

$CH_3(CH_2)_{10}COOH$ lauric acid
$CH_3(CH_2)_{12}COOH$ myristic acid
$CH_3(CH_2)_{14}COOH$ palmitic acid
$CH_3(CH_2)_{16}COOH$ stearic acid

The most important unsaturated acids have 18 carbon atoms, with a double bond located at the middle of the chain. If other double bonds are present they are more distant from the carboxyl group:

$CH_3(CH_2)_7CH{=}CH(CH_2)_7COOH$ oleic acid

$CH_3(CH_2)_4CH{=}CHCH_2CH{=}CH(CH_2)_7COOH$ linoleic acid

$CH_3CH_2CH{=}CHCH_2CH{=}CHCH_2CH{=}CH(CH_2)_7COOH$ linolenic acid

Since the glycerol molecule contains three OH groups, it can be esterified with either one, two, or three molecules of acid:

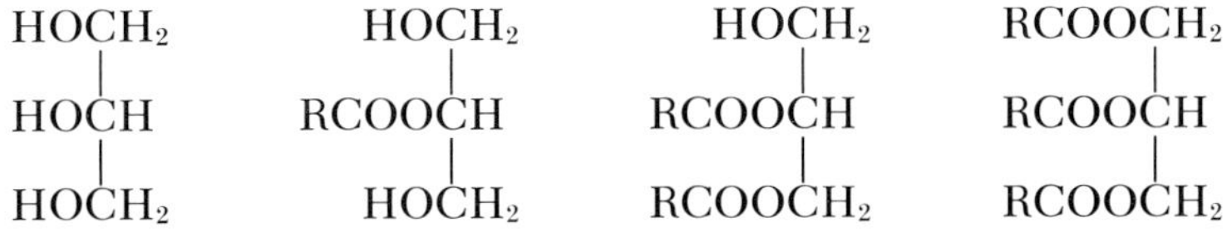

$$\begin{array}{cccc} HOCH_2 & HOCH_2 & HOCH_2 & RCOOCH_2 \\ | & | & | & | \\ HOCH & RCOOCH & RCOOCH & RCOOCH \\ | & | & | & | \\ HOCH_2 & HOCH_2 & RCOOCH_2 & RCOOCH_2 \\ \text{glycerol} & \text{a monoglyceride} & \text{a diglyceride} & \text{a triglyceride} \end{array}$$

The fats and oils are *triglycerides. Monoglycerides* and *diglycerides*, obtained by reacting triglycerides with glycerol, are useful as emulsifying agents in oleomargarine and in cosmetic creams.

A *simple triglyceride* is one in which all three OH groups have been esterified with the same acid. Examples are glyceryl tristearate (stearin) and glyceryl trioleate (olein):

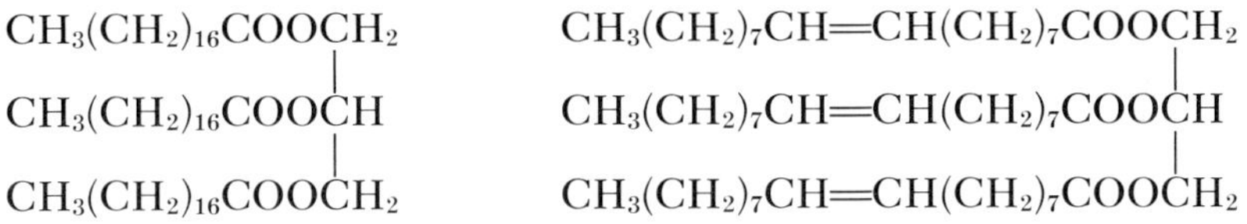

$$\begin{array}{cc} CH_3(CH_2)_{16}COOCH_2 & CH_3(CH_2)_7CH{=}CH(CH_2)_7COOCH_2 \\ | & | \\ CH_3(CH_2)_{16}COOCH & CH_3(CH_2)_7CH{=}CH(CH_2)_7COOCH \\ | & | \\ CH_3(CH_2)_{16}COOCH_2 & CH_3(CH_2)_7CH{=}CH(CH_2)_7COOCH_2 \\ \text{stearin} & \text{olein} \end{array}$$

Historically, simple triglycerides were given trivial names based on their sources, for example, *stearin* (Greek: *stear*, tallow). Common names of fatty acids were derived from the names of the fats, for example, stearic acid.

A *mixed triglyceride* is one in which the OH groups of the glycerol molecule have been esterified with two or three different acids.

All fats are mixtures of different triglycerides. Furthermore, most of the triglyceride molecules in a fat are of the mixed glyceride type. The different RCOO groups in a fat appear to be distributed at random. Unless more than two-thirds of the RCOO groups are of one kind, simple glycerides are not present in quantity.

As esters, the glycerides can be hydrolyzed to give the parent acids and the parent alcohol (glycerol):

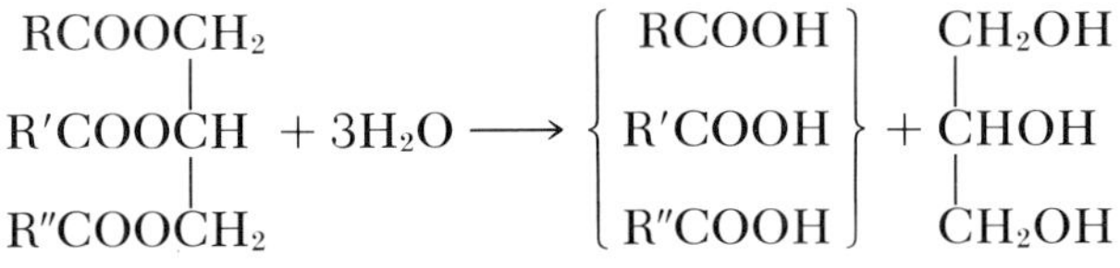

$$\begin{array}{c} RCOOCH_2 \\ | \\ R'COOCH \\ | \\ R''COOCH_2 \\ \text{a mixed glyceride} \end{array} + 3H_2O \longrightarrow \begin{array}{c} \left\{ \begin{array}{c} RCOOH \\ R'COOH \\ R''COOH \end{array} \right\} \\ \text{fatty acids} \end{array} + \begin{array}{c} CH_2OH \\ | \\ CHOH \\ | \\ CH_2OH \\ \text{glycerol} \end{array}$$

To occur at a substantial rate, this reaction must be catalyzed with acid, alkali, or certain digestive enzymes called lipases. With alkali, the metal salts of the acids are obtained. The reaction of alkali with any ester is called *saponification* (Section 14-10). The following is an example of saponification of a glyceride:

$$\begin{array}{l} CH_3(CH_2)_{16}COOCH_2 \\ \quad\quad\quad\quad\quad\quad | \\ CH_3(CH_2)_{16}COOCH \\ \quad\quad\quad\quad\quad\quad | \\ CH_3(CH_2)_{16}COOCH_2 \\ \text{stearin (a fat)} \end{array} + 3Na^+OH^- \longrightarrow \underset{\substack{\text{sodium stearate} \\ \text{(a soap)}}}{3C_{17}H_{35}COO^-Na^+} + \begin{array}{c} CH_2OH \\ | \\ CHOH \\ | \\ CH_2OH \\ \text{glycerol} \end{array}$$

15-12 Chemical Composition and the Properties of Fats

The melting points of the saturated straight-chain monocarboxylic acids increase with chain length from −8°C for butyric acid (four carbon atoms) to 70°C for stearic acid (18 carbon atoms). Double bonds tend to lower the melting point. Thus, oleic acid (18 carbon atoms, one double bond) melts at 14°C; linoleic acid (18 carbon atoms, two double bonds) melts at −5°C; and the melting point of linolenic acid (18 carbon atoms, three double bonds) is −111°C.

The effects of double bonds on the melting points of the fatty acids are reflected in the melting points of the glycerides formed from them. In general, unsaturated acid chains predominate in the vegetable fats and saturated groups in animal fats. The melting points of glycerides are also affected by molecular weight. Butter contains only small amounts of unsaturated acid groups, but it is nevertheless a low-melting fat. This is due in part to the low molecular weights of some of the acids from which butter fats are formed. Butter softens gradually over a range of temperature because it contains a large number of different glycerides.

Coconut oil and palm oil are examples of fats that are referred to as oils despite the fact that they are solids in the temperate zones. They are liquid in the tropics where they are obtained.

The composition and properties of animal fats are affected by the diet of the animals. As an example, lard from hogs fed on corn has a higher melting point than lard from hogs fed on peanuts. However, beef tallow has a higher melting point than hog lard, regardless of the diet of the animals. The first stomach (rumen) of a cud-chewing animal (ruminant) has an enzyme that reduces the double bonds of unsaturated fats to single bonds.

The hydrogenation of oils is a large-scale technical operation carried out for the purpose of raising the melting point. An example is the conversion of olein to stearin:

$$
\begin{array}{l}
CH_3(CH_2)_7CH{=}CH(CH_2)_7COOCH_2 \\
\qquad\qquad\qquad\qquad\qquad\quad | \\
CH_3(CH_2)_7CH{=}CH(CH_2)_7COOCH \\
\qquad\qquad\qquad\qquad\qquad\quad | \\
CH_3(CH_2)_7CH{=}CH(CH_2)_7COOCH_2 \\
\qquad\qquad\quad \text{olein}
\end{array}
+ 3H_2 \xrightarrow[\text{catalyst}]{\text{nickel}}
\begin{array}{l}
CH_3(CH_2)_{16}COOCH_2 \\
\qquad\qquad\qquad\quad | \\
CH_3(CH_2)_{16}COOCH \\
\qquad\qquad\qquad\quad | \\
CH_3(CH_2)_{16}COOCH_2 \\
\qquad\quad \text{stearin}
\end{array}
$$

Hydrogen gas is bubbled through the hot oil containing finely divided nickel in suspension. Any desired consistency in the product may be obtained by controlling the amount of hydrogen added. It is usually desirable to hydrogenate only part of the double bonds.

The original reason for converting vegetable oils to solid fats, a process often referred to as the "hardening" of fats, lay in the unsuitability of the natural oils for use as less expensive substitutes for butter and lard. Since 1953 the production of oleomargarine has exceeded that of butter. Also, solid fats are preferred for making soap.

Glycerides of fatty acids containing two or more double bonds are known as *drying oils* because of the polymerization they undergo in the presence of oxygen, forming a tough film. Examples of natural products containing such glycerides are linseed oil and tung oil, which are important constituents of paints and varnishes. In the "drying" of an oil-base paint, the evaporation of a volatile solvent (such as turpentine) is only a minor part of the process. More important is the chemical reaction in which the unsaturated oil polymerizes to form a protective coating over the painted surface. Neither the polymerization process nor the structure of the waterproof film is well understood.

Waste or rags containing unsaturated oils are a fire hazard unless air is excluded or enough ventilation is provided to remove the heat generated by the oxidation of the oils. If the temperature rises, the rate of oxidation increases. This in turn causes a further rise in temperature. When the kindling temperature is reached, the material bursts into flame. Such *spontaneous combustion* can be prevented by spreading oily rags out flat or by confining oily waste within an airproof container.

15-13 Fats in Food

When oxidized to carbon dioxide and water, fats produce more than twice as much energy as an equal weight of carbohydrate or protein (9.5 kcal per g compared to 4 kcal per g). When stored in the body, fats serve as thermal insulators, and they protect the body and its vital organs against mechanical injury from blows and falls.

Fats in their ordinary form consist of large particles that are insoluble in water. In this form they cannot be absorbed through the intestinal walls. They are acted upon in the intestine by fat-splitting enzymes (lipases), which partially hydrolyze them, forming diglycerides and monoglycerides, which act as emulsifying agents upon the unhydrolyzed fat. Emulsification is also brought about by the bile salts secreted by the liver.

Only about one-third of the fat is broken down by the lipases; the remainder, upon emulsification, is absorbed from the intestine as unchanged triglyceride. The emulsified fats are of sufficiently small particle size to be able to pass directly through the intestinal wall into the lymphatic system. From there they enter the general circulation through the thoracic duct and are transported throughout the body. Some are oxidized for energy; some are stored in fat depots from which they may be taken when required for future use; and some are converted into phospholipids, glycolipids, lipoproteins, etc.

The fats in the diet must supply either linoleic or linolenic acid to maintain a healthy condition of the skin. It is also claimed that inclusion of unsaturated glycerides in the diet lowers the cholesterol content of the blood. There has been much interest in recent years in the replacement of saturated fats in the diet by corn oil and safflower oil because of the apparent correlation between high cholesterol content of the blood and the incidence of heart disease and hardening of the arteries.

With the exception of the unsaturated fatty acids mentioned above, fats are not essential in the diet. It is well known that starches, when present in the diet in excess of the needed amounts, are converted into fats by the animal body. The fat synthesis takes place inside the cells, where glucose is degraded into active two-carbon units used to make fatty acids. By a different type of degradation, glucose gives rise to glycerol. Esterification of the fatty acids with the glycerol produces fats, which are transported to fat depots and stored like fats contained in the diet.

15-14 Phospholipids, Glycolipids, and Lipoproteins

The phospholipids, also known as phosphatides, are present in all living cells, both plant and animal. They are abundant in the brain and spinal cord, in eggs, and in soybeans. Large quantities of soybean lecithin are used to stabilize emulsified fats, as in oleomargarine and mayonnaise.

The dry matter of the human brain is about half protein and half lipid. The most abundant of the lipids present are phospholipids, glycolipids, and cholesterol.

The simplest of the phospholipids are esters of the phosphatidic acids, which are themselves fatty acid esters of glycerophosphoric acid.

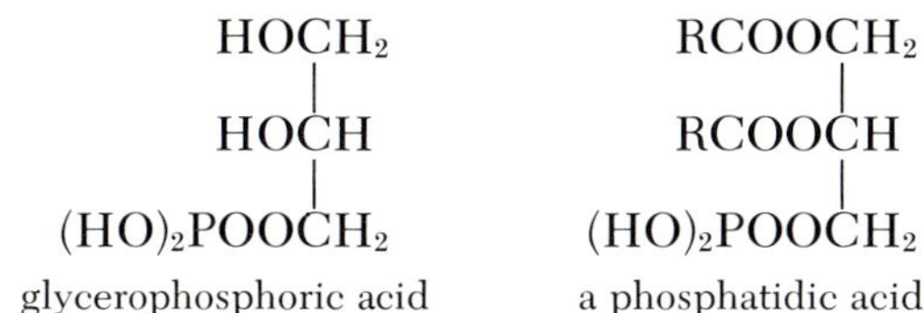

Glycolipids are compounds formed between sugars and lipids. They are often called cerebrosides because they occur in the brain and nervous tissue. The sugar unit is usually galactose.

Lipoproteins are complex substances consisting of protein linked with lipid. They occur in blood serum, brain tissue, eggs, etc. Much of the lipid in blood plasma is transported in the form of loose lipoprotein complexes.

PROTEINS

15-15 Amines and Amino Acids

Proteins are essential constituents of all living cells. Each type of cell makes its own kinds of proteins. Nerve and muscle tissue, skin, hair, nails, feathers, wool, horns, and hoofs are all composed of proteins. Hemoglobin, antibodies, viruses, enzymes, and many of the hormones are complex proteins.

Even the simplest of the proteins are far more complicated than sugars or fats. The molecular weights of proteins vary from several thousand to hundreds of thousands, depending on the source. Hydrolysis of proteins yields compounds called amino acids, which are the building blocks of protein molecules.

Amino acids are bifunctional compounds containing both amino and carboxyl groups. The *amino group*, NH_2, is the group present in *primary amines*, which are formally related to ammonia in the same way that alcohols are formally related to water.

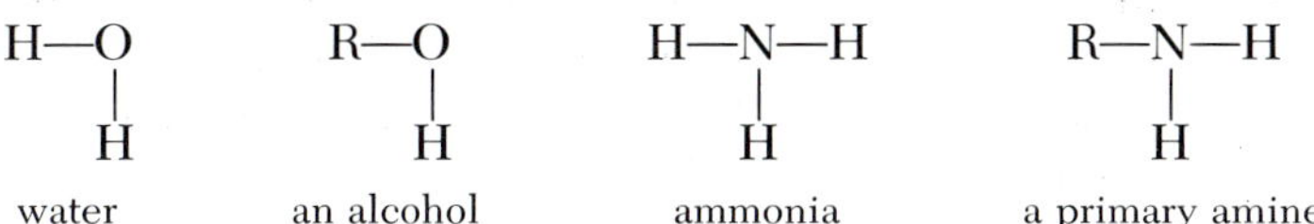

The analogy of methylamine to methyl alcohol, of ethylamine to ethyl alcohol, etc. is apparent from comparison of their formulas:

Note that the name for an amine is written as one word.

Secondary amines contain two alkyl groups, and in this respect they are analogous to ethers. Since the term *amine* is retained in naming these compounds, it is mandatory that the prefix *di-* be used if the amine contains two identical alkyl groups:

$CH_3{-}O{-}CH_3$	$CH_3{-}\underset{}{\overset{H}{\overset{\vert}{N}}}{-}CH_3$
methyl ether	dimethylamine
$CH_3{-}O{-}C_2H_5$	$CH_3{-}\overset{H}{\overset{\vert}{N}}{-}C_2H_5$
methyl ethyl ether	methylethylamine

Compounds in which a nitrogen atom is bonded to three alkyl groups are called *tertiary amines*. Examples include $(CH_3)_3N$, trimethylamine, and $(CH_3)_2NC_2H_5$, dimethylethylamine.

Like ammonia, all amines of low molecular weight are soluble in water and in most organic solvents. Molecules of primary and secondary amines are associated through hydrogen bonding. In their aqueous solutions there is hydrogen bonding between amine molecules and water molecules.

An aqueous solution of an aliphatic amine is somewhat more alkaline than an aqueous solution of ammonia. Neutralization of this solution with an acid produces a substituted ammonium salt. Just as $NH_4^+Cl^-$ is ammonium chloride, so $CH_3NH_3^+Cl^-$ is methylammonium chloride.

The simplest conceivable amino acid is carbamic acid, H_2NCOOH; but it, like carbonic acid, $CO(OH)_2$, is unstable. The simplest stable amino acid is aminoacetic acid,

$$\underset{\displaystyle NH_2}{\underset{\vert}{CH_2}}{-}COOH$$

which is usually called *glycine*. Because it contains both a basic group and an acidic group, glycine is an *amphoteric electrolyte*: it can act either as a base (proton acceptor) or as an acid (proton donor). If glycine is in aqueous solution, the most abundant species present is the one that results from the transfer of the carboxyl proton to the amino group,

$$\begin{array}{l} CH_2\text{—}COO^- \\ | \\ NH_3^+ \end{array}$$

The species resulting from this proton transfer is called a *zwitterion* or a *dipolar ion*. It is not really an ion because its net charge is zero. It is actually a highly polar molecule; the charges are written to emphasize its polar character.

In contrast to amines and carboxylic acids, which have low boiling points, the amino acids are nonvolatile crystalline solids with fairly high melting points. Furthermore, they are insoluble in nonpolar solvents, such as benzene or ether. In brief, their properties are the properties characteristic of salts. Having a zwitterion structure in the pure state, they may be regarded as *inner salts*. This simply means that viewed from one end, an amino acid zwitterion appears to be a substituted ammonium ion, while viewed from the other end, it has the appearance of a substituted acetate ion.

Since they are amphoteric electrolytes, amino acids exist in aqueous solution as equilibrium mixtures whose compositions depend upon the pH of the solution. If sufficient alkali is present to establish a pH of about 11, amino acids are present predominantly in the form of negative ions. In the case of glycine, the negative ion is the ion

$$\begin{array}{l} CH_2\text{—}COO^- \\ | \\ NH_2 \end{array}$$

If sufficient acid is present to establish a pH of about 1, the predominant species is a positive ion. In the case of glycine this is the ion

$$\begin{array}{l} CH_2\text{—}COOH \\ | \\ NH_3^+ \end{array}$$

15-16 The Peptide Linkage

The most common source of amino acids is from the hydrolysis of proteins. All the amino acids obtained from this source are α-amino acids, that is, the NH_2 group is attached to the same carbon as the COOH group.

FIGURE 15-7

Some common amino acids derived from proteins

$$\underset{\text{glycine}}{\underset{|\atop NH_2}{CH_2}-COOH} \qquad \underset{\text{alanine}}{CH_3-\underset{|\atop NH_2}{CH}-COOH} \qquad \underset{\text{leucine}}{CH_3-\underset{|\atop CH_3}{CH}-CH_2-\underset{|\atop NH_2}{CH}-COOH}$$

$$\underset{\text{tyrosine}}{HO-C_6H_4-CH_2-\underset{|\atop NH_2}{CH}-COOH} \qquad \underset{\text{aspartic acid}}{HOOC-CH_2-\underset{|\atop NH_2}{CH}-COOH}$$

$$\underset{\text{lysine}}{\underset{|\atop NH_2}{CH_2}-CH_2-CH_2-CH_2-\underset{|\atop NH_2}{CH}-COOH} \qquad \underset{\text{cystine}}{HOOC-\underset{|\atop NH_2}{CH}-CH_2-S-S-CH_2-\underset{|\atop NH_2}{CH}-COOH}$$

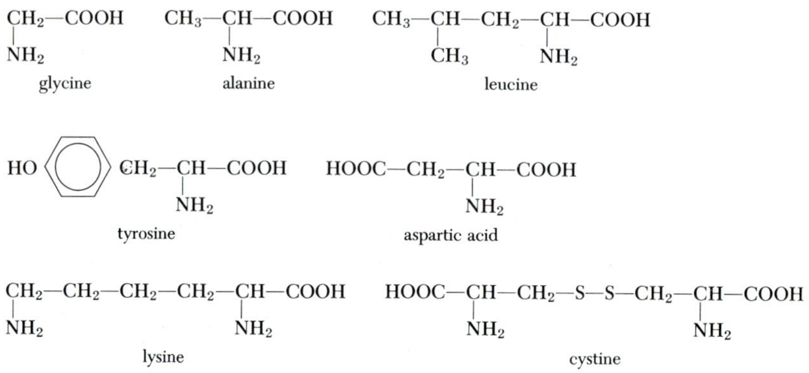

Thus, the general formula for α-amino acids is $R \cdot CH(NH_2) \cdot COOH$, where R is an alkyl or substituted alkyl group (or a H atom in the case of glycine). The structures of some common amino acids derived from proteins are given in Figure 15-7.

A total of 26 α-amino acids are known to occur in proteins. The so-called "neutral" amino acids are those that have an equal number of basic NH_2 groups and acidic COOH groups, for example, leucine and cystine. The so-called "acidic" amino acids have an additional acidic group, for example, aspartic acid. The so-called "basic" amino acids have an additional basic group, as in lysine.

The reaction between the amino group of an amino acid molecule and the carboxyl group of another amino acid molecule forms a type of compound known as a *peptide*. All peptides contain the group

$$-\overset{O\atop \|}{C}-\overset{H\atop |}{N}-$$

which is known as a *peptide linkage*. The reaction between two glycine molecules forms the *dipeptide* called glycylglycine:

$$\underset{\text{glycine}}{H_2N-CH_2-COOH} + \underset{\text{glycine}}{H_2N-CH_2-COOH} \longrightarrow$$

$$\underset{\text{glycylglycine}}{H_2N-CH_2-CO-NH-CH_2-COOH} + H_2O$$

Similarly, the reaction between two alanine molecules forms the dipeptide called alanylalanine:

$$\underset{\text{alanine}}{H_2N-\overset{\displaystyle CH_3}{\overset{|}{C}H}-COOH} + \underset{\text{alanine}}{H_2N-\overset{\displaystyle CH_3}{\overset{|}{C}H}-COOH} \longrightarrow$$

$$\underset{\text{alanylalanine}}{H_2N-\overset{\displaystyle CH_3}{\overset{|}{C}H}-CO-NH-\overset{\displaystyle CH_3}{\overset{|}{C}H}-COOH} + H_2O$$

Two different peptides could be formed by the reaction of a molecule of glycine with a molecule of alanine, depending on whether the amino group of glycine reacts with the carboxyl group of alanine or the amino group of alanine with the carboxyl group of glycine:

$$\underset{\text{alanine}}{H_2N-\overset{\displaystyle CH_3}{\overset{|}{C}H}-COOH} + \underset{\text{glycine}}{H_2N-CH_2-COOH} \longrightarrow$$

$$\underset{\text{alanylglycine}}{H_2N-\overset{\displaystyle CH_3}{\overset{|}{C}H}-CO-NH-CH_2-COOH} + H_2O$$

$$\underset{\text{glycine}}{H_2N-CH_2-COOH} + \underset{\text{alanine}}{H_2N-\overset{\displaystyle CH_3}{\overset{|}{C}H}-COOH} \longrightarrow$$

$$\underset{\text{glycylalanine}}{H_2N-CH_2-CO-NH-\overset{\displaystyle CH_3}{\overset{|}{C}H}-COOH} + H_2O$$

A dipeptide has a terminal NH_2 group capable of reacting with the COOH group of another amino acid molecule. It also has a terminal COOH group capable of reacting with the NH_2 group of another amino acid molecule. If either of these reactions occurs, the result is a *tripeptide*. A peptide containing a large number of amino acid units is called a *polypeptide*. Proteins are polypeptides of high molecular weight.

Hydrolysis of either alanylglycine or glycylalanine gives one molecule of alanine and one molecule of glycine. The first step taken in determining the chemical structure of a protein is to subject it to hydrolysis. Complete hydrolysis reveals how many amino acid units of each type are present in one molecule of the protein, provided the molecular weight of the protein has been measured. No single protein gives all 26 α-amino acids, but each protein gives several different ones. The insulin molecule is one of the smallest and simplest protein molecules known; it contains only 51 amino acid units, but even so there are 17 different kinds of units.

15-17 The Isoelectric Point

When a solution of an amino acid is placed between two oppositely charged electrodes, the amino acid may migrate toward one or the other of the electrodes, depending on whether it is positively or negatively charged. The movement of charged particles in an electric field is called *electrophoresis.* In a solution of low pH, the positive amino acid ions migrate toward the negative electrode. In a solution of high pH the negative ions migrate toward the positive electrode. At a certain intermediate pH an amino acid does not migrate toward either electrode because the amino acid is present almost entirely in the form of zwitterions that have no net charge. The pH at which this occurs is called the *isoelectric point* of the amino acid.

Each amino acid has its own characteristic isoelectric point. This point is not usually the neutral point (pH 7), because this would require that the ability of the compound to act as a proton donor be matched exactly by its ability to act as a proton acceptor. All so-called "neutral" amino acids have isoelectric points in the neighborhood of pH 6 to pH 7. The isoelectric points depend on the basic strengths of the NH_2 groups and the acid strengths of the COOH groups, each of which may be affected by the alkyl or substituted alkyl groups present in the particular acids. The isoelectric point is also affected by the presence of other functional groups that are themselves basic or acidic, such as the phenolic OH group in tyrosine. The isoelectric point tends to occur at a lower pH, of course, for amino acids that have an excess of COOH over NH_2 groups, and it tends to be higher for those that have an excess of NH_2 over COOH groups.

Since individual amino acids have different isoelectric points, they behave differently at any particular pH when a solution containing a mixture of them is placed in an electric field. A mixture of amino acids therefore can be separated by varying the pH of a solution during electrophoresis. In this manner the mixture of amino acids obtained by the hydrolysis of a protein can be analyzed.

Proteins themselves are amphoteric electrolytes, because every protein molecule must have at least one free NH_2 group and at least one free COOH group—one at one end and the other at the other end. In a sense, a protein molecule is simply a giant amino acid molecule. The isoelectric points of proteins vary from one protein to another, depending upon the kinds and relative amounts of amino acids they contain. Consequently, mixtures of proteins can be analyzed by the same electrophoretic techniques that are used for analyzing mixtures of amino acids.

Proteins are least soluble at their isoelectric points. For example, the isoelectric point of casein, the principal protein in cow's milk, is pH 4.7. The pH of cow's milk is ordinarily between pH 6.3 and pH 6.6. When milk sours, the pH approaches the isoelectric point of casein. Consequently, the casein molecules lose their electrical charges and thereby also lose their mutual electrical repulsion. Coagulation, with precipitation, follows. In other words, the milk curdles. The souring of milk is due to bacterial enzymes; lactase hydrolyzes lactose, and other enzymes convert the resulting hexoses to lactic acid:

$$C_6H_{12}O_6 \longrightarrow 2CH_3\text{—}\underset{\displaystyle \text{OH}}{\underset{|}{\text{CH}}}\text{—COOH}$$

Some proteins remain in solution at the isoelectric point but may be precipitated by adding a protein precipitant. Salts used for this purpose include NaCl, $(NH_4)_2SO_4$, and $MgSO_4$. Nonaqueous solvents that are miscible with water but are not solvents for proteins may also be used to aid the precipitation of proteins; examples are alcohol and acetone.

15-18 The Structure of Proteins

The first step taken in determining the structure of a protein is to subject a sample to complete hydrolysis, breaking all peptide linkages. The amino acids formed by the hydrolysis must be individually identified, and the relative amounts of each must be determined by analysis of the mixture. Knowing the molecular weight of each amino acid, we can calculate the number of moles of each acid obtained from the sample. From this result we can compute the relative numbers of the different amino acid units in the protein molecule. In order to compute *how many* amino acid units are present in one protein molecule, we must know the molecular weight of the protein. The molecular weights of proteins are too large to be determined from the lowering of the freezing point, but they have been determined by the measurement of osmotic pressure, rate of diffusion, rate of settling in an ultracentrifuge, and the scattering of light.

The next step is to determine the sequence of the amino acid units in the chain. This is a difficult and laborious task. It involves incomplete hydrolysis and identification of the various small peptides that are thus formed.

Even after the amino acid sequence of the peptide chains is known, the unique character of individual proteins cannot be accounted for completely without proper consideration of the ways in which peptide chains can be arranged to give the most stable conformations. Hydrogen bonding can occur between the hydrogen atom of an —NH— group and the oxygen atom of any —CO— group that it can approach at the proper distance. In many proteins a highly favorable arrangement is a coiled conformation in which the peptide chain forms a helix like a spiral staircase or a corkscrew (Figure 15-8). The helix is coiled in such a way that hydrogen bonding occurs between —NH— hydrogens and —CO— oxygens that are four peptide linkages apart. The protein coils can be arranged relative to one another in a variety of ways; for example, they may be wound into a ball, aligned beside each other in fibrils, or twisted together like the strands of a rope.

The properties of some types of proteins are sometimes changed markedly by agents that do not break any of the peptide linkages but alter the configuration of the chains. Such changes are called *denaturation*. Coagulation often follows denaturation. A striking example of denaturation occurs in the frying of an egg. High temperatures make the molecules vibrate with sufficient violence that hydrogen bonds are disrupted and the molecular configuration of the egg albumin becomes

FIGURE 15-8

A protein helix

Hydrogen bonding between different parts of the same chain holds the helix together.

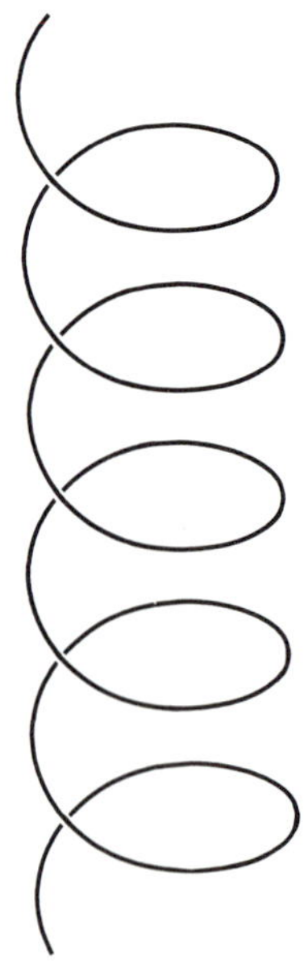

disorganized. Another example of denaturation by heat is the inactivation of enzymes at high temperatures (Section 15-2).

Proteins that yield α-amino acids exclusively when they undergo complete hydrolysis are called *simple proteins.* Others yield, in addition to α-amino acids, one or more substances of another kind and are called *conjugated proteins.* The groups of a nonprotein nature that are present in conjugated proteins are called *prosthetic groups.* In *glycoproteins* the prosthetic group is a carbohydrate; an example is mucin, which makes saliva slippery. In *lipoproteins* the prosthetic group is a lipid; examples include prothrombin and heparin, both important in connection with blood clotting. The most complex of the conjugated proteins are the *nucleoproteins*, which are present in all living cells. In these materials the prosthetic groups are nucleic acids. Nucleoproteins and nucleic acids are discussed in Section 15-20.

15-19 Proteins in Food

When proteins are ingested by an animal organism, they undergo progressive hydrolysis in the digestive tract, giving peptides and amino acids. The hydrolysis in the stomach, where the pH is between 1 and 2, is catalyzed by the enzyme *pepsin*; in the intestines, where the pH is between 6 and 8, the acting enzymes are *chymotrypsins* and *peptidases.*

The amino acids pass through the intestinal walls into the bloodstream, which carries them to the body tissues. There the amino acids are reconstituted into proteins characteristic of the particular tissues. The liver converts amino acids into plasma proteins, which can also be used as a source of amino acids for the synthesis of tissue proteins.

It is remarkable that body chemistry is so well regulated as to enable an organism to synthesize a particular protein for each purpose. This involves selecting the necessary amino acids from the bloodstream and putting them together in the proper sequence.

Organisms are also able to convert some amino acids into others. Furthermore, some of the amino acids can be synthesized in the body from ammonium salts and α-keto acids derived from carbohydrates. The amino acids that can be formed in the body in either of these ways need not be present in the diet of the organism. However, those that cannot be so formed *must* be present in the diet and are therefore called "essential" amino acids. These amino acids are usually supplied in the diet in the form of proteins containing them. Despite their name, they are *not* more important than the others.

By means of studies made with the aid of isotopes, it has been discovered that the proteins of the body are being continuously built up and torn down so that there is a fairly rapid turnover of the amino acids. The average "half-life" of proteins in the body of the adult human has been calculated to be about 80 days. This is the average time required for half of the original amino acid units in the body proteins to be replaced by other amino acid units.

NUCLEIC ACIDS

15-20 Nucleoproteins and Nucleic Acids

Nucleoproteins are present in all living cells. As the name suggests, they were once believed to occur only in the cell nucleus, but it is now known that they are also present in other parts of the cell.

The nucleoproteins are conjugated proteins in which the prosthetic groups are high polymers known as nucleic acids. Much of their biochemical importance is due to their association with the chromosomes of cell nuclei. They are believed to play an important role in the genetic mechanism of life. They are also present in the cytoplasm, where they are involved in the biosynthesis of proteins.

The structures of nucleic acids are studied in much the same way as protein structures, for example, by hydrolysis and by X-ray analysis. The nucleic acids are chemically quite different from proteins, but the two classes of substances resemble each other in consisting of long chains to which side groups are attached.

The progressive hydrolysis of a nucleoprotein occurs in the following steps:

1. Nucleoprotein $\longrightarrow$ nucleic acid + protein
2. Nucleic acid $\longrightarrow$ nucleotides
3. Nucleotide $\longrightarrow$ nucleoside + H_3PO_4
4. Nucleoside $\longrightarrow$ organic base + sugar

FIGURE 15-9

Parent compounds of the purines and pyrimidines

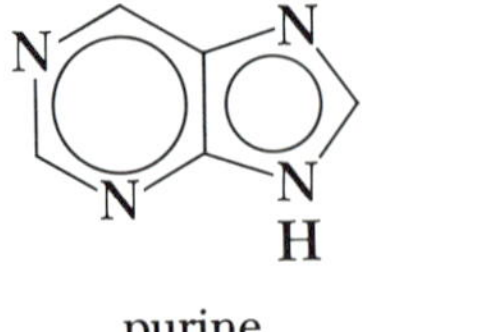

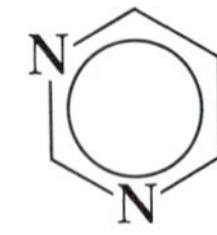

purine
$C_5H_5N_4$

pyrimidine
$C_4H_4N_2$

FIGURE 15-10

Heterocyclic bases present in nucleic acids

adenine

guanine

cytosine

uracil

thymine

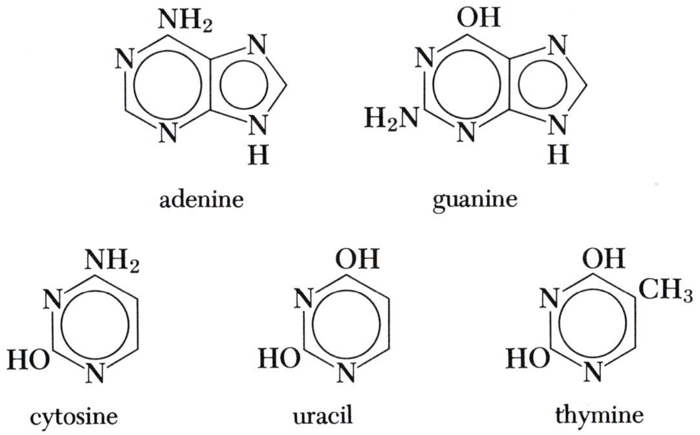

The sugars present in nucleic acids are ribose and deoxyribose. The nucleic acids containing ribose are known as ribonucleic acids (the RNA's), and those containing deoxyribose are called deoxyribonucleic acids (the DNA's). The sugar in a nucleic acid is present as a five-membered ring.

The bases present in nucleic acids are derivatives of purine and pyrimidine (Figure 15-9). They are known as heterocyclic bases because they are cyclic compounds with both carbon and nitrogen atoms present in the rings. The structures of the bases present in nucleic acids are shown in Figure 15-10. The bases present in the RNA's are adenine, guanine, cytosine, and uracil. Those present in the DNA's are adenine, guanine, cytosine, and thymine.

The combination of one molecule of a pentose (ribose or deoxyribose) with one molecule of a heterocyclic base is called a *nucleoside.* In a

FIGURE 15-11

Structure of adenosine

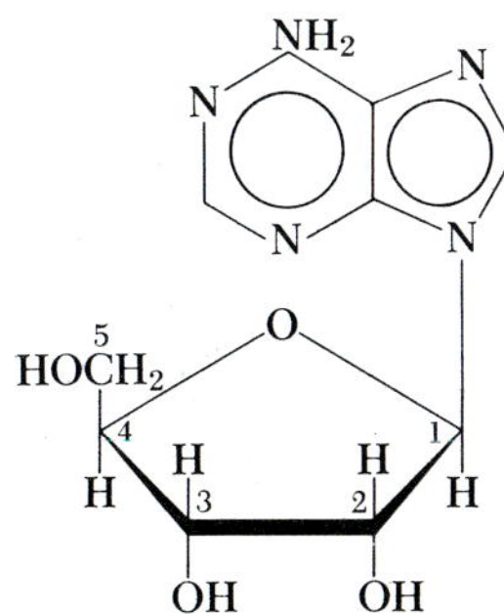

FIGURE 15-12

Structure of adenylic acid

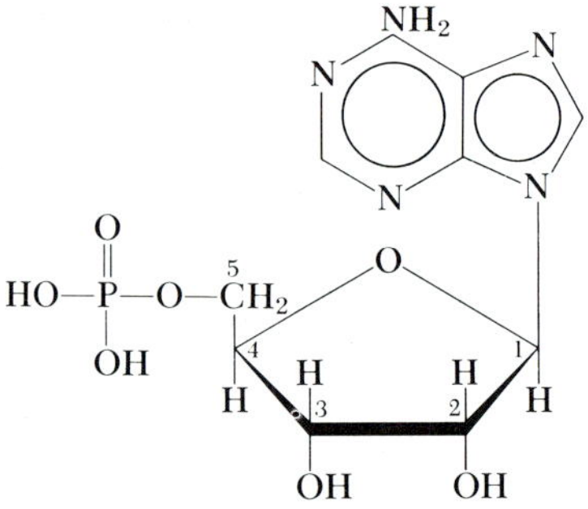

nucleoside, the pentose ring is attached through carbon-1 to a nitrogen atom of the base, as illustrated by the structure of adenosine (Figure 15-11), in which ribose is combined with adenine.

The ester formed between phosphoric acid and an OH group of a nucleoside sugar ring is called a *nucleotide.* In a nucleotide the sugar unit is joined to phosphate through carbon-5, as illustrated by the structure of adenylic acid (Figure 15-12), which is also known as adenosine monophosphate (AMP).

Nucleic acids are polynucleotides. In the same sense that proteins are polymers of amino acids, the nucleic acids are polyesters composed of nucleotide units. The sugar ring of each nucleotide unit is connected to the phosphate of another nucleotide unit through carbon-3, as indicated in Figure 15-13.

FIGURE 15-13

Portion of RNA chain

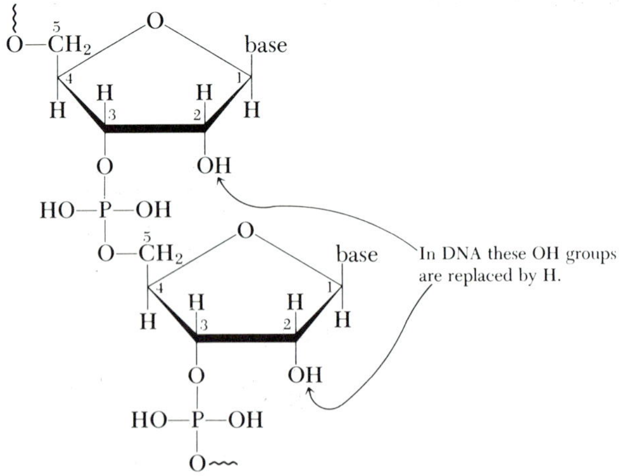

In DNA these OH groups are replaced by H.

The nucleic acids are strong acids because of the phosphoric acid groups present in each nucleotide unit. Under physiological conditions the nucleic acids are associated with basic proteins. The nucleoproteins are essentially saltlike combinations of simple basic proteins with nucleic acids. The transfer of protons from the phosphoric acid groups [illegible] the basic sites of the protein portion of the stru[illegible] mean[illegible] nucleic acids are usually present in anionic form, associated w[illegible] [illegible]teins that are in cationic form.

15-21 The Chemical Basis of Heredity

The role of DNA in living cells has been compared to that of a punched tape used for controlling the operation of an automatic machine. By this is meant that DNA is an information storage device: it supplies the information for cell development, including the synthesis of enzymes and its own self-replication. The basic structural features of DNA are believed to be the same in all organisms—but the DNA of one kind of organism cannot be exactly the same as that of another kind of organism.

The DNA molecules are quite large—their molecular weights run into the millions. X-ray diffraction studies indicate that DNA is made up of two chains twisted around each other to form a double helix (Figure 15-14).

It is a remarkable fact that the number of adenine groups in any sample of DNA is equal to the number of thymine groups and that the number of guanine groups is the same as the number of cytosine groups. This equivalence between purine and pyrimidine bases led James Watson and Francis Crick to postulate that hydrogen bonding between adenine and thymine and between cytosine and guanine holds the two strands together in the double helix. Thus, the two strands of the helix have complementary structures in which the larger purine units and the smaller pyrimidine units are enabled to fit together in a compact arrangement.

The Watson-Crick model suggests a plausible mechanism for the replication of DNA molecules. If stable bonding occurs only between adenine and thymine and between guanine and cytosine, then the sequence of bases in one strand of the helix determines that in the other strand. If, during cell division, the two strands of DNA were to separate, each strand could then serve as a template for the synthesis of a new strand. If the synthesis of the new strand were to start at one end of the template and grow by addition of one nucleotide at a time, a double strand would eventually result. At each step in the synthesis the base of the nucleotide used in building the new strand would have to be able to form hydrogen bonds with the base that lies opposite it in the template—adenine with thymine, and guanine with cytosine. The net result

FIGURE 15-14

The DNA double helix

The helixes head in opposite directions; hydrogen bonding occurs between them.

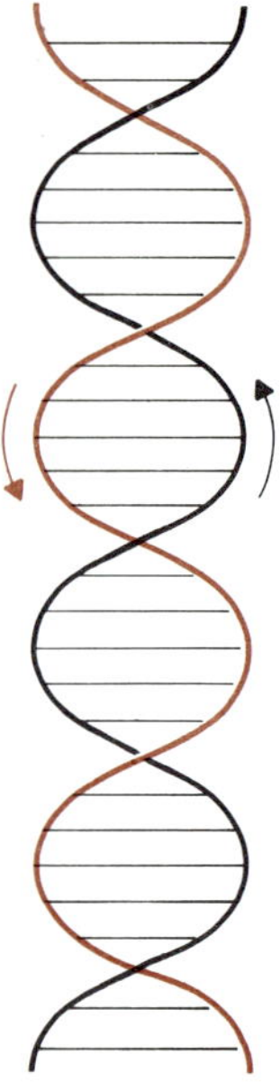

would be a new strand of DNA from each of the original strands. At the start there was one double strand; now there are two. The two original strands are no longer paired with each other—but each is now paired with a new partner that is identical with its original partner. The original DNA molecule has been replaced with two precise copies of itself. This may be the explanation of how cell division produces identical cells.

In connection with the synthesis of proteins within living cells there arises the question of how amino acid units are induced to link together in the sequence that is required for a particular type of protein. It is almost certain that DNA contains the genetic information required for this kind of synthesis. One of the first steps is the synthesis of a molecule of RNA that is complementary to a segment of one strand of DNA. Presumably this requires the temporary separation of a corresponding segment of the double-stranded DNA into two strands; after the complementary strand of RNA has been synthesized, it detaches from the template and the two separated strands of DNA can return to double-stranded form. Thus the coded information of the DNA has been transcribed to the RNA, which aligns the amino acid units for synthesis of a protein molecule in accordance with the "program" built into it by the coded sequence of nucleotide units. The synthesis involves a complex series of reactions, each under the control of a specific enzyme.

NEW TERMS

Aldose: a polyhydroxyaldehyde.

Amine: any compound conforming to any of the following general formulas: R_3N (tertiary amines), R_2NH (secondary amines), or RNH_2 (primary amines).

Amino acids: compounds that contain both NH_2 and COOH groups.

Carbohydrate: a polyhydroxyaldehyde (aldose), polyhydroxyketone (ketose), or any substance that yields such compounds (and only such compounds) upon hydrolysis.

Conjugated proteins: complex proteins that yield, upon hydrolysis, one or more substances of another kind in addition to amino acids.

Denaturation: a marked change in the properties of a protein brought about by an agent that does not break any of the peptide linkages.

Deoxyribonucleic acid (DNA): any nucleic acid containing deoxyribose.

Deoxy sugars: sugars that have H in place of one or more OH groups.

Disaccharide: a carbohydrate that can undergo hydrolysis to give two monosaccharide molecules; for example, sucrose, which hydrolyzes to give glucose and fructose.

Fats (glycerides): monocarboxylic acid esters of glycerol.

Isoelectric point: the pH at which an amino acid or protein does not migrate in an electric field.

Ketose: a polyhydroxyketone.

Lipids: the group of substances that comprise the fats and a variety of other substances whose properties are similar in some degree to those of fats.

Monosaccharide: a carbohydrate that cannot be hydrolyzed to give simpler carbohydrates.

Nucleic acids: polynucleotides, i.e., polyesters composed of nucleotide units. Nucleic acids are the prosthetic groups present in nucleoproteins.

Nucleoproteins: conjugated proteins in which the prosthetic groups are nucleic acids.

Oligosaccharide: a carbohydrate that yields only a few monosaccharide molecules upon hydrolysis.

Peptide: the type of compound formed by the reactions between the amino groups and carboxyl groups of different amino acid molecules.

Peptide linkage: the $-\overset{\overset{\displaystyle O}{\|}}{C}-\overset{\overset{\displaystyle H}{|}}{N}-$ group, which is present in all peptides.

Polypeptide: a peptide containing a large number of amino acid units and peptide linkages. Proteins are polypeptides of high molecular weight.

Polysaccharide: a carbohydrate of high molecular weight. A polysaccharide can be hydrolyzed to give, ultimately, a large number of monosaccharide molecules.

Proteins: the complex combinations of amino acids that are the essential constituents of all living cells.

Ribonucleic acid (RNA): any nucleic acid containing ribose.

Sugars: water-soluble carbohydrates that are more or less sweet.

Waxes: simple esters derived from long-chain monohydric alcohols and long-chain fatty acids.

Zwitterion: a highly polar molecule produced by the transfer of one of its own protons from an acidic site to a basic site within the molecule, as when the proton from the carboxyl group of a glycine molecule transfers to the NH_2 group.

EXERCISES

15-1 What is biochemistry? Tell how the meaning of the term *organic chemistry* has changed from its original meaning.

15-2 Tell how the meaning of the term *carbohydrate* has changed from its original meaning.

15-3 Define the following terms: sugars, monosaccharides, oligosaccharides, and polysaccharides.

15-4 The formula for formaldehyde may be written CH_2O. Does this mean that it is a carbohydrate? Explain your answer.

15-5 Describe the two methods of classifying monosaccharides. What is an aldopentose? A ketohexose? Which monosaccharides are of greatest importance?

15-6 How does galactose differ from glucose? β-Glucose from α-glucose? Fructose from glucose and galactose? Deoxyribose from ribose?

15-7 Maltose and cellobiose give the same hydrolysis products, but they are different sugars. Explain. What are the hydrolysis products of sucrose and lactose?

15-8 What are the sources of sucrose, lactose, maltose, and cellobiose?

15-9 What fundamental structural difference is there between cellulose and starch? Between amylose and amylopectin? How does the structure of glycogen compare with that of starch?

15-10 Ruminants and termites can utilize cellulose as food, but most animals cannot. Explain.

15-11 Give the chemical composition of five types of lipids.

15-12 What is beeswax? Carnauba wax?

15-13 What are fats?

15-14 What are glycerides? How do monoglycerides and diglycerides differ from triglycerides?

15-15 What is the distinction between simple triglycerides and mixed triglycerides?

15-16 Corn oil is nutritious but diesel oil is not. Explain.

15-17 Why are vegetable oils hydrogenated in the food industry? Why has there been a new emphasis in recent years on leaving the vegetable oils partially unsaturated?

15-18 What type of chemical structure is necessary in drying oils?

15-19 Explain why glycerol is a by-product of both the soap and syndet industries.

15-20 What are proteins? What are α-amino acids?

15-21 Explain why the simplest proteins are far more complicated than sugars or fats.

15-22 Define the following: peptide, peptide linkage, dipeptide, tripeptide, polypeptide.

15-23 Explain the existence of amino acids in solution as zwitterions. Explain how the charge of an amino acid in solution may be altered by changing the pH.

15-24 Explain why amino acids are crystalline substances with fairly high melting points.

15-25 Define isoelectric point and itemize some of the factors that affect its value.

15-26 Define electrophoresis and discuss its applications in protein chemistry.

15-27 Define the following: denaturation, conjugated proteins, prosthetic groups, enzyme, nucleoprotein, glycoprotein, and lipoprotein.

15-28 List some factors that affect the rates of enzyme-catalyzed reactions.

15-29 What becomes of the fats, carbohydrates, and proteins that we eat?

15-30 What is a nucleoside? A nucleotide? A nucleic acid? DNA? RNA?

15-31 Which sugars are present in nucleosides? Which bases?

15-32 Why are nucleic acids strong acids?

15-33 What is a double helix?

15-34 Describe a plausible mechanism for the self-replication of DNA molecules.

REFERENCES

Barry, J. M., *Molecular Biology: Genes and the Chemical Control of Living Cells*, Prentice-Hall, Englewood Cliffs, N.J., 1964 (paperbound). Intelligible even to readers who know only the basic facts of chemistry and biology.

Barry, J. M., and Barry, E. M., *An Introduction to the Structure of Biological Molecules*, Prentice-Hall, Englewood Cliffs, N.J., 1969 (paperbound). Discusses the structure of carbohydrates, proteins, nucleic acids, lipids, and other compounds of biological importance.

Bennett, T. P., and Frieden, E., *Modern Topics in Biochemistry*, Macmillan, New York, 1966 (paperbound). A brief text offering an authoritative modern summary for undergraduates in the lower ranks.

Green, D. E., and Goldberger, R. F., *Molecular Insights into the Living Process*, Academic Press, New York, 1967. Designed to enable the reader to savor the fascination of biochemical developments.

Holum, J., *Elements of General and Biological Chemistry*, 2nd ed., Wiley, New York, 1969. A beginning text with the primary goal of considering "the molecular basis of life."

Ihde, A. J., *The Development of Modern Chemistry*, Harper & Row, New York, 1964. Chapter 13 deals with the structure of natural products and Chapter 17 discusses the development of biological chemistry.

Jellinck, P. H., *Biochemistry: An Introduction*, Holt, Rinehart & Winston, New York, 1963. Written at an elementary level for nonspecialists.

Jevons, F. R., *The Biochemical Approach to Life*, Basic Books, New York, 1964. A well-written discussion of biochemistry in which technicalities are kept to a minimum.

Locke, D. M., *Enzymes—The Agents of Life*, Crown Publishers, New York, 1969. Written for nonscientific readers.

Rafelson, M., and Binkley, S. B., *Basic Biochemistry*, 2nd ed., Macmillan, New York, 1968 (paperbound). A brief introduction to biochemistry that assumes an adequate background in chemistry and biology.

Watson, J. D., *The Double Helix*, Atheneum, New York, 1968. A personal account of the discovery of the structure of DNA.

Watson, J. D., and Crick, F., "A Structure for Deoxyribose Nucleic Acid," *Nature*, **171**, 737 (1953); "Genetical Implications of the Structure of Deoxyribonucleic Acid," *Nature*, **171**, 964 (1953). The original articles presenting the Crick-Watson theory of the double helix.

Williams, A. L., Embree, H. D., and DeBey, H. J., *Introduction to Chemistry*, Addison-Wesley, Reading, Mass., 1968. An introductory chemistry text that includes extensive sections on organic chemistry and biochemistry.

16

Transmutation of the Elements

16-1 Discovery of Radioactivity

In 1896 Henri Becquerel, a French physicist, made the chance observation that potassium uranyl sulfate, $K_2SO_4 \cdot UO_2SO_4 \cdot 2H_2O$, emits penetrating radiation spontaneously and continuously. The radiation was found to affect photographic plates even when they were wrapped in lightproof black paper. Systematic investigation revealed that the radiation is emitted by all compounds containing uranium. Mme Marie Sklodowska Curie gave to this remarkable property the name *radioactivity.*

Becquerel's discovery of radioactivity occurred within a few months after Roentgen's discovery of X-rays. Becquerel found that the radiation from radioactive materials is similar to X-rays in penetrating materials that are opaque to ordinary light, in affecting photographic plates in much the same manner as light, and in ionizing the molecules of air in the immediate vicinity, as demonstrated by the discharge of an electroscope.

In 1898 Mme Curie undertook to make reliable quantitative comparisons of the relative activities of various radioactive materials. She found that the activities of all compounds of uranium are proportional to the quantities of uranium they contain. She also found that all compounds containing thorium are radioactive. It was fortunate that she discovered so early that radioactivity is not a property that is limited to one element. Soon she was to need this information to understand her remarkable observation that two uranium minerals, pitchblende and carnotite, are more radioactive than uranium itself. She postulated that these minerals contain another element that is more radioactive than uranium. Pierre Curie joined his wife in the search for this element.

16-2 Radioactive Elements

The Curies carried out a series of chemical separations that finally yielded a small quantity of a material 400 times as radioactive as uranium. From the nature of the chemical reactions involved in isolating this material, the Curies concluded that it was a compound of a previously unknown metal. They named the element *polonium* in honor of Mme Curie's native Poland (Latin: *Polonia*). Continuing their work, they isolated, in December 1898, another substance 900 times as radioactive as uranium. This substance, they concluded, was a compound of another hitherto unknown metal, to which they gave the name *radium* in recognition of the high intensity of its radiation.

To confirm their discoveries of the two elements, the Curies obtained a ton of pitchblende residues from which most of the uranium had been extracted. From this Mme Curie isolated 0.1 g of radium chloride, $RaCl_2$, of sufficient purity for the determination of the atomic weight of radium. This culminated what has been termed one of the supreme efforts of faith and perseverance on the part of one individual in the history of science. For this work Mme Curie received the 1911 Nobel prize in chemistry. The Curies and Becquerel had previously shared the 1903 Nobel prize in physics for the discovery and early investigation of radioactivity.

By 1911 various scientific workers had discovered about 40 different radioactive species. At the time there were only 12 positions available in the periodic table to accommodate them, and it was apparent that several radioactive species would have to occupy a single position in the table. This was equivalent to saying that different radioactive species might be different kinds of atoms of the same element, because in the periodic table there is one place for each element and presumably only one element for each place. Thus was born the concept of isotopes (Greek: *iso* + *topos*, same + place).

16-3 Three Types of Rays

After the announcement of Becquerel's discovery of radioactivity, the nature of the radiation itself had to be investigated. When a small sample of radioactive material is placed in a deep recess in a block of lead, the emerging radiation travels in a straight line as a pencil of parallel rays, all other rays being absorbed within the lead. If the rays are permitted to pass through an electric field and then impinge upon a photographic plate (Figure 16-1), it is found that the rays are of three types. These rays have been named *alpha* (α), *beta* (β), and *gamma* (γ) *rays*. This experiment was first performed by Ernest Rutherford (later Lord Rutherford) while he was a professor of physics at McGill University in Montreal.

From the directions in which they are deflected in the electric field, it is apparent that alpha rays are positively charged, that beta rays are negatively charged, and that gamma rays have no charge at all. Rutherford found that the alpha radiation consists of *particles* that are doubly charged helium ions. The beta radiation also consists of particles; beta rays are high-speed electrons and are thus similar in character to cathode rays (Section 3-5). The gamma rays are not particles but are a type of electromagnetic radiation like X-rays. Gamma rays have even greater penetrating power than X-rays because of their very short wavelengths,

FIGURE 16-1

Behavior of alpha, beta, and gamma rays in an electric field

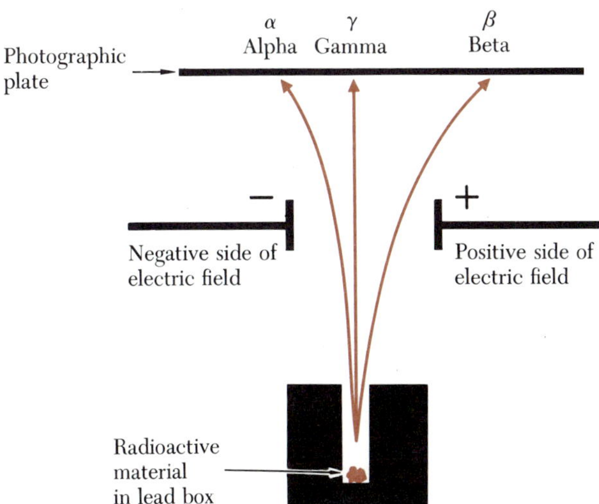

TABLE 16-1
Rays emitted from radioactive substances

Ray	*Character*	*Charge*[a]	*Mass*[b]	*Velocity*	*Penetration*
α	He^{++}	+2	4.0015	About 0.1 speed of light	Stopped by paper or human skin
β	Electron	−1	0.0005486	Up to 0.9 speed of light	Several mm in human tissue
γ	Short X-rays	0	0	Speed of light[c]	Human body or several feet of concrete

[a] The unit of charge employed here is the amount of charge of the electron itself.
[b] The mass is expressed on the atomic weight scale, defined as $^{12}C = 12$ exactly.
[c] The velocity of light in a vacuum is 3.00×10^{10} cm/sec.

10^{-8} to 10^{-10} cm. The properties of the three types of rays are summarized in Table 16-1. From this table it will be noted that the mass number (i.e., mass to the nearest integer) is 4 for the alpha particle and 0 for the beta particle.

Seldom does one radioactive species emit all three types of radiation. Nevertheless, most specimens of natural radioactive materials emit at least some of all three, because such materials are usually mixtures of alpha-emitters and beta-emitters, and gamma radiation usually accompanies emission of beta particles.

The intensity of the radiation from a sample is not affected by raising or lowering the temperature, and it is independent of the oxidation state of the radioactive element. The activity depends only on the identity of the radioactive element and the quantity that is present. The intensity decreases with time, in some cases very rapidly and in other cases so slowly as to be imperceptible by direct measurement.

16-4 Disintegration of Atoms

Rutherford and his colleague Frederick Soddy proposed a theory of radioactive disintegration to account for the many surprising facts that had come to light since Becquerel's discovery. They suggested that the atoms of radioactive species undergo spontaneous disintegration with the emission of alpha or beta particles and form atoms of a new element. This bold suggestion was directly opposed to the concept that atoms are indestructible; according to Dalton's atomic theory, such *transmutation* of one element into another is not to be expected. However, it can be demonstrated with a spectroscope that after a period of time elements are present in a sample of radioactive material that were not originally present. In most cases the new element is itself radioactive.

The rates at which radioactive disintegrations occur vary from very slow to very rapid. Half of any sample of a radioactive element disintegrates in a definite period of time known as the *half-life* of that element. For radium the half-life is 1590 years. This means that, on the average, half of the radium atoms contained in a particular sample will disintegrate in 1590 years, that half of the remaining atoms will disintegrate in the next 1590 years, etc. Half-life periods vary from 14 billion years for one of the isotopes of thorium to less than one billionth of a second for one of the isotopes of polonium. The importance of radium is a consequence of its half-life, which is neither so long as to make its radiation feeble nor so brief as to give the element only a fleeting existence.

By 1913 it was established that the nature of the emitted radiation is the factor that determines the identity of the new element. The relationship was stated by Soddy in the *group displacement law:*

1. Emission of an alpha particle reduces the mass of the atom by four atomic mass units, and the daughter element is located two groups to the left of its parent in the periodic table.
2. Emission of a beta particle causes little change in the mass of the atom, but the daughter element is located one group to the right of its parent in the periodic table.

The nuclear theory of the atom (Section 3-7) and the atomic number concept (Section 3-8) made Soddy's group displacement law understandable and made it possible for the existence of isotopes to be reconciled with the periodic table. The order of increasing atomic numbers placed cobalt and nickel, tellurium and iodine, potassium and argon, and thorium and protactinium in their correct positions in the periodic table, although the order of increasing atomic weights did not do so. The number of possible lanthanide elements could now be foretold, and the periodic table location of elements not yet discovered was definitely known for the first time. Although the recognition that radioactive disintegration involves transmutation had made the distinction between compounds and elements somewhat ambiguous, it gave rise to the concept of isotopes, which made it possible to define an element simply as a substance all of whose atoms have the same atomic number (nuclear charge).

16-5 Radioactive Disintegration Series

The radioactive disintegration of a radium nucleus consists in the ejection of an alpha particle, which is itself a helium nucleus. The radium nucleus is thereby transformed into a radon nucleus:

$$\binom{88p}{138n} \longrightarrow \binom{86p}{136n} + \binom{2p}{2n}$$

The transmutation can be represented by a *nuclear equation*:

$${}^{226}_{88}\text{Ra} \longrightarrow {}^{222}_{86}\text{Rn} + {}^{4}_{2}\text{He}$$

In this kind of equation the symbol for an element represents the nucleus of one atom of that element. The number at the lower left is the atomic number, i.e., the number of protons in the nucleus; the number at the upper left is the mass number, i.e., the mass of the nucleus expressed to the nearest whole number, which is the number of protons plus the number of neutrons. Since the daughter element produced by alpha emission has an atomic number two units less than the parent, it must occupy the position in the periodic table two places to the left of the parent, as stated in Soddy's group displacement law.

The disintegration of a thorium-234 nucleus consists in the ejection of a beta particle, which has a charge of -1 unit but has such a small mass (0.0005486 on the atomic-weight scale) that its mass number (mass to the nearest integer) is 0. The thorium-234 nucleus is therefore converted to a protactinium-234 nucleus:

$$\binom{90p}{144n} \longrightarrow \binom{91p}{143n} + {}^{0}_{-1}e$$

$${}^{234}_{90}\text{Th} \longrightarrow {}^{234}_{91}\text{Pa} + {}^{0}_{-1}e$$

Whenever a beta particle is ejected from a nucleus, the daughter nucleus has one more proton and one less neutron than the parent; evidently a neutron in the nucleus is converted to an electron that is ejected at high velocity and a proton that remains in the nucleus:

$$\underset{\text{neutron}}{{}^{1}_{0}n} \longrightarrow \underset{\text{proton}}{{}^{1}_{1}\text{H}} + \underset{\text{electron}}{{}^{0}_{-1}e}$$

The nucleus that is formed by the emission of an alpha or a beta particle may itself be radioactive, and usually is. Thus, the radon nucleus produced by the alpha decay of a radium nucleus emits an alpha particle and forms a polonium-218 nucleus:

$${}^{222}_{86}\text{Rn} \longrightarrow {}^{218}_{84}\text{Po} + {}^{4}_{2}\text{He}$$

TABLE 16-2

The uranium disintegration series

Element	*Atomic number*	*Mass number*	*Particle emitted*	*Half-life period*
Uranium	92	238	α	4.5×10^9 yr
Thorium	90	234	β	24 days
Protactinium	91	234	β	1.2 min
Uranium	92	234	α	2.5×10^5 yr
Thorium	90	230	α	8.0×10^4 yr
Radium	88	226	α	1590 yr
Radon	86	222	α	3.8 days
Polonium	84	218	α	3.05 min
Lead	82	214	β	27 min
Bismuth	83	214	β	20 min
Polonium	84	214	α	1.6×10^{-4} sec
Lead	82	210	β	19 yr
Bismuth	83	210	β	5.0 days
Polonium	84	210	α	138 days
Lead	82	206	Stable	—

Similarly, the protactinium-234 produced by the beta decay of thorium-234 emits a beta particle and forms uranium-234, which by alpha decay forms thorium-230, and so on.

Uranium-238 is the starting point for a series of radioactive disintegrations that finally ends in the formation of a nonradioactive isotope of lead, $^{206}_{82}Pb$. The successive "generations" in this *uranium disintegration series* are given in Table 16-2. Every known ore of uranium contains radium, and the uranium–radium ratio is always the same, this being a consequence of the brevity of the half-life of radium compared with that of uranium. All uranium ores also contain nonradioactive lead. This lead probably has originated as the end product of the disintegration series, because it has an atomic weight (approximately 206) that differs from the atomic weight of lead from most other sources (207.2). The atomic weight of an element is merely the average mass of its different isotopes.

From the known half-life periods of the elements involved, the time has been calculated that would be required for the formation of the lead in uranium ores. The result is of the order of 3 billion years. Presumably the earth is at least this old.

The uranium series is one of four *radioactive disintegration series*. The *thorium series* starts with thorium-232 and ends in the formation of nonradioactive lead-208. Since all members of this series have mass numbers that are divisible by four, the thorium series is known as the $4n$ series. The uranium series is the $4n + 2$ series. A third series is the $4n + 3$ or *actinium series*, which begins with uranium-235 and ends with lead-207. The fourth series is the $4n + 1$ or *neptunium series*, which involves man-made elements and was discovered during World War II in connection with the American atomic energy project. This series begins with plutonium-241 and ends with bismuth-209.

16-6 Artificial Transmutation

With transmutation taking place spontaneously in nature, man was challenged to produce it artificially. In 1919 Lord Rutherford observed that discharging alpha particles into a space containing nitrogen gas results, on rare occasions, in a collision between an alpha particle and a nitrogen nucleus that results, in turn, in the formation of a proton and an oxygen-17 nucleus:

$$^{4}_{2}He + ^{14}_{7}N \longrightarrow ^{17}_{8}O + ^{1}_{1}H$$

This was the first instance of deliberate *artificial transmutation* of one element to another. Most of the elements from boron to potassium were subsequently found to emit protons when bombarded with alpha particles. Frequently the energy of the protons was found to be higher than that of the alpha particles.

Bombardment of beryllium with alpha particles from polonium gives a highly penetrating radiation, which was proved by James Chadwick in England (in 1932) to consist of particles having no charge but a mass approximating that of the proton. The newly discovered particle was named the *neutron*:

$$^{4}_{2}He + ^{9}_{4}Be \longrightarrow ^{12}_{6}C + ^{1}_{0}n$$

Its lack of charge accounts for the high penetrating power of the neutron and for the difficulty of detecting it (and thus for the long delay in its discovery). The neutron is now regarded as a fundamental unit of atomic structure.

In 1934 Irene Curie (daughter of Pierre and Marie Curie) and her husband Frederic Joliot found that bombardment of magnesium with alpha particles causes emission of both neutrons and *positrons* (positive electrons). The emission of neutrons ceases when the source of alpha particles is removed, but positrons continue to be produced for some time. These observations were accounted for by the assumption that the transmutation of magnesium produced a radioactive isotope of silicon:

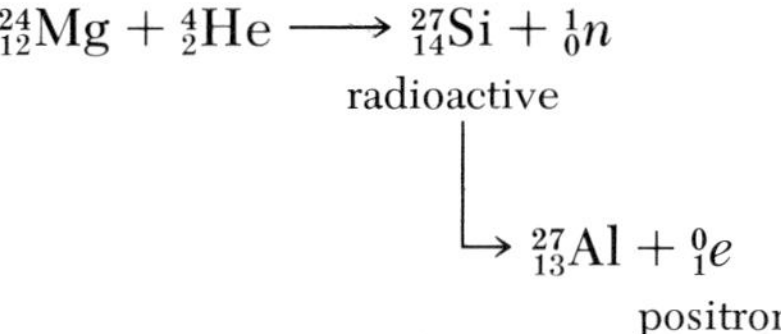

$$^{24}_{12}Mg + ^{4}_{2}He \longrightarrow \underset{\text{radioactive}}{^{27}_{14}Si} + ^{1}_{0}n$$

$$^{27}_{14}Si \longrightarrow ^{27}_{13}Al + \underset{\text{positron}}{^{0}_{1}e}$$

The first reaction ceases as soon as alpha particle bombardment is ended, but the *artificial radioactivity* continues as long as the silicon-27 is spontaneously disintegrating to form aluminum. The Curie-Joliot discovery was significant because it meant that radioactive isotopes of common elements could be obtained artificially for use in research. Radioactive carbon, nitrogen, and phosphorus have proved very useful as tracer atoms in biochemical research.

Artificial transmutations that occur instantaneously cause emission of a proton, a neutron, or some particle composed of protons and neutrons such as a deuteron (deuterium nucleus) or an alpha particle. Electrons or positrons are produced as secondary products from the spontaneous disintegration of the artificially produced radioactive isotopes. Just as electrons (beta particles) are emitted as products of the decomposition of neutrons, so positrons are attributed to the disintegration of protons:

$$\underset{\text{proton}}{^{1}_{1}H} \longrightarrow \underset{\text{neutron}}{^{1}_{0}n} + \underset{\text{positron}}{^{0}_{1}e}$$

The positron has the same mass and the same quantity of electric charge as an electron; however, the charge is positive instead of negative. (The positron was discovered in 1932 by the American physicist C. D. Anderson.)

Artificial transmutation has been achieved in recent years through the use of cyclotrons and other powerful particle accelerator machines. Such instruments as the huge synchrotron at Brookhaven National Laboratory, Upton, New York, have been built in order to use strong magnetic and electric fields to impart tremendous energies to protons, deuterons, alpha particles, and electrons. For example, bombardment of carbon-12 nuclei with protons accelerated to high velocities in a cyclotron produces radioactive nitrogen-13:

$$^{12}_{6}C + ^{1}_{1}H \longrightarrow \underset{\text{radioactive}}{^{13}_{7}N} + \text{energy}$$

$$^{13}_{7}N \xrightarrow[\text{10 min}]{\text{half-life}} \underset{\text{stable}}{^{13}_{6}C} + ^{0}_{1}e$$

Since neutrons have no charge, they are not repelled by atomic nuclei. For this reason they are extremely effective as atomic "artillery." They can be obtained by allowing alpha particles to bombard beryllium. The following are examples of nuclear reactions brought about by streams of neutrons:

$$^{200}_{80}\text{Hg} + ^{1}_{0}n \longrightarrow ^{201}_{80}\text{Hg} + \text{energy } (\gamma \text{ rays})$$

$$^{40}_{20}\text{Ca} + ^{1}_{0}n \longrightarrow ^{40}_{19}\text{K} + ^{1}_{1}\text{H}$$

16-7 Man-Made Elements

Artificial transmutation has been used not only to change many of the known elements one into another and to obtain radioactive isotopes of all known elements but to form elements not known to occur in nature. Of the 104 elements now known to exist, 15 were first obtained by artificial transmutation.

In the middle 1930's four elements with atomic numbers less than 92 (uranium) were still unknown and no elements beyond uranium were known or believed necessarily possible of existence. The four elements missing from the periodic table were those with atomic numbers 43, 61, 85, and 87.

Element 43 was first obtained in 1937 by bombardment of molybdenum with deuterons:

$$^{96}_{42}\text{Mo} + ^{2}_{1}\text{H} \longrightarrow ^{97}_{43}\text{Tc} + ^{1}_{0}n$$

It was given the name *technetium*, symbol Tc (Greek: *technetos*, artificial), because it was the first element not known on earth to be made artificially.

Element 61, the missing lanthanide, was found to be a product of uranium fission in connection with the atomic energy project during World War II. The name *promethium*, symbol Pm, which was given to the element was derived from the mythological character Prometheus who brought fire from heaven for the use of man, much as nuclear fission made atomic energy available.

Element 85, the missing halogen, was first obtained in 1940 by bombardment of bismuth with alpha particles. It was named *astatine*, symbol At (Greek: *astatos*, unstable), because it is the only halogen that has no stable isotope.

Element 87, the missing alkali metal, was discovered in 1939 (in France) and was named *francium*, symbol Fr. It occurs in nature in extremely small quantities, where it is formed by alpha particle emission from actinium-227; its longest-lived isotope has a half-life of only 21 min, which makes francium the most unstable of any element occurring naturally. Francium isotopes have been identified among products of artificial transmutation reactions.

The first transuranium elements were obtained in 1940 by bombarding uranium with high-velocity deuterons:

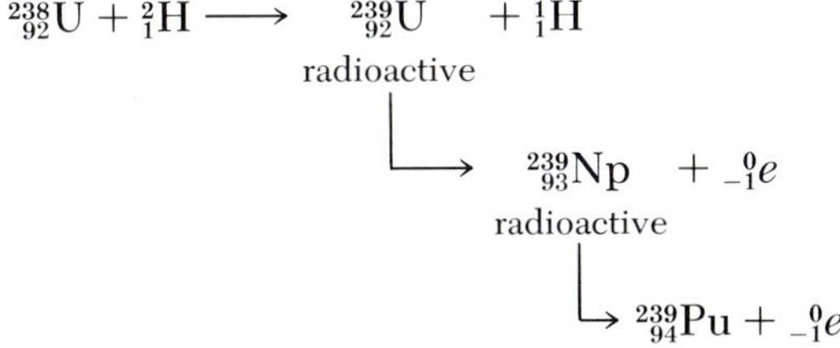

Element 93 was named *neptunium*, symbol Np, after Neptune, the planet directly beyond Uranus in the solar system. Element 94 was named *plutonium*, symbol Pu, after Pluto, the planet beyond Neptune that was discovered in 1930.

Elements 95 through 104 have all been subsequently obtained by artificial means. The actinide series was completed with the production of a very small amount of element 103, named *lawrencium*, in 1961. Its half-life is 8 sec. More recently, the preparation of element 104 has been claimed by both Russian and American scientists. At the time of this writing, scientists at the University of California's Lawrence Radiation Laboratory, Berkeley, California, expect $^{261}104$ to be the longest lived isotope of element 104, which is as yet unnamed. It is also expected that element 104 will be found to belong below hafnium in the periodic table. However, the findings of the Berkeley group are at variance with those reported by the Russians, and extensive investigations into the nature of element 104 continue, along with studies of the probable nature and properties of elements of yet higher atomic numbers and the most feasible methods of their synthesis. Fantastic as it would have seemed in the 1930s, or indeed in the 1950s, American scientists have recently postulated that man will one day synthesize element 168 as a noble element too heavy to exist as a gas.

16-8 Nuclear Fission

Before 1939 atomic scientists had little hope that atomic energy could be used for practical purposes. Although many of the artificial transmutation reactions then known released large amounts of energy, the bombardment procedures employed to produce the reactions consumed even more energy. Both the targets (atomic nuclei) and the projectiles (protons, etc.) in such procedures are so small that there is no way to aim the projectile or even to see the target. "Scoring a hit" under such circumstances is an improbable event of infrequent occurrence. Most of the energy invested in the operation is wasted on projectiles that go wide of their mark. In 1937, the year of his death, Lord Rutherford held the opinion that the outlook for gaining useful energy by artificial processes of transmutation was not promising.

In 1938 Otto Hahn and Fritz Strassmann in Germany unexpectedly discovered that bombardment of uranium with neutrons results in the formation of barium. This was impossible to account for in terms of the transmutation reactions previously known, because the atomic number of barium is only 56, that is, 36 less than the atomic number of uranium. Early in 1939 Lise Meitner (a refugee from Nazi Germany who had formerly worked with Hahn in Berlin), in association with her nephew Otto Frisch, postulated that the impact of a neutron upon a uranium nucleus sometimes causes the latter to disintegrate into portions of somewhat similar mass. The phenomenon was given the name *nuclear fission.* The following equation represents one of the many ways in which a uranium atom may undergo fission:

$$^{235}_{92}\mathrm{U} + ^{1}_{0}n \longrightarrow \underset{\text{radioactive}}{^{141}_{56}\mathrm{Ba}} + \underset{\text{radioactive}}{^{92}_{36}\mathrm{Kr}} + 3\,^{1}_{0}n + \text{energy}$$

The discovery of nuclear fission immediately changed the attitude of atomic scientists toward the atomic energy question because the process is accompanied by the liberation of large amounts of energy and because the fission, which is brought about by neutrons, is accompanied by the liberation of neutrons. The analogy with fire is clear. A fire is a self-sustaining reaction because the heat liberated by the reaction (combustion) maintains the temperature required for the reaction to continue. In similar fashion it was assumed that nuclear fission might be made self-sustaining by having neutrons liberated in the process initiate fission in other nuclei. If this could be accomplished, the practical utilization of nuclear energy would be realized.

Natural uranium is a mixture of isotopes having the approximate composition 99.3 percent U-238 and 0.7 percent U-235. Only the U-235 nuclei undergo fission. Neutrons absorbed by U-238 atoms are unavailable for bringing about fission of U-235 atoms. The utilization of the nuclear fission phenomenon in the so-called atomic bomb required that the U-235 be separated from most of the U-238 with which it occurs.

Unexpectedly, it was discovered that the absorption of neutrons by U-238 results in the formation of another fissionable species of atom, Pu-239:

$$^{238}_{92}\mathrm{U} + ^{1}_{0}n \longrightarrow ^{239}_{92}\mathrm{U}$$

$$^{239}_{92}\mathrm{U} \longrightarrow ^{239}_{93}\mathrm{Np} + ^{0}_{-1}e$$

$$^{239}_{93}\mathrm{Np} \longrightarrow ^{239}_{94}\mathrm{Pu} + ^{0}_{-1}e$$

Although the Pu-239 is itself radioactive, its half-life is of the order of 25,000 years.

16-9 Nuclear Energy

Weights of the individual isotopes of an element are expressed relative to $^{12}_{6}C = 12.0000$ exactly. Precision measurements show that the mass of each isotope is very nearly a whole number, but C-12 is the only one whose mass is exactly a whole number. The exact value for the mass of an isotope is called the *atomic mass* or *isotopic weight*, and the nearest integer is called the *mass number*. As examples, the isotopic weight of Ne-20 is 19.999 and that of Ne-22 is 21.998. The helium atom (mass number 4) has the isotopic weight 4.0026. The U-235 atom has an isotopic weight slightly in excess of 235. Atoms at the beginning and at the end of the periodic table have isotopic weights in excess of their mass numbers. Atoms in the middle of the periodic table have isotopic weights less than their mass numbers.

Nuclear fission of elements near the end of the periodic table, such as uranium, produces elements nearer the middle of the table, such as barium and krypton. The loss of mass that accompanies the fission is responsible for the formation of such enormous amounts of energy. Albert Einstein, in 1905, had derived the equation

$$E = mc^2$$

which relates the magnitude of the mass that is lost to the quantity of energy produced. In this equation, m represents the mass in grams, E is the energy in ergs, and c is the velocity of electromagnetic radiation (light, etc.) in a vacuum, which is 3.00×10^{10} cm per sec. Thus, conversion of 1 g of mass produces 9×10^{20} ergs. In other words, destruction of a small quantity of matter produces a tremendous amount of energy. In the nuclear fission of U-235 or Pu-239, the masses of the fission products are not much less than the masses of the starting materials, but even so the amount of energy produced by the fission of a small quantity of material is extremely large.

Nuclear reactors have been constructed for the utilization of nuclear energy for purposes other than explosions. In all such devices the energy is produced by fission.

The converse process, whereby nuclei of small mass undergo *nuclear fusion* to form nuclei of atoms nearer the middle of the periodic table, provides another means for the practical utilization of nuclear energy. The fusion of deuterium and tritium nuclei produces helium:

$$^{2}_{1}H + ^{3}_{1}H \longrightarrow ^{4}_{2}He + ^{1}_{0}n$$

The mass of the deuterium nucleus (2.0136 amu) plus the mass of the tritium nucleus (3.0155 amu) is 5.0291 amu. The mass of the helium-4 nucleus (4.0015 amu) plus the mass of the neutron (1.0087 amu) is 5.0102 amu. The difference between the two sums is 0.0189 amu. Thus, the formation of 1 mole of helium-4 nuclei by this process is accompanied by the loss of 0.0189 g of mass and by the production of an enormous amount of energy that can be calculated from the Einstein relationship:

$$
\begin{aligned}
E &= 0.0189 \text{ g} \times 9 \times 10^{20} \text{ ergs/g} \\
&= 1.7 \times 10^{20} \text{ ergs}
\end{aligned}
$$

This is equivalent to 4×10^{14} kcal!

All that is required to initiate the fusion reaction is an extremely high temperature. For this reason the process is called a *thermonuclear reaction.* The only known method for producing the required temperature of many millions of degrees is the explosion of a fission-type atomic bomb. Therefore, an atomic bomb (A-bomb) is used to detonate a hydrogen bomb (H-bomb). There is no theoretical limit to the possible size of an H-bomb.

Controlled fusion for peacetime purposes has proved much more difficult to achieve than controlled fission and is still in the process of development.

NEW TERMS

Artificial radioactivity: radioactivity arising from the formation, through artificial transmutation, of an atomic species that is radioactive and does not occur in nature.

Artificial transmutation: the transmutation of one element into another as the result of human action and intent (as distinguished from the "natural" spontaneous type of transmutation familiar as radioactivity).

Half-life: the time required for half of any sample of a radioactive element to disintegrate.

Nuclear equation: an equation representing a nuclear "reaction," e.g., transmutation. In a nuclear equation the chemical symbol of an element represents the nucleus of one atom of that element; the atomic number and the mass number are represented by a subscript and a superscript, respectively.

Nuclear fission: a type of transmutation characterized by the disintegration of an atomic nucleus into portions of somewhat similar mass.

Nuclear fusion: a type of transmutation in which atomic nuclei of somewhat similar mass unite to form a nucleus of greater mass than either of them.

Radioactive disintegration series: a sequence of radioactive disintegrations in which each disintegrating nucleus is itself a product of the disintegration of the preceding member of the series.

Radioactivity: the spontaneous disintegration of certain kinds of atomic nuclei, producing one or more of three kinds of radiation: alpha, beta, and gamma rays.

Thermonuclear reaction: same as nuclear fusion (so called because nuclear fusion is initiated by extremely high temperature).

EXERCISES

16-1 If you have a 16-mg sample of iodine-128 at 1:00 P.M. and only 0.25 mg remains at 3:30 P.M., what is its half-life?

16-2 How do nuclear reactions differ from chemical reactions?

16-3 If the nucleus of a bismuth-209 atom captures a neutron and the product undergoes beta decay, what is the final product?

16-4 What is the difference between radioactivity and artificial transmutation?

16-5 What distinguishes nuclear fission from nuclear fusion?

16-6 Why does uranium-235 not undergo self-sustaining nuclear fission in nature?

16-7 Why are the atomic weights of the elements not whole numbers? (Give two reasons.)

16-8 Complete the following equations:
(a) ${}^{27}_{13}\text{Al} + ? \longrightarrow {}^{24}_{11}\text{Na} + {}^{4}_{2}\text{He}$
(b) $? + {}^{2}_{1}\text{H} \longrightarrow {}^{12}_{6}\text{C} + {}^{4}_{2}\text{He}$
(c) ${}^{23}_{11}\text{Na} + {}^{2}_{1}\text{H} \longrightarrow {}^{24}_{11}\text{Na} + ?$

16-9 When would you say the Atomic Age began?

16-10 Why is so large an amount of energy produced in nuclear fission?

16-11 Discuss the Law of Conservation of Mass and the Law of Conservation of Energy in relation to nuclear reactions.

16-12 What is a thermonuclear bomb? How is such a device detonated?

16-13 Fast neutrons produced by cosmic rays in the upper atmosphere interact with nitrogen to form tritium,

$$ {}^{1}_{0}n + {}^{14}_{7}\text{N} \longrightarrow {}^{12}_{6}\text{C} + {}^{3}_{1}\text{H} $$

Consequently, there is a very small concentration of tritium oxide in rainwater. Tritium undergoes radioactive decay with a half-life of 12.5 yrs:

$$ {}^{3}_{1}\text{H} \longrightarrow {}^{3}_{2}\text{He} + {}^{0}_{-1}e $$

Suggest how the age of a vintage wine might be estimated by utilizing this information.

16-14 Some atomic nuclei undergo transmutation by capturing an electron from the space outside the nucleus. The electron is usually acquired from the $n = 1$ level and the process is known as *K*-capture. What is the effect on the atomic number of the element? Write the nuclear equation.

16-15 How many alpha particles and how many beta particles are given off in the series that begins with Th-232 and ends with Pb-208? In the series that begins with Pu-241 and ends with Bi-209? In the series that begins with U-238 and ends with Pb-206? In the series that begins with U-235 and ends with Pb-207?

REFERENCES

Garrett, A. B., *The Flash of Genius*, Van Nostrand, Princeton, N.J., 1962. Useful for accounts of discoveries in radioactivity by Becquerel, Rutherford, Curie, Soddy, Chadwick, Meitner, *et al.*

Glasstone, S., *Sourcebook on Atomic Energy*, 3rd ed., Van Nostrand, Princeton, N.J., 1967. A source of basic atomic energy information for readers with many different interests. Sponsored by the U.S. Atomic Energy Commission, this book has had worldwide acceptance.

Hewlett, R. G., and Anderson, O. E., Jr., *The New World, 1939/1946*, Pennsylvania State University Press, University Park, Pa., 1962. A complete account of the step-by-step discovery of the nature of nuclear fission and the first utilization of atomic energy. Volume I in *A History of the U.S. Atomic Energy Commission.*

Holcomb, R. W., "Heavy Elements: A Feud over 104 and a Future for 114," *Science*, **166**, 1254 (1969). Report of a Mendeleev Centennial Symposium held at Houston, November 1969. Scientists from the Lawrence Radiation Laboratory at Berkeley have proposed that element 104 be named rutherfordium, challenging the discovery claims of Russian scientists who had previously proposed that it be named kurchatovium.

Seaborg, G. T., *Man-Made Transuranium Elements*, Prentice-Hall, Englewood Cliffs, N.J., 1963. An intellectually stimulating account of the discovery and investigation of the first 11 transuranium elements. Written for use with the Chemical Education Material Study program.

Seaborg, G. T., "Some Recollections of Early Nuclear Age Chemistry," *Journal of Chemical Education*, **45**(5), 278 (1968). Personal recollections.

Seaborg, G. T., "Prospects for Further Considerable Extension of the Periodic Table," *Journal of Chemical Education*, **46**(10), 696 (1969). A paper presented at the Mendeleev Centennial Symposium at the American Chemical Society Meeting, Minneapolis, April 1969.

Seaborg, G. T., and Bloom, J. L., "The Synthetic Elements: IV," *Scientific American*, **220**(4), 57 (1969). Seaborg, a recipient of the Nobel prize for chemistry (1951) and now Chairman of the U.S. Atomic Energy Commission, and Bloom, his technical assistant, discuss the possibilities of extending the periodic table to element 168 by elemental synthesis.

Smyth, H. D., *Atomic Energy for Military Purposes*, Princeton University Press, Princeton, N.J., 1945. Complete title: "A general account of the development of methods of using atomic energy for military purposes under the auspices of the U.S. Government, 1940–45." The first book published on the subject, the Smyth report was released six days after the first use of an atomic bomb in warfare.

Wall, F., "Early Days of Radioactivity in Industry," *Chemistry*, **42**(4), 17; **42**(5), 16 (1969). Personal recollections of employment in a factory where women acquired radium poisoning.

Appendixes

Appendix 1: Alphabetical List of Symbols of the Elements

Symbol	*Name*	*Atomic number*	*Symbol*	*Name*	*Atomic number*
Ac	Actinium	89	F	Fluorine	9
Ag	Silver	47	Fe	Iron	26
Al	Aluminum	13	Fm	Fermium	100
Am	Americium	95	Fr	Francium	87
Ar	Argon	18	Ga	Gallium	31
As	Arsenic	33	Gd	Gadolinium	64
At	Astatine	85	Ge	Germanium	32
Au	Gold	79	H	Hydrogen	1
B	Boron	5	He	Helium	2
Ba	Barium	56	Hf	Hafnium	72
Be	Beryllium	4	Hg	Mercury	80
Bi	Bismuth	83	Ho	Holmium	67
Bk	Berkelium	97	I	Iodine	53
Br	Bromine	35	In	Indium	49
C	Carbon	6	Ir	Iridium	77
Ca	Calcium	20	K	Potassium	19
Cd	Cadmium	48	Kr	Krypton	36
Ce	Cerium	58	La	Lanthanum	57
Cf	Californium	98	Li	Lithium	3
Cl	Chlorine	17	Lr	Lawrencium	103
Cm	Curium	96	Lu	Lutetium	71
Co	Cobalt	27	Md	Mendelevium	101
Cr	Chromium	24	Mg	Magnesium	12
Cs	Cesium	55	Mn	Manganese	25
Cu	Copper	29	Mo	Molybdenum	42
Dy	Dysprosium	66	N	Nitrogen	7
Er	Erbium	68	Na	Sodium	11
Es	Einsteinium	99	Nb	Niobium	41
Eu	Europium	63	Nd	Neodymium	60

Symbol	*Name*	*Atomic number*	*Symbol*	*Name*	*Atomic number*
Ne	Neon	10	Sc	Scandium	21
Ni	Nickel	28	Se	Selenium	34
No	Nobelium	102	Si	Silicon	14
Np	Neptunium	93	Sm	Samarium	62
O	Oxygen	8	Sn	Tin	50
Os	Osmium	76	Sr	Strontium	38
P	Phosphorus	15	Ta	Tantalum	73
Pa	Protactinium	91	Tb	Terbium	65
Pb	Lead	82	Tc	Technetium	43
Pd	Palladium	46	Te	Tellurium	52
Pm	Promethium	61	Th	Thorium	90
Po	Polonium	84	Ti	Titanium	22
Pr	Praseodymium	59	Tl	Thallium	81
Pt	Platinum	78	Tm	Thulium	69
Pu	Plutonium	94	U	Uranium	92
Ra	Radium	88	V	Vanadium	23
Rb	Rubidium	37	W	Tungsten	74
Re	Rhenium	75	Xe	Xenon	54
Rh	Rhodium	45	Y	Yttrium	39
Rn	Radon	86	Yb	Ytterbium	70
Ru	Ruthenium	44	Zn	Zinc	30
S	Sulfur	16	Zr	Zirconium	40
Sb	Antimony	51			

Appendix 2: Names and Charges of Important Cations

Name	*Formula and charge*	*Name*	*Formula and charge*
Aluminum	Al^{3+}	Lithium	Li^{+}
Ammonium	NH_4^{+}	Magnesium	Mg^{++}
Antimony	Sb^{3+}	Manganese(II), manganous	Mn^{++}
Antimonyl	SbO^{+}	Mercury(I), mercurous	Hg_2^{++}
Barium	Ba^{++}	Mercury(II), mercuric	Hg^{++}
Bismuth	Bi^{3+}	Methylammonium	$CH_3NH_3^{+}$
Bismuthyl	BiO^{+}	Nickel	Ni^{++}
Cadmium	Cd^{++}	Potassium	K^{+}
Calcium	Ca^{++}	Silver	Ag^{+}
Chromium(III), chromic	Cr^{3+}	Sodium	Na^{+}
Cobalt(II), cobaltous	Co^{++}	Strontium	Sr^{++}
Copper(II), cupric	Cu^{++}	Tetramethylammonium	$(CH_3)_4N^{+}$
Dimethylammonium	$(CH_3)_2NH_2^{+}$	Tin(II), stannous	Sn^{++}
Hydrogen, hydronium	H^{+}, H_3O^{+}	Trimethylammonium	$(CH_3)_3NH^{+}$
Iron(II), ferrous	Fe^{++}	Uranyl	UO_2^{++}
Iron(III), ferric	Fe^{3+}	Zinc	Zn^{++}
Lead	Pb^{++}		

Appendix 3: Names and Charges of Important Anions

Name	*Formula and charge*
Acetate, CH_3COO^-	$C_2H_3O_2^-$
Aluminate	$Al(OH)_4^-$
Amide	NH_2^-
Arsenate	AsO_4^{3-}
Azide	N_3^-
Benzoate, $C_6H_5COO^-$	$C_7H_5O_2^-$
Bromate	BrO_3^-
Bromide	Br^-
Bromite	BrO_2^-
Carbonate	CO_3^{--}
Hydrogen carbonate, bicarbonate	HCO_3^-
Chlorate	ClO_3^-
Chloride	Cl^-
Chlorite	ClO_2^-
Chromate	CrO_4^{--}
Dichromate	$Cr_2O_7^{--}$
Cyanate	OCN^-
Cyanide	CN^-
Ferricyanide, hexacyanoferrate(III)	$Fe(CN)_6^{3-}$
Ferrocyanide, hexacyanoferrate(II)	$Fe(CN)_6^{4-}$
Fluoaluminate	AlF_6^{3-}
Fluoride	F^-
Fluosilicate	SiF_6^{--}
Formate, $HCOO^-$	CHO_2^-
Hydroxide	OH^-
Hypobromite	BrO^-
Hypochlorite	ClO^-
Hypoiodite	IO^-
Iodate	IO_3^-
Iodide	I^-
Lactate, $CH_3CHOHCOO^-$	$C_3H_5O_3^-$
Metaperiodate	IO_4^-
Molybdate	MoO_4^{--}
Nitrate	NO_3^-
Nitride	N^{3-}
Nitrite	NO_2^-
Oxalate	$C_2O_4^{--}$
Hydrogen oxalate, binoxalate	$HC_2O_4^-$
Oxide	O^{--}
Paraperiodate	IO_6^{5-}
Perchlorate	ClO_4^-
Permanganate	MnO_4^-
Peroxide	O_2^{--}
Phosphate	PO_4^{3-}
Monohydrogen phosphate	HPO_4^{--}
Dihydrogen phosphate	$H_2PO_4^-$
Salicylate, $HOC_6H_4COO^-$	$C_7H_5O_3^-$
Selenate	SeO_4^{--}
Selenide	Se^{--}
Silicate	SiO_3^{--}
Sulfamate	$SO_3NH_2^-$
Sulfate	SO_4^{--}
Hydrogen sulfate, bisulfate	HSO_4^-
Sulfide	S^{--}
Hydrogen sulfide, bisulfide	HS^-
Sulfite	SO_3^{--}
Hydrogen sulfite, bisulfite	HSO_3^-
Tartrate	$C_4H_4O_6^{--}$
Telluride	Te^{--}
Tetraborate	$B_4O_7^{--}$
Thioarsenate	AsS_4^{3-}
Thiocyanate	SCN^-
Thiosulfate	$S_2O_3^{--}$
Tungstate	WO_4^{--}
Uranate	UO_4^{--}
Zincate	$Zn(OH)_4^{--}$

Appendix 4: Important Acids and Their Ionization Constants, Listed in Order of Decreasing Strength

Name	*Formula*	*Ionization constant at 25°C*
Perchloric acid	$HClO_4$	Completely ionized
Hydriodic acid	HI	Completely ionized
Hydrobromic acid	HBr	Completely ionized
Sulfuric acid	H_2SO_4	Completely ionized
Hydrochloric acid	HCl	Completely ionized
Nitric acid	HNO_3	Completely ionized
Iodic acid	HIO_3	2.0×10^{-1}
Oxalic acid	$H_2C_2O_4$	5.4×10^{-2}
Sulfurous acid	H_2SO_3	1.2×10^{-2}
Hydrogen sulfate ion	HSO_4^-	1.2×10^{-2}
Phosphoric acid	H_3PO_4	7.5×10^{-3}
Ferric ion	$Fe(H_2O)_6^{3+}$	6.0×10^{-3}
Hydrotelluric acid	H_2Te	2.3×10^{-3}
Hydrofluoric acid	HF	6.7×10^{-4}
Nitrous acid	HNO_2	5.1×10^{-4}
Formic acid	$HCHO_2$ ($HCOOH$)	1.8×10^{-4}
Hydroselenic acid	H_2Se	1.7×10^{-4}
Lactic acid	$HC_3H_5O_3$	1.4×10^{-4}
Benzoic acid	$HC_7H_5O_2$ (C_6H_5COOH)	6.6×10^{-5}
Hydrogen oxalate ion	$HC_2O_4^-$	5.4×10^{-5}
Acetic acid	$HC_2H_3O_2$ (CH_3COOH)	1.8×10^{-5}
Carbonic acid	H_2CO_3	4.4×10^{-7}
Hydrosulfuric acid	H_2S	1.1×10^{-7}
Dihydrogen phosphate ion	$H_2PO_4^-$	6.3×10^{-8}
Hydrogen sulfite ion	HSO_3^-	5.6×10^{-8}
Hypochlorous acid	$HClO$	3.2×10^{-8}
Ammonium ion	NH_4^+	5.7×10^{-10}
Hydrocyanic acid	HCN	4.0×10^{-10}
Hydrogen carbonate ion	HCO_3^-	4.7×10^{-11}
Hydrogen peroxide	H_2O_2	2.4×10^{-12}
Monohydrogen phosphate ion	HPO_4^{--}	1.7×10^{-12}
Hydrogen sulfide ion	HS^-	1.0×10^{-14}

Appendix 5: Solubility Product Constants

Name	*Formula*	K_{sp} *at 25°C*
Aluminum hydroxide	$Al(OH)_3$	1.9×10^{-33}
Barium carbonate	$BaCO_3$	8.1×10^{-9}
Barium chromate	$BaCrO_4$	2.4×10^{-10}
Barium fluoride	BaF_2	1.7×10^{-6}
Barium sulfate	$BaSO_4$	1.1×10^{-10}

Name	*Formula*	K_{sp} *at 25°C*
Calcium carbonate	$CaCO_3$	4.8×10^{-9}
Calcium fluoride	CaF_2	3.9×10^{-11}
Calcium sulfate	$CaSO_4$	2.4×10^{-5}
Copper(II) sulfide	CuS	8.7×10^{-36}
Iron(III) hydroxide	$Fe(OH)_3$	1.1×10^{-36}
Lead carbonate	$PbCO_3$	1.5×10^{-13}
Lead chromate	$PbCrO_4$	1.8×10^{-14}
Lead sulfate	$PbSO_4$	1.8×10^{-8}
Lead sulfide	PbS	8.4×10^{-28}
Magnesium carbonate	$MgCO_3$	1.0×10^{-5}
Magnesium hydroxide	$Mg(OH)_2$	1.5×10^{-11}
Silver acetate	$AgC_2H_3O_2$	2.3×10^{-3}
Silver bromide	$AgBr$	5.0×10^{-13}
Silver chloride	$AgCl$	1.6×10^{-10}
Silver chromate	Ag_2CrO_4	2.4×10^{-12}
Silver iodide	AgI	1.5×10^{-16}
Strontium carbonate	$SrCO_3$	9.4×10^{-10}
Strontium sulfate	$SrSO_4$	2.8×10^{-7}
Zinc sulfide	ZnS	1.1×10^{-20}

Appendix 6: Vapor Pressure of Water

Temperature (°C)	*Pressure,* torr (mm Hg)	*Temperature* (°C)	*Pressure,* torr (mm Hg)
0	4.6	28	28.4
5	6.5	29	30.0
10	9.2	30	31.8
11	9.8	31	33.7
12	10.5	32	35.7
13	11.2	33	37.7
14	12.0	34	39.9
15	12.8	35	42.2
16	13.6	36	44.6
17	14.5	37	47.1
18	15.5	38	49.7
19	16.5	39	52.4
20	17.5	40	55.3
21	18.6	50	92.5
22	19.8	60	149.4
23	21.1	70	233.7
24	22.4	80	355.1
25	23.8	90	525.8
26	25.2	100	760.0
27	26.7	110	1074.6

Appendix 7: Units of Measurement

Any measurement is expressed by a number and a unit. The size of the unit used for a particular measurement is a matter of choice. It is desirable to express measurements with numbers that are neither fantastically large nor absurdly small. Thus, in any system of weights and measures there are multiples and submultiples of the fundamental units.

In the United States both the English system and the metric system are used. The English system employs such units as the inch for length, the acre for area, and the pound for weight. The disadvantage of the English system lies in the lack of systematic interrelationships among units. For example, the units of length are not related in a consistent manner: 1 foot = 12 inches, 1 yard = 3 feet, etc.

The metric system, less than 200 years old, derives its name from its basic unit of length, the *meter* (abbreviation: m). In this system the units are related by powers of 10. The most frequently used metric units of length are the following:

UNITS OF LENGTH (DISTANCE)

1 kilometer (km) = 1000 m (10^3 m)

1 centimeter (cm) = 0.01 m (10^{-2} m)

1 millimeter (mm) = 0.001 m (10^{-3} m)

1 angstrom (Å) = 0.0000000001 m (10^{-10} m, 10^{-8} cm)

1 nanometer (nm) = 10^{-9} m

The meter was originally intended to be one ten-millionth of the distance from the earth's equator to the North Pole (measured on a meridian passing through Paris). After this distance was measured (as accurately as possible), two fine lines were scribed on a metal bar made of an alloy of the noble metals platinum and iridium. Since the middle of the nineteenth century this metal bar has been kept in a vault at the International Bureau of Weights and Measures (at Sèvres, near Paris), and the distance between the two lines became the international standard of length. In 1875 the meter was adopted as the official standard of length by the United States, and the standard yard was defined as $\frac{3600}{3937}$ meter (from which it follows that 1 inch is approximately 2.54 cm). The requirements of present-day science and technology have necessitated increased accuracy in measurements; accordingly, the standard meter was redefined by international agreement in 1960 in terms of spectroscopic measurements. (The meter is now defined as 1,650,763.73 times the wavelength of the orange-red line of krypton-86.)

The basic unit of mass in the metric system is the *gram* (abbreviation: g), but the primary standard of mass is that of the international prototype kilogram (kg), a body of platinum-iridium alloy kept at the International Bureau of Weights and Measures.

UNITS OF MASS (WEIGHT)

1 kilogram (kg) = 1000 g (10^3 g)

1 metric ton = 1000 kg (10^3 kg, 10^6 g)

1 milligram (mg) = 0.001 g (10^{-3} g)

1 microgram (μg or γ) = 0.000001 g (10^{-6} g, 10^{-3} mg)

1 nanogram (ng) = 10^{-9} g

1 picogram (pg) = 10^{-12} g

Approximately, 1 pound is 453.59 g, 1 ounce is 28.35 g, and 1 kilogram is 2.20 lb.

The basic unit of volume in the metric system is the *liter*. In October 1964 the Twelfth General Conference on Weights and Measures defined the liter to equal a cubic decimeter, i.e., 1000 cubic centimeters. Formerly (since 1901) the liter was defined as the volume occupied by 1 kilogram of water at 4°C. This made the liter equal to 1000.028 cubic centimeters, i.e., slightly larger than a cubic decimeter. The new definition makes the milliliter the same as the cubic centimeter.

UNITS OF VOLUME

1 liter = 1 cubic decimeter (dm^3)

1 milliliter (ml) = 0.001 liter

1 liter = 1000 cubic centimeters (cm^3)

1 ml = 1 cm^3

Approximately, 1 liter is 1.057 quarts (liquid measure), and 10 liters are 2.64 United States gallons or 2.20 British Imperial gallons.

Appendix 8: Significant Figures

It is possible to count discrete objects without introducing an error, but measurements are never completely free from error because the accuracy of the procedure and the precision of the measuring instruments are limited. In the technical reporting of measurements (and calculations based on them), the convention used is to write down all digits definitely known plus one additional digit. The known digits plus the doubtful one constitute the *significant figures* of the number. Thus, a reported length of 23.42 cm is to be interpreted as being 23.42 ± 0.01 cm; that is, the first three digits are assumed to be definitely known, with the true value of the length lying between 23.41 and 23.43 cm.

The number of significant figures used to record a measurement depends on the calibration of the measuring instrument. For example, if a ruler calibrated in millimeters is used for a measurement, the result is expressed to the nearest millimeter, and an estimate is made to the nearest 0.1 mm. If a quantity is measured only to the nearest tenth, it should not be reported to the nearest hundredth, for example, 32.1 cm as against 32.10 cm. On the other hand, a length measured as 32.1000 cm should not be recorded as 32.1 cm.

Zeros may or may not be significant figures, depending on how they are used. A zero used only to locate a decimal point is not a significant figure. Thus, there is no difference among 27 mm, 2.7 cm, 0.027 m, and 0.000027 km, and each has two significant figures. However, the zeros in such numbers as 328.0 and 2.037 are counted as significant figures.

When we use numbers in a computation, we must make sure that we do not throw precision away nor appear to add to it. When a set of measurements is added or subtracted, the accuracy of the sum or difference is determined by the least accurate measurement in the set. For example, let us compute the sum of 32.1000 cm + 23.42 cm + 9.752 cm:

$$\begin{array}{r} 32.1000 \text{ cm} \\ 23.42\phantom{00 \text{ cm}} \\ 9.752\phantom{0 \text{ cm}} \\ \hline 65.2720 \text{ cm} \end{array} \qquad \begin{array}{r} 32.1000 \text{ cm} \\ 23.42??\phantom{\text{ cm}} \\ 9.752?\phantom{\text{ cm}} \\ \hline 65.27?? \text{ cm} \end{array}$$

On the left we have set up the problem in the conventional fashion for adding and have performed the addition in the naive manner of a person who regards each recorded number as a quantity known with absolute exactness. On the right we have set up the problem with question marks to indicate our lack of information about certain digits; this shows why the sum should be reported as 65.27 cm.

Sometimes the answer to a computation must be subjected to the procedure known as rounding off. For example, let us add 14.23 g and 8.649 g:

$$\begin{array}{r} 14.23 \text{ g} \\ 8.649\phantom{\text{ g}} \\ \hline 22.879 \text{ g} \end{array}$$

Clearly, there should be only four digits in the reported sum, but the number to be reported is 22.88 g. The rule in rounding off is to increase the last retained digit by 1 if the following digit to be discarded exceeds 5 (as in our example). If the following digit to be discarded is less than 5, the last retained digit is left unchanged. Thus,

1.423 rounds off to 1.42
1.427 rounds off to 1.43

If the digit to be discarded is 5, the standard procedure is to increase the preceding digit if it is odd and to leave it unchanged if it is even. Thus,

1.425 rounds off to 1.42
1.435 rounds off to 1.44

When you are multiplying or dividing measured quantities, there can be only as many significant figures in the answer as the least number of significant figures in any of the quantities used in the calculation. This means that multiplication of a number with two significant figures by a number with three significant figures gives an answer that has only two significant figures. For example, let us compute the product

$$40.2 \text{ cm} \times 0.11 \text{ cm}$$

The multiplication of 40.2 by 0.11 gives 4.422, but you should report the result as 4.4 cm^2.

There is a special problem connected with exact numbers that are not the result of measurement. Suppose 775 g of sugar is divided into 30 equal piles; how much sugar is in each pile? Because 30 is an exact number here, it has an infinite number of significant figures; that is, it could be written as 30.00000. . . . Since the least precisely known quantity in this problem is 775 g, you are justified in retaining three significant figures in the answer (25.8 g).

Appendix 9: Exponential Notation

Numbers obtained in scientific measurements and computations range from very large to extremely small. Such numbers can be expressed and manipulated most easily by the use of exponential notation, that is, by expressing them as multiples of powers of 10. For example, the mass of the earth is about 6,000,000,000,000,000,000,000,000,000 g, a quantity more conveniently written in exponential notation as 6×10^{27} g.

Numbers greater than 10 are usually written with positive powers of 10. Thus,

$$
\begin{aligned}
14.52 &= 1.452 \times 10 \\
145.2 &= 1.452 \times 10^2 \\
1452 &= 1.452 \times 10^3 \\
14520 &= 1.452 \times 10^4 \\
145200 &= 1.452 \times 10^5
\end{aligned}
$$

In each case the exponent of 10 corresponds to the number of places the decimal point is moved to the left to obtain a number between 1 and 10.

Usually, numbers less than 1 are written with negative powers of 10. For example, the mass of an electron, which is approximately 0.000,000,000,000,000,000,000,000,000,9 g, is more conveniently written as 9×10^{-28} g. This use of negative powers of 10 can be illustrated as follows:

$$
\begin{aligned}
0.1452 &= 1.452 \times 10^{-1} \\
0.01452 &= 1.452 \times 10^{-2} \\
0.001452 &= 1.452 \times 10^{-3} \\
0.0001452 &= 1.452 \times 10^{-4} \\
0.00001452 &= 1.452 \times 10^{-5}
\end{aligned}
$$

In each case the negative exponent of 10 indicates how many places the decimal point must be moved to the right to obtain a number between 1 and 10.

With a number such as 143,000 the zeros may or may not be significant figures. To avoid such ambiguity, use exponential notation. Writing 143,000 as 1.43×10^5 indicates three significant figures; as 1.430×10^5, four significant figures; as 1.4300×10^5, five significant figures; and as 1.43000×10^5, six significant figures.

Exponential numbers cannot be added or subtracted unless the powers of 10 are the same.

EXAMPLE

Add 3.00×10^2 and 4.00×10^3.

First, write 4.00×10^3 as 40.0×10^2. Then,

$$
\begin{aligned}
(3.00 \times 10^2) + (40.0 \times 10^2) &= 43.0 \times 10^2 \\
&= 4.30 \times 10^3
\end{aligned}
$$

An alternate solution is to first write 3.00×10^2 as 0.300×10^3. Then,

$$(0.300 \times 10^3) + (4.00 \times 10^3) = 4.30 \times 10^3$$

It is better to report the answer as 4.30×10^3 than as 43.0×10^2 because 4.3 is between 1 and 10.

To multiply exponential numbers, multiply the digit terms in the customary manner and also multiply the powers of 10 (by adding their exponents).

EXAMPLE

Multiply 4.8×10^5 by 2.0×10^{-7}.

$$\begin{aligned}(4.8 \times 10^5)(2.0 \times 10^{-7}) &= (4.8 \times 2.0)(10^5 \times 10^{-7}) \\ &= 9.6 \times 10^{-2}\end{aligned}$$

When one exponential number is divided by another, the digit term of the numerator is divided by the digit term of the denominator, and the powers of 10 are divided by subtracting the exponent of the denominator from the exponent of the numerator.

EXAMPLE

Divide 4.2×10^{-8} by 7×10^{-3}.

$$\begin{aligned}\frac{4.2 \times 10^{-8}}{7 \times 10^{-3}} &= \frac{42 \times 10^{-9}}{7 \times 10^{-3}} \\ &= \frac{42}{7} \times \frac{10^{-9}}{10^{-3}} \\ &= 6 \times 10^{-6}\end{aligned}$$

To raise an exponential number to a higher power, raise the digit term to the stated power by multiplying it by itself the necessary number of times, and raise the power of 10 to the higher power by multiplying the exponent by the appropriate number.

EXAMPLE 1

Square the number 9.0×10^3.

$$\begin{aligned}(9.0 \times 10^3)^2 &= (9.0)^2 \times (10^3)^2 \\ &= 81 \times 10^6 \\ &= 8.1 \times 10^7\end{aligned}$$

EXAMPLE 2

What is the cube of 4.0×10^7?

$$\begin{aligned}(4.0 \times 10^7)^3 &= (4.0)^3 \times (10^7)^3 \\ &= 64 \times 10^{21} \\ &= 6.4 \times 10^{22}\end{aligned}$$

In taking the square root of an exponential number, first make the power of 10 divisible by 2; then extract the square root of the digit term and divide the exponential term by 2.

EXAMPLE

Find the square root of 2.5×10^{-7}.

$$\begin{aligned} 2.5 \times 10^{-7} &= 25 \times 10^{-8} \\ \sqrt{25 \times 10^{-8}} &= \sqrt{25} \times \sqrt{10^{-8}} \\ &= 5.0 \times 10^{-4} \end{aligned}$$

In taking the cube root, proceed in similar fashion except that the power of 10 is made divisible by 3.

EXAMPLE

What is the cube root of 8,000,000?

$$\begin{aligned} 8{,}000{,}000 &= 8 \times 10^{6} \\ \sqrt[3]{8 \times 10^{6}} &= \sqrt[3]{8} \times \sqrt[3]{10^{6}} \\ &= 2 \times 10^{2} \end{aligned}$$

Appendix 10: Use of the Logarithm Table

Any positive number may be expressed as a unique power of any other positive number. The number 8 may be written as 2^3 and 64 may be written as 8^2 or as 2^6. The logarithm of a number to the base n is the power to which n must be raised to produce the number. The logarithm of 8 to the base 2 is 3; the logarithm of 64 to the base 8 is 2; the logarithm of 64 to the base 2 is 6; and so on.

The logarithms commonly used as an aid to multiplication, division, raising to powers, and extracting roots are called common or Briggsian logarithms. The common or Briggsian logarithm of a number is the power to which 10 must be raised to equal that number. Thus, the logarithm of 1000 is 3, and the logarithm of 0.001 is −3.

$$\begin{aligned} \log 1000 &= 3 \\ \log 0.001 &= -3 \end{aligned}$$

The first table of common logarithms was published by the English mathematician Henry Briggs in 1624. Most numbers are not integral powers of 10, and it is therefore necessary to determine their "logs" by consulting a "log" table.

EXAMPLE

What is the logarithm of 7.32?

In Appendix 11, find 73 in the first vertical column and then move across to the column headed by 2. The number found there is 8645. This means that $\log 7.32 = 0.8645$.

Appendix 11 gives only the logs of numbers from 1.00 to 9.99. To obtain the log of any number less than 1.00 or greater than 9.99, it is necessary to write the number in exponential form (Appendix 9) so that it is expressed as a number between 1 and 10 multiplied by a power of 10.

EXAMPLE 1

What is the logarithm of 732?

First, write the number as 7.32×10^2. Next, find the log of 7.32 in the table. It is 0.8645. Since the log of a product is the sum of the logs,

$$\log ab = \log a + \log b$$

we can write

$$\begin{aligned} \log 7.32 \times 10^2 &= \log 7.32 + \log 10^2 \\ &= 0.8645 + 2 \\ &= 2.8645 \end{aligned}$$

EXAMPLE 2

What is the log of 0.0732?

First, write the number as 7.32×10^{-2}. Next, find log 7.32 in the table (0.8645):

$$\begin{aligned} \log 7.32 \times 10^{-2} &= \log 7.32 + \log 10^{-2} \\ &= 0.8645 - 2 \end{aligned}$$

For some purposes (for example, in pH problems) it will be necessary to compute $0.8645 - 2 = -1.1355$. For purposes of routine computation there is no occasion for doing this.

The use of logarithms in multiplication saves time and increases accuracy. The same is true in division, raising to powers, and finding roots.

EXAMPLE 1

Compute $547 \times 25.6 \times 4.28 \times 0.0032$.

First, find the logs of these numbers in the table; then add them.

$$\begin{aligned} \log 547 &= 2.7380 \\ \log 25.6 &= 1.4082 \\ \log 4.28 &= 0.6314 \\ \log 0.0032 &= \underline{0.5051 - 3} \\ & 2.2827 \end{aligned}$$

In the table the logarithm closest to 0.2827 is 0.2833 and the number corresponding to it is 1.92. Therefore, the answer is 1.92×10^2 or 192.

EXAMPLE 2

Divide 547 by 25.6.

Subtract log 25.6 from log 547.

$$\begin{aligned} \log 547 &= 2.7380 \\ \log 25.6 &= \underline{1.4082} \\ & 1.3298 \end{aligned}$$

In the table the log closest to 0.3298 is 0.3304 and the number corresponding to it is 2.14. Therefore, the answer is $2.14 \times 10 = 21.4$.

EXAMPLE 3

Raise 13 to the seventh power.

Obtain the logarithm of 13 and multiply it by 7.

$$\begin{aligned} \log 13 &= 1.1139 \\ & \underline{7} \\ & 7.7973 \end{aligned}$$

In the table the number corresponding to the logarithm 0.7973 is 6.27. Therefore, the answer is 6.27×10^7.

EXAMPLE 4

Find the cube root of 81.

Obtain the logarithm of 81 and divide it by 3.

$$\log 81 = 1.9085$$

$$\frac{1.9085}{3} = 0.6362$$

From the log table the answer is 4.33.

EXAMPLE 5

Find the square root of 0.0348.

Obtain log 0.0348 and divide it by 2.

$$\log 0.0348 = 0.5416 - 2$$

$$\frac{0.5416 - 2}{2} = 0.2708 - 1$$

From the log table the answer is 1.87×10^{-1} or 0.187.

Appendix 11 includes a table of proportional parts that facilitates the search for the logarithm of a four-digit number. For example, what is the logarithm of 3.445? As usual, we find 3.4 in the first vertical column and then move across to the column headed by 4, where we find the number 5366. We now continue across, in the same row, to the column headed by 5 in the proportional parts table. Here the number found is 6, which we add to 5366, giving 5372. Therefore, log 3.445 = 0.5372.

Appendix 11: Logarithms

Natural numbers	0	1	2	3	4	5	6	7	8	9	*Proportional parts*								
											1	2	3	4	5	6	7	8	9
10	0000	0043	0086	0128	0170	0212	0253	0294	0334	0374	4	8	12	17	21	25	29	33	37
11	0414	0453	0492	0531	0569	0607	0645	0682	0719	0755	4	8	11	15	19	23	26	30	34
12	0792	0828	0864	0899	0934	0969	1004	1038	1072	1106	3	7	10	14	17	21	24	28	31
13	1139	1173	1206	1239	1271	1303	1335	1367	1399	1430	3	6	10	13	16	19	23	26	29
14	1461	1492	1523	1553	1584	1614	1644	1673	1703	1732	3	6	9	12	15	18	21	24	27
15	1761	1790	1818	1847	1875	1903	1931	1959	1987	2014	3	6	8	11	14	17	20	22	25
16	2041	2068	2095	2122	2148	2175	2201	2227	2253	2279	3	5	8	11	13	16	18	21	24
17	2304	2330	2355	2380	2405	2430	2455	2480	2504	2529	2	5	7	10	12	15	17	20	22
18	2553	2577	2601	2625	2648	2672	2695	2718	2742	2765	2	5	7	9	12	14	16	19	21
19	2788	2810	2833	2856	2878	2900	2923	2945	2967	2989	2	4	7	9	11	13	16	18	20
20	3010	3032	3054	3075	3096	3118	3139	3160	3181	3201	2	4	6	8	11	13	15	17	19
21	3222	3243	3263	3284	3304	3324	3345	3365	3385	3404	2	4	6	8	10	12	14	16	18
22	3424	3444	3464	3483	3502	3522	3541	3560	3579	3598	2	4	6	8	10	12	14	15	17
23	3617	3636	3655	3674	3692	3711	3729	3747	3766	3784	2	4	6	7	9	11	13	15	17
24	3802	3820	3838	3856	3874	3892	3909	3927	3945	3962	2	4	5	7	9	11	12	14	16
25	3979	3997	4014	4031	4048	4065	4082	4099	4116	4133	2	3	5	7	9	10	12	14	15
26	4150	4166	4183	4200	4216	4232	4249	4265	4281	4298	2	3	5	7	8	10	11	13	15
27	4314	4330	4346	4362	4378	4393	4409	4425	4440	4456	2	3	5	6	8	9	11	13	14
28	4472	4487	4502	4518	4533	4548	4564	4579	4594	4609	2	3	5	6	8	9	11	12	14
29	4624	4639	4654	4669	4683	4698	4713	4728	4742	4757	1	3	4	6	7	9	10	12	13
30	4771	4786	4800	4814	4829	4843	4857	4871	4886	4900	1	3	4	6	7	9	10	11	13
31	4914	4928	4942	4955	4969	4983	4997	5011	5024	5038	1	3	4	6	7	8	10	11	12
32	5051	5065	5079	5092	5105	5119	5132	5145	5159	5172	1	3	4	5	7	8	9	11	12
33	5185	5198	5211	5224	5237	5250	5263	5276	5289	5302	1	3	4	5	6	8	9	10	12
34	5315	5328	5340	5353	5366	5378	5391	5403	5416	5428	1	3	4	5	6	8	9	10	11
35	5441	5453	5465	5478	5490	5502	5514	5527	5539	5551	1	2	4	5	6	7	9	10	11
36	5563	5575	5587	5599	5611	5623	5635	5647	5658	5670	1	2	4	5	6	7	8	10	11
37	5682	5694	5705	5717	5729	5740	5752	5763	5775	5786	1	2	3	5	6	7	8	9	10
38	5798	5809	5821	5832	5843	5855	5866	5877	5888	5899	1	2	3	5	6	7	8	9	10
39	5911	5922	5933	5944	5955	5966	5977	5988	5999	6010	1	2	3	4	5	7	8	9	10
40	6021	6031	6042	6053	6064	6075	6085	6096	6107	6117	1	2	3	4	5	6	8	9	10
41	6128	6138	6149	6160	6170	6180	6191	6201	6212	6222	1	2	3	4	5	6	7	8	9
42	6232	6243	6253	6263	6274	6284	6294	6304	6314	6325	1	2	3	4	5	6	7	8	9
43	6335	6345	6355	6365	6375	6385	6395	6405	6415	6425	1	2	3	4	5	6	7	8	9
44	6435	6444	6454	6464	6474	6484	6493	6503	6513	6522	1	2	3	4	5	6	7	8	9
45	6532	6542	6551	6561	6571	6580	6590	6599	6609	6618	1	2	3	4	5	6	7	8	9
46	6628	6637	6646	6656	6665	6675	6684	6693	6702	6712	1	2	3	4	5	6	7	7	8
47	6721	6730	6739	6749	6758	6767	6776	6785	6794	6803	1	2	3	4	5	5	6	7	8
48	6812	6821	6830	6839	6848	6857	6866	6875	6884	6893	1	2	3	4	4	5	6	7	8
49	6902	6911	6920	6928	6937	6946	6955	6964	6972	6981	1	2	3	4	4	5	6	7	8
50	6990	6998	7007	7016	7024	7033	7042	7050	7059	7067	1	2	3	3	4	5	6	7	8
51	7076	7084	7093	7101	7110	7118	7126	7135	7143	7152	1	2	3	3	4	5	6	7	8
52	7160	7168	7177	7185	7193	7202	7210	7218	7226	7235	1	2	2	3	4	5	6	7	7
53	7243	7251	7259	7267	7275	7284	7292	7300	7308	7316	1	2	2	3	4	5	6	6	7
54	7324	7332	7340	7348	7456	7364	7372	7380	7388	7396	1	2	2	3	4	5	6	6	7

Natural numbers	0	1	2	3	4	5	6	7	8	9	Proportional parts								
											1	2	3	4	5	6	7	8	9
55	7404	7412	7419	7427	7435	7443	7451	7459	7466	7474	1	2	2	3	4	5	5	6	7
56	7482	7490	7497	7505	7513	7520	7528	7536	7543	7551	1	2	2	3	4	5	5	6	7
57	7559	7566	7574	7582	7589	7597	7604	7612	7619	7627	1	2	2	3	4	5	5	6	7
58	7634	7642	7649	7657	7664	7672	7679	7686	7694	7701	1	1	2	3	4	4	5	6	7
59	7709	7716	7723	7731	7738	7745	7752	7760	7767	7774	1	1	2	3	4	4	5	6	7
60	7782	7789	7796	7803	7810	7818	7825	7832	7839	7846	1	1	2	3	4	4	5	6	6
61	7853	7860	7868	7875	7882	7889	7896	7903	7910	7917	1	1	2	3	4	4	5	6	6
62	7924	7931	7938	7945	7952	7959	7966	7973	7980	7987	1	1	2	3	3	4	5	6	6
63	7993	8000	8007	8014	8021	8028	8035	8041	8048	8055	1	1	2	3	3	4	5	5	6
64	8062	8069	8075	8082	8089	8096	8102	8109	8116	8122	1	1	2	3	3	4	5	5	6
65	8129	8136	8142	8149	8156	8162	8169	8176	8182	8189	1	1	2	3	3	4	5	5	6
66	8195	8202	8209	8215	8222	8228	8235	8241	8248	8254	1	1	2	3	3	4	5	5	6
67	8261	8267	8274	8280	8287	8293	8299	8306	8312	8319	1	1	2	3	3	4	5	5	6
68	8325	8331	8338	8344	8351	8357	8363	8370	8376	8382	1	1	2	3	3	4	4	5	6
69	8388	8395	8401	8407	8414	8420	8426	8432	8439	8445	1	1	2	2	3	4	4	5	6
70	8451	8457	8463	8470	8476	8482	8488	8494	8500	8506	1	1	2	2	3	4	4	5	6
71	8513	8519	8525	8531	8537	8543	8549	8555	8561	8567	1	1	2	2	3	4	4	5	5
72	8573	8579	8585	8591	8597	8603	8609	8615	8621	8627	1	1	2	2	3	4	4	5	5
73	8633	8639	8645	8651	8657	8663	8669	8675	8681	8686	1	1	2	2	3	4	4	5	5
74	8692	8698	8704	8710	8716	8722	8727	8733	8739	8745	1	1	2	2	3	4	4	5	5
75	8751	8756	8762	8768	8774	8779	8785	8791	8797	8802	1	1	2	2	3	3	4	5	5
76	8808	8814	8820	8825	8831	8837	8842	8848	8854	8859	1	1	2	2	3	3	4	5	5
77	8865	8871	8876	8882	8887	8893	8899	8904	8910	8915	1	1	2	2	3	3	4	4	5
78	8921	8927	8932	8938	8943	8949	8954	8960	8965	8971	1	1	2	2	3	3	4	4	5
79	8976	8982	8987	8993	8998	9004	9009	9015	9020	9026	1	1	2	2	3	3	4	4	5
80	9031	9036	9042	9047	9053	9058	9063	9069	9074	9079	1	1	2	2	3	3	4	4	5
81	9085	9090	9096	9101	9106	9112	9117	9122	9128	9133	1	1	2	2	3	3	4	4	5
82	9138	9143	9149	9154	9159	9165	9170	9175	9180	9186	1	1	2	2	3	3	4	4	5
83	9191	9196	9201	9206	9212	9217	9222	9227	9232	9238	1	1	2	2	3	3	4	4	5
84	9243	9248	9253	9258	9263	9269	9274	9279	9284	9289	1	1	2	2	3	3	4	4	5
85	9294	9299	9304	9309	9315	9320	9325	9330	9335	9340	1	1	2	2	3	3	4	4	5
86	9345	9350	9355	9360	9365	9370	9375	9380	9385	9390	1	1	2	2	3	3	4	4	5
87	9395	9400	9405	9410	9415	9420	9425	9430	9435	9440	0	1	1	2	2	3	3	4	4
88	9445	9450	9455	9460	9465	9469	9474	9479	9484	9489	0	1	1	2	2	3	3	4	4
89	9494	9499	9504	9509	9513	9518	9523	9528	9533	9538	0	1	1	2	2	3	3	4	4
90	9542	9547	9552	9557	9562	9566	9571	9576	9581	9586	0	1	1	2	2	3	3	4	4
91	9590	9595	9600	9605	9609	9614	9619	9624	9628	9633	0	1	1	2	2	3	3	4	4
92	9638	9643	9647	9652	9657	9661	9666	9671	9675	9680	0	1	1	2	2	3	3	4	4
93	9685	9689	9694	9699	9703	9708	9713	9717	9722	9727	0	1	1	2	2	3	3	4	4
94	9731	9736	9741	9745	9750	9754	9759	9763	9768	9773	0	1	1	2	2	3	3	4	4
95	9777	9782	9786	9791	9795	9800	9805	9809	9814	9818	0	1	1	2	2	3	3	4	4
96	9823	9827	9832	9836	9841	9845	9850	9854	9859	9863	0	1	1	2	2	3	3	4	4
97	9868	9872	9877	9881	9886	9890	9894	9899	9903	9908	0	1	1	2	2	3	3	4	4
98	9912	9917	9921	9926	9930	9934	9939	9943	9948	9952	0	1	1	2	2	3	3	4	4
99	9956	9961	9965	9969	9974	9978	9983	9987	9991	9996	0	1	1	2	2	3	3	3	4

Appendix 12: Answers to Selected Exercises

1-11 7.2 g/cm^3
1-12 0.790 g/ml
1-13 367 g; 367 lb
1-14 45; 3; 200
1-15 NO_2; N_2O; N_2O_5

2-2 14
2-3 31.810; 200.41; 32.065
2-5 31 g; 50 g; 2.9 g; 2.2×10^{-14} g
2-6 2.821 moles; 2.821 moles; 7.2175 moles
2-9 117 moles; 279 moles
2-10 2.923 moles
2-14 $C_2H_6O_2$
2-17 (a) 10.0 moles; (b) 62.33 moles; (c) 192 g; (d) 997.3 g; (e) 245.6 g

3-5 KNO_3; $Be(NO_3)_2$; BF_3; SiS_2; H_2S; CF_4
3-9 50 protons; 50 electrons; the mass number of the isotope
3-10 18; 32
3-15 2.17×10^{14} g/cm^3; 2.08×10^{13}

4-2 18; 32; 8
4-5 5; 4; 9
4-6 6; 14; 2; 8
4-8 6 paired, 2 unpaired; nickel
4-15 Element 119

5-4 No, no, no, no, yes, no, yes, yes
5-12 1000; 2000; 6.023×10^{23}
5-15 Carbon monoxide, carbon dioxide; iron(II) chloride, iron(III) chloride; tin(II) chloride, tin(IV) chloride; phosphorus(III) oxide, phosphorus(V) oxide.

6-2 28.8 amu
6-3 Heavier; lighter
6-4 Same
6-9 307 cm^3
6-11 1.60 liters
6-13 2.647 g
6-15 61.7 g/mole
6-22 65 liters
6-23 258 ml of O_2

7-4 346 ml
7-7 12.6 g
7-8 117 g
7-11 1.6×10^7 cal

8-4 2.88×10^3 kcal; 8.16×10^4 kcal
8-7 160 g
8-8 $KClO_3$
8-16 $LiAlH_4$
8-19 51.16%
8-20 $Na_2B_4O_7 \cdot 10H_2O$
8-23 0.5 kg of ice and 1.5 kg of water at 0°C
8-27 48.21%

9-3 $Ca_3(PO_4)_2$, $MgCl_2$, Na_2Se, BaTe, Ag_3AsO_4
9-7 P_4O_{10} (empirical formula P_2O_5), N_2O_5, SeO_3, Cl_2O_7, Cl_2O_5
9-15 Strong acids: $HClO_4$, HI, HBr, H_2SO_4, HCl, HNO_3, H_3O^+. Their conjugate bases (all very weak bases): ClO_4^-, I^-, Br^-, HSO_4^-, Cl^-, NO_3^-, H_2O.

10-2 17.6 ml
10-3 Molecular weight of solvent
10-5 $0.185m$; 265 g/mole
10-7 100.241°C
10-9 174 g/mole; $C_6H_{12}O_6$
10-13 99.9%
10-17 $0.149M$

11-5 0.51
11-8 6.8×10^{-4}
11-10 $2.9 \times 10^{-11}M$
11-13 0.62 g
11-14 2.4×10^{-12}
11-15 8×10^{-5} g

12-2 6.242×10^4 electrons
12-4 613 min
12-6 38.0 g
12-8 202 ml STP; 1.445 g

13-2 −92 kcal
13-4 +27.44 kcal
13-6 −15.00 kcal
13-8 (a) −123.0 kcal; (b) −20.2 kcal; (c) −795.5 kcal
13-12 −86.7 kcal; spontaneous
13-13 Iron anode, cadmium cathode

14-12 Skeletal isomerism: the two butanes
Positional isomerism: 1-butene and 2-butene
Geometric isomerism: *cis*-2-butene and *trans*-2-butene
Functional isomerism: methyl ether and ethyl alcohol
14-17 The six C—C bonds are of equal length. The conclusion is confirmed by the nonexistence of isomeric 1,2-dibromobenzenes, etc.

15-13 Fats are mixtures of triglycerides; most of the triglyceride molecules in fats are of the mixed glyceride type.

15-28 Temperature, pH, concentration of enzyme, concentration of substrate

15-31 (a) Ribose and deoxyribose; (b) adenine, guanine, cytosine, uracil, and thymine

15-32 Each nucleotide unit contains a phosphoric acid group.

15-33 The DNA double helix consists of two DNA strands, headed in opposite directions, twisted around each other, and held together by hydrogen bonds

16-3 Polonium-210

16-8 (a) ${}^{1}_{0}n$; (b) ${}^{14}_{7}N$; (c) ${}^{1}_{1}H$

16-15 (a) 6 α, 4 β; (b) 8 α, 5 β; (c) 8 α, 6 β; (d) 7 α, 4 β

Index

Italicized page numbers refer to the "New Terms" sections.

B C D E F G H I J
0 1 2 3 4 5 6 7 8

atomic weight*	102.905	acid-base properties†
electronegativity	2.2	
symbol	Rh	
melting point (°C)	1960	
boiling point (°C)‡	3900	
atomic number	45	
atomic radius (Å)	1.25	

* Based on carbon-12. ( ) indicates the most stable or best-known isotope.

† For representative oxides of group. Oxide is acidic if color is rust, basic if color is blue, and amphoteric if both colors are shown. Intensity of color indicates relative strength.

‡ s. indicates sublimation.

IA

1
1.0080 2.1 H −259.2 −252.8 1 0.37

IIA

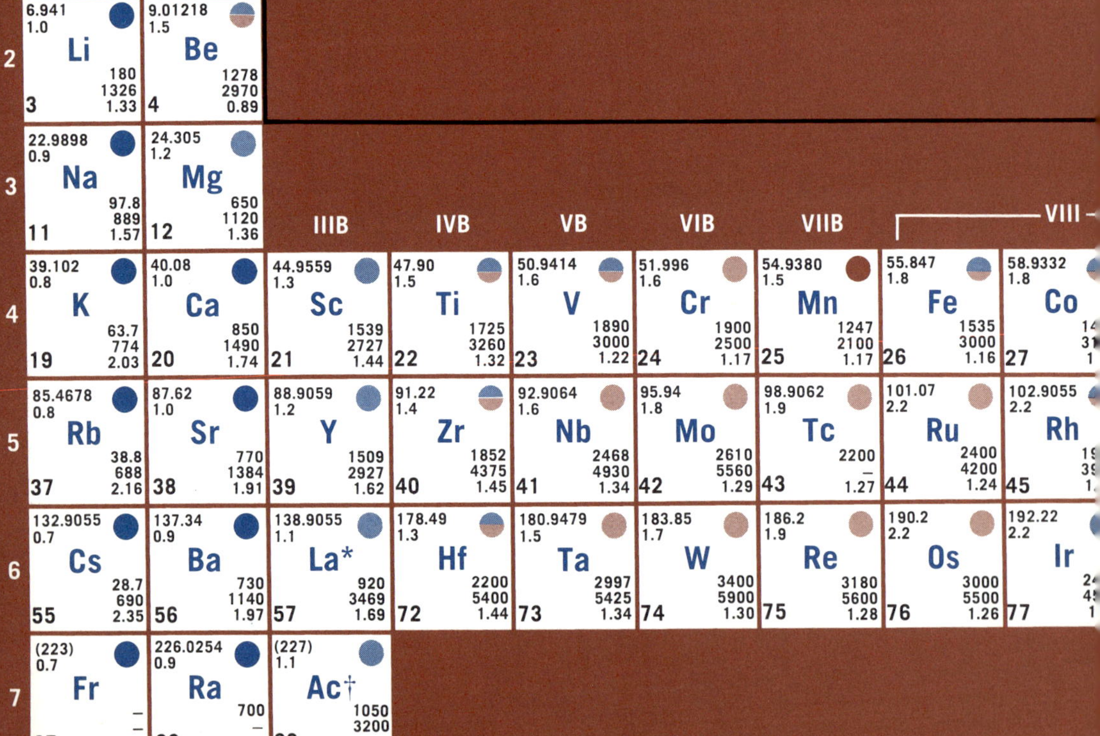

* Lanthanides

140.12 1.1 Ce 795 3468 58 1.65
140.9077 1.1 Pr 935 3127 59 1.65
144.24 1.2 Nd 1024 3027 60 1.64
(147) – Pm 1035 2730 61 –
150.4 1.2 Sm 62

† Actinides

232.0381 1.3 Th 1750 4000 90 1.65
231.0359 1.5 Pa – – 91 –
238.029 1.7 U 1132 3818 92 1.42
237.0482 1.3 Np 640 – 93 –
(242) – Pu 94